Stochastic Averaging

Wei-Qiu Zhu · Mao-Lin Deng · Guo-Qiang Cai

Stochastic Averaging

Methods and Applications, Volume 1

Science Press
Beijing

Springer

Wei-Qiu Zhu
Department of Mechanics
Zhejiang University
Hangzhou, Zhejiang, China

Mao-Lin Deng
Department of Mechanics
Zhejiang University
Hangzhou, Zhejiang, China

Guo-Qiang Cai
Florida Atlantic University
Boca Raton, FL, USA

ISBN 978-981-96-3831-4 ISBN 978-981-96-3829-1 (eBook)
https://doi.org/10.1007/978-981-96-3829-1

Jointly published with Science Press
The print edition is not for sale in China (Mainland). Customers from China (Mainland) please order the print book from: Science Press.

Translation from the Chinese language edition: "The Stochastic Averaging Method and Its Applications, Volume 1" by Wei-Qiu Zhu et al., © Science Press 2023. Published by Science Press. All Rights Reserved.

This Springer imprint is published by the registered company Springer Nature Singapore Pte Ltd.
The registered company address is: 152 Beach Road, #21-01/04 Gateway East, Singapore 189721, Singapore

Preface

Nonlinear stochastic dynamic systems are prevalent in the natural, technical sciences, and social sciences. The study on nonlinear stochastic dynamics began in the early 1960s. So far, it has been developed to a well-established discipline. Several exact, approximate analytical and numerical solution methods have been proposed and developed. The only exact solution method is based on that the excitation is Gaussian white noise and the response is Markov diffusion process. The exact probability density and statistics of system response are obtained from the formulation and solution of Fokker-Planck-Kolmogorov (FPK) equation. However, the exact solution method has limited application since real nonlinear stochastic dynamical system has no exact solution and the FPK equation is very difficult to solve. The approximate analytical solution methods include equivalent (statistical) linearization, equivalent nonlinear system method, moment equation and its truncation scheme, stochastic averaging method, etc. The numerical methods include Monte Carlo simulation, cell mapping and path integral methods, etc. Each method has its advantages and drawbacks. Relatively, the stochastic averaging methods are the most powerful and widespread applied approximate analytical methods. These methods have stochastic averaging principle as their sound mathematical foundation. They extend the application range of exact solution method. They simplify the system and reduce the dimension of system while retaining the essential nonlinearity of original system. Physically, the stochastic averaging methods transform the study of nonlinear stochastic dynamical system into the study of amplitude or energy of system or its subsystems, and the obtained results for amplitude or energy can be translated back into the probability and statistics for the original system. Stochastic averaging methods can be used to predict the response, to study the stability, reliability, and stochastic optimal control.

Starting from the early 1990s, the first author and his collaborators and students developed the stochastic averaging methods of quasi-Hamiltonian systems under Gaussian white noise excitation and then extend them to non-Gaussian, non-white noise excitation, to quasi-generalized Hamiltonian systems and some cases involving non-Gaussian and non-white noise excitations, etc. Meanwhile, Prof. Y. K. Lin and the second author at the Center for Applied Stochastics Research of Florida Atlantic

University, USA, derived the non-smooth and smooth averaged equations for general nonlinear stochastic dynamical systems and applied them to numerous single-degree-of-freedom and two-dimensional nonlinear stochastic systems, particularly ecological systems. Although most research results have been published as technical papers, to write a book is still necessary for the readers to understand easily and systematically. Chapters 2, 3, 4 in Volume 1 and Chap. 4 in Volume 2 were drafted by the third author while the remaining chapters (except Chap. 1 in Volume 1) were drafted by the second author. The first author revised and finalized the manuscript.

We are grateful to the National Natural Science Foundation of China for funding our research. We also extend gratitude to our colleagues Dr. Y. Q. Yang, Prof. Z. L. Huang, Prof. Z. G. Ying, Prof. Z. H. Liu, Prof. R. H. Huan, Prof. L. C. Chen, Assoc. Profs. Y. Wang, Y. Zeng, W. T. Jia, Y. J. Wu, Q. F. Lü, J. J. Sun for their invaluable contribution in developing the stochastic averaging methods. The support and help of Science Press and the financial support for publication of the book from China Classics International Publishing Company and Institute of Applied Mechanics of Zhejiang University are greatly appreciated.

Hangzhou, China — Wei-Qiu Zhu
Hangzhou, China — Mao-Lin Deng
Boca Raton, USA — Guo-Qiang Cai

Contents

Chapter 1
Introduction

Nonlinear stochastic dynamics have been studied for over sixty years and are theoretically well-developed, but many challenges still remain. The achievements in nonlinear stochastic dynamics are mainly based on the research results of Markov process and its related theory of stochastic differential equation in mathematics. The only case in nonlinear stochastic dynamics that can be exactly solved is that excitation is white noise and the system response is Markov process with transition probability density governed by Fokker–Planck-Kolmogorov (FPK) equation. For a nonlinear dynamical system excited by white noise, establishing and solving FPK equation to obtain probability density and statistics of the response is the only method of exact solution. However, white noise and Markov process are merely mathematical concepts and do not exist in reality. As a matter of fact, all random excitations in reality are colored noise. One question is under what conditions can colored noise be approximated by white noise, and then the above-mentioned exact solution method is applicable. On the other hand, solving high-dimensional FPK equations is extremely challenging. The other question is whether and how the dimension of the FPK equation can be reduced to lower the difficulty or reduce computational complexity. Stochastic averaging methods provide a solution for both these two questions. Stochastic averaging principle states that when the correlation time of a colored noise is much shorter than system's relaxation time, the colored noise can be approximated by an equivalent white noise. When a system contains both slowly-varying and fast-varying processes, the original system can be approximated by a lower dimensional equivalent system obtained from time averaging of the slowly-varying processes. When the degenerated conservative system is ergodic on some sub-manifold, time averaging can be replaced by spatial averaging with respect to fast-varying processes, eliminating fast-varying processes and then reducing dimension of FPK equation. As mentioned in the preface of this book, the stochastic averaging methods are powerful approximate analytical techniques in nonlinear stochastic dynamics with many advantages and have been widely applied.

W.-Q. Zhu et al., *Stochastic Averaging*, https://doi.org/10.1007/978-981-96-3829-1_1

The stochastic averaging methods were first proposed by Stratonovich (1963, 1967) based on physical concepts and applied to the noise problems in communication systems. Later, Khasminskii (1966) provided a rigorous mathematical formulation and proof. Papanicolaou and Kohler (1974) extended the mathematical foundation, and Blankenskip and Papanicolaou (1978) further proved that averaged equations can approximate to original system as time goes to infinity. Khasminskii (1968) proposed the averaging principle for Itô stochastic differential equations with both slowly-varying and fast-varying processes. In the 1970s and 1980s, the stochastic averaging methods were applied by many researchers such as Ariaratnam, Dimentberg, Roberts, Spanos, etc., to study the response, stability, and reliability of single and multiple degree-of-freedom (DOF) quasi-linear and single-DOF strongly nonlinear random vibration systems. Their research results have been summarized in review articles (Roberts and Spanos 1986; Zhu 1988, 1996) and monographs (Dimentberg 1988; Zhu 1992; Lin and Cai 1995).

In the mid-1990s, Lin and Cai (1995) explicitly distinguish between stochastic averaging and time averaging, derived the stochastic averaging equations for general nonlinear stochastic systems, including non-smooth and smooth averaged equations and applied stochastic averaging methods to various single-DOF and two-dimensional nonlinear stochastic dynamical systems, especially to systematic study of ecological systems (Cai and Lin 2007). At the same time, Zhu and his research group (Zhu 2003, 2006) proposed and developed the stochastic averaging methods of quasi-Hamiltonian systems excited by Gaussian white noise and extended them to quasi-generalized Hamiltonian systems and quasi-integrable Hamiltonian systems with genetic effect forces, as well as to quasi-Hamiltonian systems excited by both Gaussian and Poisson white noises, fractional Gaussian noise, and colored noise. This book provides a comprehensive summary of research results on the stochastic averaging methods and their applications obtained by these two teams since the mid-1990s.

The entire book, consisting of Volume 1 and Volume 2, is divided into three parts. The first part includes Chaps. 2 and 3 of vol. 1, providing the preliminary knowledge for the book. The second part consists of Chaps. 4, 5, 6, 7 in vol. 1 and Chaps. 1, 2, 3 in vol. 2, which are the main contents of this book and present the stochastic averaging methods for various nonlinear dynamical systems under several random excitations. The third part comprises Chaps. 4, 5, 6 of vol. 2, giving several applications of stochastic averaging methods in natural and technical sciences. In the following, the main contents and key points of each chapter are introduced.

In Chap. 2 of vol. 1, the stochastic processes involved in this book and their related stochastic differential equations and stochastic differential rules are introduced, including Gaussian white noise, Wiener process, diffusion process, and their related Itô stochastic differential equation, Itô differential rule and FPK equation; Poisson white noise, Markov process, stochastic differential equation with compound Poisson process and its equivalent stochastic differential integral equation, Di Paola-Falsone differential rule; fractional Gaussian noise, fractional stochastic differential equation, and fractional stochastic differential rule.

In Chap. 3 of vol. 1, the three mathematical models for the nonlinear stochastic dynamical systems involved in these books, namely, the general equation, the stochastically excited and dissipated Lagrangian equation and the stochastically excited and dissipated Hamiltonian equation are introduced. Since the third model is most frequently used in these books and many readers may not be very familiar with it, the definitions, basic properties, and the classification of Hamiltonian system and generalized Hamiltonian system based on integrability and resonance, i.e., nonintegrable, (completely) integrable and nonresonant, (completely) integrable and resonant, partially integrable and nonresonant, partially integrable and resonant are briefly explained. The ergodicity on some sub-manifolds of the five classes of Hamiltonian systems are emphasized, which provides a basis for the operation of replacing time averaging with spatial averaging. Finally, the mathematical models for genetic effective forces, including hysteresis force, viscoelastic force, and fractional derivative damping force are introduced.

In Chap. 4 of vol. 1, the concepts of stochastic averaging and time averaging are introduced firstly and the non-smooth and smooth stochastic averaged equations for general stochastic dynamical systems are derived. Stochastic averaging is using equivalent Gaussian white noise to replace a stochastic excitation with its correlation time much shorter than system relaxation time. Time averaging can be used when system contains periodic coefficients or both fast-varying and slowly-varying processes. The time averaged equations of slowly-varying processes are used to approximate the original system, reducing the system dimension. Then, the detailed derivations of amplitude-envelope and energy-envelope stochastic averaged equations and their associated FPK equations for single-DOF nonlinear dynamic equations under excitations such as Gaussian white noise, wideband noise, harmonic and wideband noise, Poisson white noise, fractional Gaussian noise, etc. are given. How to decouple viscoelastic forces into elastic restoring forces and viscous damping forces, and then to apply stochastic averaging method are explained. Finally, the aforementioned stochastic averaging methods cannot be applied to the saddle point and homoclinic orbit of system with two-potential wells is pointed out. Alternatively, the stochastic averaging methods have to be applied to one potential well at each time. It is recommended that readers who are new to stochastic averaging method read this chapter carefully to gain a deep understanding and a solid foundation for reading the following chapters.

Starting from Chap. 5 of vol. 1, the stochastic averaging methods for multi-DOF nonlinear systems, especially strongly nonlinear systems, under various stochastic excitations are presented. Since the Hamiltonian formulation is essential for explaining the global relationships between degrees-of-freedom in multi-DOF strongly nonlinear systems, the stochastic averaging methods of five classes of quasi-Hamiltonian systems, including nonintegrable, (completely) integrable and nonresonant, (completely) integrable and resonant, partially integrable and nonresonant, partially integrable and resonant quasi-Hamiltonian systems under various stochastic excitations, are presented. In Chap. 5 of vol. 1, the stochastic averaging methods of five classes of quasi-Hamiltonian systems under Gaussian white noise excitation are presented. Since the excitation is white noise, the stochastic averaging is not required. Instead, it is needed to identify the slowly-varying processes in five classes of quasi Hamiltonian systems and to derive averaged Itô stochastic differential equations via time averaging over drift and diffusion coefficients. Here, one challenge is how to use the ergodicity of Hamiltonian systems on some sub-manifolds to replace time averaging with spatial averaging with respect to fast-varying processes. Another challenge is how to obtain the approximate probability density of the original system from the stationary probability density obtained from solving averaged FPK equation. Notably, for any quasi-nonintegrable Hamiltonian system with finite number of DOF, the averaged equation is one-dimensional, and a unified expression for stationary probability density can be obtained. This result is surprisingly simple, similar to the power spectral density method for all linear stochastic systems. However, it comes at the price of challenging in integral calculation. For quasi-nonintegrable Hamiltonian systems with more than two DOFs, the multi-dimensional domain integrals in averaged drift and diffusion coefficients are very difficult. To overcome this difficulty, a two-step generalized elliptic coordinate transformation method to convert $(2n\text{-}1)$-fold domain integral into n-fold integral is introduced. In Sect. 5.4, an example of a two-DOF vibration-impact system is used to illustrate how to properly select the stochastic averaging methods according to the intensity of system nonlinearity (nonintegrability). As an extension, the stochastic averaging method of quasi-nonintegrable Hamiltonian systems with Markov jumping parameters is presented in the last section.

As a typical example of system excited by both continuous and impulsive random excitations, the stochastic averaging method of quasi-Hamiltonian systems excited by both Gaussian and Poisson white noises are developed in Chap. 6 of vol. 1. Similar to Chap. 5 of vol. 1, the systems are divided into five classes based on the integrability and resonance of Hamiltonian systems, and the stochastic averaging is not needed and only time averaging of slowly-varying processes is required. Furthermore, the ergodicity of the Hamiltonian system on sub-manifold can be used to replace time averaging with spatial averaging with respect to fast-varying processes. The formula for converting the probability density of the averaged system into that of the original system remains the same as those in Chap. 5 of vol. 1. Chapter 6 of vol. 1 differs from Chap. 5 of vol. 1 in that Poisson white noise is a kind of non-Gaussian white noise and can be considered as the derivative process of compound Poisson process. The quasi-Hamiltonian systems excited by both Gaussian and Poisson white noises are transformed into stochastic differential equations with additional Wong-Zakai and Di Paola-Falsone correction terms. These equations can be rewritten as equivalent stochastic differential integral equations with Poisson stochastic integrals. The derivation of the averaged stochastic differential integral equations for slowly-varying processes needs using the stochastic jump-diffusion chain rule. The resulting averaged FPK equation is of infinite order and truncation is needed. In Chap. 6 of vol. 1, truncation is performed at the fourth order, and perturbation method is used to solve the truncated averaged FPK equation. The results obtained from the stochastic averaging methods are validated via comparing with the results from Monte Carlo simulations of the original system, and the effects of Gaussian white noise and Poisson white noise on the same system are compared. One notable feature of Chap. 6 of vol. 1 is that lengthy and complicated expressions require the readers to refer to relevant literature and to be patient.

Fractional Gaussian noise is essentially a colored noise, which can be regarded as derivative process of fractional Brownian motion. Quasi-Hamiltonian systems excited by fractional Gaussian noise can be transformed into fractional stochastic differential equations. In Chap. 7 of vol. 1, the stochastic averaging methods of quasi-Hamiltonian systems excited by fractional Gaussian noise are developed. Although fractional Gaussian noise is not white noise, here only time averaging is required and the time averaging can be replaced by spatial averaging with respect to fast-varying processes. The formula of transforming probability density of averaged system into that of original system also remains unchanged. The main differences are that when a quasi-Hamiltonian system excited by fractional Gaussian noise is transformed into fractional stochastic differential equation, there is no correction term and the process governed by the averaged fractional stochastic differential equation is non-Markovian, and its probability density and statistics can be obtained only via Monte Carlo simulation. The advantage of the stochastic averaging methods in this chapter is significantly saving computation time compared with that for Monte Carlo simulation of original system, while yielding fairly close results.

Almost all real random excitations are colored noises, which can be either wideband or narrowband. At the same time, a colored noise can exhibit wideband character in some frequency band, and narrowband character in another frequency band. In Chap. 1 of vol. 2, the stochastic averaging methods of quasi-integrable Hamiltonian systems excited by wideband and narrowband noise are developed. The stochastic averaging method of quasi-integrable Hamiltonian systems under wideband noise excitation developed in Sect. 1.1 of vol. 2 can be viewed as extending of stochastic averaging method of quasi-linear systems under wideband noises excitation to multi-DOF strongly nonlinear systems. For the case of wideband noise excitation, both stochastic averaging and time averaging are necessary. Differed from Chaps. 5, 6, 7 of vol. 1, in Chap. 1 of vol. 2 it is assumed that the integrable Hamiltonian system undergoes generalized harmonic motion, while the quasi-integrable Hamiltonian system undergoes randomly periodic motion. Slowly-varying and fast-varying processes are separated for the cases of non-internal resonance or internal resonance, stochastic averaging and time averaging are used to derive the averaged Itô differential equations and associated FPK equation. Spatial averaging with respect to fast-varying processes can used to replace time averaging. The relationship between stationary probability density of original system and that of averaged system is the same as those in Chaps. 5, 6, 7 of vol. 1.

Consider a quasi-integrable Hamiltonian system excited by fractional Gaussian noise. When the natural frequencies of the system are higher than certain value and the fractional Gaussian noise in this frequency band can be considered wideband noise, then the stochastic averaging method presented in Sect. 1.1 of vol. 2 can be applied. In Sect. 1.2 of vol. 2, the quasi-integrable Hamiltonian system excited by fractional Gaussian noise is studied by using the stochastic averaging method of quasi-integrable Hamiltonian systems under wideband excitations. The results from

Monte Carlo simulation confirm the validity of this approach over a wide range of Hurst indices.

In Sect. 1.3 of vol. 2, the stochastic averaging method of quasi-integrable Hamiltonian systems excited by both harmonic and wideband noise is developed. Since harmonic excitation has significant influence on the system only in the case of external resonance, only the following two cases are considered, i.e., external resonance only and simultaneous internal and external resonance. Deriving Itô stochastic differential equations for slowly-varying processes, performing stochastic averaging and time averaging, establishing and solving averaged FPK equations yield the stationary probability density of the system. The combination of harmonic and wideband noise excitation is equivalent to narrowband noise excitation. As an example, the random jump and its bifurcations of Duffing oscillator under such narrowband noise excitation are studied. The method described in this section is also applicable to quasi-integrable Hamiltonian systems excited by both harmonic and Gaussian white noise.

Randomized harmonic process can be wideband or narrowband. In Sect. 1.4 of vol. 2, it is treated as a narrowband noise to excite a quasi-integrable Hamiltonian system. The response of the system is randomized periodic process. Only the cases of external and both external and internal resonances are considered. In these cases, the equation for amplitude or energy is deterministic while the equation for the combination of angles is stochastic. The two equations are coupled. Only time averaging is necessary. It is still necessary to solve multi-dimensional FPK equation. The random jump and its bifurcation of Duffing oscillator under randomized harmonic process excitation will also occur.

In a quasi-Hamiltonian system, restoring force and damping force are usually uncoupled. However, in reality, some forces such as hysteresis force, viscoelastic force, fractional derivative damping force, and time-delay forces exhibit genetic effect containing coupled restoring and damping forces. Before applying stochastic averaging method to quasi-integrable Hamiltonian systems with genetic effective forces, it is necessary to decouple these forces into elastic restoring forces and viscous damping forces. Therefore, Chap. 2 of vol. 2 focus on how to decouple these forces into elastic restoring forces and viscous damping forces. A unified approach is generalized harmonic balance technique, which is applicable to all four aforementioned genetic effective forces. For hysteresis force, there is also a method to determine elastic restoring force and viscous damping force by calculating the potential and dissipative energies of the hysteresis loop. After decoupling, the quasi-integrable Hamiltonian systems with genetic effective force become equivalent quasi-integrable Hamiltonian systems with elastic restoring forces and viscous damping forces dependent on amplitude or energy. These quasi-integrable Hamiltonian systems under various random excitations can be carried out stochastic averaging by using the methods presented in Chaps. 5, 6, 7 of vol. 1 and Chap. 1 of vol. 2.

Quasi-Hamiltonian systems are even-dimensional. To study the stochastic averaging method for odd-dimensional stochastic systems, quasi-generalized Hamiltonian systems are especially introduced. The main distinction between generalized

Hamiltonian systems and Hamiltonian systems is that the former has Casimir functions which serve as first-order integrals. In Chap. 3 of vol. 2, the stochastic averaging methods of quasi-generalized Hamiltonian systems excited by Gaussian white noise are presented. The only difference between the stochastic averaging methods in Chap. 3 of vol. 2 and those in Chap. 5 of vol. 1 is that the former methods contain averaged equations for Casimir processes, which increase the dimension of the averaged FPK equations and make them more difficult to be solved. Additionally, the formula of converting the stationary probability density of the averaged system into that of the original system includes Jacobian determinant of transforming original variables into Casimir functions.

As a stochastic averaging method for two-dimensional dynamical systems with first integral, the stochastic averaging method for predator–prey ecological systems is presented in Chap. 4 of vol. 2. Classical Lotka-Volterra model and its modified versions, involving predator saturation, predator competition, time delays, and complex environments are introduced. Various noises such as Gaussian white noise and different colored noises are considered as excitations. The dynamical properties, probability densities and statistics are obtained and verified by using results from Monte Carlo simulation. This chapter serves as a successful application of stochastic averaging methods presented in Chap. 4 of vol. 1.

Nonlinear stochastic dynamical systems widely exist in many fields of natural sciences such as physics, chemistry and biology. In Chap. 5 of vol. 2, the applications of the stochastic averaging methods to five different problems in natural sciences, including motion of active Brownian particles, reaction rate theory, Fermi resonance, thermal denaturation of DNA molecules, and conformational transition of biomolecules are presented in detail. The results obtained using the stochastic averaging methods are compared with those from Monte Carlo simulations, demonstrating a high degree of satisfaction.

Many engineering structures are subjected to various random loadings, leading to numerous nonlinear stochastic dynamical systems in technical sciences. Stochastic averaging methods have already been utilized in various studies. In Chap. 6 of vol. 2, the applications of stochastic averaging methods in vortex-induced vibration of structures in wind engineering, multi-machine power systems, ship rolling and capsizing, random stability, and nonlinear stochastic optimal control are presented. This chapter aims to inspire more applications of stochastic averaging methods in technical sciences.

In conclusion, although abundant research results on the stochastic averaging methods and their applications in nonlinear dynamical systems under various random excitations are presented in these books, there is still room for further developments of the methods and their wider applications. The authors eagerly anticipate the continued advancements and dissemination of stochastic averaging methods and their applications.

References

Blankenship G, Papanicolaou G C. 1978. Stability and control of stochastic systems with wide-band noise disturbances. SIAM Journal of Applied Mathematics, 34(3): 437–476.

Cai G Q, Lin Y K. 2007. Stochastic analysis of predator-prey type ecosystems. Ecological Complexity, 4: 241–249.

Dimentberg M F. 1988. Statistical Dynamics of Nonlinear and Time-Varying Systems. Research Studies Press, Taunton, U.K., Distributed by Wiley, Chichester.

Khasminskii R Z. 1966. A limit theorem for the solution of differential equations with random right hand sides. Theory of Probability and Application, 11(3): 390–405.

Khasminskii R Z. 1968. On the averaging principle for Itô stochastic differential equations. Kibernetika, 3(4): 260–279. (in Russian).

Lin Y K, Cai G Q. 1995. Probabilistic Structural Dynamics: Advanced Theory and Applications. New York: McGraw-Hill.

Papanicolaou G C, Kohler W. 1974. Asymptotic theory of mixing stochastic ordinary differential equations. Communication on Pure and Applied Mathematics 27: 641–668.

Roberts J B, Spanos P D. 1986. Stochastic averaging: an approximate method of solving random vibration problems. International Journal of Non-Linear Mechanics, 21: 111–134.

Stratonovich R L. 1963. Topics in the Theory of Random Noise, Vol. 1, New York: Gordon and Breach.

Stratonovich R L. 1967. Topics in the Theory of Random Noise, Vol. 2, New York: Gordon and Breach.

Zhu W Q. 1988. Stochastic averaging method in random vibration. ASME Applied Mechanics Reviews, 41(5): 189–199.

Zhu W Q. 1992. Random Vibration. Beijing: Science Press. (in Chinese).

Zhu W Q. 1996. Recent developments and applications of the stochastic averaging method in random vibration. ASME Applied Mechanics Reviews, 49(10): s72–s80.

Zhu W Q. 2003. Nonlinear Stochastic Dynamics and Control – Hamilton's Theoretical Framework. Beijing: Science Press. (in Chinese).

Zhu W Q. 2006. Nonlinear stochastic dynamics and control in Hamiltonian formulation. ASME Applied Mechanics Reviews, 59(4): 230–248.

Chapter 2
Stochastic Processes

Stochastic averaging methods are powerful techniques to predict responses, analyze stability and bifurcation, estimate reliability, and design optimal controller for nonlinear stochastic dynamic systems. The excitations and responses of nonlinear stochastic dynamic systems are stochastic processes. As a prerequisite, in this chapter the fundamentals of stochastic processes and relevant stochastic differential equations involved in this book are briefly introduced. Section 2.1 gives the definition of a stochastic process, its description, stationarity, ergodicity, and spectral analysis. Section 2.2 specifies the Gaussian stochastic processes and their properties. Section 2.3 describes the Markov diffusion processes, the associated Fokker-Plank-Kolmogorov (FPK) equation, and Itô stochastic differential equation. Section 2.4 describes Poisson white noise, stochastic differential-integral equation and FPK equation. Section 2.5 defines the fractional Gaussian noise, fractional Brownian motion, fractional stochastic calculus and fractional stochastic differential equation. Section 2.6 introduces colored noises generated from linear and nonlinear filters, and randomized harmonic noises.

2.1 Fundamentals

Consider a physical phenomenon that evolves with respect to time in random fashion, such as vibration of a building during an earthquake, the motion of a ship in a sea, etc. Denoting $X(t)$ as a physical quantity investigated in a random phenomenon, then at different time instants, $X(t_1)$, $X(t_2)$, $\cdots$ are random variables. Such a random phenomenon can be investigated by introducing a concept called stochastic process, or random process, which is defined below:

A stochastic process $X(t)$ is a family of random variables with t as a parameter belonging to an index set T, denoted by $\{X(t), t \in T\}$. Although the index set can be of various types, conventionally and also in this book, $X(t)$ is referred as stochastic

W.-Q. Zhu et al., *Stochastic Averaging*, https://doi.org/10.1007/978-981-96-3829-1_2

process only for the case in which the index set is time. Another common situation, although not involved in this book, is that the parameter t is a spatial variable, in which case X(t) is called random field.

A stochastic process is, in strict sense, a function of two arguments, $\{X(t,\omega); t \in T, \omega \in \Omega\}$, where Ω is the sample space. For a fixed time t, $X(t,\omega)$ is a random variable defined in the sample space Ω; while each possible $X(t,\omega)$ function with respect to time t is called a sample function.

In general, the sample space Ω can be discrete or continuous. For example, a continuous stochastic process means a stochastic process with a continuous sample space. Except specified otherwise, only continuous stochastic processes are considered.

Figure 2.1 illustrates three sample functions of a stochastic process schematically. At a specific time instant, say t_1 or t_2, the values of the sample functions constitute a random variable.

Fig. 2.1 Sample functions of a stochastic process

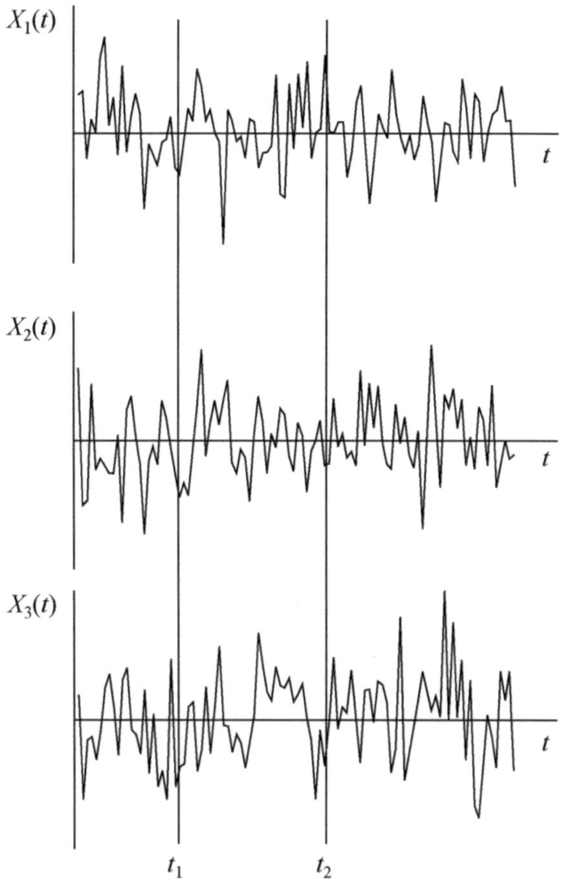

2.1.1 Descriptions of Stochastic Processes

As defined, a stochastic process is a family of random variables; thus, its specification is the same as that of jointly distributed random variables, except the number of random variables may be infinite or even uncountable. Therefore, a complete description of a stochastic process may be only possible for very limited cases. Practically, the most important features, although incomplete, may be sufficient for analysis in various applications, such as in engineering, biology, ecology, finance, etc.

Consider a stochastic process $X(t)$. Its first-order, second-order, and up to nth-order PDFs are denoted respectively as

$$p(x,t), p(x_1,t_1;x_2,t_2), \ldots ,p(x_1,t_1;x_2,t_2;\ldots;x_n,t_n). \tag{2.1}$$

In the above notation, x is the state variable of stochastic process $X(t)$, i.e. the possible value of $X(t)$, x_i is the state variable of stochastic process $X(t_i)$. The convention of using a capital letter to denote a random quantity, such as a random variable or a stochastic process, and a low case letter to represent its corresponding state variable will be followed throughout the book.

Since a lower-order PDF can be obtained from a higher-order one by integration, the higher-order one contains more information than the lower-order ones.

A stochastic process can also be described by its moment functions

$$\begin{aligned}
&E[X(t)] = \int xp(x,t)dx,\\
&E[X(t_1)X(t_2)] = \iint x_1x_2p(x_1,t_1;x_2,t_2)dx_1dx_2,\\
&\cdots\cdots\cdots\\
&E[X(t_1)X(t_2)\cdots\cdots X(t_n)]\\
&\quad = \int\cdots\int x_1x_2\cdots x_np(x_1,t_1;x_2,t_2;\cdots;x_n,t_n)dx_1dx_2\cdots dx_n.
\end{aligned} \tag{2.2}$$

The first and second moment functions, called the mean function and autocorrelation function respectively, are

$$\mu_X(t) = E[X(t)], \quad R_{XX}(t_1,t_2) = E[X(t_1)X(t_2)]. \tag{2.3}$$

The autocorrelation function is non-negative definite, i.e., it satisfies

$$R_{XX}(t_1,t_2)h(t_1)h^*(t_2) \geq 0 \quad \text{for any } t_1 \text{ and } t_2, \tag{2.4}$$

where $h(t)$ is an arbitrary function, and the asterisk "*" denotes the complex conjugate. The proof is given in (Lin 1967).

The auto-covariance function is defined as the second central moment functions, i.e.,

$$\begin{aligned}\kappa_{XX}(t_1,t_2)&=E\{[X(t_1)-\mu_X(t_1)][X(t_2)-\mu_X(t_2)]\}\\&=R_{XX}(t_1,t_2)-\mu_X(t_1)\mu_X(t_2).\end{aligned}\quad(2.5)$$

The variance function is the special case of the auto-covariance function when $t_1 = t_2$, i.e.,

$$\sigma_X^2(t)=E\{[X(t)-\mu_X(t)]^2\}.\quad(2.6)$$

The autocorrelation coefficient function is defined as

$$\rho_{XX}(t_1,t_2)=\frac{\kappa_{XX}(t_1,t_2)}{\sigma_X(t_1)\sigma_X(t_2)}.\quad(2.7)$$

Since $\rho_{XX}(t_1,t_2)$ is in fact the correlation coefficient of two random variables $X(t_1)$ and $X(t_2)$, it is known that $\rho_{XX}(t_1,t_2)\le 1$ for any t_1 and t_2 (Cai and Zhu 2016).

The autocorrelation or auto-covariance function is a measure of correlation of a stochastic process at two different time instants. A larger value of the autocorrelation function indicates a closer relationship of the process at two different time instants.

The mean and variance functions are the first-order statistical properties of the stochastic process since it is related to the first-order probability distribution involving only one time instant t. The autocorrelation, auto-covariance and autocorrelation coefficient functions are the second-order properties involving two time instants t_1 and t_2. Although they may be incomplete description for a stochastic process, they are among the most important properties in practical applications.

Consider two stochastic processes $X_1(t)$ and $X_2(t)$. Analogous to the case of a single stochastic process, the cross-correlation, cross-covariance, and cross-correlation coefficient functions are defined to describe the relationship of the two processes. They are

$$R_{X_1X_2}(t_1,t_2)=E[X_1(t_1)X_2(t_2)],\quad(2.8)$$

$$\begin{aligned}\kappa_{X_1X_2}(t_1,t_2)&=E\{[X_1(t_1)-\mu_{X_1}(t_1)[X_2(t_2)-\mu_{X_2}(t_2)]]\}\\&=R_{X_1X_2}(t_1,t_2)-\mu_{X_1}(t_1)\mu_{X_2}(t_2),\end{aligned}\quad(2.9)$$

$$\rho_{X_1X_2}(t_1,t_2)=\frac{\kappa_{X_1X_2}(t_1,t_2)}{\sigma_{X_1}(t_1)\sigma_{X_2}(t_2)}.\quad(2.10)$$

It is clear that the autocorrelation and the cross-correlation functions are both symmetric in the following sense,

$$R_{XX}(t_1,t_2)=R_{XX}(t_2,t_1),\quad R_{X_1X_2}(t_1,t_2)=R_{X_2X_1}(t_2,t_1).\quad(2.11)$$

The statistical properties of the derivative process $\dot{X}(t)$ can be obtained from those of the original process $X(t)$. For the mean function $\mu_{\dot{X}}(t)$,

$$\begin{aligned}\mu_{\dot{X}}(t) &= E\big[\dot{X}(t)\big] = E\left[\underset{\Delta t\to 0}{\text{l.i.m.}}\frac{X(t+\Delta t)-X(t)}{\Delta t}\right]\\ &= \lim_{\Delta t\to 0}\frac{E[X(t+\Delta t)]-E[X(t)]}{\Delta t} = \frac{\mathrm{d}\mu_X(t)}{\mathrm{d}t} = \dot{\mu}_X(t),\end{aligned} \tag{2.12}$$

where l.i.m denotes the mean-square limit (Lin and Cai 1995). In deriving (2.12), use has been made of the commutability between operations of the expectation and the limiting process.

Similarly, we have

$$R_{\dot{X}X}(t_1,t_2) = E\big[\dot{X}(t_1)X(t_2)\big] = \frac{\partial}{\partial t_1}R_{XX}(t_1,t_2), \tag{2.13}$$

$$R_{\dot{X}\dot{X}}(t_1,t_2) = E\big[\dot{X}(t_1)\dot{X}(t_2)\big] = \frac{\partial^2}{\partial t_1\partial t_2}R_{XX}(t_1,t_2). \tag{2.14}$$

More generally,

$$\mu_{X^{(n)}}(t) = E\big[X^{(n)}(t)\big] = \frac{\mathrm{d}^n\mu_X(t)}{\mathrm{d}t^n}, \tag{2.15}$$

$$\begin{aligned}R_{X^{(n)}X^{(m)}}(t_1,t_2) &= E\big[X^{(n)}(t_1)X^{(m)}(t_2)\big]\\ &= \frac{\mathrm{d}^{n+m}}{\mathrm{d}t_1^n\mathrm{d}t_2^m}R_{XX}(t_1,t_2).\end{aligned} \tag{2.16}$$

2.1.2 *Stationarity and Ergodicity*

Stochastic processes can be classified according to different criteria. One of the criteria is whether the probabilistic and statistical properties are independent of a shift of the time, i.e., when $t \to t+\tau$. A stochastic process is said to be strongly stationary, or stationary in strict sense, if its complete probability structure is invariant under the shift, i.e., the following equations hold for any τ,

$$\begin{aligned}&p(x_1,t_1;x_2,t_2;\cdots;x_n,t_n) = p(x_1,t_1+\tau;x_2,t_2+\tau;\ldots;x_n,t_n+\tau),\\ &n = 1,2,\ldots.\end{aligned} \tag{2.17}$$

Equation (2.17) indicates that the first-order PDF is independent of the time, i.e., $p(x,t+\tau) = p(x,t) = p(x)$, and the higher-order PDFs only depend on the time difference τ. When Eq. (2.17) only holds for $n = 1$ and 2, the stochastic process is said to be weakly stationary, or stationary in wide (or weak) sense. In most practical problems, only weakly stationary processes are involved; therefore, we may omit the term "weakly" for simplicity.

For a weakly stationary process, the first-order properties are independent of the time; thus $E[X^n(t)] = E[X^n]$, $\mu_X(t) = \mu_X$, and $\sigma_X^2(t) = \sigma_X^2$. The second-order properties only depend on the time difference, i.e.

$$R_{XX}(t_1, t_2) = R_{XX}(\tau), \quad \kappa_{XX}(t_1, t_2) = \kappa_{XX}(\tau),$$
$$\rho_{XX}(\tau) = \frac{\kappa_{XX}(\tau)}{\sigma_X^2}, \quad \tau = t_2 - t_1, \tag{2.18}$$

$$R_{XX}(0) = E\left[X^2(t)\right] = E\left[X^2\right], \quad \kappa_{XX}(0) = \sigma_X^2. \tag{2.19}$$

The physical meaning of the autocorrelation or auto-covariance function is better revealed for the stationary processes. For a given time difference, a larger value of the autocorrelation function indicates a closer relationship of the process at two different time instants. On the other hand, a smaller autocorrelation indicates that the process changes very rapidly in stochastic fashion. With an increasing time difference, the correlation function generally becomes smaller.

Two stochastic processes are jointly stationary if each is stationary, and the following holds for any t_1 and t_2,

$$R_{X_1X_2}(t_1, t_2) = R_{X_1X_2}(t_2 - t_1) = R_{X_1X_2}(\tau). \tag{2.20}$$

From the definitions of autocorrelation and cross-correlation functions, we have

$$R_{XX}(\tau) = R_{XX}(-\tau), \quad R_{X_1X_2}(\tau) = R_{X_2X_1}(-\tau) \tag{2.21}$$

Using the following inequalities

$$E\left\{\left[\frac{X_1(t_1)}{\sqrt{R_{X_1X_1}(0)}} \pm \frac{X_2(t_2)}{\sqrt{R_{X_2X_2}(0)}}\right]^2\right\} = 2 \pm 2\frac{R_{X_1X_2}(\tau)}{\sqrt{R_{X_1X_1}(0)R_{X_2X_2}(0)}} \geq 0, \tag{2.22}$$

we obtain

$$\left|R_{X_1X_2}(\tau)\right| \leq \sqrt{R_{X_1X_1}(0)R_{X_2X_2}(0)} = \sqrt{E\left[X_1^2\right]E\left[X_2^2\right]}, \tag{2.23}$$

$$|R_{XX}(\tau)| \leq R_{XX}(0) = E\left[X^2\right]. \tag{2.24}$$

Inequality (2.24) shows that the autocorrelation function reaches the maximum at $\tau = 0$. It makes sense intuitively since the correlation of a random variable and itself is of course the strongest. Similar properties are present for the auto-covariance and cross-covariance functions

$$\left|\kappa_{X_1X_2}(\tau)\right| \le \sqrt{\kappa_{X_1X_1}(0)\kappa_{X_2X_2}(0)} = \sigma_{X_1}\sigma_{X_2}, \; |\kappa_{XX}(\tau)| \le \sigma_X^2. \tag{2.25}$$

A parameter called correlation time is defined as a quantitative measure for the correlation of a stationary process (Cai and Zhu 2016)

$$\tau_0 = \int_0^\infty |\rho_{XX}(\tau)| \mathrm{d}\tau, \tag{2.26}$$

where $\rho_{XX}(\tau)$ is the autocorrelation coefficient given in (2.10). When the process is totally uncorrelated, $\rho_{XX}(0) = 1$ and $\rho_{XX}(\tau) = 0$ for $\tau \neq 0$, the correlation time $\tau_0 = 0$. In this case, the process at two different time instants is not correlated no matter how the two time instants are closed with each other. When $\rho_{XX}(\tau) \neq 0$ even for $\tau \to \infty$, τ_0 is infinite, indicating a very long correlation time.

For the properties of the derivative process of stationary processes, (2.12), (2.13) and (2.14) lead to

$$\mu_{\dot{X}}(t) = \frac{\mathrm{d}\mu_X(t)}{\mathrm{d}t} = 0, \tag{2.27}$$

$$R_{\dot{X}X}(t_1, t_2) = R_{\dot{X}X}(\tau) = -\frac{\mathrm{d}}{\mathrm{d}\tau} R_{XX}(\tau), \quad \tau = t_2 - t_1, \tag{2.28}$$

$$R_{\dot{X}\dot{X}}(t_1, t_2) = R_{\dot{X}\dot{X}}(\tau) = -\frac{\mathrm{d}^2}{\mathrm{d}\tau^2} R_{XX}(\tau), \quad \tau = t_2 - t_1. \tag{2.29}$$

Since $R_{XX}(\tau)$ is an even function, (2.28) results in

$$R_{\dot{X}X}(0) = E\left[X(t)\dot{X}(t)\right] = -\frac{\mathrm{d}}{\mathrm{d}\tau} R_{XX}(\tau)\bigg|_{\tau=0} = 0. \tag{2.30}$$

Equation (2.30) indicates that a stationary process $X(t)$ and its derivative process $\dot{X}(t)$ are uncorrelated.

For a stochastic process to be modeled and analyzed in practical applications, its characteristics need to be found; at least, its mean and correlation functions need to be estimated. Assume that N sample functions $x_i(t)$ ($i = 1, 2, \ldots, N$) are obtained from measurements, and then the mean and correlation functions can be calculated from the ensemble averaging as follows

$$\mu_X(t) = E[X(t)] = \frac{1}{N} \sum_{i=1}^{N} x_i(t), \tag{2.31}$$

$$R_{XX}(t_1, t_2) = E[X(t_1)X(t_2)] = \frac{1}{N}\sum_{i=1}^{N} x_i(t_1)x_i(t_2). \tag{2.32}$$

The accuracy of the estimations depends on the number of sample functions. The more the sample functions are, the more reliable the estimations are. However, for many physical stochastic processes, the number of sample functions obtained from measurements is inadequate to provide reliable estimations.

The difficulty may be overcome for stationary processes for which the first-order properties are independent of the time and the higher-order properties only depend on the time shift. For such a case, a single sample function in a sufficient long time period may be used to obtain the characteristics of the stochastic process. Consider a record (sample function) $x(t)$ of a stochastic process $X(t)$ in the time period $[0, T]$ where T is sufficiently large. The time average of $X(t)$ is defined as

$$\langle X(t)\rangle_t = \lim_{T\to\infty}\frac{1}{T}\int_0^T x(t)\mathrm{d}t, \tag{2.33}$$

where the angular bracket $\langle[\cdot]\rangle_t$ denotes the time averaging. If the time average is the same for all sample functions, and it is the same as the ensemble average, i.e. $\langle X(t)\rangle_t = E[X(t)] = \mu_X$,, we say that the process is ergodic in the mean.

Ergodicity of a stochastic process can be defined on different levels. A process is ergodic in the mean square if the following is satisfied

$$\left\langle X^2(t)\right\rangle_t = \lim_{T\to\infty}\frac{1}{T}\int_0^T x^2(t)\mathrm{d}t = E\left[X^2(t)\right]. \tag{2.34}$$

Ergodicity in the correlation is defined as

$$\begin{aligned}\langle X(t)X(t+\tau)\rangle_t &= \lim_{T\to\infty}\frac{1}{T}\int_0^T x(t)x(t+\tau)\mathrm{d}t\\ &= E[X(t)X(t+\tau)] = R_{XX}(\tau),\end{aligned} \tag{2.35}$$

which is equivalent to ergodicity in the covariance defined as

$$\begin{aligned}\langle[X(t)-\mu_X][X(t+\tau)-\mu_X]\rangle_t &= \lim_{T\to\infty}\frac{1}{T}\int_0^T x(t)x(t+\tau)\mathrm{d}t - \left[\langle X(t)\rangle_t\right]^2\\ &= E\{[X(t)-\mu_X][X(t+\tau)-\mu_X]\} = \kappa_{XX}(\tau).\end{aligned} \tag{2.36}$$

It is obvious that ergodicity in a higher-order statistics implies ergodicity in the lower orders, and that ergodicity in correlation or covariance implies a weakly stationary process. However, the reverse of either of the above conclusions is not necessarily true.

Ergodicity in the correlation is often assumed for a physical stationary process, in which case the mean value, the mean square value, and correlation function can be estimated from the time averaging shown in (2.33), (2.34), and (2.35) from one sample function in a very long time period. It greatly reduces the time and effort in analytical and numerical investigations.

2.1.3 Spectral Analysis

The autocorrelation function is a second-order statistical property of a stochastic process since it involves two different time instants, and is related to the second-order PDF function. Equivalent to the autocorrelation function is another second-order statistical property, called as power spectral density, which is one of the most important characteristic of a stochastic process in practice.

In this section, we only consider stationary stochastic processes with zero mean such that the autocorrelation function is the same as the auto-covariance function, and both depend only on the time difference.

Consider a stationary stochastic processes $X(t)$ with zero mean. The power spectral density function of $X(t)$ is defined as the Fourier transform of its autocorrelation function, i.e.,

$$S_{XX}(\omega) = \frac{1}{2\pi}\int_{-\infty}^{\infty} R_{XX}(\tau)e^{-i\omega\tau}\mathrm{d}\tau. \tag{2.37}$$

Since $R_{XX}(\tau)$ is non-negative definite, as shown in (2.4), its Fourier transform $S_{XX}(\omega)$ is also non-negative according to Bochner (1959). The inverse of (2.37) is

$$R_{XX}(\tau) = \int_{-\infty}^{\infty} S_{XX}(\omega)e^{i\omega\tau}\mathrm{d}\omega. \tag{2.38}$$

The Fourier transformation pair of (2.37) and (2.38) is referred to as the well-known Wiener-Khintchine theorem.

To reveal the physical meaning of the power spectral density, let $\tau = 0$ in (2.38) to have

$$R_{XX}(0) = E\left[X^2(t)\right] = \int_{-\infty}^{\infty} S_{XX}(\omega)\mathrm{d}\omega. \tag{2.39}$$

Equation (2.39) shows that $S_{XX}(\omega)$ describes the distribution of the mean-square value over the entire frequency domain. In many cases, the mean-square value is a measure of energy, such as the case where $X(t)$ is the displacement of a mechanical system and $X^2(t)$ is proportional to the elastic potential energy. In such cases, the power spectral density $S_{XX}(\omega)$ represents the energy distribution of $X(t)$ in the frequency domain. Thus, it is named as power spectral density.

For two joint stationary processes $X_1(t)$ and $X_2(t)$, the cross-spectral density is defined as the Fourier transform of the cross-correlation function,

$$S_{X_1X_2}(\omega) = \frac{1}{2\pi}\int_{-\infty}^{\infty} R_{X_1X_2}(\tau)e^{-i\omega\tau}\mathrm{d}\tau, \tag{2.40}$$

$$R_{X_1X_2}(\tau) = \int_{-\infty}^{\infty} S_{X_1X_2}(\omega)e^{i\omega\tau}\mathrm{d}\omega. \tag{2.41}$$

Using the symmetrical properties of the autocorrelation and cross-correlation functions, as shown in (2.21), we obtain

$$\begin{aligned} &S_{XX}(-\omega) = S_{XX}(\omega) \geq 0, \\ &S_{X_1X_2}(\omega) = S^*_{X_2X_1}(\omega), \quad S_{X_1X_2}(-\omega) = S^*_{X_2X_1}(\omega). \end{aligned} \tag{2.42}$$

Equation (2.42) indicates that the power spectral density functions are real, nonnegative, and even, and the cross-spectral densities are generally complex, with even real parts and odd imaginary parts.

From (2.38), we obtain

$$\frac{d^n}{d\tau^n}R_{XX}(\tau) = i^n\int_{-\infty}^{\infty} \omega^n S_{XX}(\omega)e^{i\omega\tau}\mathrm{d}\omega. \tag{2.43}$$

Using (2.43), (2.13) and (2.14) lead to

$$R_{\dot{X}X}(\tau) = -\frac{\mathrm{d}}{\mathrm{d}\tau}R_{XX}(\tau) = \int_{-\infty}^{\infty} (-i\omega)S_{XX}(\omega)e^{i\omega\tau}\mathrm{d}\omega, \tag{2.44}$$

$$R_{\dot{X}\dot{X}}(\tau) = -\frac{\mathrm{d}^2}{\mathrm{d}\tau^2}R_{XX}(\tau) = \int_{-\infty}^{\infty} \omega^2 S_{XX}(\omega)e^{i\omega\tau}\mathrm{d}\omega. \tag{2.45}$$

(2.44) and (2.45) indicate that

$$S_{\dot{X}X}(\omega) = (-i\omega)S_{XX}(\omega), \quad S_{\dot{X}\dot{X}}(\omega) = \omega^2 S_{XX}(\omega). \tag{2.46}$$

More generally,

$$S_{X^{(m)}X^{(n)}}(\omega) = (-1)^m i^{m+n} \omega^{m+n} S_{XX}(\omega). \tag{2.47}$$

Thus, the spectral density functions of a stationary process and its derivative processes can be calculated with one another. It is quite useful for a displacement process, its velocity process, and its acceleration processes, if the spectral density of one process is known.

Energy distribution over the entire frequency domain is an important characteristic of a stochastic process. For some processes, the energy is concentrated in a small frequency band, and they are referred as narrowband processes. On the contrary, a process is called wideband or wideband process if its power spectrum has significant values over a large frequency band. Figure 2.2 illustrates schematically the correlation functions and spectral density functions of narrowband and wideband processes, Fig. 2.3 depicts three sample functions for a narrowband process, and Fig. 2.4 depicts three sample functions for a wideband process.

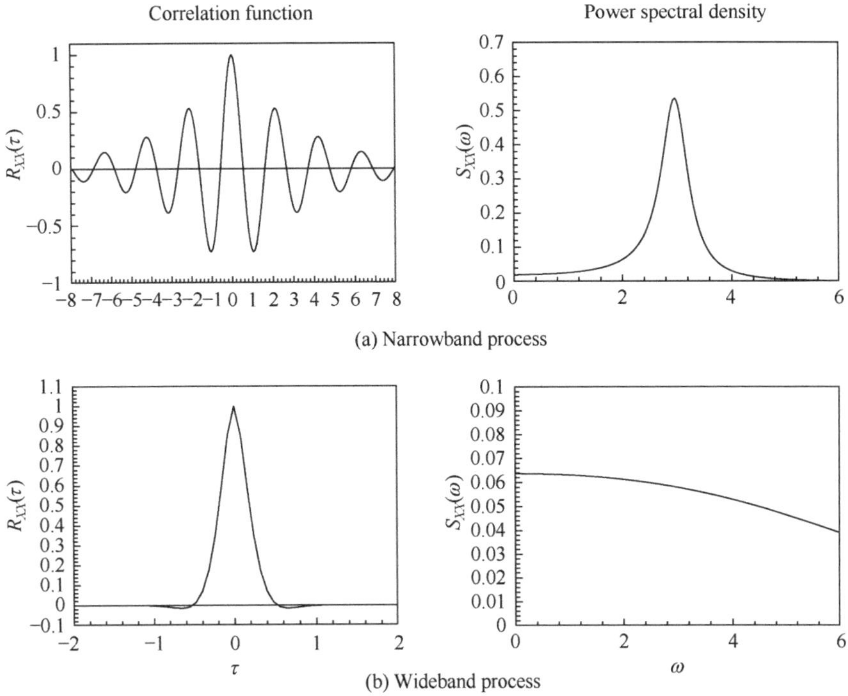

Fig. 2.2 Schematics of narrowband and wideband processes

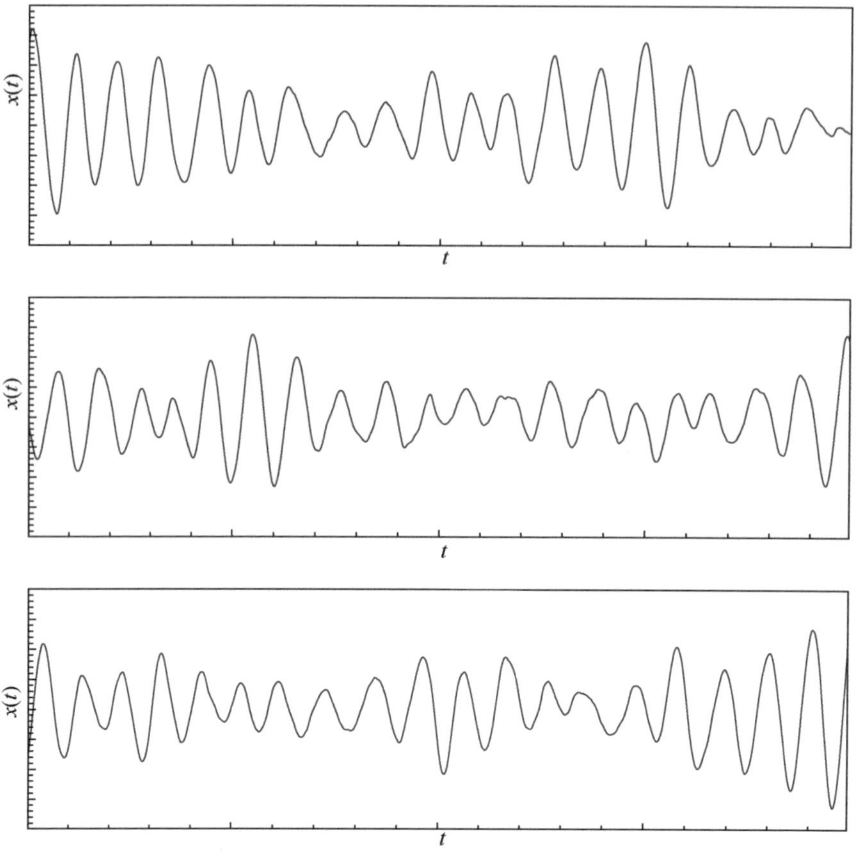

Fig. 2.3 Sample functions of a narrowband process

Example 2.1 Gaussian white noise

A stochastic process is called a Gaussian white noise if it is Gaussian distributed, with zero mean and a constant power spectral density over the entire frequency range $(-\infty, \infty)$. Denoting it as $W_g(t)$, its power spectral density and autocorrelation function are

$$S_{W_g W_g}(\omega) = K, \quad R_{W_g W_g}(\tau) = 2\pi K\delta(\tau) = 2D\delta(\tau), \tag{2.48}$$

where D is called the intensity of the noise. The second equation in (2.48) shows that $\sigma^2_{W_g} = R_{W_g W_g}(0) = \infty$ and $R_{W_g W_g}(\tau) = 0$ for any $\tau \neq 0$. The mean square $\sigma^2_{W_g} = \infty$ indicates that the process has infinite amount of energy, and $R_{W_g W_g}(\tau) = 0$ for $\tau \neq 0$ tells that the process changes infinitely fast. Obviously, such a process does not exist in reality. Nevertheless, it has been used frequently to approximate many physical processes with a broad spectrum due to (i) its simplicity in mathematical treatment,

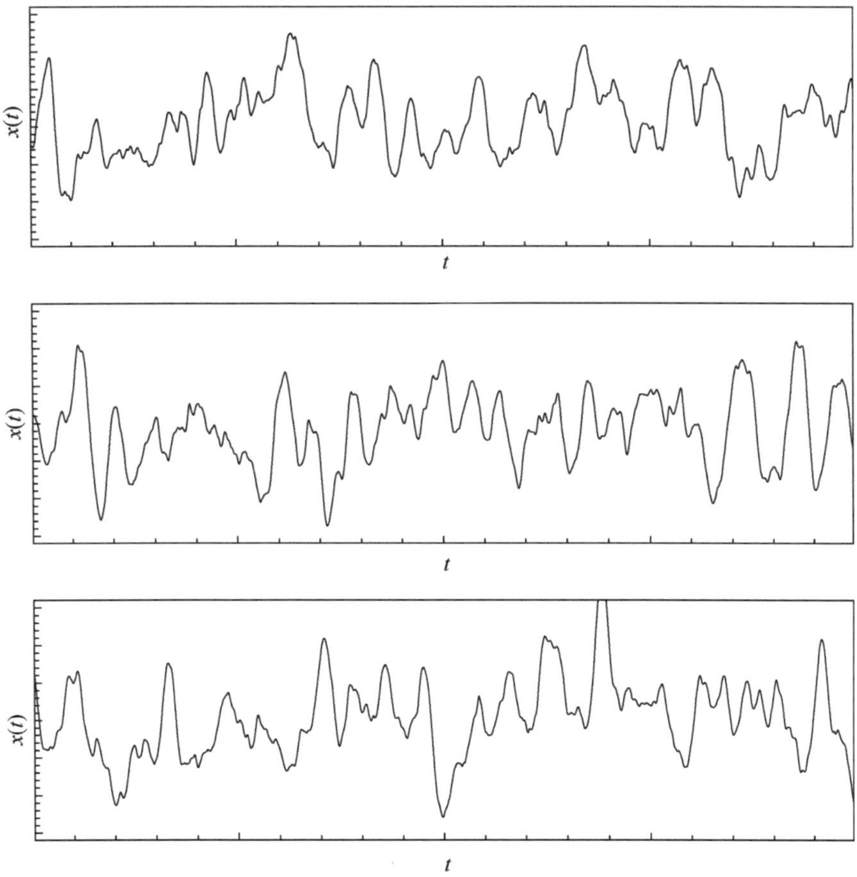

Fig. 2.4 Sample functions of a wideband process

and (ii) rapid decay of system responses to the white noise excitation in the high frequency range.

Example 2.2 Banded white noise

To make the white-noise process more practical, the banded white noise is introduced with the power spectral density in positive ω domain as

$$S_{XX}(\omega) = \begin{cases} S_0, & \omega_0 - \frac{B}{2} \leq |\omega| \leq \omega_0 + \frac{B}{2}, 0 < B \leq 2\omega_0 \\ 0, & \text{otherwise} \end{cases}, \tag{2.49}$$

and the associated correlation function as

$$R_{XX}(\tau) = \frac{4S_0}{\tau} \cos(\omega_0 \tau) \sin\left(\frac{1}{2} B\tau\right). \tag{2.50}$$

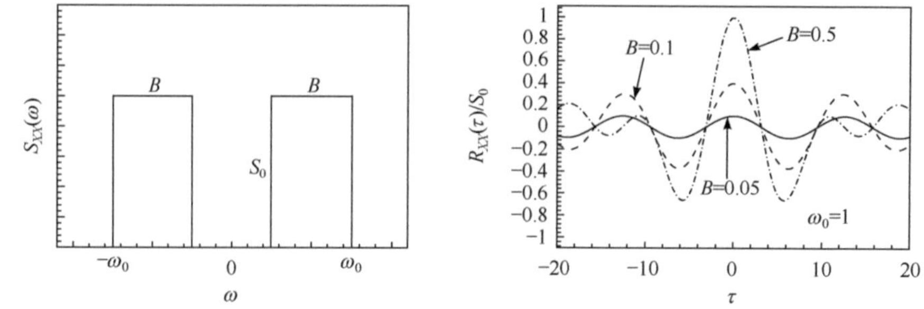

Fig. 2.5 Spectral density and correlation functions of the banded white noise

Figure 2.5 shows the spectral density and correlation function of the banded white noise for the cases of $\omega_0 = 1$ and three different values, 0.05, 0.1 and 0.5, of the band-width parameter B. It is seen that, for the case of a broader band, $B = 0.5$, the correlation function has higher values in the vicinity of $\tau = 0$, and decays rapidly.

Example 2.3 Consider a stationary process $X(t)$ with a power spectral density and an autocorrelation as follows

$$S_{XX}(\omega) = \frac{\sigma_X^2 \alpha}{\pi\left(\omega^2 + \alpha^2\right)}, \quad R_{XX}(\tau) = \sigma_X^2 e^{-\alpha|\tau|}. \tag{2.51}$$

Figure 2.6 shows the spectral density and correlation functions for several different α values. A larger α value corresponds to a broader band spectrum. Since the central frequency is zero, the processes are called the low-pass processes.

For non-stationary processes, the correlation functions depend on two time instants, not just their difference, as shown in Eqs. (2.3) and (2.8). There are two different ways to carry out the spectral analysis. One way is to define the power spectral density and cross spectral density as follows

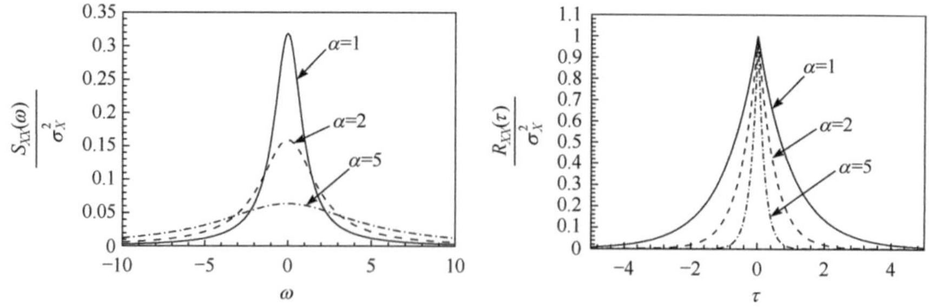

Fig. 2.6 Spectral density and correlation functions of the low-pass processes

$$S_{XX}(\omega_1, \omega_2) = \frac{1}{(2\pi)^2} \int_{-\infty}^{\infty} \int_{-\infty}^{\infty} R_{XX}(t_1, t_2) e^{-i(\omega_1 t_1 - \omega_2 t_2)} \mathrm{d}t_1 \mathrm{d}t_2, \tag{2.52}$$

$$S_{X_1X_2}(\omega_1, \omega_2) = \frac{1}{(2\pi)^2} \int_{-\infty}^{\infty} \int_{-\infty}^{\infty} R_{X_1X_2}(t_1, t_2) e^{-i(\omega_1 t_1 - \omega_2 t_2)} \mathrm{d}t_1 \mathrm{d}t_2. \tag{2.53}$$

Corresponding to the two time instants t_1 and t_2, two frequencies ω_1 and ω_2 are used to specify the frequency domain property of the stochastic processes.

The power spectral density and cross spectral density can also be defined in another way as

$$S_{XX}(\omega, t) = \frac{1}{2\pi} \int_{-\infty}^{\infty} R_{XX}(t, t + \tau) e^{-i\omega\tau} \mathrm{d}\tau, \tag{2.54}$$

$$S_{X_1X_2}(\omega, t) = \frac{1}{2\pi} \int_{-\infty}^{\infty} R_{X_1X_2}(t, t + \tau) e^{-i\omega\tau} \mathrm{d}\tau. \tag{2.55}$$

Intuitively, the definitions (2.54) and (2.55) have clearer physical meaning in the sense that it describes the time-varying energy distribution over the frequency domain. The spectral densities defined in (2.52) through (2.55) are called the general spectral densities (Lin 1967).

2.2 Gaussian Stochastic Processes

A stochastic process $X(t)$ is a family of random variables with t as a parameter. A stochastic process $X(t)$ is said to be Gaussian distributed if all random variables at different time instants are jointly Gaussian distributed. The first-order PDF of a Gaussian stochastic process is

$$p(x, t) = \frac{1}{\sqrt{2\pi}\sigma_X(t)} \exp\left\{-\frac{[x - \mu_X(t)]^2}{2\sigma_X^2(t)}\right\}. \tag{2.56}$$

The second-order PDF is

$$p(x_1, t_1; x_2, t_2) = \frac{1}{2\pi\sigma_1\sigma_2\sqrt{1 - \rho^2}} \times$$
$$\exp\left[-\frac{\sigma_2^2(x_1 - \mu_1)^2 - 2\sigma_1\sigma_2\rho(x_1 - \mu_1)(x_2 - \mu_2) + \sigma_1^2(x_2 - \mu_2)^2}{2\sigma_1^2\sigma_2^2(1 - \rho^2)}\right], \tag{2.57}$$

where $\mu_1 = \mu_X(t_1), \mu_2 = \mu_X(t_2), \sigma_1 = \sigma_X(t_1), \sigma_2 = \sigma_X(t_2)$, and $\rho = \rho_{XX}(t_1, t_2)$.

If a Gaussian stochastic process is weakly stationary, its mean function $\mu_X(t)$ is a constant and covariance function $\kappa_{XX}(t_1,t_2)$ only depends on the time difference $\tau = t_2 - t_1$. Since a Gaussian stochastic process is completely specified by its mean function and covariance function, the weakly stationary property implies strongly stationary property.

Similar to the case of Gaussian random variables, a linear algebraic operation of a Gaussian stochastic process results in another Gaussian stochastic process. This conclusion can be extended to non-algebraic linear operations of differentiation and integration (Lin 1967).

2.3 Markov Processes

A type of stochastic processes is known as Markov process, which is especially important in stochastic dynamics due to (i) its capability of modeling many practical stochastic processes, (ii) the possibility of applying existing mathematical theories of Markov processes to various stochastic problems in different fields, and (iii) the convenience of generating and simulating Markov processes.

Denote a Markov Process by $X(t)$. If both the value of the process X and the parameter t are discrete, it is called Markov chain. If the value of X is continuous, but the parameter t is discrete, it is called the Markov series. In many applications, both the X value and the parameter t are continuous, and it is conventionally called the Markov process, which will be dealt with in this book exclusively.

2.3.1 Markov Processes and Chapman-Kolmogorov-Smoluwski Equation

When a stochastic process is related to a system with short memory, the process at the current instant may be affected only by the most recent history. Such a process belongs to a general category of Markov process. A stochastic process $X(t)$ is said to be Markovian if its conditional probability satisfies

$$\begin{aligned}&\text{Prob}\left[X(t_n) \le x_n | X(t_{n-1}) \le x_{n-1}, \ldots, X(t_1) \le x_1\right]\\&= \text{Prob}\left[X(t_n) \le x_n | X(t_{n-1}) \le x_{n-1}\right]\end{aligned} \tag{2.58}$$

where $t_1 < t_2 < \cdots < t_n$, and Prob[$\cdot$ | $\cdot$] denotes the conditional probability. A sufficient condition for a stochastic process $X(t)$ to be Markovian is that its increments in any two non-overlapping time intervals are independent, i.e., $X(t_2) - X(t_1)$ and $X(t_4) - X(t_3)$ are independent if $t_1 < t_2 \le t_3 < t_4$. If $X(t)$ is a Gaussian process, then the sufficient condition is

$$E\{[X(t_2) - X(t_1)][X(t_4) - X(t_3)]\} = 0, \quad t_1 < t_2 \le t_3 < t_4. \tag{2.59}$$

It is obvious that a Markov process is a mathematical idealization of a real stochastic process. Nevertheless, a great number of stochastic processes can be represented by Markov process. The Brownian motion in Physics is a Markov process. Many noise and signal processes in various fields, such as engineering, communication, ecology, biology etc., are frequently modeled as Markov processes or in terms of Markov processes.

For a Markov process, the definition, Eq. (2.58), can be expressed in terms of the PDF function, i.e.,

$$p(x_n, t_n | x_{n-1}, t_{n-1}; \ldots; x_1, t_1) = p(x_n, t_n | x_{n-1}, t_{n-1}). \tag{2.60}$$

Using (2.60) and the property of the conditional probability density, we have

$$\begin{aligned} p(x_1, t_1; x_2, t_2; \ldots; x_n, t_n) &= p(x_n, t_n | x_{n-1}, t_{n-1}) \\ & p(x_{n-1}, t_{n-1} | x_{n-2}, t_{n-2}) \cdots p(x_1, t_1). \end{aligned} \tag{2.61}$$

It suggests that a higher order PDF can be obtained from the initial PDF and the conditional probability density. In other words, a Markov process is completely characterized by its conditional PDF and the initial probability density. The latter includes the special case in which the initial sate is fixed, i.e., the initial PDF is a Dirac delta function. Therefore, the conditional PDF $p\big(x_k, t_k | x_j, t_j\big)$ is the most important for quantifying a Markov process $X(t)$. It is also called the transition PDF in literature. A Markov process is said to be stationary if its transition PDF is invariant with a time shift, i.e. for any Δt,

$$p\big(x_k, t_k | x_j, t_j\big) = p\big(x_k, t_k + \Delta t | x_j, t_j + \Delta t\big) = p(x_\tau, \tau | x_0, 0), \tag{2.62}$$

where $\tau = t_k - t_j$, and x_τ and x_0 are the state variables of $X(\tau)$ and $X(0)$, respectively. In this case, the stationary PDF may be obtained by letting the transition time interval approach infinity, i.e.,

$$p(x) = \lim_{\tau \to \infty} p(x, \tau | x_0). \tag{2.63}$$

The above concept of a scalar Markov process is readily extended to a vector Markov process. Let $\boldsymbol{X}(t) = [X_1(t), X_2(t), \ldots, X_n(t)]^{\mathrm{T}}$ be a n-dimensional Markov vector process. It satisfies

$$p(\boldsymbol{x}_n, t_n | \boldsymbol{x}_{n-1}, t_{n-1}; \ldots; \boldsymbol{x}_1, t_1) = p(\boldsymbol{x}_n, t_n | \boldsymbol{x}_{n-1}, t_{n-1}). \tag{2.64}$$

It is important to note that a component of Markov vector process may or may not be a scalar Markov process.

Consider three time instants $t_1 < t < t_2$. Equation (2.64) implies

$$p(\boldsymbol{x}_2, t_2; \boldsymbol{y}, t|\boldsymbol{x}_1, t_1) = p(\boldsymbol{x}_2, t_2|\boldsymbol{y}, t)p(\boldsymbol{y}, t|\boldsymbol{x}_1, t_1). \tag{2.65}$$

Integrating (2.65) with respect to $\boldsymbol{y}$, we obtain

$$p(\boldsymbol{x}_2, t_2|\boldsymbol{x}_1, t_1) = \int p(\boldsymbol{x}_2, t_2|\boldsymbol{y}, t)p(\boldsymbol{y}, t|\boldsymbol{x}_1, t_1)d\boldsymbol{y}. \tag{2.66}$$

Equation (2.66) is known as the Chapman-Kolmogorov-Smoluwski equation, which can be considered as a governing integral equation for the transition probability density.

2.3.2 *Markov Diffusion Processes and Fokker–Planck-Kolmogorov (FPK) Equation*

For more convenient analysis, the integral Eq. (2.66) can be converted to an equivalent form of differential equation, known as Fokker–Planck-Kolmogorov (FPK) equation shown below

$$\begin{aligned} &\frac{\partial}{\partial t}p + \sum_{j=1}^{n}\frac{\partial}{\partial x_j}(a_j p) - \frac{1}{2}\sum_{j,k=1}^{n}\frac{\partial^2}{\partial x_j \partial x_k}(b_{jk}p) \\ &+ \frac{1}{3!}\sum_{j,k,l=1}^{n}\frac{\partial^3}{\partial x_j \partial x_k \partial x_l}(c_{jkl}p) - \cdots = 0, \end{aligned} \tag{2.67}$$

where $p = p(\boldsymbol{x}, t|\boldsymbol{x}_0, t_0)$ is the transition probability density, and

$$\begin{aligned} a_j &= a_j(\boldsymbol{x}, t) = \lim_{\Delta t \to 0}\frac{1}{\Delta t}E\big[X_j(t+\Delta t) - X_j(t)|\boldsymbol{X}(t) = \boldsymbol{x}\big], \\ b_{jk} &= b_{jk}(\boldsymbol{x}, t) = \lim_{\Delta t \to 0}\frac{1}{\Delta t}E\big\{\big[X_j(t+\Delta t) - X_j(t)\big][X_k(t+\Delta t) - X_k(t)]|\boldsymbol{X}(t) = \boldsymbol{x}\big\} \\ c_{jkl} &= c_{jkl}(\boldsymbol{x}, t) = \lim_{\Delta t \to 0}\frac{1}{\Delta t}E\big\{\big[X_j(t+\Delta t) - X_j(t)\big]\big\} \\ &\quad\times [X_k(t+\Delta t) - X_k(t)][X_l(t+\Delta t) - X_l(t)]|\boldsymbol{X}(t) = \boldsymbol{x} \end{aligned} \tag{2.68}$$

The functions a_j, b_{jk}, c_{jkl}, ..., are called the derivate moments, describing the change rate of moments of various increments in $\boldsymbol{X}(t)$ at time t, on the condition that $\boldsymbol{X}(t) = \boldsymbol{x}$. The derivation of the FPK Eq. (2.67) from the Chapman-Kolmogorov-Smoluwski Eq. (2.66) is given in detail in the book (Lin and Cai 1995).

In many practical problems, the derivate moments of orders higher than two vanish or can be neglected, then Eq. (2.67) reduces to

$$\frac{\partial}{\partial t}p+\sum_{j=1}^{n}\frac{\partial}{\partial x_j}(a_jp)-\frac{1}{2}\sum_{j,k=1}^{n}\frac{\partial^2}{\partial x_j\partial x_k}(b_{jk}p)=0. \tag{2.69}$$

Conventionally, the name of FPK equation refers to Eq. (2.69). In this case, the Markov process $\boldsymbol{X}(t)$ is called a Markov diffusion process, or simply a diffusion process.

The FPK Eq. (2.69) can be rewritten as

$$\frac{\partial}{\partial t}p+\sum_{j=1}^{n}\frac{\partial}{\partial x_j}G_j=0, \tag{2.70}$$

where

$$G_j=a_jp-\frac{1}{2}\sum_{k=1}^{n}\frac{\partial}{\partial x_k}(b_{jk}p). \tag{2.71}$$

Equation (2.70) is analogous to the continuity equation in fluid mechanics for the conservation of mass in a fluid flow. Thus, Eq. (2.70) may be interpreted as conservation of probability, and G_j as the jth component of the probability flow vector $\boldsymbol{G}(\boldsymbol{x}, t|\boldsymbol{x}_0, t_0)$.

Equation (2.69) is a partial differential equation with first-order derivative in time t, and first and second derivatives in the state variable $\boldsymbol{x}$. For a practical problem, the derivate moments $a_j(\boldsymbol{x},t)$ and $b_{jk}(\boldsymbol{x},t)$ are determined from the system equations of motion. In order to solve the FPK equation, proper initial condition and boundary conditions are required, and they should be established based on the underlined physical problem. In many practical problems, the initial state is fixed, and then the initial condition is

$$p(\boldsymbol{x}, t_0|\boldsymbol{x}_0, t_0)=\delta(\boldsymbol{x}-\boldsymbol{x}_0)=\prod_{j=1}^{n}\delta(x_j-x_{j0}). \tag{2.72}$$

The boundary conditions depend on the sample behavior of the system. For boundaries not at infinity, there are several typical cases: reflective boundary, absorbing boundary, and periodic boundary. For many engineering problems, a boundary at infinity is of importance. At infinity boundary, the probability flow must vanish, i.e.,

$$\lim_{x_j\to\pm\infty}\boldsymbol{G}(\boldsymbol{x}, t|\boldsymbol{x}_0, t_0)=0. \tag{2.73}$$

Moreover, since the total probability is finite, we also have

$$\lim_{x_j\to\pm\infty}p(\boldsymbol{x}, t|\boldsymbol{x}_0, t_0)=0, \tag{2.74}$$

and it approaches zero at least as fast as $|x_j|^{-\alpha}$, where $\alpha > 1$ and is determined from the specific system.

If a Markov diffusion process reaches the stationary state, its stationary PDF is the limit of the transition probability density, as shown in Eq. (2.61). Then the time derivative term in Eq. (2.67) vanishes, resulting in the so-called reduced FPK equation,

$$\sum_{j=1}^{n} \frac{\partial}{\partial x_j}(a_j p) - \frac{1}{2} \sum_{j,k=1}^{n} \frac{\partial^2}{\partial x_j \partial x_k}(b_{jk} p) = 0, \tag{2.75}$$

where $p = p(\boldsymbol{x})$, the stationary probability density, and a_j and b_{jk} are independent of t. Equation (2.75) can also be written as

$$\sum_{j=1}^{n} \frac{\partial}{\partial x_j} G_j = 0, \; G_j = a_j p - \frac{1}{2} \sum_{k=1}^{n} \frac{\partial}{\partial x_k}(b_{jk} p). \tag{2.76}$$

2.3.3 *Wiener Processes and Gaussian White Noise*

The simplest Markov diffusion process is the Wiener process, also known as the Brownian motion process, denoted by $B(t)$. A stochastic process $B(t)$ is said to be a Wiener process if the followings are satisfied: (i) $B(t)$ is a Gaussian process, (ii) $B(0) = 0$, (iii) $E[B(t)] = 0$, and (iv)

$$E[B(t_1)B(t_2)] = \sigma^2 \min(t_1, t_2), \tag{2.77}$$

where σ^2 is known as the intensity of the Wiener process. (2.77) shows that the Wiener process is not a stationary process. Letting $t_1 < t_2 \leq t_3 < t_4$, it follows from (2.77) that

$$\begin{aligned} & E\{[B(t_2) - B(t_1)][B(t_4) - B(t_3)]\} \\ & = E[B(t_2)B(t_4) - B(t_1)B(t_4) - B(t_2)B(t_3) + B(t_1)B(t_3)] \\ & = \sigma^2(t_2 - t_1 - t_2 + t_1) = 0 \end{aligned} \tag{2.78}$$

According to the sufficient condition (2.59), $B(t)$ is a Markov process.

Additional properties can be derived for Wiener process. First, it is continuous in L_2 sense (Cai and Zhu 2016) since its correlation function is continuous along the diagonal $t_1 = t_2$, as shown from (2.76). Furthermore, use Eq. (2.14) to find the correlation function of the derivative $\dot{B}(t)$ as

$$
\begin{aligned}
E\big[\dot{B}(t_1)\dot{B}(t_2)\big] &= \frac{\partial^2}{\partial t_1 \partial t_2} E[B(t_1)B(t_2)] \\
&= \sigma^2 \frac{\partial^2}{\partial t_1 \partial t_2} \min(t_1, t_2) = \sigma^2 \frac{\partial H(t_2 - t_1)}{\partial t_2} = \sigma^2 \delta(t_2 - t_1),
\end{aligned} \tag{2.79}
$$

where $H(t)$ is a Heaviside unit step function

$$
H(t) = \begin{cases} 1, & t > 0 \\ 0, & t < 0 \end{cases}. \tag{2.80}
$$

In deriving (2.79), use has been made of

$$
\frac{\partial}{\partial t_1} \min(t_1, t_2) = \begin{cases} 1, & t_1 < t_2 \\ 0, & t_1 > t_2 \end{cases} = H(t_2 - t_1) \tag{2.81}
$$

Equation (2.79) shows that $B(t)$ is not differentiable in L_2 sense since the mixed second derivative of its correlation function is unbounded along $t_1 = t_2$ (Lin and Cai 1995).

Denote the differential increment of $B(t)$ as

$$
\mathrm{d}B(t) = B(t + \mathrm{d}t) - B(t). \tag{2.82}
$$

Using (2.77), we have

$$
E[\mathrm{d}B(t_1)\mathrm{d}B(t_2)] = \begin{cases} \sigma^2 \mathrm{d}t, & t_1 = t_2 \\ 0, & t_1 \neq t_2 \end{cases}. \tag{2.83}
$$

and hence,

$$
E[B(t + \mathrm{d}t) - B(t)] = 0, \quad E\big\{[B(t + \mathrm{d}t) - B(t)]^2\big\} = \sigma^2 \mathrm{d}t. \tag{2.84}
$$

According to (2.68), the first and second derivate moments are

$$
a = 0, \quad b = \sigma^2. \tag{2.85}
$$

Since the Wiener process $B(t)$ is Gaussian, it can be proved that all derivate moments of orders higher than two vanish. Thus, it is a diffusion process, and the FPK equation is

$$
\frac{\partial}{\partial t} p - \frac{1}{2} \sigma^2 \frac{\partial^2 p}{\partial z^2} = 0, \tag{2.86}
$$

where z is the state variable of $B(t)$, and $p = p(z, t | z_0, t_0)$ is the transition probability density. With the initial condition and boundary conditions given as

$$\lim_{t \to t_0} p(z, t|z_0, t_0) = \delta(z - z_0), \quad \lim_{z \to \pm\infty} p(z, t|z_0, t_0) = 0,$$
$$\lim_{z \to \pm\infty} \frac{\partial}{\partial z} p(z, t|z_0, t_0) = 0. \tag{2.87}$$

the solution of (2.86) is

$$p(z, t|z_0, t_0) = \frac{1}{\sqrt{2\pi(t - t_0)}\sigma} \exp\left[-\frac{(z - z_0)^2}{2\sigma^2(t - t_0)}\right]. \tag{2.88}$$

As expected, $B(t)$ is a Gaussian process with mean z_0 and standard deviation $\sigma\sqrt{t - t_0}$.

It is shown that (Cai and Zhu 2016)

$$dB(t_1)dB(t_2) = \begin{cases} \sigma^2 dt, & t_1 = t_2 \\ 0, & t_1 \neq t_2 \end{cases}, \tag{2.89}$$

which is a much stronger statement than (2.83). (2.89) indicates that d $B(t)$ is of the order of $(\mathrm{d}\, t)^{1/2}$. Thus, it is proved again that d $B(t)$/ d t is unbounded as d $t \to 0$, and process $B(t)$ is not differentiable.

Besides that $B(t)$ is not differentiable, it can also be proved that $B(t)$ has unbounded variation within any finite time interval. Therefore, the Wiener process is only a mathematical idealization of a type of physical processes.

Consider the following equation

$$\frac{dX(t)}{dt} = W_g(t), \quad X(0) = 0, \tag{2.90}$$

where $W_g(t)$ is a Gaussian white noise with a spectral density K, i.e.,

$$E\left[W_g(t)\right] = 0, \quad E\left[W_g(t)W_g(t + \tau)\right] = 2\pi K\delta(\tau) \tag{2.91}$$

In terms of (2.90), $X(t)$ can be expressed as

$$X(t) = \int_0^t W_g(u)du. \tag{2.92}$$

Thus, we have

$$E[X(t)] = \int_0^t E\left[W_g(u)\right]du = 0, \tag{2.93}$$

$$E[X(t_1)X(t_2)] = \int_0^{t_1}\int_0^{t_2} E\big[W_g(u)W_g(v)\big]\mathrm{d}u\mathrm{d}v.$$

$$= 2\pi K \int_0^{t_1}\int_0^{t_2} \delta(u-v)\mathrm{d}u\mathrm{d}v \tag{2.94}$$

Without loss of generality, assume $t_1 < t_2$, and the last integral in (2.94) can be calculated as.

$$\int_0^{t_1}\int_0^{t_2} \delta(u-v)\mathrm{d}u\mathrm{d}v = \int_0^{t_1}\int_0^{t_1} \delta(u-v)\mathrm{d}u\mathrm{d}v + \int_0^{t_1}\int_{t_1}^{t_2} \delta(u-v)\mathrm{d}u\mathrm{d}v = \int_0^{t_1} \mathrm{d}v = t_1 \tag{2.95}$$

Thus, we have

$$E[X(t_1)X(t_2)] = 2\pi K \min(t_1, t_2). \tag{2.96}$$

According to the definition of the Wiener process, $X(t)$ is a Wiener process, and we may write

$$\frac{\mathrm{d}B(t)}{\mathrm{d}t} = W_g(t), \tag{2.97}$$

with an intensity σ^2 of the Wiener process and the spectral density K of the white noise related as

$$\sigma^2 = 2\pi K. \tag{2.98}$$

It is noted that the relationship (2.97) is only a formal one since the Wiener process $B(t)$ is not differentiable.

The validity of (2.97) can also be substantiated by comparing Eqs. (2.79) and (2.48), which shows that the correlation functions of $\dot{B}(t)$ and the white noise $W(t)$ are both the delta function.

2.3.4 Itô Stochastic Differential Equations

As the simplest Markov diffusion process, the Wiener process $B(t)$ can be used to construct other Markov diffusion processes through stochastic differential equations. According to Itô (1951a), a scalar Markov diffusion process $X(t)$ may be generated from

$$\mathrm{d}X(t) = m(X, t)\mathrm{d}t + \sigma(X, t)\mathrm{d}B(t), \tag{2.99}$$

where $B(t)$ is a unit Wiener process, i.e.,

$$E[B(t_1)B(t_2)] = \min(t_1, t_2), \quad E[\mathrm{d}B(t_1)\mathrm{d}B(t_2)] = \begin{cases} 0, & t_1 \neq t_2 \\ \mathrm{d}t, & t_1 = t_2 = t \end{cases}. \tag{2.100}$$

In (2.99), functions m and σ may depend on $X(t)$, and may also depend explicitly on t. Equation (2.99) has the formal solution

$$X(t) = X(0) + \int_0^t m(X, \tau)\mathrm{d}\tau + \int_0^t \sigma(X, \tau)\mathrm{d}B(\tau). \tag{2.101}$$

The last term in (2.101) is a Stieltjes integral (Cai and Zhu 2016), i.e.

$$\int_0^t \sigma[X(\tau), \tau]\mathrm{d}B(\tau) = \lim_{\substack{n\to\infty \\ \Delta_n\to 0}} \sum_{j=1}^n \sigma\left[X\left(\tau_j'\right), \tau_j'\right]\left[B\left(\tau_{j+1}\right) - B\left(\tau_j\right)\right], \tag{2.102}$$

where $0 = \tau_1 < \tau_2 < \cdots < \tau_n < \tau_{n+1} = t$, $\Delta_n = \max(\tau_{j+1} - \tau_j)$, and $\tau_j \le \tau_j' \le \tau_{j+1}$. Since $B(t)$ is a very unusual stochastic process, being not differentiable and of unbounded variation within any finite time interval, this Stieltjes integral must be interpreted properly. The key issue is the choice of τ_j'. Two different choices were proposed, one is the Itô type and another is the Stratonovich type.

The Itô integral selects $\tau_j' = \tau_j$, and the Stieltjes integral (2.102) becomes

$$\int_0^t \sigma[X(\tau), \tau]\mathrm{d}B(\tau) = \lim_{\substack{n\to\infty \\ \Delta_n\to 0}} \sum_{j=1}^n \sigma\left[X\left(\tau_j\right), \tau_j\right]\left[B\left(\tau_{j+1}\right) - B\left(\tau_j\right)\right]. \tag{2.103}$$

In terms of Itô integral, the differential Eq. (2.99) is known as the Itô stochastic differential equation (Itô 1951a). The stochastic process $X(t)$ is a diffusion process, and functions m and σ in (2.99) are called the drift and diffusion coefficients, respectively. In the Itô integral (2.103), the difference $B(\tau_{j+1}) - B(\tau_j)$ is taken in a forward time interval following the time instant τ_j at which function $\sigma[X(\tau), \tau]$ is evaluated; thus, ensuring that $dB(t)$ is independent of $X(t)$ in (2.99).

To obtain the first and second derivate moments, consider a very small Δt and write

$$\begin{aligned} X(t + \Delta t) - X(t) &= \int_t^{t+\Delta t} m[X(\tau), \tau]\mathrm{d}\tau + \int_t^{t+\Delta t} \sigma[X(\tau), \tau]\mathrm{d}B(\tau) \\ &\approx m[X(t), t]\Delta t + \sigma[X(t), t][B(t + \Delta t) - B(t)]\cdot \end{aligned} \tag{2.104}$$

From (2.68), we have

$$a(x,t) = \lim_{\Delta t \to 0} \frac{1}{\Delta t} E[X(t+\Delta t) - X(t) | X(t) = x] = m(x,t), \tag{2.105}$$

$$b(x,t) = \lim_{\Delta t \to 0} \frac{1}{\Delta t} E\{[X(t+\Delta t) - X(t)]^2 | X(t) = x\} = \sigma^2(x,t). \tag{2.106}$$

Therefore, the first and second derivate moments in the FPK equation can be obtained directly from the drift and diffusion coefficients, respectively. However, it should be noticed that the first and second derivate moments are functions of the state variable x, while the drift and diffusion coefficients are functions of the stochastic process $X(t)$.

The above analysis can be extended to n-dimensional case. A n-dimensional diffusion vector process may be generated from the following set of Itô stochastic differential equations

$$\mathrm{d}X_j(t) = m_j(\boldsymbol{X},t)\mathrm{d}t + \sum_{l=1}^{m} \sigma_{jl}(\boldsymbol{X},t)\mathrm{d}B_l(t), \quad j = 1, 2, \cdots, n, \tag{2.107}$$

where $B_l(t)$, $l = 1, 2, \ldots, m$, are assumed to be independent unit Wiener process. The first and second derivate moments for the corresponding FPK equation are obtained as

$$a_j(\boldsymbol{x},t) = m_j(\boldsymbol{X},t)\big|_{\boldsymbol{X}=\boldsymbol{x}}, \quad b_{jk}(\boldsymbol{x},t) = \sum_{l=1}^{m} \sigma_{jl}(\boldsymbol{X},t)\sigma_{kl}(\boldsymbol{X},t)\Big|_{\boldsymbol{X}=\boldsymbol{x}}, \tag{2.108}$$

which can be written as a matrix form

$$\boldsymbol{a}(\boldsymbol{x},t) = \boldsymbol{m}(\boldsymbol{X},t)|_{\boldsymbol{X}=\boldsymbol{x}}, \quad \boldsymbol{b}(\boldsymbol{x},t) = \boldsymbol{\sigma}(\boldsymbol{X},t)\boldsymbol{\sigma}^T(\boldsymbol{X},t)|_{\boldsymbol{X}=\boldsymbol{x}}, \tag{2.109}$$

where the superscript T denotes matrix transpose.

Consider a function of the diffusion vector process $\boldsymbol{X}(t)$, say $F(\boldsymbol{X},t)$, differentiable with respect to t and twice differentiable with respect to the components of $\mathbf{X}(t)$. The differential of $F(\mathbf{X}, t)$ is obtained as

$$\mathrm{d}F(\boldsymbol{X},t) = \frac{\partial F}{\partial t}\mathrm{d}t + \sum_{j=1}^{n} \frac{\partial F}{\partial X_j}\mathrm{d}X_j + \frac{1}{2}\sum_{j,k=1}^{n} \frac{\partial^2 F}{\partial X_j \partial X_k}\mathrm{d}X_j\mathrm{d}X_k + \cdots. \tag{2.110}$$

Substituting $\mathrm{d}X_j$ in (2.107) into (2.110), using the Wiener process property (2.89), and keeping terms of the order $\mathrm{d}t$ and $\mathrm{d}B_l(t)$, we have

$$\mathrm{d}F(X,t) = \left(\frac{\partial F}{\partial t} + \sum_{j=1}^{n} m_j \frac{\partial F}{\partial X_j} + \frac{1}{2} \sum_{j,k=1}^{n} \sum_{l=1}^{m} \sigma_{jl}\sigma_{kl} \frac{\partial^2 F}{\partial X_j \partial X_k} \right) \mathrm{d}t + \sum_{j=1}^{n} \sum_{l=1}^{m} \sigma_{jl} \frac{\partial F}{\partial X_j} \mathrm{d}B_l(t) \tag{2.111}$$

Equation (2.111) is known as Itô's differential rule or Itô's lemma (Itô 1951b). For the case of one dimensional system subjected to one Wiener process excitation, (2.111) reduces to

$$\mathrm{d}F(X,t) = \left(\frac{\partial F}{\partial t} + m \frac{\mathrm{d}F}{\mathrm{d}X} + \frac{1}{2}\sigma^2 \frac{\mathrm{d}^2 F}{\mathrm{d}X^2} \right) dt + \sigma \frac{\mathrm{d}F}{\mathrm{d}X} \mathrm{d}B(t). \tag{2.112}$$

The Itô's lemma shows that the Itô equation for a function of a Markov diffusion process can be derived quite directly and simply. This property is an advantage of using the Itô stochastic differential equation to describe a Markov diffusion process.

Example 2.4 Consider the Itô differential equation

$$\mathrm{d}X(t) = KX(t)\mathrm{d}B(t), \tag{2.113}$$

where K is a constant, and $B(t)$ is a unit Wiener process. Let $Y(t) = \ln X(t)$. According to the Itô differential rule (2.112), we have

$$\mathrm{d}Y(t) = -\frac{1}{2}K^2 \mathrm{d}t + K\mathrm{d}B(t). \tag{2.114}$$

Equation (2.114) is the Itô differential equation for $Y(t)$.

2.3.5 Responses of Systems Under Gaussian White-Noise Excitations

In many engineering systems, the excitations are Gaussian distributed and of wide-band nature, which may be modeled as Gaussian white noises. The simplest of such systems is a one-dimensional system governed by

$$\frac{\mathrm{d}}{\mathrm{d}t} X(t) = f(X,t) + g(X,t)W_g(t), \tag{2.115}$$

where f and g are deterministic functions, and $W_g(t)$ is a Gaussian white noise with a spectral density K, i.e. $E[W_g(t)W_g(t+\tau)] = 2\pi K\delta(\tau)$. From (2.115), we have

$$X(t+\Delta t)-X(t)=\int_t^{t+\Delta t} f[X(u),u]\mathrm{d}u+\int_t^{t+\Delta t} g[X(u),u]W_g(u)\mathrm{d}u. \tag{2.116}$$

Expand $f[X(u),u]$ and $g[X(u),u]$ at $u=t$,

$$\begin{aligned} f[X(u),u] =& f[X(t),t]+(u-t)\frac{\partial}{\partial t}f[X(t),t] \\ &+[X(u)-X(t)]\frac{\partial}{\partial X}f[X(t),t]+\cdots, \end{aligned} \tag{2.117}$$

$$\begin{aligned} g[X(u),u] =& g[X(t),t]+(u-t)\frac{\partial}{\partial t}g[X(t),t] \\ &+[X(u)-X(t)]\frac{\partial}{\partial X}g[X(t),t]+\cdots. \end{aligned} \tag{2.118}$$

In (2.117) and (2.118), we use (2.116) to substitute $X(u)-X(t)$, i.e.,

$$X(u)-X(t)=\int_t^{u} f[X(v),v]\mathrm{d}v+\int_t^{u} g[X(v),v]W_g(v)\mathrm{d}v. \tag{2.119}$$

Combining (2.116) through (2.119) and keeping only the leading terms, we have

$$\begin{aligned} X(t+\Delta t)-X(t) =& f[X(t),t]\Delta t+g[X(t),t]\int_t^{t+\Delta t} W_g(u)\mathrm{d}u \\ &+\left[\frac{\partial}{\partial t}g[X(t),t]\right]\int_t^{t+\Delta t}(u-t)W_g(u)\mathrm{d}u \\ &+\left[\frac{\partial}{\partial X}g[X(t),t]\right]\int_t^{t+\Delta t} W_g(u)\mathrm{d}u\int_t^{u} g[X(v),v]W_g(v)\mathrm{d}v+\cdots. \end{aligned} \tag{2.120}$$

Since $E[W_g(t)]=0$, $E[W_g(t)W_g(t+\tau)]=2\pi K\delta(\tau)$, (2.120) results in

$$\begin{aligned} &\lim_{\Delta t\to 0}\frac{1}{\Delta t}E[X(t+\Delta t)-X(t)] \\ &=f[X(t),t]+\frac{2\pi K}{\Delta t}\left[\frac{\partial}{\partial X}g[X(t),t]\right]\times \\ &\int_t^{t+\Delta t}\mathrm{d}u\int_t^{u} g[X(v),v]\delta(u-v)\mathrm{d}v+\mathrm{O}(\Delta t) \end{aligned} \tag{2.121}$$

To calculate the integral in (2.121), let $\tau = u - v$, and change the order of the double integration to obtain

$$\int_t^{t+\Delta t} \mathrm{d}u \int_t^u g[X(v), v]\delta(u-v)\mathrm{d}v = \int_{-\Delta t}^{0} \delta(\tau)\mathrm{d}\tau \int_{t-\tau}^{t+\Delta t} g[X(u+\tau), u+\tau]\mathrm{d}u \quad (2.122)$$

Substituting (2.122) into (2.121), and considering (i) $\delta(\tau) = 0$ for $\tau \neq 0$, and (ii) $\int_{-\Delta t}^{0} \delta(\tau)\mathrm{d}\tau = \frac{1}{2}$, Eq. (2.121) reduces to

$$\lim_{\Delta t \to 0} \frac{1}{\Delta t} E[X(t+\Delta t) - X(t)] = f(X, t) + \pi K g(X, t)\frac{\partial}{\partial X} g(X, t). \quad (2.123)$$

Using (2.116) again, we obtain

$$\begin{aligned} &\lim_{\Delta t \to 0} \frac{1}{\Delta t} E\left\{[X(t+\Delta t) - X(t)]^2\right\} \\ &= \lim_{\Delta t \to 0} \frac{1}{\Delta t} \int_t^{t+\Delta t} \mathrm{d}u \int_t^{t+\Delta t} g[X(u), u] g[X(v), v] E\left[W_g(u) W_g(v)\right]\mathrm{d}v \\ &= 2\pi K \lim_{\Delta t \to 0} \frac{1}{\Delta t} \int_t^{t+\Delta t} \mathrm{d}u \int_t^{t+\Delta t} g[X(u), u] g[X(v), v]\delta(u-v)\mathrm{d}v \\ &= 2\pi K g^2(X, t). \end{aligned} \quad (2.124)$$

The first and second derivate moments are then obtained from (2.123) and (2.124) as

$$a = f(x, t) + \pi K g(x, t)\frac{\partial}{\partial x} g(x, t), \quad b = 2\pi K g^2(x, t), \quad (2.125)$$

which lead to the Itô equation

$$\mathrm{d}X(t) = \left[f(X, t) + \pi K g(X, t)\frac{\partial}{\partial X} g(X, t)\right]\mathrm{d}t + \sqrt{2\pi K} g(X, t)\mathrm{d}B(t), \quad (2.126)$$

and FPK equation

$$\frac{\partial}{\partial t} p + \frac{\partial}{\partial x}\left[\left(f + \pi K g\frac{\partial g}{\partial x}\right)p\right] - \pi K\frac{\partial^2}{\partial x^2}\left(g^2 p\right) = 0. \quad (2.127)$$

The above analysis can be extended to multi-dimensional cases. Consider a stochastic vector process $\boldsymbol{X}(t) = [X_1(\mathrm{t}), X_1(\mathrm{t}), \cdots, X_n(t)]^{\mathrm{T}}$, governed by

$$\frac{\mathrm{d}}{\mathrm{d}t}X_j(t) = f_j(\boldsymbol{X}, t) + \sum_{l=1}^{m} g_{jl}(\boldsymbol{X}, t)W_{gl}(t), \quad j = 1, 2, \cdots, n, \tag{2.128}$$

where $W_l\ (t)$, $l = 1, 2, \ldots, m$, are Gaussian white noises with the correlations

$$E\left[W_{gl}(t)W_{gs}(t+\tau)\right] = 2\pi K_{ls}\delta(\tau). \tag{2.129}$$

The FPK equation corresponding to (2.128) is

$$\frac{\partial}{\partial t}p + \sum_{j=1}^{n}\frac{\partial}{\partial x_j}\left(a_j p\right) - \frac{1}{2}\sum_{j,k=1}^{n}\frac{\partial^2}{\partial x_j \partial x_k}\left(b_{jk}p\right) = 0, \tag{2.130}$$

where the first and second derivate moments are obtained as

$$a_j(\boldsymbol{x}, t) = f_j(\boldsymbol{x}, t) + \pi\sum_{r=1}^{n}\sum_{l,s=1}^{m} K_{ls}g_{rs}(\boldsymbol{x}, t)\frac{\partial}{\partial x_r}g_{jl}(\boldsymbol{x}, t), \tag{2.131}$$

$$b_{jk}(\boldsymbol{x}, t) = 2\pi\sum_{l,s=1}^{m} K_{ls}g_{jl}(\boldsymbol{x}, t)g_{ks}(\boldsymbol{x}, t). \tag{2.132}$$

Equation (2.132) shows that the Itô equation corresponding to the physical system (2.127) is not unique as long as the diffusion coefficientσ_{jl} satisfy Eq. (2.132). The equivalency of these Itô equations is in the sense that they generate the same FPK equation.

Example 2.5 Consider a single-degree-of-freedom oscillator subjected to random excitations, governed by

$$\ddot{X} + h\left(X, \dot{X}\right) = XW_{g1}(t) + \dot{X}W_{g2}(t) + W_{g3}(t), \tag{2.133}$$

where $h\left(X, \dot{X}\right)$ represents the damping force and restoring force, and $W_{gl}(t)$, $l = 1$, 2, 3, are assumed to be independent Gaussian white noises with the correlations

$$E\left[W_{gl}(t)W_{gs}(t+\tau)\right] = 2\pi K_{ls}\delta_{ls}\delta(\tau), \tag{2.134}$$

where δ_{ls} is the Kronecker delta, i.e., $\delta_{ls} = 1$ if $l = s$ and $\delta_{ls} = 0$ if $l \neq s$. By denoting $X_1 = X$ and $X_2 = \dot{X}$, Eq. (2.133) can be converted to two equations in the state space

$$\begin{aligned}\dot{X}_1 &= X_2,\\ \dot{X}_2 &= -h(X_1, X_2) + X_1W_{g1}(t) + X_2W_{g2}(t) + W_{g3}(t),\end{aligned} \tag{2.135}$$

Using (2.131) and (2.132), we obtain the first and second derivate moments

$$a_1 = x_2, \quad a_2 = -h(x_1, x_2) + \pi K_{22} x_2,$$
$$b_{11} = 0, \quad b_{12} = 0, \quad b_{21} = 0, \quad b_{22} = 2\pi K_{11} x_1^2 + 2\pi K_{22} x_2^2 + 2\pi K_{33}\cdot \tag{2.136}$$

Thus, the FPK equation is

$$\begin{aligned} &\frac{\partial}{\partial t} p + \frac{\partial}{\partial x_1}(x_2 p) + \frac{\partial}{\partial x_2}\{[(-h(x_1, x_2)) + \pi K_{22} x_2]p\} \\ &- \pi \frac{\partial^2}{\partial x_2^2}\left[(K_{11} x_1^2 + K_{22} x_2^2 + K_{33})p\right] = 0. \end{aligned} \tag{2.137}$$

The corresponding Itô equation can be derived from (2.136) and (2.132) as follows

$$\begin{aligned} \mathrm{d}X_1 &= X_2 \mathrm{d}t, \\ \mathrm{d}X_2 &= [-h(X_1, X_2) + \pi K_{22} X_2]\mathrm{d}t, \\ &+ \sqrt{2\pi K_{11}} X_1 \mathrm{d}B_1(t) + \sqrt{2\pi K_{22}} X_2 \mathrm{d}B_2(t) + \sqrt{2\pi K_{33}} \mathrm{d}B_3(t) \end{aligned} \tag{2.138}$$

or

$$\begin{aligned} \mathrm{d}X_1 &= X_2 \mathrm{d}t \\ \mathrm{d}X_2 &= [-h(X_1, X_2) + \pi K_{22} X_2]\mathrm{d}t \\ &+ \sqrt{2\pi \left(K_{11} X_1^2 + K_{22} X_2^2\right)} \mathrm{d}B_1(t) + \sqrt{2\pi K_{33}} \mathrm{d}B_2(t), \end{aligned} \tag{2.139}$$

or

$$\begin{aligned} \mathrm{d}X_1 &= X_2 \mathrm{d}t, \\ \mathrm{d}X_2 &= [-h(X_1, X_2) + \pi K_{22} X_2]\mathrm{d}t \\ &+ \sqrt{2\pi \left(K_{11} X_1^2 + K_{22} X_2^2 + K_{33}\right)} \mathrm{d}B(t). \end{aligned} \tag{2.140}$$

Although (2.138), (2.139), and (2.140) have different forms, they yield the same FPK Eq. (2.137); therefore, they are equivalent.

2.4 Poisson White Noise Processes

2.4.1 Poisson Processes

The Poisson distribution for a type of discrete random variables represents the occurrence of some random events. The Poisson distribution is

$$P_X(n) = \text{Prob}[N = n] = \frac{\mu^n}{n!} e^{-\mu}, \quad n = 0, 1, 2, \cdots\cdots, \tag{2.141}$$

where random variable N is defined in the domain of non-negative integers, and μ is a parameter. In practical problems, a counting process is often required to record the number of events which occur repeatedly and randomly, such as the number of passengers arriving at an airport or a train station, or a sequence of impact forces from sea waves. The Poisson process is defined for such purpose.

Let $N(t)$ denote the number of events arriving (occurring) in the time interval [0, t), and $P_N(n, t)$ denote the probability of the event $N(t) = n$. The counting process is said to be a homogeneous Poisson process, or Poisson process with stationary increments, if the following conditions are satisfied:

1. Independent arrival. Each arrival of the event is independent of the arrivals in the past.
2. Stationary arrival. The probability of one arrival in an interval $(t, t + \text{dt}]$ equals to λdt where λ is the average arrival rate of positive constant. It depends only on the interval length dt, analogous to the stationary case.
3. Isolated arrival. In an infinitesimal interval $(t, t + \text{dt}]$, only one arrival with a PDF λdt is considered, and the probability of two or more arrivals is negligible.

It is shown that the Poisson process has the probability (Lin 1967)

$$P_N(n, t) = \frac{(\lambda t)^n}{n!} e^{-\lambda t}. \tag{2.142}$$

Comparing (2.142) and (2.141), we find that (2.142) is the Poisson distribution with mean value $\mu_N(t) = \text{E}[N(t)] = \lambda t$ and variance $\sigma_N^2(t) = \text{E}[N^2(t)] = \lambda t$. The constant λ has the physical meaning of average arrival rate. The correlation function and the covariance function are (Sun 2006)

$$R_N(t_1, t_2) = \lambda \min(t_1, t_2) + \lambda^2 t_1 t_2, \kappa_N(t_1, t_2) = \lambda \min(t_1, t_2). \tag{2.143}$$

The mean value of the Poisson process increases with the time, and the correlation function depends on both time instants; therefore, it is a non-stationary process although the process is called the Poisson process with stationary increments.

2.4.2 Poisson White Noise

The Poisson process can be used to construct a type of stochastic processes, known as impulsive noise processes, as follows

$$W_p(t) = \sum_{j=1}^{N(T)} Y_j \delta(t - \tau_j), \tag{2.144}$$

where $N(t)$ is a Poisson process, Y_j, $j = 1, 2, \ldots$, are zero-mean independent and identically distributed random variables, and $N(t)$ and Y_j are independent. The mth cumulant function of $W_p(t)$ is (Lin 1967)

$$\kappa_m\left[W_p(t_1), W_p(t_2), \ldots, W_p(t_m)\right] = D_m\delta(t_2 - t_1)\delta(t_3 - t_1)\cdots\delta(t_m - t_1), \quad (2.145)$$

where

$$D_m = \lambda E\left[Y^m\right]. \quad (2.146)$$

It is shown that the impulsive noise $W_p(t)$ is a stationary process, and its correlation function can be obtained from (2.145) as (Cai and Zhu 2016)

$$R_p(\tau) = \kappa_2\left[W_p(t_1), W_p(t_2)\right] = D_2\delta(\tau), \quad \tau = t_2 - t_1. \quad (2.147)$$

The spectral density is then

$$S_p(\omega) = K = \frac{D_2}{2\pi} = \frac{1}{2\pi}\lambda E\left[Y^2\right]. \quad (2.148)$$

Equations (2.147) and (2.148) show that the impulsive noise is a white noise; thus, it is also called the Poisson white noise. Since its spectral density must be finite, $D_2 = \lambda E[Y^2]$ is finite. Consider the case in which the average arrival rate λ of the counting Poisson process $N(t)$ approaches infinite. Thus, $E[Y^2]$ is of order of λ^{-1}, $E[Y^m]$ is at most of order $\lambda^{-m/2}$, and all D_m ($m \geq 3$) approach zero. In this case, the cumulants of order higher than two vanish, and process $W_p(t)$ is Gaussian white noise. On the other hand, if λ is small, it is a non-Gaussian white noise.

The impulsive noise process (2.144) is a mathematical idealization for real stochastic processes in which the duration of the pulses is much shorter than a characteristic time of the investigated dynamical system, known as the relaxation time (Cai and Zhu 2016).

The Poisson white noise can be formally written as the derivative of the compounded Poisson noise $C(t)$,

$$X(t) = \frac{\mathrm{d}}{\mathrm{d}t}C(t), \quad C(t) = \sum_{j=1}^{N(T)} Y_j U\left(t - \tau_j\right), \quad (2.149)$$

where $U(t)$ is a unit step function, i.e.

$$U(t) = \begin{cases} 1, & t \geq 0 \\ 0, & t < 0 \end{cases}. \quad (2.150)$$

It can be shown that

$$E\big[(\mathrm{d}C^n)\big] = D_n\mathrm{d}t = \lambda E\big[Y^n\big]\mathrm{d}t. \tag{2.151}$$

The compounded Poisson noise $C(t)$ is a process with independent increments. The rth-order correlation function of its increment $\mathrm{d}C(t) = C(t + \mathrm{d}t) - C(t)$ is

$$R^{(r)}\big[\mathrm{d}C(t_1), \mathrm{d}C(t_2), \cdots, \mathrm{d}C(t_r)\big]= \lambda E[Y_1^r]\delta(t_2 - t_1)\cdots\delta(t_r - t_1)\mathrm{d}t_1\mathrm{d}t_2\cdots\mathrm{d}t_r. \tag{2.152}$$

Considering the relationship between the correlation function and moment function, the rth-order moment function of $C(t)$ is obtained as

$$E\big[(\mathrm{d}C(t))^r\big] = \lambda E\big[Y_1^r\big]\mathrm{d}t. \tag{2.153}$$

In deriving (2.153), terms with orders higher than $\mathrm{d}t$ are neglected.

Consider r independent compounded Poisson noise $C_1(t), C_2(t), \cdots, C_r(t)$. Similar to the derivation of (2.152) and (2.153), we have the joint moment of $C_{k_1}(t), C_{k_2}(t), \cdots, C_{k_s}(t)$ $(s \le r)$ as follows

$$\begin{aligned} &E\big[\mathrm{d}C_{k_1}(t)\mathrm{d}C_{k_2}(t)\cdots\mathrm{d}C_{k_s}(t)\big] \\ &= \begin{cases} \lambda_k E\big[Y_k^s\big]\mathrm{d}t, & \text{if } k_1 = k_1 = \cdots = k_s = k \\ 0, & \text{if } k_i \ne k_j \text{ for any} i \text{and } j \end{cases} \end{aligned} \tag{2.154}$$

Similar to (2.153), terms with orders higher than $\mathrm{d}t$ are neglected in (2.154). The result of (2.154) can be used to simplify the Itô stochastic differential equations of systems excited by Poisson white noises.

The compounded Poisson noe $C(t)$ can be written as (Hanson 2007; Di Paola and Vasta 1997)

$$C(t) = \int_0^t \int_Y Y\mathcal{P}(\mathrm{d}t, \mathrm{d}Y). \tag{2.155}$$

where $\mathcal{P}(\mathrm{d}t, \mathrm{d}Y) = \mathcal{P}[(t, t + \mathrm{d}t), (Y, Y + \mathrm{d}Y)]$ is Poisson random measure, the integration domain Y is the Poisson mark space. In terms of (2.155), the Poisson counting process $N(t)$ can be expressed as

$$N(t) = \int_0^t \int_Y \mathcal{P}(\mathrm{d}t, \mathrm{d}Y). \tag{2.156}$$

Due to the independence of increments of Poisson random measure (Hanson 2007), $\mathcal{P}(\mathrm{d}t_i, \mathrm{d}Y_k)$ and $\mathcal{P}\big(\mathrm{d}t_j, \mathrm{d}Y_l\big)$ are independent for non-overlapping time and space intervals $(t_i, t_i + \Delta t_i)$,$(Y_k, Y_k + \Delta Y_k]$ and $\big(t_j, t_j + \Delta t_j\big)$,$(Y_l, Y_l + \Delta Y_l)$. The mean and covariance of Poisson random measure are respectively

$$E[\mathcal{P}(\mathrm{d}t, \mathrm{d}Y)] = p_Y(y)\lambda \mathrm{d}y\mathrm{d}t, \tag{2.157}$$

$$\begin{aligned} &\mathrm{Cov}[\mathcal{P}(\mathrm{d}s_1, \mathrm{d}Y_1), \mathcal{P}(\mathrm{d}s_2, \mathrm{d}Y_2)] = \lambda\delta(s_2 - s_1)\mathrm{d}s_1\mathrm{d}s_2, \\ &p_Y(y_1)\delta(y_1 - y_2)\mathrm{d}y_1\mathrm{d}y_2, \end{aligned} \tag{2.158}$$

where $p_Y(y)$ denotes the PDF of Y.

Since it is assumed that the random variables Y_k are independent, the integration form of the compounded Poisson noise $C(t)$ can be written as

$$(\mathrm{d}C(t))^j = \int_Y Y^j \mathcal{P}(\mathrm{d}t, \mathrm{d}Y), \quad j = 1, 2, \cdots, \tag{2.159}$$

Based on the mean and covariance of the Poisson random measure (2.157) and (2.158), the ensemble average of $(\mathrm{d}C(t))^j$ can be calculated as

$$E\big[(\mathrm{d}C(t))^j\big] = \int_Y Y^j E[\mathcal{P}(\mathrm{d}t, \mathrm{d}Y)] = \lambda E\big[Y^j\big]\mathrm{d}t. \tag{2.160}$$

It can be seen from (2.153) and (2.160) that the moments derived from the differential form and integration form in terms of Poisson random measure are the same (Di Paola and Vasta 1997).

It is noted that the Poisson white noise is a stochastic jump process which is not a continuous process. When its average arrival rate approaches infinity, it becomes a Gaussian white process, a continuous process.

2.4.3 Stochastic Differential-Integral Equation and FPK Equation

The Poisson white noise is a type of random impulses with both the impulse magnitudes and impulse arrival times being random. Before studying stochastic dynamical systems, it is necessary to establish the corresponding stochastic differential rule. Since 1990's, two different differential rules for systems excited by Poisson white noises have been proposed. The one proposed by Di Paola and Falsone (1993a, b) is for non-Gaussian δ-correlated stochastic processes including Poisson white noise by confirming the existence of Wong-Zakai correction terms. On contrary, Hu (1993, 1994, 1995) and Grigoriu (1998) proposed, respectively, a set of different rules for Poisson white noise, treating the parametric and external excitations in the same way without adding the Wong-Zakai correction terms. Gough (1999) and Zygadlo (2003) verified the correctness of the one by Di Paola and Falsone. This book will use the stochastic differential rule of Di Paola and Falsone. The results in Chap. 6 will prove the validity of the rule.

Assume a system under both the Gaussian white noises and Poisson white noises, governed by (Jia and Zhu 2014a, b)

$$\dot{\boldsymbol{X}} = \boldsymbol{f}(\boldsymbol{X}, t) + \boldsymbol{g}(\boldsymbol{X}, t)\boldsymbol{W}_g(t) + \boldsymbol{h}(\boldsymbol{X}, t)\boldsymbol{W}_p(t), \tag{2.161}$$

where $\boldsymbol{X} = [X_1, X_2, \ldots, X_n]^{\mathrm{T}}$ is a n-dimensional vector of stochastic responses, $\boldsymbol{f} = [f_1, f_2, \ldots, f_n]^{\mathrm{T}}$, $\boldsymbol{g} = [g_{ij}]_{n\times n_g}$ is a $n \times n_g$ matrix, $\boldsymbol{h} = [h_{ij}]_{n\times n_p}$ is a $n \times n_p$ matrices,$\boldsymbol{W}_g(t) = [W_{g1}, W_{g2}, \ldots, W_{gn_g}]^{\mathrm{T}}$ is a n_g—dimensional vector of Gaussian white noises with

$$E\big[W_{gk}(t)\big] = 0, \quad E\big[W_{gk}(t)W_{gl}(t+\tau)\big] = 2D_{kl}\delta(\tau), \tag{2.162}$$

and $\boldsymbol{W}_p(t) = [W_{p1}, W_{p2}, \ldots, W_{pn_p}]^{\mathrm{T}}$ is a n_p-dimensional vector of Poisson white noises with all components independent.

Equation (2.161) can be transformed into the following Stratonovich stochastic differential equation

$$\mathrm{d}\boldsymbol{X} = \boldsymbol{f}(\boldsymbol{X}, t)\mathrm{d}t + \boldsymbol{\sigma}(\boldsymbol{X}, t)\mathrm{d}^{\circ}\boldsymbol{B}(t) + \boldsymbol{h}(\boldsymbol{X}, t)\mathrm{d}^{\circ}\boldsymbol{C}(t), \tag{2.163}$$

where $\sigma_{ik} = [\boldsymbol{gL}]_{ik}$ with $\boldsymbol{LL}^{\mathrm{T}} = 2\boldsymbol{D}$, the notation "$d^{\circ}$" indicates a differential operation using the Stratonovich differential rule, $\boldsymbol{C}(t)$ is a n_p-dimensional vector of compound Poisson processes corresponding to $\boldsymbol{W}_p(t)$. The scalar form of Eq. (2.163) is

$$\begin{aligned} \mathrm{d}X_i &= f_i(\boldsymbol{X}, t)\mathrm{d}t + \sum_{k=1}^{n_g} \sigma_{ik}(\boldsymbol{X}, t)\mathrm{d}^{\circ}B_k(t) + \sum_{l=1}^{n_p} h_{il}(\boldsymbol{X}, t)\mathrm{d}^{\circ}C_l(t), \\ i &= 1, 2, \ldots, n \end{aligned} \tag{2.164}$$

By adding the Wong-Zakai correction terms for both the Gaussian and Poisson white noises (Wong and Zakai, 1965; Di Paola and Falsone, 1993a; 1993b), equations in (2.164) are re-written as the Itô stochastic differential equations

$$\begin{aligned} \mathrm{d}X_i &= m_i(\boldsymbol{X}, t)\mathrm{d}t + \sum_{k=1}^{n_g} \sigma_{ik}(\boldsymbol{X}, t)\mathrm{d}B_k(t) \\ &+ \sum_{l=1}^{n_p}\sum_{j=1}^{\infty}\left[\frac{1}{j!}H_{il}^{(j)}(\boldsymbol{X}, t)(\mathrm{d}C_l(t))^j\right], \\ i &= 1, 2, \ldots, n \end{aligned} \tag{2.165}$$

where are independent unit Wiener processes, are independent compound Poisson processes, and

$$m_i(\boldsymbol{X},t)=f_i(\boldsymbol{X},t)+\frac{1}{2}\sum_{j=1}^{n}\sum_{k=1}^{n_g}\sigma_{jk}(\boldsymbol{X},t)\frac{\partial\sigma_{ik}}{\partial X_j}(\boldsymbol{X},t), \tag{2.166}$$

$$H_{il}^{(j)}(\boldsymbol{X},t)=\sum_{s=1}^{n}H_{sl}(\boldsymbol{X},t)\frac{\partial H_{il}^{(j-1)}}{\partial X_s}(\boldsymbol{X},t),\ H_{il}^{(1)}(\boldsymbol{X},t)=h_{il}(\boldsymbol{X},t). \tag{2.167}$$

Compared with (2.164), $\frac{1}{2}\sum\limits_{j=1}^{n}\sum\limits_{k=1}^{n_g}\sigma_{jk}\frac{\partial\sigma_{ik}}{\partial X_j}$ in (2.166) is the Wong-Zakai correction terms. The terms $\sum\limits_{l=1}^{n_p}\sum\limits_{j=1}^{\infty}\left[\frac{1}{j!}H_{il}^{(j)}(\mathrm{d}C_l(t))^j\right]$ in (2.165) are the correction terms due to compound Poisson white noises.

In terms of the integral form of compound Poisson white noises in (2.155), equations in (2.165) can be written as the following stochastic differential-integral equations

$$\begin{aligned}\mathrm{d}X_i&=m_i(\boldsymbol{X},t)\mathrm{d}t+\sum_{k=1}^{n_g}\sigma_{ik}(\boldsymbol{X},t)\mathrm{d}B_k(t)\\&\quad+\sum_{l=1}^{n_p}\int_Y\boldsymbol{\gamma}(\boldsymbol{X},Y_l,t)\mathcal{P}_l(\mathrm{d}t,\mathrm{d}Y_l)\\ i&=1,2,\ldots,n\end{aligned} \tag{2.168}$$

where $\boldsymbol{\gamma}(\boldsymbol{X},Y_l,t)=[\gamma_{il}]_{n\times n_p}$, $\gamma_{il}=\sum\limits_{j=1}^{\infty}\frac{1}{j!}H_{il}^{(j)}(\boldsymbol{X},t)Y_l^j$, and $\mathcal{P}_l$ $(l=1,2,\ldots,n_p)$ are independent Poisson random measure corresponding to the compound Poisson process, satisfying (2.157) through (2.160). Equation (2.168) is the stochastic differential-integral equation in the Itô sense corresponding to the Stratonovich stochastic differential equation (2.164).

The process satisfies C_l corresponding stochastic jump-diffusion chain rule (Hanson 2007; Di Paola and Vasta 1997). Assume that $F(\boldsymbol{X}(t),t)$ is a function of $\boldsymbol{X}(t)$. Based on the stochastic jump-diffusion chain rule, the stochastic differential equation of $F(\boldsymbol{X}(t),t)$ can be derived from those of $\boldsymbol{X}(t)$ in (2.168) as

$$\begin{aligned}\mathrm{d}F(\boldsymbol{X}(t),t)&=\left(\frac{\partial F}{\partial t}+\sum_{j=1}^{n}f_j\frac{\partial F}{\partial X_j}+\frac{1}{2}\sum_{j=1}^{n}\sum_{k=1}^{n}\sum_{l=1}^{n_g}\sigma_{kl}\sigma_{jl}\frac{\partial^2F}{\partial X_j\partial X_k}\right)\mathrm{d}t\\&\quad+\sum_{j=1}^{n}\sum_{l=1}^{n_g}\sigma_{jl}\frac{\partial F}{\partial X_j}\mathrm{d}B_l(t)\\&\quad+\sum_{l=1}^{n_p}\int_{Y_l}\left[F\left(\boldsymbol{X}(t)+\hat{\boldsymbol{\gamma}}_l,t\right)-F(\boldsymbol{X}(t),t)\right]\mathcal{P}_l(\mathrm{d}t,\mathrm{d}Y_l)\end{aligned} \tag{2.169}$$

where $\hat{\boldsymbol{\gamma}}_l$ is the lth column of the matrix $\boldsymbol{\gamma}$. It is seen that a jump term is added in (2.169) compared with the chain rule in (2.111) for the case of Gaussian white noise excitations.

According to (Hanson 2007; Zeng and Zhu 2011), the FPK equation governing the PDF of $\boldsymbol{X}(t)$, corresponding to stochastic differential-integral equation Eq. (2.168), is

$$\frac{\partial p(\boldsymbol{x},t)}{\partial t} = \sum_{k=1}^{\infty}(-1)^k \sum_{\substack{s_1+s_2+\cdots s_n=k \\ s_i \in N, i=1,2,\ldots,n}} \left[\prod_{j=1}^{n} \frac{1}{s_j!}\frac{\partial^{s_j}}{\partial x_j^{s_j}}\right]\left[A_{s_1,s_2,\cdots,s_n}(\boldsymbol{x},t)p(\boldsymbol{x},t)\right], \tag{2.170}$$

where $\mathbb{N}$ is the set of natural numbers, $p(\boldsymbol{x}, t)$ is the PDF or transition probability density, and

$$\begin{aligned} A_{s_1,s_2,\cdots,s_n}(\boldsymbol{x},t) &= \lim_{\Delta t\to 0}\frac{1}{\Delta t}E\left[\prod_{j=1}^{n}[X_i(t+\Delta t)-X_i(t)]^{s_j}|\boldsymbol{X}(t)=\boldsymbol{x}_0\right] \\ &= \frac{1}{\mathrm{d}t}E\left[\prod_{j=1}^{n}(\mathrm{d}X_i)^{s_j}|\boldsymbol{X}(t)=\boldsymbol{x}_0\right]. \end{aligned} \tag{2.171}$$

Example 2.6 Find the FPK equation of the following one-dimensional system subjected to a Poisson white noise excitation

$$\frac{\mathrm{d}}{\mathrm{d}t}X(t) = f(X,t) + h(X,t)W_p(t), \tag{2.172}$$

where f and h are deterministic functions, and $W_p(t)$ is a Poisson white noise given in Eq. (2.144) with its cumulant functions given by (2.145), i.e.

$$\kappa_m\left[W_p(t_1), W_p(t_2), \cdots, W_p(t_m)\right] = D_m\delta(t_2-t_1)\delta(t_3-t_1)\cdots\delta(t_m-t_1), \tag{2.173}$$

where $D_m = \lambda E[Y^m]$ shown in (2.146).

In terms of the compounded Poisson process $C(t)$ defined in (2.149), (2.162) is first written as the Stratonovich stochastic differential equation

$$\mathrm{d}X(t) = f(X,t)\mathrm{d}t + h(X,t)\mathrm{d}^{\circ}C(t), \tag{2.174}$$

and then transferred to the following Itô stochastic differential equation according to the stochastic differential rule of Di Paola and Falsone,

$$dX = f(X,t)dt + h(X,t)dC(t) + \sum_{k=2}^{\infty} \frac{1}{k!} H_k(X,t)[dC(t)]^k, \tag{2.175}$$

where

$$H_k(X,t) = \frac{\partial H_{k-1}(X,t)}{\partial X} h(X,t), \quad H_1(X,t) = h(X,t). \tag{2.176}$$

From (2.176),

$$\begin{aligned} H_2(X,t) &= h\frac{\partial h}{\partial X}, H_3(X,t) = h^2\frac{\partial^2 h}{\partial X^2} + h\left(\frac{\partial h}{\partial X}\right)^2, \\ H_4(X,t) &= h^3\frac{\partial^3 h}{\partial X^3} + 4h^2\frac{\partial h}{\partial X}\frac{\partial^2 h}{\partial X^2} + h\left(\frac{\partial h}{\partial X}\right)^3, \ldots\ldots. \end{aligned} \tag{2.177}$$

Then (2.175) is specified as

$$\begin{aligned} dX(t) &= f\,dt + h dC(t) + \frac{1}{2}h\frac{\partial h}{\partial X}[dC(t)]^2 \\ &+ \frac{1}{3!}\left[h^2\frac{\partial^2 h}{\partial X^2} + h\left(\frac{\partial h}{\partial X}\right)^2\right][dC(t)]^3. \\ &+ \frac{1}{4!}\left[h^3\frac{\partial^3 h}{\partial X^3} + 4h^2\frac{\partial h}{\partial X}\frac{\partial^2 h}{\partial X^2} + h\left(\frac{\partial h}{\partial X}\right)^3\right][dC(t)]^4 + \ldots \end{aligned} \tag{2.178}$$

Assuming that the probability distribution of random magnitude Y is symmetric, then $E[C^k(t)] = 0$ for $k =$ odd numbers. Noticing Eq. (4.11), i.e., $E\{[dC(t)]^n\} = D_n dt$, and neglecting D_n for $n > 4$, the first four derivate moments can be obtained directly from (2.178) as follows

$$\begin{aligned} a &= \lim_{\Delta t \to 0} \frac{1}{\Delta t} E[X(t+\Delta t) - X(t)|X(t) = x] \\ &= f + \frac{D_2}{2}h\frac{\partial h}{\partial x} + \frac{D_4}{4!}\left[h^3\frac{\partial^3 h}{\partial x^3} + 4h^2\frac{\partial h}{\partial x}\frac{\partial^2 h}{\partial x^2} + h\left(\frac{\partial h}{\partial x}^3\right)\right] \end{aligned} \tag{2.179}$$

$$\begin{aligned} b &= \lim_{\Delta t \to 0} \frac{1}{\Delta t} E\{[X(t+\Delta t) - X(t)]^2 |X(t) = x\} \\ &= D_2 h^2 + D_4\left[\frac{1}{3}h^3\frac{\partial^2 h}{\partial x^2} + \frac{7}{12}h^2\left(\frac{\partial h}{\partial x}^2\right)\right], \end{aligned} \tag{2.180}$$

$$c = \lim_{\Delta t \to 0} \frac{1}{\Delta t} E\{[X(t+\Delta t) - X(t)]^3 |X(t) = x\} = \frac{3D_4}{2}h^3\frac{\partial h}{\partial x}, \tag{2.181}$$

$$d = \lim_{\Delta t \to 0} \frac{1}{\Delta t} E\{[X(t + \Delta t) - X(t)]^4 | X(t) = x\} = D_4 h^4. \tag{2.182}$$

The FPK equation is then

$$\begin{aligned}
&\frac{\partial}{\partial t} p + \frac{\partial}{\partial x}\left\{\left(f + \frac{D_2}{2} h \frac{\partial h}{\partial x} + \frac{D_4}{4!}\left[h^3 \frac{\partial^3 h}{\partial x^3} + 4h^2 \frac{\partial h}{\partial x}\frac{\partial^2 h}{\partial x^2} + h\left(\frac{\partial h}{\partial x}\right)^3\right]\right) p\right\} \\
&\quad - \frac{1}{2!}\frac{\partial^2}{\partial x^2}\left\{\left(D_2 h^2 + D_4\left[\frac{1}{3} h^3 \frac{\partial^2 h}{\partial x^2} + \frac{7}{12} h^2 \left(\frac{\partial h}{\partial x}\right)^2\right]\right) p\right\} \\
&\quad + \frac{1}{3!} D_4 \frac{\partial^3}{\partial x^3}\left(\frac{3}{2} h^3 \frac{\partial h}{\partial x} p\right) - \frac{1}{4!} D_4 \frac{\partial^4}{\partial x^4}(h^4 p) + \cdots\cdots = 0.
\end{aligned} \tag{2.183}$$

(i) Consider the case of $h(X, t) = 1$, i.e. only an external excitation is present,

$$\frac{\mathrm{d}}{\mathrm{d}t} X(t) = f(X) + W_p(t). \tag{2.184}$$

The corresponding stochastic differential equation is

$$\mathrm{d}X(t) = f(X)\mathrm{d}t + \mathrm{d}C(t), \tag{2.185}$$

and the first four derivate moments are

$$a = f, \quad b = D_2, \quad c = 0, \quad d = D_4, \tag{2.186}$$

(ii) Consider another case of $h(X, t) = X$, i.e. there is only a parametric excitation

$$\frac{\mathrm{d}}{\mathrm{d}t} X(t) = f(X) + X W_p(t). \tag{2.187}$$

The corresponding Itô stochastic differential equation is

$$\begin{aligned}
\mathrm{d}X(t) &= f(X)\mathrm{d}t + X\mathrm{d}C(t) + \frac{1}{2!} X[\mathrm{d}C(t)]^2 \\
&\quad + \frac{1}{3!} X[\mathrm{d}C(t)]^3 + \frac{1}{4!} X[\mathrm{d}C(t)]^4 + \cdots,
\end{aligned} \tag{2.188}$$

and the first four derivate moments are

$$a = f + \left(\frac{D_2}{2} + \frac{D_4}{24}\right)x, \quad b = \left(D_2 + \frac{7D_4}{12}\right)x^2, \quad c = \frac{3D_4}{2} x^3, \quad d = D_4 x^4. \tag{2.189}$$

Similar to the case of Gaussian white noises, the analysis can be extended to multi-dimensional cases.

Example 2.7 Consider a SDOF oscillator

$$\ddot{X} + f\left(X, \dot{X}\right) = h\left(X, \dot{X}\right) W_p(t), \tag{2.190}$$

where $h\left(X, \dot{X}\right)$ represents the damping force and restoring force, and $W_p(t)$ is a Poisson white noise of zero mean. By denoting $X_1 = X$ and $X_2 = \dot{X}$, Eq. (2.189) can be converted to two equations in the state space

$$\dot{X}_1 = X_2, \quad \dot{X}_2 = -f(X_1, X_2) + h(X, X_1) W_p(t). \tag{2.191}$$

According to (2.168), we have the Itô differential equations

$$\begin{aligned}
\mathrm{d}X_1 &= X_2 \mathrm{d}t, \\
\mathrm{d}X_2 &= -f \mathrm{d}t + h \mathrm{d}C(t) + \frac{1}{2} h \frac{\partial h}{\partial X_2} [\mathrm{d}C(t)]^2 \\
&+ \frac{1}{3!}\left[h^2 \frac{\partial^2 h}{\partial X_2^2} + h\left(\frac{\partial h}{\partial X_2}\right)^2\right][\mathrm{d}C(t)]^3 \\
&+ \frac{1}{4!}\left[h^3 \frac{\partial^3 h}{\partial X_2^3} + 4h^2 \frac{\partial h}{\partial X_2} \frac{\partial^2 h}{\partial X_2^2} + h\left(\frac{\partial h}{\partial X_2}\right)^3\right][\mathrm{d}C(t)]^4 + \cdots\cdots,
\end{aligned} \tag{2.192}$$

According to (2.68) and neglecting D_n for $n > 4$, the first four derivate moments are

$$\begin{aligned}
a_1 &= x_2, \\
a_2 &= -f + \frac{D_2}{2} h \frac{\partial h}{\partial x_2} + \frac{D_4}{4!}\left[h^3 \frac{\partial^3 h}{\partial x_2^3} + 4h^2 \frac{\partial h}{\partial x_2} \frac{\partial^2 h}{\partial x_2^2} + h\left(\frac{\partial h}{\partial x_2}\right)^3\right],
\end{aligned} \tag{2.193}$$

$$b_{22} = D_2 h^2 + D_4\left[\frac{1}{3} h^3 \frac{\partial^2 h}{\partial x_2^2} + \frac{7}{12} h^2\left(\frac{\partial h}{\partial x_2}\right)^2\right], \tag{2.194}$$

$$c_{222} = \frac{3D_4}{2} h^3 \frac{\partial h}{\partial x_2}, \quad d_{2222} = D_4 h^4. \tag{2.195}$$

Other b_{ij}, c_{ijk}, and d_{ijkl} are zeros. Thus, the FPK equation is

$$\begin{aligned}
&\frac{\partial}{\partial t}p + \frac{\partial}{\partial x_1}(x_2 p) \\
&+ \frac{\partial}{\partial x_2}\left\{\left(-f + \frac{D_2}{2}h\frac{\partial h}{\partial x_2} + \frac{D_4}{4!}\left[h^3\frac{\partial^3 h}{\partial x_2^3} + 4h^2\frac{\partial h}{\partial x_2}\frac{\partial^2 h}{\partial x_2^2} + h\left(\frac{\partial h}{\partial x_2}\right)^3\right]p\right)\right\} \\
&- \frac{1}{2!}\frac{\partial^2}{\partial x_2^2}\left\{\left(D_2 h^2 + D_4\left[\frac{1}{3}h^3\frac{\partial^2 h}{\partial x_2^2} + \frac{7}{12}h^2\left(\frac{\partial h}{\partial x_2}\right)^2\right]\right)p\right\} \\
&+ \frac{1}{3!}D_4\frac{\partial^3}{\partial x_2^3}\left(\frac{3}{2}h^3\frac{\partial h}{\partial x_2}p\right) - \frac{1}{4!}D_4\frac{\partial^4}{\partial x_2^4}(h^4 p) + \cdots\cdots = 0.
\end{aligned} \tag{2.196}$$

2.5 Fractional Gaussian Processes

2.5.1 Fractional Calculus

The fractional calculus is a theory of integrals and derivatives of arbitrary positive order, which is an extension of common notions of integrals and derivatives, i.e., differentiation and integration of positive integer order.

Assume that n is a positive integer, and consider a sequence of nth-order derivatives

$$\frac{\mathrm{d}f(t)}{\mathrm{d}t}, \frac{\mathrm{d}^2 f(t)}{\mathrm{d}t^2}, \cdots\cdots, \frac{\mathrm{d}^n f(t)}{\mathrm{d}t^n}, \tag{2.197}$$

and a sequence of n-fold integrals

$$\begin{aligned}
&\int_a^t f(\tau)\mathrm{d}\tau, \\
&\int_a^t \mathrm{d}\tau_1 \int_a^{\tau_1} f(\tau_2)\mathrm{d}\tau_2 = \int_a^t (t-\tau)f(\tau)\mathrm{d}\tau, \\
&\int_a^t \mathrm{d}\tau_1 \int_a^{\tau_1} \mathrm{d}\tau_2 \int_a^{\tau_2} f(\tau_3)\mathrm{d}\tau_3 = \frac{1}{2}\int_a^t (t-\tau)^2 f(\tau)\mathrm{d}\tau, \\
&\cdots\cdots\cdots \\
&\int_a^t \mathrm{d}\tau_1 \int_a^{\tau_1} \mathrm{d}\tau_2 \ldots \int_a^{\tau_{n-1}} f(\tau_n)\mathrm{d}\tau_n = \frac{1}{(n-1)!}\int_a^t (t-\tau)^{n-1} f(\tau)\mathrm{d}\tau.
\end{aligned} \tag{2.198}$$

The fractional integral of an arbitrary order β, denoted by ${}_aI_t^\beta f(t)$, $\beta > 0$, is defined as

$$ {}_aI_t^\beta f(t) = \frac{1}{\Gamma(\beta)} \int_a^t (t-\tau)^{\beta-1} f(\tau)\mathrm{d}\tau \tag{2.199}$$

where $\Gamma(\cdot)$ is the Gamma function. It is consistent with the positive integer-order integral, i.e., the last equation in (2.198) since $\Gamma(n) = (n-1)!$.

The fractional derivative of an arbitrary order β is denoted by a similar notation, i.e., ${}_aD_t^\beta f(t)$, $\beta > 0$, with the following definition

$$\begin{aligned} {}_aD_t^\beta f(t) &= \frac{\mathrm{d}^n}{\mathrm{d}t^n}\left[{}_aI_t^{n-\beta} f(t)\right] \\ &= \frac{1}{\Gamma(n-\beta)} \frac{\mathrm{d}^n}{\mathrm{d}t^n} \int_a^t (t-\tau)^{n-\beta-1} f(\tau)\mathrm{d}\tau, \quad n-1 \le \beta < n. \end{aligned} \tag{2.200}$$

If β is an integer k, (2.200) leads to

$$ {}_aD_t^k f(t) = \frac{1}{\Gamma(1)} \frac{\mathrm{d}^{k+1}}{\mathrm{d}t^{k+1}} \int_a^t f(\tau)\mathrm{d}\tau = \frac{\mathrm{d}^k}{\mathrm{d}t^k} f(t). \tag{2.201}$$

Therefore, definition (2.200) is also consistent with integer-order differentiations.

It is noted that the fractional integration in (2.199) is easy to calculate, but calculation of the fractional derivative in (2.200) is not straightforward. An integer n is needed to determine first, and then it can be calculated by carrying out the integration and differentiation. An example is

$$ {}_0D_t^{1/2} t = \frac{1}{\Gamma(1/2)} \frac{\mathrm{d}}{\mathrm{d}t} \int_0^t (t-\tau)^{-1/2} \tau \mathrm{d}\tau = 2\sqrt{\frac{t}{\pi}}. \tag{2.202}$$

The definitions of the fractional integral in (2.199) and derivative in (2.200) are the well-known Riemann–Liouville representation, although other equivalent definitions have also been proposed. Theories and applications about the fractional calculus are given in many books, for example, the two volumes by Uchaikin (2012a, b).

2.5.2 *Fractional Brownian Motion*

It is known that the Brownian motion can be represented as the integral of the Gaussian white noise as shown in Sect. 2.3.3,

$$B(t) = \int_0^t \mathrm{d}B(\tau) = {}_0I_t^1 W_g(t) = \int_0^t W_g(\tau)\mathrm{d}\tau, \tag{2.203}$$

where use has been made of $\mathrm{d}B(t) = W_g(t)\mathrm{d}t$. Naturally, the fractional Brownian motion can be introduced as the fractional integral of Gaussian white noise according to the Riemann–Liouville representation (Levy 1953), i.e.,

$${}_0I_t^\beta W_g(t) = \frac{1}{\Gamma(\beta)} \int_0^t (t-\tau)^{\beta-1} W_g(\tau)\mathrm{d}\tau = \frac{1}{\Gamma(\beta)} \int_0^t (t-\tau)^{\beta-1} \mathrm{d}B(\tau). \tag{2.204}$$

Letting $\mathcal{H} = \beta - 1/2$, known as the Hurst index, Eq. (2.204) can be written as another form

$$B_H(t) = {}_0I_t^{\mathcal{H}+1/2} W_g(t) = \frac{1}{\Gamma(\mathcal{H}+1/2)} \int_0^t (t-\tau)^{\mathcal{H}-1/2} \mathrm{d}B(\tau). \tag{2.205}$$

$B_H(t)$ is known as the fractional Brownian motion (fbm) of Hurst index $\mathcal{H}$. A more popular one is (Mandelbrot and van Ness 1968),

$$B_H(t) = C_H \left[\int_{-\infty}^t (t-\tau)^{\mathcal{H}-1/2} \mathrm{d}B(\tau) - \int_{-\infty}^0 (-\tau)^{\mathcal{H}-1/2} \mathrm{d}B(\tau) \right], \quad t > 0 \tag{2.206}$$

where the Hurst index is required to be $0 < \mathcal{H} < 1$, and C_H is a normalization constant. The requirement of $0 < \mathcal{H} < 1$ is due to two reasons. Under this condition, the increment of $B_H(t)$ and the fractional Gaussian noise are stationary, see Eqs. (2.211) and (2.215). Another reason is that $B_H(t)$ is $\mathcal{H}$ self-similar, namely, $a^{\mathcal{H}} B_H(t)$ has the same probability distribution as $B_H(at)$ (Uchaikin 2012a). These two properties are essential in many practical applications.

It is known that the fractional Brownian motion $B_H(t)$ defined in (2.206) is a Gaussian process, and possesses the following properties (Mishura 2008).

(i) $B_H(0) = 0$,
(ii) $E[B_H(t)] = 0, t \geq 0$,
(iii)

$$E[B_H(t)B_H(s)] = \frac{1}{2}\left(t^{2\mathcal{H}} + s^{2\mathcal{H}} - |t-s|^{2\mathcal{H}}\right), \quad t, s \geq 0. \tag{2.207}$$

According to (2.207), the normalization constant C_H is determined as

$$C_H = \frac{\sqrt{2\mathcal{H}\Gamma(2\mathcal{H})\sin(\mathcal{H}\pi)}}{\Gamma(\mathcal{H}+1/2)}. \tag{2.208}$$

When $\mathcal{H} = 1/2$, (2.207) leads to

$$E[B_H(t)B_H(s)] = \min(t, s), \quad \mathcal{H} = 1/2. \tag{2.209}$$

By comparing (2.209) and (2.77), it is known that $B_H(t)$ ($\mathcal{H} = 1/2$) is a standard Brownian motion with a unit intensity.

From (2.207), we also know

$$E\left[B_H^2(t)\right] = t^{2\mathcal{H}}. \tag{2.210}$$

Indicating that $B_H(t)$ has an anomalous diffusion behavior. It is known as sub-diffusion if $0 < \mathcal{H} < 1/2$, normal diffusion if $\mathcal{H} = 1/2$, super-diffusion if $1/2 < \mathcal{H} < 1$, and trajectory diffusion if $\mathcal{H} = 1$.

It can be shown that the increment of $B_H(t)$, i.e.

$$\Delta B_H(t) = B_H(t + \Delta t) - B_H(t), \tag{2.211}$$

possesses the following property (Mishura 2008)

$$E\left[|\Delta B_H(t)|^n\right] = \frac{2^{n/2}}{\sqrt{\pi}}\Gamma\left(\frac{n+1}{2}\right)|\Delta t|^{n\mathcal{H}}, \quad n \geq 1. \tag{2.212}$$

Therefore, the increment process is stationary. For $n = 2$, Eq. (2.211) leads to

$$E\left[|\Delta B_H(t)|^2\right] = |\Delta t|^{2\mathcal{H}}, \quad E\left[|\mathrm{d}B_H(t)|^2\right] = |\mathrm{d}t|^{2\mathcal{H}}. \tag{2.213}$$

The fractional Brownian motion has applications in communication, physics, finance, bioengineering, etc. (Bassingthwaighte and Raymond 1995; Rangarajan and Ding 2003; Dieker 2004; Mishura 2008; Biagini et al. 2008).

2.5.3 *Fractional Gaussian Noises*

Similar to the relationship of the Brownian motion and Gaussian white noise, the fractional Gaussian noise $W_H(t)$ is defined as the derivative of the fractional Brownian motion $B_H(t)$, i.e.,

$$W_H(t) = \frac{\mathrm{d}B_H(t)}{\mathrm{d}t}, \quad B_H(t) = \int_0^t W_H(t)\mathrm{d}t. \tag{2.214}$$

Since the increment of $B_H(t)$ is stationary, then $W_H(t)$ is also stationary. Note that we restrict $0 < \mathcal{H} < 1$. Letting $t = s$,

$$E[W_H(t)W_H(t)] = E\left\{\left[\lim_{\Delta t\to 0}\frac{B_H(t+\Delta t)-B_H(t)}{\Delta t}\right]^2\right\} = \lim_{\Delta t\to 0}(\Delta t)^{2\mathcal{H}-2} = \infty. \tag{2.215}$$

If $t \neq s$, we calculate from (2.207)

$$\begin{aligned} E[W_H(t)W_H(s)] &= \frac{\partial^2 E[B_H(t)B_H(s)]}{\partial t\partial s} \\ &= \frac{1}{2}\frac{\partial^2}{\partial t\partial s}\left(t^{2\mathcal{H}} + s^{2\mathcal{H}} - |t-s|^{2\mathcal{H}}\right) \\ &= -\frac{1}{2}\frac{\partial^2}{\partial t\partial s}|t-s|^{2\mathcal{H}} = \mathcal{H}(2\mathcal{H}-1)|t-s|^{2\mathcal{H}-2}. \end{aligned} \tag{2.216}$$

Combining (2.215) and (2.216), we have the correlation function of the unit fractional Gaussian white noise $W_H(t)$,

$$R(\tau) = E[W_H(t+\tau)W_H(t)] = \begin{cases} \mathcal{H}(2\mathcal{H}-1)|\tau|^{2\mathcal{H}-2}, & \tau \neq 0 \\ \infty, & \tau = 0 \end{cases}. \tag{2.217}$$

It is found that the mean square value of a fractional Gaussian noise $W_H(t)$ ($0 < H < 1$) is infinity. Therefore, it is not a real noise similar to the case of Gaussian white noise.

An important concept known as the long-range dependence is introduced to measure the correlation of a stationary random process. The long-range dependence index μ is defined from the equation

$$\lim_{\tau\to\infty}\frac{R(\tau)}{\tau^{-\mu}} = C, \quad C > 0, \quad 0 < \mu < 1, \tag{2.218}$$

where C is a constant. When (2.218) holds, we say that the process has long-range dependence, or long-memory. Equation (2.218) indicates that the decay rate of the correlation function $R(\tau)$ is compared with the decay rate of $\tau^{-\mu}$. The smaller the index μ is, the slower pace of decay the correlation function has. It can be seen from (2.217) that when $1/2 < \mathcal{H} < 1, 0 < \mu = 2 - 2\mathcal{H} < 1$, then the process has long-range dependence. For the Gaussian white noise, i.e., $\mathcal{H} = 1/2$, $\mu = 0$; thus, it does not have the long-range dependence. In fact, it has no dependence at any $\tau \neq 0$.

The power spectral density of $W_H(t)$ does not exist if $\mathcal{H} \in (0, 1/2)$. If $\mathcal{H} \in (1/2, 1)$, the power spectral density of the fractional Gaussian noise $W_H(t)$ with unit intensity is obtained as (Rangarajan and Ding 2003; Uchaikin 2012a)

$$S(\omega) = \frac{1}{2\pi}\int_{-\infty}^{\infty} R(\tau)e^{-i\omega\tau}\mathrm{d}\tau = \frac{1}{\pi}\mathcal{H}\Gamma(2\mathcal{H})\sin(\mathcal{H}\pi)|\omega|^{1-2\mathcal{H}}. \tag{2.219}$$

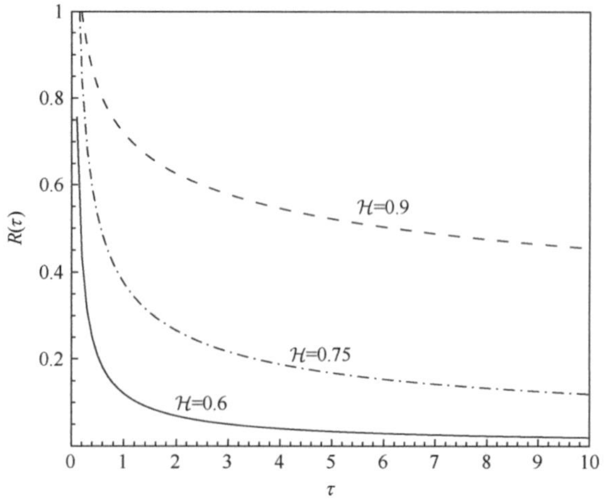

Fig. 2.7 Correlation functions of fractional Gaussian noises

For the special case of $\mathcal{H} = 1/2$, $W_H(t)$ is a Gaussian white noise $W(t)$ with

$$R(\tau) = E[W(t+\tau)W(t)] = \delta(\tau), \quad S(\omega) = \frac{1}{2\pi} \tag{2.220}$$

It is noticed from (2.219) that $\lim_{\mathcal{H}\to 1} S(\omega) = \delta(\omega)\big/2$. Thus, $W_H(t)$ is approaching a Gaussian random variable when $\mathcal{H} \to 1$. When $\mathcal{H} \in (1/2, 1)$, $W_H(t)$ is a colored noise between Gaussian white noise and Gaussian random variable.

Figures 2.7 and 2.8 depict the correlation and spectral density functions of fractional Gaussian noises for three values of Hurst index. The long-range dependence of the fractional Gaussian noise can be observed in Fig. 2.7. Except in the low frequency range, the fractional Gaussian noise has the characteristic of broad band.

It is clear that the fractional Brownian motion and fractional Gaussian noise are meaningful and have practical applications only if the Hurst index $\mathcal{H}$ is in the range of $1/2 \le \mathcal{H} < 1$. Without specified, the Hurst index of the fractional Brownian motions and fractional Gaussian noises are restricted in the range of $1/2 < \mathcal{H} < 1$ in this book.

2.5.4 Stochastic Integration with Respect to Fractional Brownian Motion and Fractional Stochastic Differential Equations

Stochastic integration with respect to fractional Brownian motion can be defined in different ways. The pathwise integration is the most simple and natural, and widely used (Biagini et al., 2008). Its definition is as follows

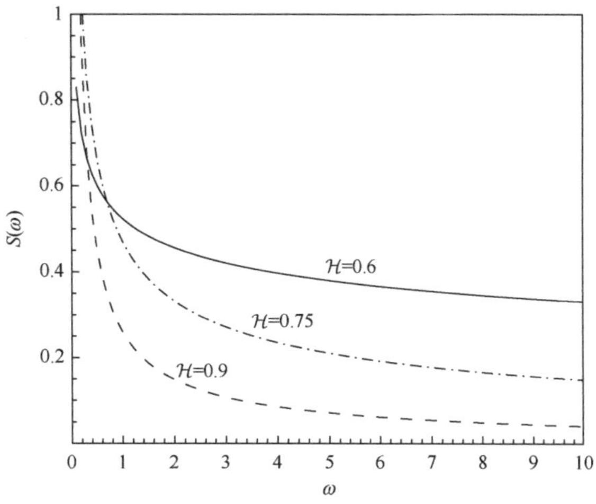

Fig. 2.8 Spectral density functions of fractional Gaussian noises

$$\int_{t_0}^{t} f(s)\mathrm{d}B_H(s) = \lim_{n\to\infty}\sum_{i=1}^{n} f(t_i')[B_H(t_i) - B_H(t_{i-1})], \tag{2.221}$$

where $t_i = i\Delta t$, $\Delta t = (t - t_0)/n$, $t_{i\text{-}1} \le t_i' \le t_i$. Depending on the t_i' value, there are three pathwise integrals: forward integral if $t_i' = t_{i\text{-}1}$, symmetric integral if $t_i' = (t_i - t_{i\text{-}1})/2$, and backward integral if $t_i' = t_i$. For the case of $\mathcal{H} = 1/2$, the forward integral is the Itô integral and the symmetric integral is the Stratonovich integral.

Consider the following stochastic differential equation

$$\mathrm{d}X(t) = \alpha(X, t)\mathrm{d}t + \beta(X, t)\mathrm{d}^{-}B_H(t). \tag{2.222}$$

It is known that when establishing Eq. (2.222), the associated stochastic integral must be defined. Here in (2.222), the notation $d^{\,-}$ in the last term is defined as the forward pathwise integral. It is noted that, although (2.222) has the same form as an Itô differential equation, $X(t)$ is not a Markov diffusion process. Since there are different pathwise integrals, the fractional stochastic differential equations may have different interpretations. Consider the difference between the symmetric integral and forward integral

$$\int_{t_0}^{t} \beta[X(s), s]\mathrm{d}^{o}B_H(s) - \int_{t_0}^{t} \beta[X(s), s]\mathrm{d}^{-}B_H(s)$$
$$= \begin{cases} \frac{1}{2}\int_{t_0}^{t} \frac{\partial\beta[x(s),s]}{\partial x}\beta[x(s), s]\mathrm{d}s, & \mathcal{H} = 1/2 \\ 0, \;\; \mathcal{H} > 1/2 \end{cases}. \tag{2.223}$$

where the symbol d° denotes the symmetric integral. Equation (2.223) shows that when $\mathcal{H} = 1/2$, the difference is the Wong-Zakai correction term. When $\mathcal{H} = 1/2$ as assumed in this book, the difference disappears, indicating that the forward and symmetric integrals are equivalent in the fractional stochastic differential equations.

The fractional stochastic differential equations considered in this book use the forward integral. For convenience, symbol d^- is simplified as d.

Similar to the Itô differential rule, the fractional stochastic differential equation for the function of $X(t)$, $Y = F\,[X(t), t]$ can be found as

$$\begin{aligned} \mathrm{d}Y &= \frac{\partial F}{\partial t}\mathrm{d}t + \frac{\partial F}{\partial X}\mathrm{d}X + \frac{1}{2}\frac{\partial^2 F}{\partial X^2}(\mathrm{d}X)^2 + \ldots \\ &= \left[\frac{\partial F}{\partial t} + \frac{\partial F}{\partial X}\alpha(X,t)\right]\mathrm{d}t + \frac{\partial F}{\partial X}\beta(X,t)\mathrm{d}B_H(t) + \ldots. \end{aligned} \tag{2.224}$$

According to (2.212), $[\mathrm{d}B_H(t)]^n$ $(n \geq 2)$ has the same order as $(\mathrm{d}t)^{n\mathcal{H}}$. When $\mathcal{H} > 1/2$ and $n \geq 2$, $[\mathrm{d}B_H(t)]^n$ has an order higher than dt; therefore, all terms containing $(\mathrm{d}X)^n$ $(n \geq 2)$ are negligible. It is different from the Itô differential rule (2.112) for Wiener processes in which the term of $\mathrm{d}^2F/\mathrm{d}X^2$ is kept since $[\mathrm{d}B(t)]^2$ has the same order of dt.

Now consider n- dimensional fractional stochastic differential equation

$$\mathrm{d}\boldsymbol{X} = \boldsymbol{\alpha}(\boldsymbol{X},t)\mathrm{d}t + \boldsymbol{\beta}(\boldsymbol{X},t)\mathrm{d}\boldsymbol{B}_H(t), \tag{2.225}$$

where $B_H(t) = \left[B_{H_1}(t), B_{H_2}(t), \cdots, B_{H_m}(t)\right]^T$ are m independent unit fractional Brownian motions with Hurst indexes $1/2 < \mathcal{H}_1, \mathcal{H}_2, \cdots, \mathcal{H}_m < 1$, respectively;

$$\begin{aligned} &\boldsymbol{X} = [X_1, X_2, \ldots, X_n]^T, \boldsymbol{\alpha} = [\alpha_1(\boldsymbol{X},t), \alpha_2(\boldsymbol{X},t), \ldots, \alpha_n(\boldsymbol{X},t)]^T, \\ &\boldsymbol{\beta} = \begin{bmatrix} \beta_{11}(\boldsymbol{X},t) & \beta_{12}(\boldsymbol{X},t) & \ldots & \beta_{1m}(\boldsymbol{X},t) \\ \beta_{21}(\boldsymbol{X},t) & \beta_{21}(\boldsymbol{X},t) & \ldots & \beta_{2m}(\boldsymbol{X},t) \\ \vdots & \vdots & \ddots & \vdots \\ \beta_{n1}(\boldsymbol{X},t) & \beta_{n1}(\boldsymbol{X},t) & \ldots & \beta_{nm}(\boldsymbol{X},t) \end{bmatrix}, \mathrm{d}\boldsymbol{B}_H = \begin{bmatrix} \mathrm{d}B_{H_1}(t) \\ \mathrm{d}B_{H_2}(t) \\ \vdots \\ \mathrm{d}B_{H_m}(t) \end{bmatrix}. \end{aligned} \tag{2.226}$$

where

$$\left(\mathrm{d}B_{H_i}\right)\left(\mathrm{d}B_{H_j}\right) = \begin{cases} 0, & i \neq j \\ (\mathrm{d}t)^{2\mathcal{H}_i}, & i = j \end{cases}. \tag{2.227}$$

Note that the order of $(\mathrm{d}t)^{2\mathcal{H}_i}$ is higher than dt and have to be neglected, the stochastic differential equation of function $Y = F\,[\boldsymbol{X}(t), t]$ is then

$$\mathrm{d}Y = \frac{\partial F}{\partial t}\mathrm{d}t + (F_X)^T\mathrm{d}X = \left[\frac{\partial F}{\partial t} + (F_X)^T\alpha\right]\mathrm{d}t + (F_X)^T\beta\mathrm{d}B_H, \tag{2.228}$$

where

$$F_X = \left[\frac{\partial F}{\partial X_1}, \frac{\partial F}{\partial X_2}, \cdots, \frac{\partial F}{\partial X_n} \right]^{\mathrm{T}}. \tag{2.229}$$

The scalar form of (2.227) is

$$\mathrm{d}Y = \left[\frac{\partial F}{\partial t} + \sum_{i=1}^{n} \alpha_i \frac{\partial F}{\partial X_i} \right] \mathrm{d}t + \sum_{i=1}^{n} \sum_{j=1}^{m} \beta_{ij} \frac{\partial F}{\partial X_i} \mathrm{d}B_{H_j}. \tag{2.230}$$

It is noticed that the correlation between a pair of increments, $\mathrm{d}B_{H_i}$ and $\mathrm{d}B_{H_j}(i \neq j)$, has no effect on Eq. (2.230).

2.5.5 Response of Linear Systems Excited by Fractional Gaussian Noises

Response of a linear system under excitations of fractional Gaussian noises is a Gaussian process. Analytical methods can be applied to obtain correlation functions and power spectral density functions.

Consider a n-DOF linear vibration system

$$\boldsymbol{M}\ddot{\boldsymbol{X}} + \boldsymbol{C}\dot{\boldsymbol{X}} + \boldsymbol{K}\boldsymbol{X} = \boldsymbol{R}\boldsymbol{W}_H(t), \tag{2.231}$$

where $\boldsymbol{X} = [X_1, X_2, .., X_n]^{\mathrm{T}}$ is the displacement vector, $\boldsymbol{M}$, $\boldsymbol{C}$, and $\boldsymbol{K}$ are $n \times n$ mass, damping and stiffness matrices respectively, $W_H(t) = \left[W_{H_1}(t), W_{H_2}(t), \cdots, W_{H_m}(t) \right]^T$ is a vector of m independent unit fractional Gaussian noises with Hurst indexes $\mathcal{H}_1, \mathcal{H}_2, \cdots, \mathcal{H}_{\mathrm{m}} \in (1/2, 1)$, $\boldsymbol{R}$ is a $n \times m$ matrix specifying the excitation intensities. Applying method of spectral analysis, we obtain $n \times n$ matrix of power spectral densities of displacements

$$\boldsymbol{S}_X(\omega) = \boldsymbol{H}^*(\omega)\boldsymbol{S}_W(\omega)\boldsymbol{H}^{\mathrm{T}}(\omega), \tag{2.232}$$

where $\boldsymbol{S}_W(\omega)$ is a $m \times m$ power spectral density matrix of excitation vector $\boldsymbol{W}_H(t)$; $\boldsymbol{H}(\omega)$ is a $n \times m$ frequency response matrix of system (2.231)

$$\boldsymbol{H}(\omega) = \left(-\omega^2 \boldsymbol{M} + i\omega \boldsymbol{C} + \boldsymbol{K} \right)^{-1} \boldsymbol{R} \tag{2.233}$$

$\boldsymbol{H}^*(\omega)$ is the complex conjugate of $\boldsymbol{H}(\omega)$. The response matrices of mean-squares of displacements and velocities can be calculated from integrations of (2.232)

$$E\left[XX^T\right] = \int_{-\infty}^{\infty} S_X(\omega)\mathrm{d}\omega, \quad E\left[\dot{X}\dot{X}^{\mathrm{T}}\right] = \int_{-\infty}^{\infty} \omega^2 S_X(\omega)\mathrm{d}\omega. \tag{2.234}$$

Consider the following n-dimensional linear fractional stochastic differential equation

$$\mathrm{d}\boldsymbol{X}(t) = \boldsymbol{A}\boldsymbol{X}(t)\mathrm{d}t + \boldsymbol{Q}\mathrm{d}\boldsymbol{B}_H(t), \quad \boldsymbol{X}(0) = \boldsymbol{X}_0, \tag{2.235}$$

where $\boldsymbol{B}_H(t) = \left[B_{H_1}(t), \cdots, B_{H_m}(t)\right]^{\mathrm{T}}$ is a vector of m independent unit fractional Brownian motions with Hurst indexes $\mathcal{H}_1, \mathcal{H}_2, \cdots, \mathcal{H}_{\mathrm{m}} \in (1/2, 1)$, $\boldsymbol{A}$ is a $n \times n$ constant matrix, and $\boldsymbol{Q}$ is a $n \times m$ matrix specifying the excitation magnitude.

Introducing vector function $\boldsymbol{Y}(t) = e^{-\boldsymbol{A}t}\boldsymbol{X}(t)$ and applying fractional stochastic differential rule (2.228), we have

$$\mathrm{d}\boldsymbol{Y}(t) = -\boldsymbol{A}e^{-\boldsymbol{A}t}\boldsymbol{X}(t)\mathrm{d}t + e^{-\boldsymbol{A}t}\mathrm{d}\boldsymbol{X}(t). \tag{2.236}$$

Substituting (2.235) into (2.236) and integrating both sides, we obtain the solution of fractional stochastic differential equation (2.235)

$$\boldsymbol{X}(t) = e^{\boldsymbol{A}t}\boldsymbol{X}_0 + \int_0^t e^{\boldsymbol{A}(t-s)}\boldsymbol{Q}\mathrm{d}\boldsymbol{B}_H(s), \tag{2.237}$$

where $e^{\boldsymbol{A}t}\boldsymbol{X}_0$ is the transient term, which approaches zero for a dissipative system after long time period. The stationary solution is then

$$\boldsymbol{X}(t) = \int_{-\infty}^t e^{\boldsymbol{A}(t-s)}\boldsymbol{Q}\mathrm{d}\boldsymbol{B}_H(s). \tag{2.238}$$

Example 2.8 Consider the following 2-DOF linear system excited by fractional Gaussian noises

$$\ddot{\boldsymbol{X}} + \boldsymbol{C}\dot{\boldsymbol{X}} + \boldsymbol{K}\boldsymbol{X} = \boldsymbol{W}_H(t), \tag{2.239}$$

where

$$\boldsymbol{X} = \begin{bmatrix} X_1 \\ X_2 \end{bmatrix}, \quad \boldsymbol{C} = \begin{bmatrix} \gamma & 0 \\ 0 & 0 \end{bmatrix}, \quad \boldsymbol{K} = \begin{bmatrix} k_1 & -k_s \\ -k_s & k_2 \end{bmatrix}, \quad \boldsymbol{W}_H(t) = \begin{bmatrix} \sqrt{2D}W_H(t) \\ 0 \end{bmatrix}. \tag{2.240}$$

and $W_H(t)$ in (2.240) is a unit fractional Gaussian noise. Application of (2.232) and (2.233) results in the power spectral density matrix of displacement responses

$$S_X(\omega) = \frac{2DS_W(\omega)}{Z}\begin{bmatrix} (k_2-\omega^2)^2 & k_2(k_2-\omega^2) \\ k_2(k_2-\omega^2) & k_s^2 \end{bmatrix},$$
$$Z = (k_1k_2-k_s^2)^2 + \left[k_2(2k_s^2-2k_1^2+k_2\gamma^2) - 2k_1(k_2^2-k_s^2)\right]\omega^2$$
$$+\left[(k_1+k_2)^2 + 2k_2(k_1-\gamma^2) - 2k_s^2\right]\omega^4 + \left[\gamma^2 - 2(k_1+k_2)\right]\omega^6 + \omega^8, \quad (2.241)$$

where $S_W(\omega)$ is the power spectral density of $W_H(t)$. Then the mean-square values of the displacements can be calculated from (2.234),

$$E[X^2] = \int_{-\infty}^{\infty} \frac{2D(k_2-\omega^2)^2 S_W(\omega)}{Z} d\omega, \quad E[X^2] = \int_{-\infty}^{\infty} \frac{2Dk_s^2 S_W(\omega)}{Z} d\omega. \quad (2.242)$$

Example 2.9 Consider a SDOF linear system subjected to a fractional Gaussian noise excitation.

$$\ddot{X} + \gamma\dot{X} + kX = \sqrt{2D}W_H(t). \quad (2.243)$$

According to the mean-square Eq. (2.234) and the power spectral density of $W_H(t)$ given in (2.223), the mean-square of the displacement

$$E[X^2] = \frac{2D\mathcal{H}(2\mathcal{H})\sin(\mathcal{H}\pi)}{\pi}\int_{-\infty}^{\infty} \frac{|\omega|^{1-2\mathcal{H}}}{(k-\omega^2)^2 + \gamma^2\omega^2} d\omega. \quad (2.244)$$

It is shown that the analytical result of (2.244) is (Deng and Zhu 2015)

$$E[X^2] = \frac{D\Gamma(1+2\mathcal{H})\sin[\mathcal{H}\pi - \mathcal{H}\arccos(1-2\zeta^2)]}{2k^{\mathcal{H}+1}\zeta\sqrt{1-\zeta^2}}, \quad (2.245)$$

where $\zeta = \frac{\gamma}{(2\sqrt{k})}$ is the damping ratio. (2.245) is valid for all case of $0 < \zeta < \infty$ if calculation is performed in complex plane.

Let the damping ratio $\zeta \to \infty$, (2.245) leads to

$$E[X^2] \to \frac{2D}{\gamma^2}\mathcal{H}\Gamma(2\mathcal{H})\left(\frac{k}{\gamma}\right)^{-2\mathcal{H}}, \quad \text{as } \zeta \to \infty. \quad (2.246)$$

Result (2.246) is the same as that for the Ornstein–Uhlenbeck system under fractional Gaussian noise excitation (Kaarakka and Salminen 2011).

Based on the analytical solution (2.246), more analyses can be carried out for the stochastic system (2.244). One example is to find the ratio of the mean kinetic energy to the mean total energy, denoted by r_k, as follows

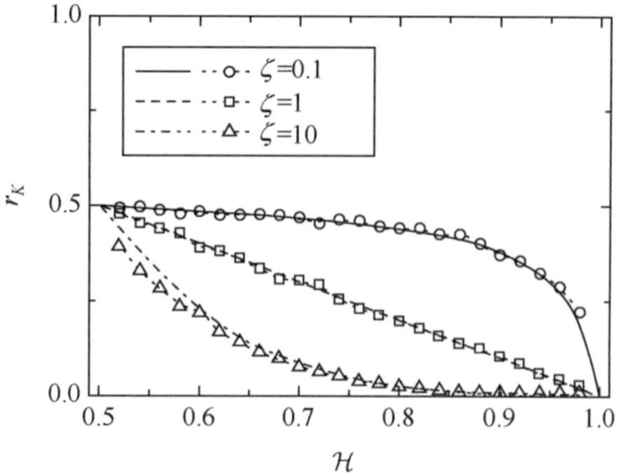

Fig. 2.9 The ratio r_k of the kinetic energy to the total energy of system (2.243)

$$
\begin{aligned}
r_K =& \frac{1}{2} - \zeta \frac{\left[\pi - \arccos(1 - 2\zeta)^2\right]}{2\sqrt{1-\zeta^2}} \left(\mathcal{H} - \frac{1}{2}\right) \\
& - \zeta \frac{\left[\pi - \arccos\left(1 - 2\zeta^2\right)\right]^3}{6\sqrt{1-\zeta^2}} \left(\mathcal{H} - \frac{1}{2}\right)^3 + \cdots .
\end{aligned} \tag{2.247}
$$

When $\mathcal{H} = 1/2$, $W_H(t)$ is a white noise, $r_k = 1/2$ indicating the kinetic energy is half of the total energy and the other half is potential energy. Figure 2.9 depicts the ration r_k with changing $\mathcal{H}$. In the figure, lines represent the analytical results, symbols □, △, and ○ represent the Monte Carlo simulation results. The good agreement of the results indicates the validity of the analytical solution. Figure 2.9 also shows that when $\mathcal{H} \to 1$, the system kinetic energy vanishes because $W_H(t)$ is degenerated to a Gaussian variable and the excitation becomes a static load.

Under a fractional Gaussian noise excitation, the system response is not a Markov process. The displacement process has the characteristic of long-range dependence, which can be measured by index μ defined in (2.217). The stationary sample solution of Eq. (2.243) can be obtained from (2.238) as

$$
X(t) = \sqrt{\frac{D}{2k\left(\zeta^2 - 1\right)}} \int_{-\infty}^{t} \left(e^{\alpha(s-t)} - e^{\beta(s-t)}\right) \mathrm{d}B_H(s), \tag{2.248}
$$

which leads to the correlation function of $X(t)$

$$
\begin{aligned}
R_X(\tau) &= E[X(t)X(t+\tau)] = E[X(0)X(\tau)] \\
&= \frac{D}{2k\left(\zeta^2 - 1\right)} E\left[\int_{-\infty}^{0} \left(e^{\alpha u} - e^{\beta u}\right) \mathrm{d}B_H(u) \int_{-\infty}^{t} \left(e^{\alpha(v-t)} - e^{\beta(v-t)}\right) \mathrm{d}B_H(v) \right]
\end{aligned} \tag{2.249}
$$

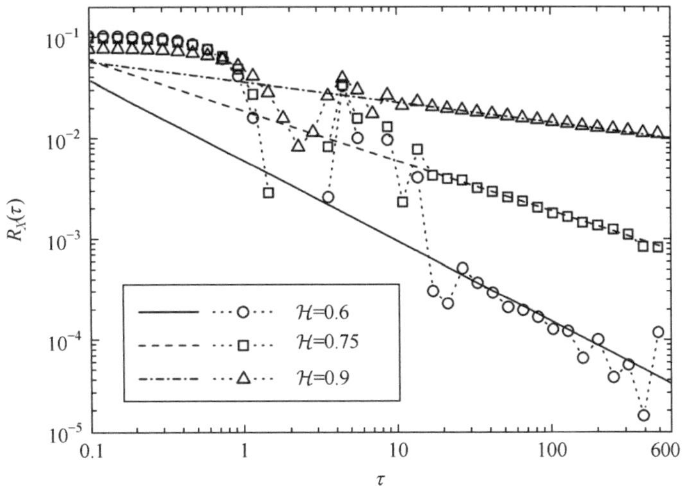

Fig. 2.10 Correlation function $R_X(\tau)$ of response displacement of system (2.243)

It is shown that when $\tau \to \infty$, the correlation function has the following limit (Deng and Zhu 2015),

$$R_X(\tau) = \frac{2D\mathcal{H}(2\mathcal{H} - 1)}{k^2}\tau^{-(2-2\mathcal{H})}, \quad \tau \to \infty. \tag{2.250}$$

Figure 2.10 depicts the correlation functions with respect to τ for three different Hurst indexes. For large τ values, results (denoted by lines) from (2.250) match those (denoted by symbols) from Mote Carlo simulations quite well.

From the definition (2.217), it is inferred that the long-range dependence index of the response displacement process $X(t)$ is $\mu = 2 - 2\mathcal{H}$. It is the same as that of the excitation process $W_H(t)$; therefore, the response displacement prosses is also a type of fractional Gaussian noise in nature. Further research (Deng and Zhu 2015) shows that the long-range dependence index of the response velocity process $\dot{X}(t)$ is $\mu = 4 - 2\mathcal{H} > 1$, and process $\dot{X}(t)$ no longer has the long-range dependence.

2.6 Colored Noises

It is known that a white noise process has some unrealistic characteristics. Its spectral density is a constant over entire frequency band, i.e., $(-\infty, \infty)$, and its mean square value is infinity, indicating the process has infinite amount of energy. Its correlation function is a delta function, meaning that it changes infinitely fast. Therefore, it is only a mathematical idealization. Obviously, such a process does not exist in reality, and any real stochastic process must have a limited frequency band, called colored noise in general. In this section, several types of colored noises are modeled and introduced. A common advantage of these colored noises is that they are modeled analytically in terms of mathematical expressions, such as algebraic equations or differential equations, so that they can be applied for analysis conveniently.

2.6.1 Noises Generated from Linear Filters

A type of colored noises is modeled as the responses of linear systems, governed by linear differential equations with constant coefficients, subjected to Gaussian white noise excitations. Such colored noises are known as linearly filtered noises. Their probability distribution is still Gaussian, but spectral densities are no longer constant, but decay fast with increasing frequency. Depending on the nature of the linear systems, the filtered noises exhibit different spectral properties. The most popular filters are the first- and second-order linear filters.

The first-order linear filter is

$$\dot{X} + \alpha X = W_g(t), \tag{2.251}$$

where $W_g(t)$ is a Gaussian white noise with a spectral density K. The power spectral density and correlation function of $X(t)$ are, respectively,

$$S(\omega) = \frac{K}{\omega^2 + \alpha^2}, \tag{2.252}$$

$$R(\tau) = \frac{\pi K}{\alpha} e^{-\alpha|\tau|}. \tag{2.253}$$

Equations (2.252) and (2.253) are the same as (2.51) since $\sigma_X^2 = \pi K / \alpha$. A larger K corresponds to a stronger process intensity, while a larger α indicates a broader band, or equivalently, a shorter correlation time. The process is called the low-pass noise since the spectrum peak is at zero frequency ($\omega = 0$). Figure 2.11 depicts the power spectral density and correlation functions of the low-pass process with three different α values. It is seen that the low-pass processes are broad band processes when parameter α is large.

Now consider the second-order linear filter governed by

$$\ddot{X} + 2\zeta\omega_0\dot{X} + \omega_0^2 X = W_g(t). \tag{2.254}$$

The spectral densities of $X(t)$ and $\dot{X}(t)$ are given by, respectively,

$$S_{XX}(\omega) = \frac{K}{\left(\omega^2 - \omega_0^2\right)^2 + 4\zeta^2\omega_0^2\omega^2}, \tag{2.255}$$

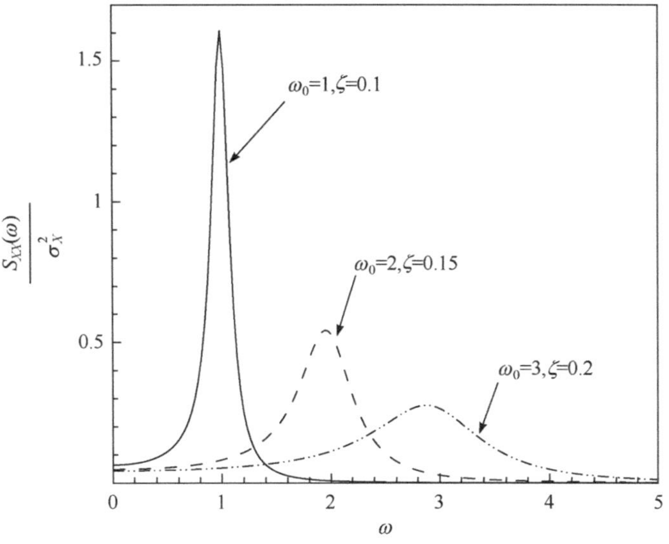

Fig. 2.11 Spectral density functions of $X(t)$ from 2nd-order linear filters

$$S_{\dot{X}\dot{X}}(\omega) = \frac{K\omega^2}{\left(\omega^2 - \omega_0^2\right)^2 + 4\zeta^2\omega_0^2\omega^2}. \tag{2.256}$$

In each of (2.255) and (2.256), the spectral density function has single peak. Parameter ζ controls the band width, ω_0 determines the peak location, and K specifies the magnitude. If the damping is weak, the peak is close to ω_0, and the band width is narrow. The difference of the two spectral densities lies in their different values at very low frequency band, as shown in the equations. The spectral values of $\dot{X}(t)$ is very small at low frequencies, and hence it can be used to effectively model random wave excitation on a ship (Chai el al. 2015, 2016). The two spectra are shown in Figs. 2.11 and 2.12 for several sets of ω_0 and ζ values. Note that $\sigma_{\dot{X}}^2 = \pi K/\left(2\zeta\omega_0^3\right)$ and $\sigma_{\dot{X}}^2 = \pi K/(2\zeta\omega_0)$. It is seen that the damping ration ζ controls the band width, while ω_0 determines the peak location.

Many practical stochastic processes can be represented by the processes generated from the first- and second-order linear filters. Since the filters are linear, it is convenient for mathematical treatment.

If a process has more than one peak, a higher-order filter may be required, but more parameters in the filter need to be identified.

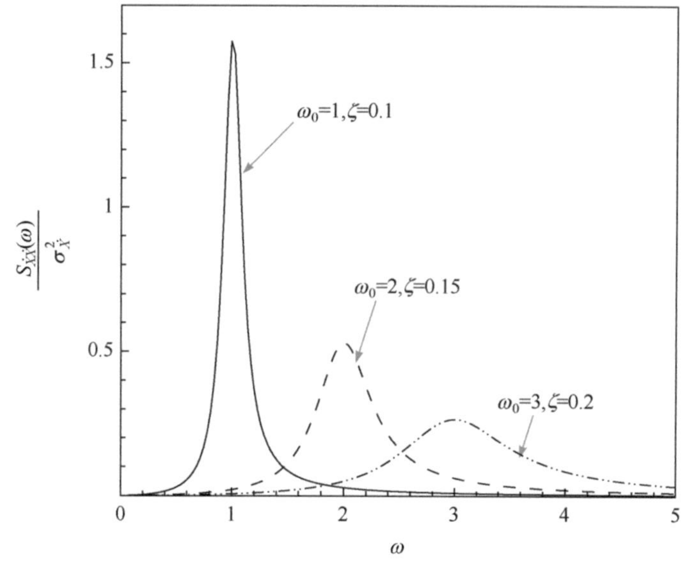

Fig. 2.12 Spectral density functions of $\dot{X}(t)$ from 2nd-order linear filters

2.6.2 Noises Generated from Nonlinear Filters

The noises from linear filters have Gaussian distributions and unlimited range $(-\infty, \infty)$. To model non-Gaussian noises in limited or unlimited ranges, nonlinear filters may be used. Consider a diffusion process $X(t)$ defined in an interval between x_l and x_r, and governed by the Itô equation

$$\mathrm{d}X = -\alpha X\,\mathrm{d}t + D(X)\mathrm{d}B(t), \tag{2.257}$$

where α is a positive constant, and $B(t)$ is a unit Wiener process. Without loss of generality, it is assumed that $X(t)$ has zero mean; thus, $x_l < 0$ and $x_r > 0$. Equation (2.257) represents a first-order nonlinear filter with function $D(X)$ being non-constant.

Multiplying both sides of (2.257) by $X(t\text{-}\tau)$ and taking ensemble average, we obtain

$$\frac{\mathrm{d}}{\mathrm{d}t}R_{XX}(\tau) = -\alpha R_{XX}(\tau), \tag{2.258}$$

where $R_{XX}(\tau)$ is the correlation function, i.e., $R_{XX}(\tau) = E[X(t\text{-}\tau)X(t)]$. In deriving (2.258), the property of the Itô equation is applied, i.e., $\mathrm{d}B(t)$ is independent of $X(t)$ and $X(t\text{-}\tau)$. The initial condition for Eq. (2.258) is

$$R_{XX}(0) = \sigma_X^2, \tag{2.259}$$

where σ_X^2 is the mean square of $X(t)$. Then the correlation function is solved from (2.258) as

$$R_{XX}(\tau) = \sigma_X^2 \exp(-\alpha|\tau). \tag{2.260}$$

The corresponding spectral density of $X(t)$ is the Fourier transform of (2.6.10), found to be

$$S_{XX}(\omega) = \frac{\alpha\sigma_X^2}{\pi\left(\omega^2 + \alpha^2\right)} \tag{2.261}$$

The correlation functions and spectral densities are exactly the same as those of the noises from the first-order linear filters, shown in Sect. 2.6.1. It is noted that the diffusion coefficient $D(X)$ in (2.257) does not directly affect the spectral density.

The stationary PDF of $X(t)$ is governed by the reduced FPK equation

$$\frac{\mathrm{d}}{\mathrm{d}x}G = -\frac{\mathrm{d}}{\mathrm{d}x}\left\{\alpha x p(x) + \frac{1}{2}\frac{\mathrm{d}}{\mathrm{d}x}\left[D^2(x)p(x)\right]\right\} = 0, \tag{2.262}$$

where G is the probability flow. For the present one dimensional case, G must vanish at the two boundaries, and hence, must vanish everywhere. Then (2.262) reduces to

$$\alpha x p(x) + \frac{1}{2}\frac{\mathrm{d}}{\mathrm{d}x}\left[D^2(x)p(x)\right] = 0 \tag{2.263}$$

Integration of (2.263) results in

$$D^2(x)p(x) = -2\alpha \int_{x_l}^{x} up(u)\mathrm{d}u + C, \tag{2.264}$$

where C is an integration constant. It is proved that $C = 0$ (Cai and Zhu 2016), and (2.264) reduces to

$$D^2(x) = -\frac{2\alpha}{p(x)} \int_{x_l}^{x} up(u)\mathrm{d}u. \tag{2.265}$$

Function $D^2(x)$ in (2.265) is nonnegative, as it should be, since $p(x) \geq 0$ and the mean of $X(t)$ is zero. Equation (2.265) shows that, for any valid PDF $p(x)$, the diffusion coefficient $D(X)$ can be determined accordingly. Therefore, the Itô equation (2.267) can be used to model stochastic processes with a low-pass spectral density (2.261) and an arbitrary probability distribution.

Example 2.10 If the process to be generated has a uniform distribution, i.e.,

$$p(x) = \frac{1}{2\Delta}, \quad -\Delta \le x \le \Delta \tag{2.266}$$

From (2.265) and (2.257),

$$D^2(x) = \alpha(\Delta^2 - x^2), \quad dX = -\alpha X dt + \sqrt{\alpha(\Delta^2 - X^2)} dB(t). \tag{2.267}$$

Example 2.11 Assume that $X(t)$ has the following distribution

$$p(x) = C(\Delta^2 - x^2)^\delta, \quad -\Delta \le x \le \Delta \text{ for } \delta \ge 0$$
$$\text{or } -\Delta < x < \Delta \text{ for } -1 < \delta < 0, \tag{2.268}$$

where C is a normalization constant. Applying (2.265), we have

$$D^2(x) = \frac{\alpha}{\delta+1}(\Delta^2 - x^2), \quad dX = -\alpha X dt + \sqrt{\frac{\alpha}{\delta+1}(\Delta^2 - X^2)} dB(t). \tag{2.269}$$

The PDFs (2.268) are depicted in Fig. 2.13 for several δ values. It is shown that the shapes of the PDFs diverse for different δ values. The case of $\delta = 0$ corresponds to a uniform distribution. For a fixed α value and different δ values, the processes have different probability distributions, yet they may share the similar spectral density (2.261).

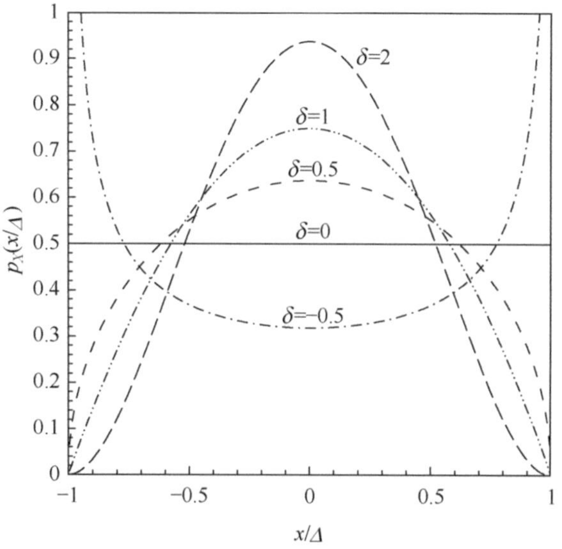

Fig. 2.13 Stationary PDFs of $X(t)$ given in (2.268)

Example 2.12 Let $X(t)$ be governed by an exponential distribution.

$$p(x) = \lambda \exp(-\lambda x), \quad \lambda > 0, \quad 0 \leq x < \infty. \tag{2.270}$$

Since its mean is $1/\lambda$, we define a process with zero mean as $Y(t) = X(t) - 1/\lambda$. The PDF of $Y(t)$ is

$$p(y) = \lambda \exp\left[-\lambda\left(y + \frac{1}{\lambda}\right)\right], \quad -\frac{1}{\lambda} \leq y < \infty \tag{2.271}$$

From Eq. (2.265),

$$D^2(y) = \frac{2\alpha}{\lambda}\left(y + \frac{1}{\lambda}\right) \tag{2.272}$$

The Itô equation for $Y(t)$ is

$$\mathrm{d}Y = -\alpha Y \mathrm{d}t + \frac{1}{\lambda}\sqrt{2\alpha(\lambda Y + 1)}\mathrm{d}B(t), \tag{2.273}$$

and the Itô equation for $X(t)$ is

$$\mathrm{d}X = -\alpha\left(X - \frac{1}{\lambda}\right)\mathrm{d}t + \sqrt{\frac{2\alpha}{\lambda}X}\,\mathrm{d}B(t). \tag{2.274}$$

Note that the spectral density of $X(t)$ contains a delta function $(1/\lambda^2)\delta(\omega)$ due to the nonzero mean $1/\lambda$.

The above examples show that the first-order nonlinear filter (2.257) can generate a low-pass stochastic process defined in a finite or infinite interval and with an arbitrary probability distribution.

Now we try to use second-order filters to generate other colored noises. Extend Eq. (2.257) from one-dimensional to two-dimensional, i.e.,

$$\begin{aligned} \mathrm{d}X_1 &= (-a_{11}X_1 - a_{12}X_2)\mathrm{d}t + D_1(X_1, X_2)\mathrm{d}B_1(t), \\ \mathrm{d}X_2 &= (-a_{21}X_1 - a_{22}X_2)\mathrm{d}t + D_2(X_1, X_2)\mathrm{d}B_2(t), \end{aligned} \tag{2.275}$$

where a_{ij} are parameters, and $B_1(t)$ and $B_2(t)$ are independent unit Wiener processes. Multiplying the two equations in (2.276) by $X_1(t-\tau)$, taking the ensemble average, and denoting $R_{ij}(\tau) = E[X_i(t-\tau)X_j(t)]$, we obtain

$$\begin{aligned} \frac{\mathrm{d}}{\mathrm{d}\tau}R_{11}(\tau) &= -a_{11}R_{11}(\tau) - a_{12}R_{12}(\tau), \\ \frac{\mathrm{d}}{\mathrm{d}\tau}R_{12}(\tau) &= -a_{21}R_{11}(\tau) - a_{22}R_{12}(\tau), \end{aligned} \tag{2.276}$$

subjected to initial conditions

$$R_{11}(0) = E\left[X_1^2\right] = \sigma^2, \quad R_{12}(0) = E[X_1 X_2] \tag{2.277}$$

Equation set (2.276) can be solved for the correlation functions. In modeling a stochastic process, its spectral density is usually of interest. It can be calculated from the correlation function using the Fourier transformation. However, it can be derived using an alternative procedure (Cai and Zhu 2016) described below. Define the following integral transformation

$$\overline{S}_{ij}(\omega) = \Im\left[R_{ij}(\tau)\right] = \frac{1}{\pi}\int_0^\infty R_{ij}(\tau)e^{-i\omega\tau}\mathrm{d}\tau. \tag{2.278}$$

The spectral density functions can be derived in terms of $\overline{S}_{ij}(\omega)$ as follows

$$\begin{aligned} S_{ii}(\omega) &= S_{X_iX_i}(\omega) = \mathrm{Re}[\overline{S}_{ii}(\omega)], \\ S_{ij}(\omega) &= S_{X_iX_j}(\omega) = \frac{1}{2}\left[\overline{S}_{ij}(\omega) + \overline{S}^*_{ji}(\omega)\right]. \end{aligned} \tag{2.279}$$

Since $R_{ij}(\tau) \to 0$ as $\tau \to \infty$, it can be shown by using integration by parts that

$$\Im\left[\frac{\mathrm{d}R_{ij}(\tau)}{\mathrm{d}\tau}\right] = i\omega\Im[R_{ij}(\tau)] - \frac{1}{\pi}R_{ij}(0) = i\omega\overline{S}_{ij}(\omega) - \frac{1}{\pi}m_{ij}. \tag{2.280}$$

Using (2.278) and (2.280), (2.276) can be transformed to

$$\begin{aligned} i\omega\overline{S}_{11} - \frac{1}{\pi}E\left[X_1^2\right] &= -a_{11}\overline{S}_{11} - a_{12}\overline{S}_{12}, \\ i\omega\overline{S}_{12} - \frac{1}{\pi}E[X_1X_2] &= -a_{21}\overline{S}_{11} - a_{22}\overline{S}_{12}. \end{aligned} \tag{2.281}$$

Solutions are readily obtained from complex linear algebraic equation set (2.281), leading to

$$S_{11}(\omega) = \frac{\left(a_{11}\omega^2 + A_2 a_{22}\right)E\left[X_1^2\right] + a_{12}\left(\omega^2 - A_2\right)E[X_1X_2]}{\pi\left[\left(A_2 - \omega^2\right)^2 + A_1^2\omega^2\right]} \tag{2.282}$$

where $A_1 = a_{11} + a_{22}$ and $A_2 = a_{11}a_{22} - a_{12}a_{21}$. By adjusting parameters $a_{ij,}$ (2.282) can represent a spectral density with a peak at a specified location and a given band width.

The FPK equation for the joint stationary PDF $p_{x_1x_2}(x_1, x_2)$ of $X_1(t)$ and $X_2(t)$ corresponding to (2.275) is given by

$$\begin{aligned} &\frac{\partial}{\partial x_1}[(-a_{11}x_1 - a_{12}x_2)p] + \frac{\partial}{\partial x_2}[(-a_{21}x_1 - a_{22}x_2)p] \\ &-\frac{1}{2}\frac{\partial^2}{\partial x_1^2}[D_1^2(x_1, x_2)p] - \frac{1}{2}\frac{\partial^2}{\partial x_2^2}[D_2^2(x_1, x_2)p] = 0. \end{aligned} \tag{2.6.33}$$

Equation (2.283) is satisfied if the following three conditions are met

$$-a_{12}x_2\frac{\partial p}{\partial x_1} - a_{21}x_1\frac{\partial p}{\partial x_2} = 0, \tag{2.284}$$

$$-a_{11}x_1 p - \frac{1}{2}\frac{\partial}{\partial x_1}[D_1^2(x_1, x_2)p] = 0, \tag{2.285}$$

$$-a_{22}x_2 p - \frac{1}{2}\frac{\partial}{\partial x_2}[D_2^2(x_1, x_2)p] = 0, \tag{2.286}$$

indicating that the system belongs to the case of detailed balance (Lin and Cai 1995). The general solution for Eq. (2.284) is given by

$$p(x_1, x_2) = \rho(\lambda), \quad \lambda = k_1x_1^2 + k_2x_2^2 \tag{2.287}$$

where ρ is an arbitrary function of λ, k_1 and k_2 are positive constants satisfying the following condition

$$k_1a_{12} + k_2a_{21} = 0. \tag{2.288}$$

Substituting (2.287) into (2.285) and (2.286), we obtain

$$D_1^2(x_1, x_2) = -\frac{2a_{11}}{p_{X_1X_2}(x_1, x_2)}\int_{x_{1l}}^{x_1} u p_{X_1X_2}(u, x_2)\mathrm{d}u = \frac{a_{11}}{k_1\rho(\lambda)}\int_{\lambda}^{\lambda_1}\rho(\lambda)\mathrm{d}\lambda, \tag{2.289}$$

$$D_2^2(x_1, x_2) = -\frac{2a_{22}}{p_{X_1X_2}(x_1, x_2)}\int_{x_{2l}}^{x_2} v p_{X_1X_2}(x_1, v)\mathrm{d}v = \frac{a_{22}}{k_2\rho(\lambda)}\int_{\lambda}^{\lambda_2}\rho(\lambda)\mathrm{d}\lambda, \tag{2.290}$$

where λ_1 is the λ value at $x_1 = x_{1l}$ for a fixed x_2, and λ_2 is the λ value at $x_2 = x_{2l}$ for a fixed x_1. Equation (2.287) shows that, if the PDF of the stochastic process $X_1(t)$ is of a form of $p(x_1) = p(x_1^2)$, function $\rho(\lambda)$ can be constructed, two functions $D_1(x_1, x_2)$ and $D_2(x_1, x_2)$ can be calculated from (2.289) and (2.290), and (2.275) can be used to generate stochastic process $X_1(t)$.

The second-order filter (2.275) can be used to generate stochastic process $X_1(t)$ defined in an infinite domain $(-\infty, \infty)$, as well as in a finite range $(-\Delta, \Delta)$.

Example 2.13 Consider the case in which x_1 and x_2 are bounded by.

$$k_1 x_1^2 + k_2 x_2^2 \leq k_1 \Delta^2, \tag{2.291}$$

and $\rho(\lambda) = C\big(k_1\Delta^2 - \lambda\big)^{\delta - \frac{1}{2}}, \delta > -\frac{1}{2}$

$$\rho(\lambda) = C(k_1\Delta^2 - \lambda)^{\delta - \frac{1}{2}}, \quad \delta > -\frac{1}{2} \tag{2.292}$$

The joint stationary PDF is

$$p_{X_1X_2}(x_1, x_2) = C\big(k_1\Delta^2 - k_1x_1^2 - k_2x_2^2\big)^{\delta - \frac{1}{2}}, \tag{2.293}$$

and the marginal PDF of X_1 is then obtained as

$$p_{X_1}(x_1) = 2\int_0^{\sqrt{k_1(\Delta^2 - x_1^2)/k_2}} p_{X_1X_2}(x_1, x_2)\mathrm{d}x_2 = C_1\big(\Delta^2 - x_1^2\big)^{\delta}, \tag{2.294}$$

where C_1 is normalization constant. Substituting (2.293) into (2.289) and (2.290), we obtain

$$D_1^2(x_1, x_2) = \frac{2a_{11}}{k_1(2\delta + 1)}\big(k_1\Delta^2 - k_1x_1^2 - k_2x_2^2\big), \tag{2.295}$$

$$D_2^2(x_1, x_2) = \frac{2a_{22}}{k_2(2\delta + 1)}\big(k_1\Delta^2 - k_1x_1^2 - k_2x_2^2\big). \tag{2.296}$$

The PDF (2.294) has the same form as (2.268), but with a more restrictive range for parameter δ due to the validity of the joint PDF (2.293) and the positivity requirement of (2.295) and (2.296). Thus, equation set (2.285), with $D_1(X_1, X_2)$ and $D_1(X_1, X_2)$ given by (2.295) and (2.296) respectively, can be used to generate a stochastic process $X_1(t)$ with a spectral density (2.282) and a PDF (2.294). Parameters a_{ij} $(i, j = 1, 2)$ are used to adjust the spectral density, Δ is determined by the allowable range of process $X_1(t)$, and δ is used to match the shape of its probability distribution.

As a special case, let the parameters in system (2.275) be

$$\begin{aligned} &a_{11} = 0, \quad a_{12} = -1, \quad a_{21} = \omega_0^2, \quad a_{22} = 2\zeta\omega_0, \\ &D_1^2 = 0, \quad D_2^2 = \frac{4\zeta\omega_0^3}{2\delta + 1}\left(\Delta^2 - X_1^2 - \frac{1}{\omega_0^2}X_2^2\right). \end{aligned} \tag{2.297}$$

Then we have the spectral density from (2.282) as

$$S_{XX}(\omega) = \frac{2\zeta\omega_0^3\sigma_{XX}^2}{\pi\left[\left(\omega^2 - \omega_0^2\right)^2 + 4\zeta^2\omega_0^2\omega^2\right]} \tag{2.298}$$

Equation (2.298) is the same as (2.265) noting that $\sigma_{XX}^2 = \pi K / \left(2\zeta\omega_0^3\right)$.
Another special case is to let the parameter in system (2.275) be

$$\begin{aligned} &a_{11} = 2\zeta\omega_0, \quad a_{12} = \omega_0^2, \quad a_{21} = -1, \quad a_{22} = 0, \\ &D_1^2 = \frac{4\zeta\omega_0}{2\delta + 1}\left(\Delta^2 - X_1^2 - \omega_0^2 X_2^2\right), \quad D_2^2 = 0. \end{aligned} \tag{2.299}$$

Then the spectral density obtained from (2.282) is

$$S_{XX}(\omega) = \frac{2\zeta\omega_0\omega^2\sigma_{XX}^2}{\pi\left[\left(\omega^2 - \omega_0^2\right)^2 + 4\zeta^2\omega_0^2\omega^2\right]}. \tag{2.300}$$

In this case, $X(t)$ in (2.275) plays the role of derivative process in system (2.264), and Eq. (2.300) is the same as (2.266) since $\sigma_{\dot{X}\dot{X}}^2 = \pi K / (2\zeta\omega_0)$.

In both cases, ζ and ω_0 can be used to adjust the spectral density, and Δ and δ are used to match the probability density.

2.6.3 Randomized Harmonic Process

A type of stochastic processes, known as randomized harmonic processes, is modeled as

$$X(t) = A\sin[\omega_0 t + \sigma B(t) + U], \tag{2.301}$$

where A is a positive constant specifying the intensity of the process, ω_0 and σ are also positive constants representing the mean frequency and the randomness level in the phase, respectively, $B(t)$ is a unit Wiener process, and U is a random variable uniformly distributed in $[0,2\pi)$ and independent of $B(t)$. Physically, the introduction of the random variable U in (2.301) means that the initial phase is random. This randomized harmonic process was proposed independently by Dimentberg and Wedig (Dimentberg 1988; Wedig 1989), and has been used in a variety of engineering problems (e.g., Naess et al. 2008).

Taking into consideration of the periodicity of $X(t)$ with respect to U, we have

$$E[X(t)] = \int_{-\infty}^{\infty} p_B(b)\mathrm{d}b \int_0^{2\pi} \frac{A}{2\pi} \sin(\omega_0 t + \sigma b + u)\mathrm{d}u = 0, \tag{2.302}$$

$$E\left[X^2(t)\right] = \int_{-\infty}^{\infty} p_B(b)\mathrm{d}b \int_0^{2\pi} \frac{A^2}{2\pi} \sin^2(\omega_0 t + \sigma b + u)\mathrm{d}u = \frac{1}{2}A^2, \tag{2.303}$$

$$\begin{aligned} E[X(t_1)X(t_2)] &= A^2 E\{\sin[\omega_0 t_1 + \sigma B(t_1) + U]\sin[\omega_0 t_2 + \sigma B(t_2) + U]\} \\ &= \frac{A^2}{2} E\{\cos[\omega_0(t_2 - t_1) + \sigma B(t_2) - \sigma B(t_1)]\}. \end{aligned} \tag{2.304}$$

Denote the increment of $B(t)$ as

$$Z = B(t_2) - B(t_1). \tag{2.305}$$

According to (4.2.29), its mean and variance are

$$E[Z(t_1, t_2)] = 0, \quad E\left[Z^2(t_1, t_2)\right] = E\left\{[B(t_2) - B(t_1)]^2\right\} = t_2 - t_1, \quad t_2 \geq t_1 \tag{2.306}$$

Since the Wiener process $B(t)$ is Gaussian distributed, its increment Z is also Gaussian distributed with a probability density

$$p_Z(z) = \frac{1}{\sqrt{2\pi(t_2 - t_1)}} \exp\left[-\frac{z^2}{2(t_2 - t_1)}\right]. \tag{2.307}$$

Continuing the calculation of (2.304), we have

$$\begin{aligned} &E[X(t_1)X(t_2)] \\ &= \frac{A^2}{2} E\{\cos[\omega_0(t_2 - t_1)]\cos(\sigma Z) - \sin[\omega_0(t_2 - t_1)]\sin(\sigma Z)\} \\ &= \frac{1}{2}A^2 \cos[\omega_0(t_2 - t_1)] \int_{-\infty}^{\infty} \frac{\cos(\sigma z)}{\sqrt{2\pi(t_2 - t_1)}} \exp\left[-\frac{z^2}{2(t_2 - t_1)}\right]\mathrm{d}z \\ &= \frac{1}{2}A^2 \cos(\omega_0 \tau)\exp\left(-\frac{1}{2}\sigma^2 \tau\right), \quad \tau = t_2 - t_1 \geq 0. \end{aligned} \tag{2.308}$$

Equation (2.308) shows that $X(t)$ is a weakly stationary process with an autocorrelation function

$$R_{XX}(\tau) = \frac{1}{2}A^2 \cos(\omega_0 \tau)\exp\left(-\frac{1}{2}\sigma^2|\tau|\right). \tag{2.309}$$

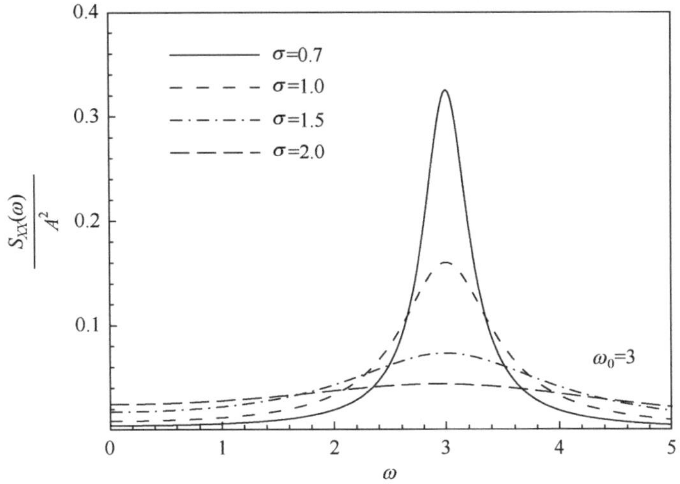

Fig. 2.14 Spectral density functions of randomized harmonic process $X(t)$ for $\omega_0 = 3$ and different σ values

Figure 2.14 depicts the spectral densities in the positive ω range for the case of $\omega_0 = 3$ and several values of σ. It is seen that the spectral densities reach their peaks near $\omega = \omega_0$, and exhibit different band widths for different σ values. In the case of $\sigma = 0$, the stochastic process $X(t)$ reduces to a pure harmonic process with a random initial phase. As σ increases, the band width of the process becomes broader, indicating an increasing randomness.

As a special case of $\omega_0 = 0$, the correlation function and spectral density are, respectively,

$$R_{XX}(\tau) = \frac{1}{2}A^2 \exp\left(-\frac{1}{2}\sigma^2|\tau|\right) \tag{2.310}$$

$$S_{XX}(\omega) = \frac{A^2\sigma^2}{4\pi\left(\omega^2 + \sigma^4/4\right)}. \tag{2.311}$$

It is noted that Eqs. (2.310) and (2.311) have the same forms as those in Eqs. (2.252) and (2.253), and the process is known as the low-pass process. However, the PDF of the randomized harmonic process is different as those generated from a linear or nonlinear filter, that is (Cai and Zhu 2016)

$$p_X(x) = p_{\Theta_1}(\theta_1)\left|\frac{d\theta_1}{dx}\right| = \frac{1}{\pi\sqrt{A^2 - x^2}}, \quad -A < x < A. \tag{2.312}$$

Figure 2.15 depicts the PDF of $X(t)$. It has very large values near the two boundaries. Note that the probability distribution only depends on A, to be determined according to the physical boundary of the underlined phenomenon. The parameters

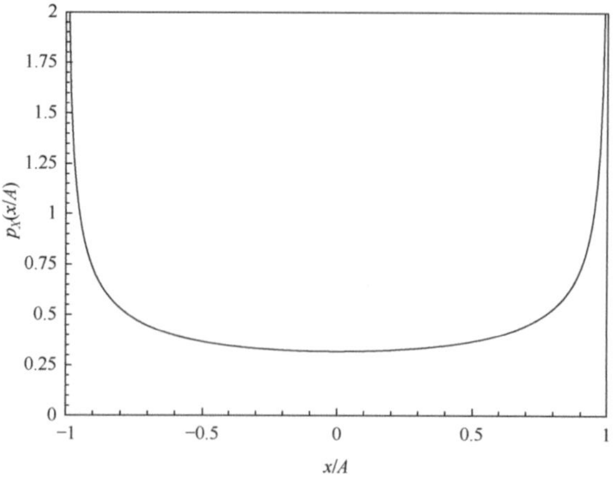

Fig. 2.15 PDF of randomized harmonic process $X(t)$

ω_0 and σ have no effect on the probability distribution; however, they can be adjusted to match the spectral density of $X(t)$ to be modeled.

There are two advantages to use the randomized harmonic process (2.301) to model a practical stochastic process: (i) it is more realistic due to its boundedness, and (ii) the spectral density can be matched by adjusting ω_0 and σ according to the peak magnitude, peak location, and band width.

References

Bassingthwaighte J B, Raymond G M. 1995. Evaluation of the dispersional analysis method for fractional time series. Annals of Biomedical Engineering, 23: 491-505.

Biagini F, Hu Y, Øksendal B, Zhang T. 2008. Stochastic Calculus for Fractional Brownian Motion and Applications. London: Springer-Verlag.

Bochner S. 1959. Lectures on Fourier Integrals. Princeton, New Jersey: Princeton University Press.

Cai G Q, Zhu W Q. 2016. Elements of Stochastic Dynamics. Singapore: World Scientific Publisher.

Chai W, Naess A, Leira B J. 2015. Filter models for prediction of stochastic ship roll response. Probabilistic Engineering Mechanics, 41: 104-114.

Chai W, Naess A, Leira B J. 2016. Stochastic nonlinear ship rolling in random beam seas by path integration method, Probabilistic Engineering Mechanics, 44: 43-52.

Deng M L, Zhu W Q. 2015. Responses of linear and nonlinear oscillators to fractional Gaussian noise with Hurst index between 1/2 and 1. ASME Journal of Applied Mechanics, 82(10): 101008.

Di Paola M, Vasta M. 1997. Stochastic integro-differential and differential equations of non-linear systems excited by parametric Poisson pulses. International Journal of Non-Linear Mechanics, 32(5): 855-862.

Di Paola M, Falsone G. 1993a. Itô and Stratonovich integrals for delta-correlated processes. Probabilistic Engineering Mechanics, 8(3): 197–208.

Di Paola M, Falsone G. 1993b. Stochastic dynamics of nonlinear systems driven by non-normal delta-correlated processes. ASME Journal of Applied Mechanics, 60(1): 141–148.

Dieker T. 2004. Simulation of fractional Brownian motion. M. Sc. Thesis, University of Twente, The Netherlands.

Dimentberg M F. 1988. Statistical Dynamics of Nonlinear and Time-Varying Systems. Taunton, U.K.: Research Studies Press.
Gough J. 1999. Asymptotic stochastic transformations for nonlinear quantum dynamical systems. Reports on Mathematical Physics, 44(3): 313-338.
Grigoriu M. 1998. Simulation of stationary non-Gaussian translation processes. ASCE Journal of Engineering Mechanics, 124(2): 121-126.
Hanson F B. 2007. Applied stochastic processes and control for Jump-diffusions: modeling, analysis, and computation. Philadelphia: Siam.
Hu S L J. 1993. Responses of dynamic-systems excited by non-Gaussian pulse processes. ASCE Journal of Engineering Mechanics, 119(9): 1818-1827.
Hu S L J. 1994. Responses of dynamic-systems excited by non-Gaussian pulse processes-closure. ASCE Journal of Engineering Mechanics, 120(11): 2473-2474.
Hu S L J. 1995. Parametric random vibrations under non-Gaussian delta-correlated processes. J. Engng. Mech. Div. ASCE, 121(12): 1366-1371.
Itô K. 1951a. On stochastic differential equations. Memoirs of the American Mathematical Society, 4: 289–302.
Itô K. 1951b. On a formula concerning stochastic differentials. Nagoya Mathematical Journal, 3: 55–65.
Jia W T, Zhu W Q. 2014a. Stochastic averaging of quasi-integrable and non-resonant Hamiltonian systems under combined Gaussian and Poisson white noise excitations. Nonlinear Dynamics, 76(2): 1271–1289.
Jia W T, Zhu W Q. 2014b. Stochastic averaging of quasi-partially integrable Hamiltonian systems under combined Gaussian and Poisson white noise excitations. Physica A: Statistical Mechanics and its Applications, 398: 125–144.
Kaarakka T, Salminen P. 2011. On fractional Ornstein–Uhlenbeck processes. Communications on Stochastic Analysis, 5(1): 121–133.
Lévy P. 1953. Random functions: General theory with special reference to Laplacian random functions. University California Publications in Statistics, 1: 331-390.
Lin Y K. 1967. Probabilistic Theory of Structural Dynamics. New York: McGraw Hill, Reprint R. Melbourne, FL: E. Krieger, 1976.
Lin Y K, Cai G Q. 1995. Probabilistic structural dynamics: advanced theory and applications. New York: McGraw-Hill.
Mandelbrot B B, Van Ness J W. 1968. Fractional Brownian motions, fractional noises and applications. SIAM Rev. 10: 422–437.
Mishura Y S. 2008. Stochasic Calculus for Fractional Brownian Motion and Related Processes. Berlin: Springer-Verlag.
Naess A, Dimentberg M F, Gaidai O. 2008. Lotka-Volterra systems in environments with randomly disordered temporal periodicity. Physical Review E, 78: 021126.
Rangarajan G, Ding M. 2003. Processes with Long Range Correlations: Theory and Applications. Berlin: Springer-Verlag.
Sun J Q. 2006. Stochastic Dynamics and Control. Netherlands: Elsevier.
Uchaikin V V. 2012a. Fractional Derivatives for Physicists and Engineers: Vol. I, Background and Theory, Berlin: Springer-Verlag.
Uchaikin V V. 2012b. Fractional Derivatives for Physicists and Engineers: Vol. II, Applications, Berlin: Springer-Verlag.
Wedig W V. 1989. Analysis and simulation of nonlinear stochastic systems. Nonlinear Dynamics in Engineering Systems, 337–344, IUTAM Symposium, Berlin: Spring-Verlag.
Wong E, Zakai M. 1965. On the relation between ordinary and stochastic equations. International Journal of Engineering Sciences, 47(1): 150-154.
Zeng Y, Zhu W Q. 2011. Stochastic averaging of quasi-nonintegrable-Hamiltonian systems under Poisson white noise excitation. ASME Journal of Applied Mechanics, 78: 021002.
Zygadlo R. 2003. Martingale integrals over Poissonian processes and the Itô-type equations with white shot noise. Physical Review E, 68(4): 046117.

Chapter 3
Nonlinear Stochastic Dynamical Systems

As another prerequisite, in this chapter the mathematical expressions of equations of motion for nonlinear stochastic dynamical systems used in this book are introduced. Section 3.1 describes commonly used three forms: the most general mathematical expression, the Lagrange equations used in many mechanical and structural systems, and the Hamiltonian format developed recently for MDOF strongly nonlinear dissipated systems subjected to stochastic excitations. The last form, namely the Hamiltonian systems, and their classification based on the integrability and resonance characteristics are presented in Sect. 3.2. The Hamiltonian systems in Sect. 3.2 are classical ones. The generalized Hamiltonian systems and their classification are described in Sect. 3.3. In above mentioned systems, the restoring force and dissipated force are separated. In some practical situations, the two types of forces may be coupled, called forces with genetic effect, including hysteretic forces, visco-elastic forces, damping forces with fractional derivatives, etc. These three forces will be described in Sect. 3.4.

3.1 Modeling of Nonlinear Stochastic Dynamical Systems

A stochastic dynamical system may be described by the following stochastic differential equations

$$\frac{\mathrm{d}}{\mathrm{d}t}X_j(t) = f_j[\boldsymbol{X}(t), t] + \sum_{l=1}^{m} g_{jl}[\boldsymbol{X}(t), t]\xi_l(t), \quad j = 1, 2, \ldots, n, \tag{3.1}$$

where $\boldsymbol{X}(t) = [X_1(t), X_2(t), \ldots, X_n(t)]^{\mathrm{T}}$ is a vector of system responses, also known as state variables, $\xi_l(t)$ are excitations among which at least one is stochastic process. Note that the capital letters used in (3.1) for the state variables signify that they are

W.-Q. Zhu et al., *Stochastic Averaging*, https://doi.org/10.1007/978-981-96-3829-1_3

random or stochastic quantities. Functions f_j and g_{jl} represent system properties which may or may not depend on time explicitly. An excitation $\xi_l(t)$ is called a parametric (or multiplicative) one if the associated function g_{jl} depends on $\boldsymbol{X}$; otherwise, it is known as an external (additive) excitation.

If all functions f_j are linear functions of $\boldsymbol{X}$ and functions g_{jl} are all constants, the system is linear. If all functions f_j and g_{jl} are linear functions of $\boldsymbol{X}$, it is known as parametrically excited linear system, although it is essentially nonlinear since the superposition principle is no longer applicable. If at least one of functions f_j and g_{jl} is nonlinear, it is a nonlinear system. For the case of $n = 1$, it is a one-dimensional system. Otherwise, it is called a multi-dimensional system. A continuous system governed by a partial differential equation can be discretized to a multi-dimensional system using schemes such as finite-element procedure.

The stochasticity (randomness) may occur in system properties, in which case some parameters in functions f_j and g_{jl} are not precisely known in advance, and they can be modeled as random variables. It may also occur in excitations, namely, some of excitations $\xi_l(t)$ in Eq. (3.1) are stochastic processes. In this book, only the latter case is considered, and systems properties represented by functions f_j and g_{jl} are assumed to be deterministic.

Equations of motion of many mechanical and structural systems are usually established by means of the Newton's second law or the Lagrange equations according to the physical nature. The governing equations often appear in the following form

$$\ddot{Z}_j + h_j(\mathbf{Z}, \dot{\mathbf{Z}}) + u_j(\mathbf{Z}) = \sum_{l=1}^{m} g_{jl}(\mathbf{Z}, \dot{\mathbf{Z}})\xi_l(t), \quad j = 1, 2, \ldots, n, \tag{3.2}$$

where $\mathbf{Z} = [Z_1, Z_2, \ldots, Z_n]^{\mathrm{T}}$ and $\dot{\mathbf{Z}} = [\dot{Z}_1, \dot{Z}_2, \ldots, \dot{Z}_n]^{\mathrm{T}}$ are vectors of displacements and velocities, respectively, and $h_j(\mathbf{Z}, \dot{\mathbf{Z}})$ and $u_j(\mathbf{Z})$ represent damping forces and restoring forces, respectively. Letting $X_{2j-1} = Z_j$, $X_{2j} = \dot{Z}_j$, and $\boldsymbol{X} = [X_1, X_2, \ldots, X_{2n}]^T$, system (3.2) is transformed to

$$\begin{aligned} \dot{X}_{2j-1} &= X_{2j}, \\ \dot{X}_{2j} &= -h_j(\boldsymbol{X}) - u_j(\boldsymbol{X}) + \sum_{l=1}^{m} g_{jl}(\boldsymbol{X})\xi_l(t). \end{aligned} \tag{3.3}$$

By comparison of (3.3) and (3.1), it can be seen that equation set (3.3) is a special case of system (3.1).

A stochastic dynamical systems can also be formulated as stochastically excited and dissipated Hamiltonian system, governed by the equations

$$\dot{Q}_j = \frac{\partial H}{\partial P_j},$$

$$\dot{P}_j = -\frac{\partial H}{\partial Q_j} - \sum_{k=1}^{n} c_{jk}(\boldsymbol{Q}, \boldsymbol{P})\frac{\partial H}{\partial P_k} + \sum_{l=1}^{m} g_{jl}(\boldsymbol{Q}, \boldsymbol{P})\xi_l(t). \tag{3.4}$$

where Q_j and P_j are generalized displacements and momenta, respectively, $\boldsymbol{Q} = [Q_1, Q_2, \ldots, Q_n]^{\mathrm{T}}, \boldsymbol{P} = [P_1, P_2, \ldots, P_n]^{\mathrm{T}}$, and $H = H(\boldsymbol{Q}, \boldsymbol{P})$ is a Hamiltonian function. Equation set (3.2) can be transformed to the form of (3.4) by using the Legendre transform. It can be seen that the stochastically excited and dissipated Hamiltonian system (3.4) is also a special case of system (3.1).

Conventionally, systems (3.2) and (3.4) are known as n-degree-of-freedom systems, which are equivalent to two $2n$-dimensional systems (3.1). These two technical terms will be followed throughout this book. A single-degree-of-freedom (SDOF) system is a two-dimensional system, and a n degrees-of-freedom system is a $2n$-dimensional system.

Mathematically, equations of motion of (3.1) are more general than (3.2) and (3.4) since the latter two equation sets can be transformed to the former equation set. However, for many engineering systems, equations (3.2) are usually derived directly from Lagrange equations, and then transformed to (3.4). They describe the relationships between different degrees of freedom, and possess more explicit physical meanings. Therefore, the methods and procedures introduced in the book, although applicable to the general system (3.1), are especially suitable for systems (3.2) and (3.4).

The vectors $\boldsymbol{X}(t) = [X_1(t), X_2(t), \ldots, X_n(t)]^{\mathrm{T}}$ in system (3.1), $\boldsymbol{Z} = [Z_1, Z_2, \ldots, Z_n]^{\mathrm{T}}$ and $\dot{\boldsymbol{Z}} = [\dot{Z}_1, \dot{Z}_2, \ldots, \dot{Z}_n]^T$ in system (3.2), and $\boldsymbol{Q} = [Q_1, Q_2, \ldots, Q_n]^{\mathrm{T}}$ and $\boldsymbol{P} = [P_1, P_2, \ldots, P_n]^{\mathrm{T}}$ in system (3.4) are known as system responses. Moreover, their functions, such as the system Hamiltonian, the amplitude envelope of a single response, the system total energy, etc. also belong to the category of system responses. Although the systems considered in this book are deterministic, the system responses are stochastic processes due to the stochastic excitations, as illustrated in Fig. 3.1.

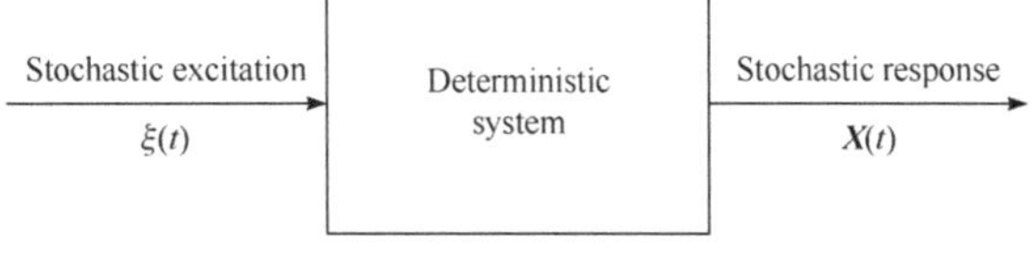

Fig. 3.1 System excitations and responses

3.2 Hamiltonian Systems and Their Classification

3.2.1 Hamilton Equation

The constraints restricting the configuration and motion of a dynamical system are called geometric constraints and motion constraints, respectively. A dynamical system contains only geometric constraints, including the case in which the motion constraints can be converted into geometric constraints, is called holonomic system. The least number of independent variables for describing a holonomic system is called the number of degree-of-freedom. For example, a system of k particles containing s holonomic constraints, the number of degree-of-freedom is $n = 3k - s$. The independent variables necessary for determining the configuration of a dynamical system are called generalized coordinates or generalized displacements (e.g., linear displacements, angle displacements, etc.). For holonomic systems, the number of generalized coordinates or generalized displacements is equal to the number of degree-of-freedom.

Let q_i $(i = 1, 2, \ldots, n)$ denote the generalized displacements of a holonomic dynamical system. For conservative systems involving only potential forces, the following Lagrange equations can be derived from the universal equations of dynamics, which denote vanishing of the summation of generalized potential forces and generalized inertial forces,

$$\frac{\mathrm{d}}{\mathrm{d}t}\left(\frac{\partial T}{\partial \dot{q}_i}\right) - \frac{\partial T}{\partial q_i} + \frac{\partial U}{\partial q_i} = 0, \quad i = 1, 2, \ldots, n, \tag{3.5}$$

where $T = T(\boldsymbol{q}, \dot{\boldsymbol{q}})$ and $U = U(\boldsymbol{q})$ are kinetic energy and potential energy, respectively; $\boldsymbol{q}$ and $\dot{\boldsymbol{q}}$ are vectors of generalized displacements and generalized velocities, respectively; n is the number of degree-of freedom.

Introduce Lagrange function

$$L = T - U. \tag{3.6}$$

Equation (3.5) becomes

$$\frac{\mathrm{d}}{\mathrm{d}t}\left(\frac{\partial L}{\partial \dot{q}_i}\right) - \frac{\partial L}{\partial q_i} = 0, \quad i = 1, 2, \ldots, n. \tag{3.7}$$

Equation (3.7) can also be derived from the following Hamilton principle

$$\delta \int_{t_1}^{t_2} L(\boldsymbol{q}, \dot{\boldsymbol{q}}, t)\mathrm{d}t = 0. \tag{3.8}$$

The solutions of Eq. (3.7) are the trajectories in n-dimensional configuration space. Since infinite trajectories may pass through any point in the configuration space, it is inconvenient to use Lagrange equation (3.7) in the theoretical study of dynamical systems. It is possible to introduce the generalized velocities as independent variables besides the generalized displacements and then converted Eq. (3.7) into state equations. The generalized displacements and generalized velocities constitute the phase space, and the above-mentioned situation will be avoided. However, a better way is to introduce the generalized momenta as independent variables besides the generalized displacements. They together constitute the phase space. There are advantages to transfer Eq. (3.7) to Hamilton equations.

The generalized momenta are defined as

$$p_i = \frac{\partial L}{\partial \dot{q}_i}, \quad i = 1, 2, \ldots, n, \tag{3.9}$$

which are the Legendre transformations of the Lagrange function L. Assume that the determinant of Hesse matrix of L with respect to $\dot{q}_i$ does not vanish, i.e.,

$$\det\left[\frac{\partial^2 L}{\partial \dot{q}_i \partial \dot{q}_j}\right] \neq 0, \quad i, j = 1, 2, \ldots, n. \tag{3.10}$$

Then the transformation (3.10) is not singular and inversive, and the inverse transformation is also a Legendre transformation. According to the theorem for inverse Legendre transformation, the generating function of the inverse Legendre transformation of transformation (3.9) is

$$\sum_{i=1}^{n} (p_i \dot{q}_i - L)_{\dot{q}_i \to p_i} = H(\boldsymbol{q}, \boldsymbol{p}, t), \tag{3.11}$$

where $\boldsymbol{p} = [p_1, p_2, \ldots, p_n]^{\mathrm{T}}$. The inverse Legendre transformation of Eq. (3.9) is

$$\dot{q}_i = \frac{\partial H}{\partial p_i}, \quad i = 1, 2, \ldots, n. \tag{3.12}$$

The generating functions of Legendre transformation and its inverse are related as follows:

$$\frac{\partial H}{\partial q_i} = -\frac{\partial L}{\partial q_i}, \quad i = 1, 2, \ldots, n. \tag{3.13}$$

The following equations can be deduced from Eqs. (2.61), (2.63) and (2.66)

$$\dot{p}_i = -\frac{\partial H}{\partial q_i}, \quad i = 1, 2, \ldots, n. \tag{3.14}$$

Finally, the following Hamilton equations with q_i and $p_i (i = 1, 2, \ldots, n)$ as independent variables can be obtained by combining Eqs. (3.12) and (3.14):

$$\dot{q}_i = \frac{\partial H}{\partial p_i}, \quad \dot{p}_i = -\frac{\partial H}{\partial q_i}; \quad i = 1, 2, \ldots, n. \tag{3.15}$$

Equation (3.15) is equivalent to Eq. (3.7). $q_i, p_i (i = 1, 2, \ldots, n)$ are called canonical coordinates. The state space formed by q_i, p_i is called canonical phase space. $H(\boldsymbol{q}, \boldsymbol{p}, t)$ is called Hamilton function or Hamiltonian, and the dynamical system governed by Eq. (3.15) is called Hamiltonian system.

Let $\boldsymbol{z} = [\boldsymbol{q}^{\mathrm{T}}\boldsymbol{p}^{\mathrm{T}}]^{\mathrm{T}}$ denote $2n$-dimensional canonical vector, $H(\boldsymbol{z}, t)$ the Hamiltonian, and $\boldsymbol{D} = [\frac{\partial}{\partial z_1} \frac{\partial}{\partial z_2} \cdots \frac{\partial}{\partial z_{2n}}]^{\mathrm{T}}$ the vector of gradient operator. The Hamiltonian Eq. (3.15) can be rewritten as

$$\dot{\boldsymbol{z}} = \boldsymbol{J}\boldsymbol{D}H(\boldsymbol{z}, t), \tag{3.16}$$

where

$$\boldsymbol{J} = \begin{bmatrix} 0 & \boldsymbol{I}_n \\ -\boldsymbol{I}_n & 0 \end{bmatrix} \tag{3.17}$$

is a symplectic matrix, and $\boldsymbol{I}_n$ is a $n \times n$ unit matrix. It can be proved that the symplectic matrix $\boldsymbol{J}$ has the following properties:

$$\boldsymbol{J}^{\mathrm{T}} = \boldsymbol{J}^{-1} = -\boldsymbol{J}, \quad |\boldsymbol{J}| = 1. \tag{3.18}$$

The duality of Eqs. (3.15) and (3.16) makes the Hamiltonian systems with symplectic structure.

Example 3.1 As an example, consider a pendulum attached to a vibrating mass as shown in Fig. 3.2. Denote the displacement of mass M and the angle of pendulum by q_1 and q_2, respectively.

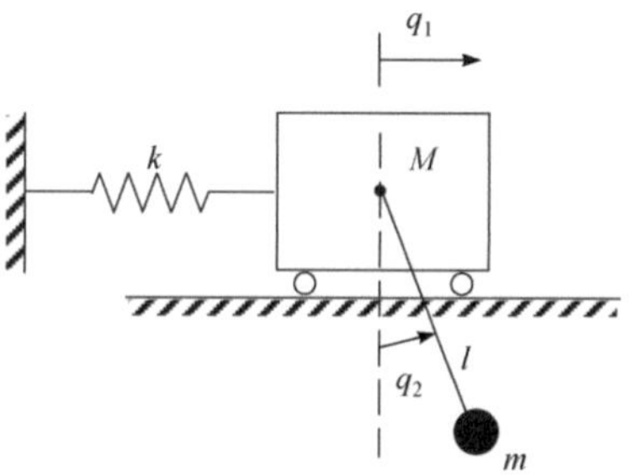

Fig. 3.2 Coupled system of vibrating mass and pendulum

The kinetic energy and potential energy of the system are

$$T = \frac{1}{2}(M + m)\dot{q}_1^2 + ml\dot{q}_1\dot{q}_2 \cos q_2 + \frac{1}{2}ml^2\dot{q}_2^2, \tag{3.19}$$

$$V = \frac{1}{2}kq_1^2 + mgl(1 - \cos q_2). \tag{3.20}$$

According to Eq. (3.6), the Lagrange function of the system is

$$\begin{aligned} L = T - V = &\frac{1}{2}(M + m)\dot{q}_1^2 + ml\dot{q}_1\dot{q}_2\cos q_2 \\ &+ \frac{1}{2}ml^2\dot{q}_2^2 - \frac{1}{2}kq_1^2 - mgl(1 - \cos q_2). \end{aligned} \tag{3.21}$$

The Lagrange equations of the system is obtained by substituting Eq. (3.21) into Eq. (3.7) as follows

$$\begin{aligned} &(M + m)\ddot{q}_1 + ml(\ddot{q}_2 \cos q_2 - \dot{q}_2^2 \sin q_2) + kq_1 = 0, \\ &ml^2\ddot{q}_2 + ml\ddot{q}_1 \cos q_2 + mgl \sin q_2 = 0. \end{aligned} \tag{3.22}$$

Following Eq. (3.9), the generalized momenta of system are

$$\begin{aligned} p_1 &= \frac{\partial L}{\partial \dot{q}_1} = (M + m)\dot{q}_1 + ml\dot{q}_2\cos q_2, \\ p_2 &= \frac{\partial L}{\partial \dot{q}_2} = ml\dot{q}_1 \cos q_2 + ml^2\dot{q}_2. \end{aligned} \tag{3.23}$$

Note that the determinant of Hesse matrix of L with respect to $\dot{q}_i (i = 1, 2)$ is not zero, i.e.,

$$\begin{vmatrix} M + m & ml\cos q_2 \\ ml\cos q_2 & ml^2 \end{vmatrix} \neq 0. \tag{3.24}$$

The inverse Legendre transformation of Eq. (3.23) exists and it is

$$\begin{aligned} \dot{q}_1 &= \frac{lp_1 - p_2\cos q_2}{l(M + m \sin^2 q_2)}, \\ \dot{q}_2 &= \frac{(M + m)p_2 - mlp_1\cos q_2}{ml^2(M + m \sin^2 q_2)}. \end{aligned} \tag{3.25}$$

The Hamiltonian of the system is obtained by substituting Eq. (3.25) into Eq. (3.21) and then into Eq. (3.11) as follows

$$\begin{aligned} H &= (p_1\dot{q}_1 + p_2\dot{q}_2 - L)|_{\dot{q}_1,\dot{q}_2 \to p_1,p_2} \\ &= Ap_1^2 + Bp_1p_2 + Cp_2^2 + \frac{1}{2}kq_1^2 + mgl(1 - \cos q_2), \end{aligned} \tag{3.26}$$

where

$$\begin{aligned} A &= \frac{1}{2\left(M + m\sin^2 q_2\right)}, B = -\frac{cosq_2}{l\left(M + m\sin^2 q_2\right)}, \\ C &= \frac{M + m}{2ml^2\left(M + m\sin^2 q_2\right)}. \end{aligned} \tag{3.27}$$

Substituting Eq. (3.26) into Eq. (3.15) leads to the following Hamilton equations of the system

$$\begin{aligned} \dot{q}_1 &= 2Ap_1 + Bp_2, \\ \dot{p}_1 &= -kq_1, \\ \dot{q}_2 &= Bp_1 + 2Cp_2, \end{aligned}$$

$$\dot{p}_2 = -mgl\sin q_2 + A'p_1^2 - B'p_1p_2 + C'p_2^2, \tag{3.28}$$

where

$$\begin{aligned} A' &= \frac{m\sin 2q_2}{2(M + m\sin^2 q_2)^2}, \quad B' = \frac{\left[M + m\left(2 - \sin^2 q_2\right)\right]\sin q_2}{l(M + m\sin^2 q_2)^2}, \\ C' &= \frac{(M + m)\sin 2q_2}{2l^2(M + m\sin^2 q_2)^2}. \end{aligned} \tag{3.29}$$

It is seen from Eq. (3.28) that the Hamiltonian system shown in Fig. 3.2 is nonlinear. It can be linearized in the vicinity of the equilibrium point $q_1 = q_2 = p_1 = p_2 = 0$ by letting $A' = B' = C' = 0$, that is, when the angle of the pendulum is very small.

3.2.2 *Poisson Bracket*

Assume that $F = F(\boldsymbol{q}, \boldsymbol{p}, t)$ and $G = G(\boldsymbol{q}, \boldsymbol{p}, t)$ are continuous and differentiable dynamical quantities defined in canonical phase space $(\boldsymbol{q}, \boldsymbol{p})$. The Poisson bracket of F and G is defined as

$$[F, G] = \sum_{i=1}^{n} \frac{\partial F}{\partial p_i}\frac{\partial G}{\partial q_i} - \frac{\partial F}{\partial q_i}\frac{\partial G}{\partial p_i}. \tag{3.30}$$

It is shown that the Poisson bracket has the following properties.

(a) Anti-symmetry

$$[F, G] = -[G, F]. \tag{3.31}$$

(b) Bi-linearity

$$[aF + bG, K] = a[F, K] + b[G, K], \tag{3.32}$$

where K is a dynamical quantity, and a and b are constants.

(c) 'Leibnitz rule

$$[F \cdot G, K] = F[G, K] + G[F, K], \tag{3.33}$$

where "·" denotes a dot product..

(d) Jacobi identity

$$[F, [G, K]] + [G, [K, F]] + [K, [F, G]] = 0. \tag{3.34}$$

(e) Non-degeneration. If $\boldsymbol{z} = [\boldsymbol{q}^{\mathrm{T}}\boldsymbol{p}^{\mathrm{T}}]^{\mathrm{T}}$ is not a critical point of F, namely, $\boldsymbol{D}F(\boldsymbol{z}) \neq 0$, then a continuously differentiable function G exists so that $[F, G] \neq 0$. In other words, if $[F, G] = 0$ holds for all continuously differentiable function G, F must be a motion constant, namely F remains as a constant in the entire motion process of the Hamiltonian system.

Using Poisson bracket, Hamilton equation (3.15) can be rewritten as

$$\dot{q}_i = \left[q_i, H\right], \quad \dot{p}_i = \left[p_i, H\right], \quad i = 1, 2, \ldots, n. \tag{3.35}$$

The time change rate of any dynamical quantity $F(\boldsymbol{q}, \boldsymbol{p}, t)$ of a Hamiltonian system can be written as

$$\mathrm{d}F/\mathrm{d}t = [F, H] + \partial F/\partial t. \tag{3.36}$$

If F is independent of t explicitly, then

$$\mathrm{d}F/\mathrm{d}t = [F, H]. \tag{3.37}$$

For autonomous Hamiltonian systems, if

$$[F, H] = 0, \tag{3.38}$$

then F is called the first integral (motion constant, or conservative quantity). Specially,

$$[H, H] = 0. \tag{3.39}$$

Thus, an autonomous Hamiltonian system is a conservative system. During the motion process, the total energy of the system does not change.

3.2.3 Phase Flow

For $2n$-dimensional autonomous Hamiltonian systems, equation

$$\dot{z} = \boldsymbol{f}(z) = \boldsymbol{J}DH(z) = \left[(\partial H/\partial \boldsymbol{p})^{\mathrm{T}}, (-\partial H/\partial \boldsymbol{q})^{\mathrm{T}}\right]^{\mathrm{T}} \tag{3.40}$$

defines a vector field $\boldsymbol{f}(z)$ in $2n$-dimensional canonical phase space. Assume that the solutions of Eq. (3.40) may extend over the entire time axis $(-\infty, \infty)$ and denote the solution of Eq. (3.40) by $(\boldsymbol{q}(t), \boldsymbol{p}(t))$ with initial state $(\boldsymbol{q}(0), \boldsymbol{p}(0))$. The phase flow g^t of a Hamiltonian system in $2n$-dimensional phase space is defined as

$$g^t : (\boldsymbol{q}(0), \boldsymbol{p}(0)) \mapsto (\boldsymbol{q}(t), \boldsymbol{p}(t)). \tag{3.41}$$

where "$\mapsto$" denotes that system moves from point $(\boldsymbol{q}(0), \boldsymbol{p}(0))$ to point $(\boldsymbol{q}(t), \boldsymbol{p}(t))$ in canonical phase space as time goes from 0 to t.

The divergence of vector field $\boldsymbol{f}(z)$ is

$$\mathrm{div}\boldsymbol{f}(z) = \sum_{i=1}^{n}\left(\frac{\partial \dot{q}_i}{\partial q_i} + \frac{\partial \dot{p}_i}{\partial p_i}\right) = \sum_{i=1}^{n}\left[\frac{\partial}{\partial q_i}\left(\frac{\partial H}{\partial p_i}\right) + \frac{\partial}{\partial p_i}\left(-\frac{\partial H}{\partial q_i}\right)\right] = 0, \tag{3.42}$$

which implies that Hamiltonian phase flow is incompressible. In other words, the volume of Hamiltonian phase flow is constant, i.e.,

$$\iint_D \mathrm{d}\boldsymbol{q}\mathrm{d}\boldsymbol{p} = \iint_{g^t D} \mathrm{d}\boldsymbol{q}\mathrm{d}\boldsymbol{p}, \tag{3.43}$$

which is called Liouville theorem.

One inference of Liouville theorem is that the equilibrium states of Hamiltonian systems can be centers, saddle points, simple closed orbits, homoclinic orbits, and heteroclinic orbits. It cannot be focuses.

3.2.4 Canonical Transformation

A transformation from a set of canonical coordinates in a phase space to a set of canonical coordinates in another phase space with the same dimension is called canonical transformation or symplectic transformation. Assume that

$$Q_i = Q_i(\boldsymbol{q}, \boldsymbol{p}, t), \quad P_i = P_i(\boldsymbol{q}, \boldsymbol{p}, t), \quad i, = 1, 2, \ldots, n \tag{3.44}$$

is the canonical transformation from canonical coordinates q_i, p_i to another canonical coordinates Q_i, P_i. The following properties of canonical transformation can be readily proved.

(a) For all continuous and differentiable functions F and G,

$$[F, G]_{\boldsymbol{Q},\boldsymbol{P}} = [F, G]_{\boldsymbol{q},\boldsymbol{p}}, \tag{3.45}$$

where the left side and right side of the equation are the Poisson brackets in canonical coordinates $\boldsymbol{Q}, \boldsymbol{P}$ and $\boldsymbol{q}, \boldsymbol{p}$, respectively.

(b) The Jacobi matrix of a canonical transformation $\boldsymbol{T} = \frac{\partial(\boldsymbol{Q},\boldsymbol{P})}{\partial(\boldsymbol{q},\boldsymbol{p})}$ satisfies the following relations

$$\boldsymbol{T}^T \boldsymbol{J} \boldsymbol{T} = \boldsymbol{J}, \tag{3.46}$$

$$\boldsymbol{T} \boldsymbol{J}^{-1} \boldsymbol{T}^T = \boldsymbol{J}^{-1}, \tag{3.47}$$

$$\boldsymbol{T} \boldsymbol{J} \boldsymbol{T}^T = \boldsymbol{J}, \tag{3.48}$$

where $\boldsymbol{J}$ is the symplectic matrix defined in Eq. (3.17).

(c) The volume of the phase space is unchanged during canonical transformation, i.e.,

$$\iint_{TD} \mathrm{d}\boldsymbol{Q}\mathrm{d}\boldsymbol{P} = \iint_{D} \mathrm{d}\boldsymbol{q}\mathrm{d}\boldsymbol{p}, \tag{3.49}$$

which is the Liouville theorem with respect to canonical transformation. Thus, the Jacobi determinant of a canonical transformation is unit, i.e.,

$$\det(\boldsymbol{T}) = \det\left(\boldsymbol{T}^{-1}\right) = 1. \tag{3.50}$$

(d) The form of Hamilton equations is unchanged after a canonical transformation, i.e., Eq. (3.15) of the Hamilton system with canonical coordinates $\boldsymbol{q}, \boldsymbol{p}$ is

$$\dot{Q}_i = \frac{\partial K}{\partial P_i}, \quad \dot{P}_i = -\frac{\partial K}{\partial Q_i}, \quad i = 1, 2, \ldots, n, \tag{3.51}$$

where K is the Hamiltonian with canonical coordinates $\boldsymbol{Q}, \boldsymbol{P}$. It is noted that this property can not be treated as the definition of canonical transformation. For example, transformation $Q = q, P = 2p$ will not change the form of Hamilton equations. However, it is in fact not a canonical transformation. It will be seen later that for completely integrable Hamiltonian system, the transformation from Cartesian coordinates to polar coordinates is not a canonical transformation while that from Cartesian coordinates to action-angle variables is canonical one.

3.2.5 Completely Integrable Hamiltonian System

Liouville proved that a Hamiltonian system of n-degree-of-freedom is completely integrable if there exist n independent and in involution first integrals $H_1, H_2, \ldots, H_n$. Here independence means that $dH_1, dH_2, \ldots, dH_n$ are not linearly related and in involution implies that the Poisson bracket of any two first integrals vanishes, i.e.,

$$\left[H_i, H_j\right] = 0, \quad i, j = 1, 2, \ldots, n. \tag{3.52}$$

Completely integrable means that the integral solution of the Hamiltonian system can be obtained in terms of limited times of algebraic operations and integrations of known functions.

For completely integrable autonomous Hamiltonian systems, the following canonical transformation

$$I_i = I_i(\boldsymbol{q}, \boldsymbol{p}), \quad \theta_i = \theta_i(\boldsymbol{q}, \boldsymbol{p}), \quad i = 1, 2, \ldots, n \tag{3.53}$$

can be introduced to convert the Hamilton Eq. (3.15) into

$$\frac{\mathrm{d}I_i}{\mathrm{d}t} = 0, \quad \frac{\mathrm{d}\theta_i}{\mathrm{d}t} = \omega_i(\boldsymbol{I}), \quad i = 1, 2, \ldots, n, \tag{3.54}$$

where I_i and θ_i are called action variables and angle variables, respectively. They constitute a set of canonical coordinates. $\omega_i(\boldsymbol{I})$ are the system frequencies. The solution of Eq. (3.54) is

$$I_i(t) = I_i(0), \quad \theta_i(t) = \omega_i(\boldsymbol{I}(0))t + \theta_i(0), \quad i = 1, 2, \ldots, n, \tag{3.55}$$

where $I_i(0), \theta_i(0)$ are the initial coordinates at $t = 0$.

Any dynamical quantity $f(\boldsymbol{I}, \boldsymbol{\theta})$ of a completely integrable autonomous Hamiltonian system can be expressed by using the following multi-fold Fourier series

$$f(\boldsymbol{I}, \boldsymbol{\theta}) = f_0(\boldsymbol{I}) + \sum_{r=1}^{\infty} \sum_{|\boldsymbol{k}|=r} \left[f_{\boldsymbol{k}}^c(\boldsymbol{I}) \cos(\boldsymbol{k}, \boldsymbol{\theta}) + f_{\boldsymbol{k}}^s(\boldsymbol{I}) \sin(\boldsymbol{k}, \boldsymbol{\theta})\right],$$
$$i = 1, 2, \ldots, n, \tag{3.56}$$

where $\boldsymbol{k} = [k_1 k_2 \ldots k_n]^{\mathrm{T}}$ is an integer vector, $|\boldsymbol{k}| = \sum_{i=1}^{n} |k_i|$, $(\boldsymbol{k}, \boldsymbol{\theta}) = \sum_{i=1}^{n} k_i\theta_i$, $i = 1, 2, \ldots, n$. For a given set of initial coordinates $\boldsymbol{I}(0)$, all orbits of the system are on a n-dimensional torus T^n.

If $\boldsymbol{\omega} = [\omega_1, \omega_2, \ldots, \omega_n]^{\mathrm{T}}$ satisfies the following non-resonant condition

$$(\boldsymbol{k}, \boldsymbol{\omega}) \neq 0, \boldsymbol{k} \in \boldsymbol{Z}^n \backslash \{0\}, \tag{3.57}$$

where $\boldsymbol{Z}^n$ is a set of n-dimensional vectors of integers except zero and $\boldsymbol{k}$ is a member of $\boldsymbol{Z}^n$, then the completely integrable autonomous Hamiltonian system is non-resonant and the motion of the system is almost periodic.

For any Rieman integrable function f on T^n,

$$\bar{f} = \frac{1}{(2\pi)^n} \int_0^{2\pi} f(\boldsymbol{\theta}) \mathrm{d}\boldsymbol{\theta} \tag{3.58}$$

is called space average of f on T^n, while

$$f^*(\boldsymbol{\theta}_0) = \lim_{T \to \infty} \frac{1}{T} \int_0^T f(\boldsymbol{\theta}_0 + \boldsymbol{\omega} t) \mathrm{d}t \tag{3.59}$$

is called time average. By substituting Eq. (3.56) into Eqs. (3.58) and (3.59), it can be readily proved that

$$\bar{f} = f^*. \tag{3.60}$$

That is, the space average is equal to the time average, which is called average theorem. In other words, completely integrable non-resonant Hamiltonian system is ergodic on n-dimensional torus T^n, and any orbit will cover the torus uniformly.

If $\boldsymbol{\omega}$ satisfies the following resonant condition

$$(\boldsymbol{k}^u, \boldsymbol{\omega}) = 0 \quad \text{or} \quad \sum_{i=1}^{n} k_i^u \omega_i = 0, \quad i = 1, 2, \ldots, n,$$
$$\boldsymbol{k}^u \in \boldsymbol{Z}^n \backslash \{0\}, \quad u = 1, 2, \ldots, \alpha, \tag{3.61}$$

then the completely integrable Hamiltonian system is resonant. Note that it is internal resonance between various degree-of-freedom of the system. Physically, internal resonance implies energy exchange between various degree-of- freedom of system involved in resonance. $|\boldsymbol{k}^u| = \sum_{i=1}^{n} |k_i^u|$ is called order of resonance. For n degree-of-freedom completely integrable Hamiltonian systems, the largest possible number α of resonant relations is $n-1$. In this case, the system is called completely resonant and the orbits of the system on T^n are closed. If $1 \le \alpha < n-1$, the system is called partially resonant, and $(n-\alpha)$ frequencies are independent of each other. The orbits of the system are closed on the $(\alpha+1)$-dimensional subtorus $T^{\alpha+1}$.

For completely integrable resonant Hamiltonian systems, the following combinations of angle variables can be introduced

$$\Psi_u = \sum_{i=1}^{n} k_i^u \theta_i = \sum_{i=1}^{n} k_i^u \theta_i(0), \quad u = 1, 2, \ldots, \alpha, \tag{3.62}$$

Then the motion Eq. (3.54) becomes

$$\frac{\mathrm{d}\boldsymbol{I}}{\mathrm{d}t} = 0, \quad \frac{\mathrm{d}\Psi_u}{\mathrm{d}t} = 0, \quad \frac{\mathrm{d}\theta_j}{\mathrm{d}t} = \omega_j(\boldsymbol{I}), \quad u = 1, 2, \ldots, \alpha; \quad j = \alpha+1, \ldots, n, \tag{3.63}$$

and the solution is

$$\boldsymbol{I}(t) = \boldsymbol{I}(0), \quad \Psi_u(t) = \Psi_u(0), \quad \theta_j(t) = \omega_j(\boldsymbol{I})t + \theta_j(0),$$
$$u = 1, 2, \ldots, \alpha; \quad j = \alpha+1, \ldots, n. \tag{3.64}$$

Obviously, there are $n+\alpha$ constants of motion in this case.

For 2-degree-of-freedom completely integrable Hamiltonian systems, the ratio of the two frequencies

$$\mu = \frac{\omega_1}{\omega_2} \tag{3.65}$$

is called rotation number. When μ is a rational number, i.e., $\mu = r/s$, r and s are relatively prime integers, the system is resonant, and the motion of the system is periodic and the orbit is closed on 2-dimensional torus T^2.

For n-degree-of-freedom completely integrable Hamiltonian systems, if the n-dimensional frequency vector $\boldsymbol{\omega}$ satisfies the following condition

$$\det\left(\frac{\partial \boldsymbol{\omega}(\boldsymbol{I})}{\partial \boldsymbol{I}}\right) = \det\left(\frac{\partial^2 H(\boldsymbol{I})}{\partial \boldsymbol{I}^2}\right) \neq 0, \tag{3.66}$$

the system is called non-degenerate. For such a system, $\boldsymbol{\omega}$ is different for different torus T^n and the system is nonlinear. As $\boldsymbol{I}$ varies, there is an infinite number of tori. Among them, some tori are covered with periodic or partially periodic orbits, and are called resonant tori. The other tori are covered with non-periodic orbits and called non-resonant tori. The measure of resonant tori is zero while that of non-resonant tori is one. That is, for nonlinear completely integral Hamiltonian systems almost all tori are non-resonant. If the system satisfies the following relation

$$\det\left(\frac{\partial \boldsymbol{\omega}(\boldsymbol{I})}{\partial \boldsymbol{I}}\right) = \det\left(\frac{\partial^2 H(\boldsymbol{I})}{\partial \boldsymbol{I}^2}\right) = 0, \tag{3.67}$$

then the system is called intrinsically degenerate, e.g., linear Hamiltonian systems.

It is very hard to find the first integrals other than Hamiltonian for a given Hamiltonian system although some methods have been proposed in the past century. In the following, some examples of completely integrable Hamiltonian systems are given.

(a) The Hamiltonian of two coupled Duffing oscillators is of the form

$$H = \frac{1}{2}\left(p_1^2 + p_2^2 + Aq_1^2 + Bq_2^2\right) + \frac{1}{4}\left(q_1^4 + \sigma q_2^4 + 2\rho q_1^2 q_2^2\right). \tag{3.68}$$

Except trivial case of $\rho = 0$, the system is completely integrable in the following two cases.

Case 1. $A = B, \sigma = \rho = 1$. Let $q_1 = r\cos\varphi, q_2 = r\sin\varphi$. Then H is independent of φ, and φ is a cyclic coordinate. The angular momentum $r^2\dot{\varphi}$ is a constant. After eliminating $\dot{\varphi}$ from H, the system becomes one degree-of-freedom and thus is completely integrable.

Case 2. $A = B, \sigma = 1, \rho = 3$. In this case, another first integral is

$$H_2 = p_1 p_2 + Aq_1 q_2 + q_1 q_2\left(q_1^2 + q_2^2\right), \tag{3.69}$$

which is independent of and in involution with the Hamiltonian. Introduce the following transformation

$$Q_1 = q_1 + q_2, \quad Q_2 = q_1 - q_2, \quad P_1 = p_1, \quad P_2 = p_2. \tag{3.70}$$

Then H can be separated, i.e.,

$$\begin{aligned} H &= H_1 + H_2, \\ H_1 &= \frac{1}{2}P_1^2 + \frac{1}{4}AQ_1^2 + \frac{1}{8}Q_1^4, \\ H_2 &= \frac{1}{2}P_2^2 + \frac{1}{4}AQ_2^2 + \frac{1}{8}Q_2^4. \end{aligned} \tag{3.71}$$

Thus, the system is completely integrable and separable in coordinates $\boldsymbol{Q}, \boldsymbol{P}$.

(b) The Hamiltonian of two nonlinearly coupled diffusion oscillators is

$$H = \frac{1}{2}(p_1^2 + p_2^2 + Aq_1^2 + Aq_2^2) + \frac{1}{2}(q_1 - q_2)^2 + \frac{1}{4}D(q_1 - q_2)^4. \quad (3.72)$$

Introducing transformation (3.70), the Hamiltonian is separated as

$$\begin{aligned} H &= H_1 + H_2, \\ H_1 &= \frac{1}{2}P_1^2 + \frac{1}{4}AQ_1^2, \\ H_2 &= \frac{1}{2}P_2^2 + \frac{1}{4}AQ_2^2 + \frac{1}{2}CQ_2^2 + \frac{1}{4}DQ_2^4. \end{aligned} \quad (3.73)$$

Thus, the system is completely integrable and separable in coordinates $\boldsymbol{Q}, \boldsymbol{P}$.

(c) The Hamiltonian of Hénon-Heiles oscillator with adjustable coefficients is

$$H = \frac{1}{2}(p_1^2 + p_2^2 + Aq_1^2 + Bq_2^2) + Cq_1^2q_2 - \frac{1}{3}Dq_2^3. \quad (3.74)$$

The system is completely integrable in the following four cases.

Case 1. $C = 0$. The Hamiltonian is separable.

Case 2. $A = B$, $C/D = -1$. Introducing transformation (3.70), the Hamiltonian is separated as

$$\begin{aligned} H &= H_1 + H_2, \\ H_1 &= \frac{1}{2}P_1^2 + \frac{1}{4}AQ_1^2 + \frac{1}{6}CQ_1^3, \\ H_2 &= \frac{1}{2}P_2^2 + \frac{1}{4}AQ_2^2 - \frac{1}{6}CQ_2^3. \end{aligned} \quad (3.75)$$

Case 3. $C/D = -1/6$. In this case, the other first integral which is independent of and in involution with the Hamiltonian is

$$H_2 = q_1^4 + 4q_1^2q_2^2 + 4p_1(p_1q_2 - p_2q_1) - 4Aq_1^2q_2 + (4A - B)\left(p_1^2 + Aq_1^2\right). \quad (3.76)$$

The motion equations are separable in parabolic coordinate.

Case 4. $B = 16A$, $D = -16C$. In this case, the other first integral which is independent of and in involution with the Hamiltonian is

$$\begin{aligned} H_2 = &\frac{1}{4}p_1^4 + \left(\frac{1}{2}q_1^2 + 4q_1^2q_2\right)p_1^2 - \frac{4}{3}q_1^3p_1p_2 \\ &+ \frac{1}{4}q_1^4 - \frac{4}{3}q_1^4q_2 - \frac{8}{9}q_1^6 - \frac{16}{3}q_1^4q_2^2. \end{aligned} \quad (3.77)$$

Thus, the system is completely integrable.

3.2.6 Non-Integrable Hamiltonian System

A Hamiltonian system of n-degree-of-freedom ($n \geq 2$) is called non-integrable if except the Hamiltonian there is no other first integrals which are independent of and in involution with the Hamiltonian. It is very difficult to study non-integrable Hamiltonian systems directly. In nonlinear science, usually only nearly integrable Hamiltonian systems are studied, i.e., the effects of small perturbation of non-integrable Hamiltonian on integrable Hamiltonian systems are studied. Historically, the study leads to a theory of canonical perturbation. The basic idea of the theory is to approximate the solution of a nearly integrable Hamiltonian system as the sum of the exact solution of its integrable part and small correction of its non-integrable part. However, a difficulty of small denominate occurs in the development of the theory. This problem was solved in the sixties of the last century and the result was KAM theorem briefly escribed below.

Let $\boldsymbol{I}$, $\boldsymbol{\theta}$ be the action-angle vectors of an unperturbed integrable Hamiltonian system with Hamiltonian $H_0(\boldsymbol{I})$. The equations of system are

$$\dot{\boldsymbol{I}} = 0, \quad \dot{\boldsymbol{\theta}} = \boldsymbol{\omega}(\boldsymbol{I}_0) = \left.\frac{\partial H_0(\boldsymbol{I})}{\partial \boldsymbol{I}}\right|_{\boldsymbol{I}=\boldsymbol{I}_0}. \tag{3.78}$$

Assume that the system is subjected to a small perturbation $\varepsilon H_1(\boldsymbol{\theta}, \boldsymbol{I})$, where H_1 is a periodic function of $\boldsymbol{\theta}$ with period 2π. The equations of the perturbed Hamiltonian system are

$$\dot{\boldsymbol{I}} = -\varepsilon \frac{\partial H_1}{\partial \boldsymbol{\theta}}, \quad \dot{\boldsymbol{\theta}} = \boldsymbol{\omega}(\boldsymbol{I}) + \varepsilon \frac{\partial H_1}{\partial \boldsymbol{I}}, \tag{3.79}$$

where $\boldsymbol{I}$ and $\boldsymbol{\theta}$ are slowly and rapidly varying vectors, respectively. KAM theorem implies that if Hamiltonian $H(\boldsymbol{\theta}, \boldsymbol{I}) = H_0(\boldsymbol{I}) + \varepsilon H_1(\boldsymbol{\theta}, \boldsymbol{I})$ is a smooth analytical function, the unperturbed Hamiltonian system is non-degenerate and does not satisfies resonant condition exactly or approximately. Then after a small non-integrable Hamiltonian perturbation, the non-resonant torus of the unperturbed Hamiltonian system still exists and has only small deformation. However, the resonant torus of the unperturbed Hamiltonian system will be destroyed under non-integrable Hamiltonian perturbation. The destroyed torus is the seeds of the chaos in non-integrable Hamiltonian system. As the non-integrable Hamiltonian perturbation increases, the perturbed Hamiltonian system will change from regular motion (periodic or almost periodic) to chaotic motion. It can be seen from the following example.

The Hénon-Heiles oscillator is given in example (c) of last section. The oscillator with the Hamiltonian in Eq. (3.74) and $A = B = C = D = 1$ is a 2-degree-of-freedom non-integrable Hamiltonian system. The motion of the system is bounded

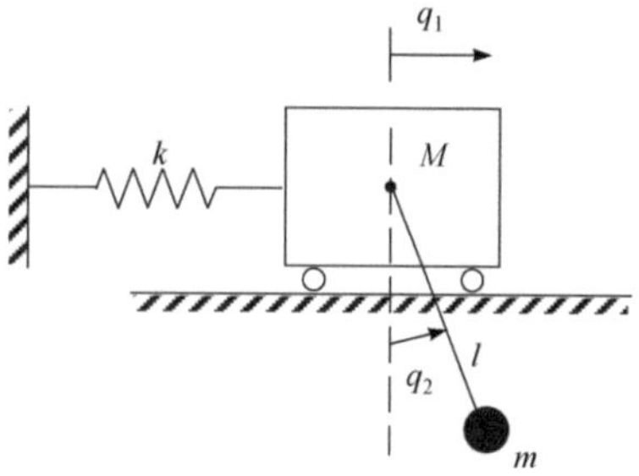

Fig. 3.3 Poincaré section of Hénon-Heiles oscillator

if its potential function $V(q_1, q_2) \leq 1/6$. (Hénon and Heiles 1964) computed the Poincaré sections on plan (q_2, p_2) as shown in Fig. 3.3.

When $H \leq 1/12$, the system is essentially integrable and all initial conditions lead to curves, among which a self-intersection curve is separation boundary. When $H = 1/8$, some curves remain while the others break. The self-intersection curve disappears and becomes randomly distributed points, which means chaos occurs. When $H = 1/6$, all curves disappear and become distributed points. The points generated from a single trajectory fill the most part of equi-energy surface. The motion of the system is essentially chaotic. Thus, with the increase of system energy, the system gradually becomes non-integrable from integrable, and its motion goes to chaotic from regular.

In the following, it is considered for simplicity that the motion of non-integrable Hamiltonian systems is chaotic, and the system states are distributed over the equi-energy surface with equi-probability, namely ergodic over the equi-energy surface..

3.2.7 *Partially Integrable Hamiltonian System*

A Hamiltonian system of n-degree-of-freedom ($n \geq 3$) having $r(1 < r < n)$ independent first integrals in involution is called partially integrable Hamiltonian system. In Hamilton dynamics, there is no concept of partially integrable since in such case it is possible to use $r - 1$ first integrals excluding the energy integral to reduce the original system into $(n - r + 1)$-degree-of-freedom non-integrable Hamiltonian system. However, in this book we will study the randomly excited and dissipated Hamiltonian systems (see Sect. 3.2.9) and it is impossible to do such reduction. Thus, we introduce the concept of partially integrable Hamiltonian systems.

Theoretically, it is always possible to separate a partially integrable Hamiltonian system into a completely integrable Hamiltonian sub-system and a non-integrable Hamiltonian sub-system via canonical transformation. If the completely integrable Hamiltonian sub-system can be further separated, then the Hamiltonian of the partially integrable Hamiltonian system will be of the form

$$H=\sum_{\eta=1}^{r-1}H_\eta\left(q_\eta,p_\eta\right)+H_r(q_r,\ldots,q_n;p_r,\ldots,p_n), \tag{3.80}$$

or

$$H=\sum_{\eta=1}^{r-1}H_\eta\left(I_\eta\right)+H_r(q_r,\ldots,q_n;p_r,\ldots,p_n), \tag{3.81}$$

where H_η are the $r-1$ independent first integrals in involution except the energy integral. Since the completely integrable Hamiltonian sub-system can be resonant or non-resonant, the partially integrable Hamiltonian systems can also be resonant or non-resonant. A partially integrable Hamiltonian system contains both regular (periodic or almost periodic) motion and chaotic motion.

3.2.8 Ergodicity of Hamiltonian Systems

Let z be state vector, g' be system phase flow, $f(z)$ be a dynamical quantity. Similar to equations (3.58) and (3.59),

$$f^*(z_0)=\lim_{T\to\infty}\frac{1}{T}\int_0^T f\left(g'(z_0)\right)\mathrm{d}t \tag{3.82}$$

is called the time average of f, while

$$\overline{f}=\int_\Omega f(z)\mathrm{d}\mu,\quad \mathrm{d}\mu=p(z)\mathrm{d}z \tag{3.83}$$

is called space average of f. Ω may be the phase space or some sub-manifold in the phase space. μ is invariant measure and p is the density of the invariant measure. For Hamiltonian systems,$p(z)=1$. The system is called ergodic on Ω if

$$f^*(z_0)=\overline{f} \tag{3.84}$$

for almost all z_0. That is, for ergodic systems the time average is independent of initial value z_0. Since the initial value can be chosen arbitrarily, the system is ergodic only if its orbit can reach anywhere of the phase space or its sub-manifold.

It is seen from above definition that the ergodicity of a system is related to the phase space or its sub-manifold. For single degree-of-freedom autonomous Hamiltonian system, the equi-energy surface and torus are the same one-dimensional manifold and the system is ergodic both on the equi-energy surface and on torus. For multi-degree-of-freedom completely integrable autonomous Hamiltonian system, it is not ergodic on its equi-energy surface since the $n - 1$ independent in involution first integrals exclude the energy integral and render the system orbit not to reach all parts of equi-energy surface. The system is ergodic on the non-resonant torus and not ergodic on the resonant torus. For nearly integrable Hamiltonian system, it is ergodic on KAM torus if it exists, and ergodic on the equi-energy surface if KAM torus disappears. In classical statistical mechanics, there is an ergodic hypothesis: a general nonlinear Hamiltonian system is ergodic on the equi-energy surface, i.e., the system orbit can reach any point on the equi-energy surface with the same probability. Obviously, this hypothesis holds if the Hamiltonian system is non-integrable and the magnitude of its Hamiltonian is large enough (see Fig. 3.3). For a partially integrable non-resonant Hamiltonian system, the integrable sub-system is ergodic on the (r-1)-dimensional torus and the non-integrable sub-system is ergodic on the curved surface with H_r being constant. For a partially integrable resonant Hamiltonian system, the integrable sub-system is ergodic on the (r-1-β)-dimensional torus where β is the number of resonant relations, and the non-integrable sub-system is ergodic on the curved surface with H_r being constant.

It is noted that the ergodicity of Hamiltonian system is a key concept in the development of the stochastic averaging method for quasi Hamiltonian systems in the sense that the time average can be replaced by space average. For detail see the derivation of averaged equations in Chaps. 5, 6, 7 of vol. 1 and Chaps. 1, 2 of vol. 2.

3.2.9 Stochastically Excited and Dissipated Hamiltonian Systems

In the derivation of Lagrange equations (3.7) for conservative systems, only generalized potential (conservative) forces and generalized inertial forces are included. If non-conservative generalized forces are also included, then the Lagrange equations is of the form

$$\frac{\mathrm{d}}{\mathrm{d}t}\frac{\partial L}{\partial \dot{q}_i} - \frac{\partial L}{\partial q_i} = \overline{F}_i, \quad i = 1, 2, \cdots, n, \tag{3.85}$$

where $\overline{F}_i = F_i(\boldsymbol{q}, \dot{\boldsymbol{q}}, t)$ are non-conservative generalized forces. Equations (3.85) are converted into the following Hamilton equation via Legendre transformation (3.9)

$$\dot{q}_i = \frac{\partial H}{\partial p_i}, \quad \dot{p}_i = -\frac{\partial H}{\partial q_i} + F_i, \quad i = 1, 2, \cdots, n, \tag{3.86}$$

where

$$F_i = F_i(\boldsymbol{q}, \boldsymbol{p}, t) = \overline{F}_i(\boldsymbol{q}, \dot{\boldsymbol{q}}, t)|_{\dot{\boldsymbol{q}} \to \boldsymbol{p}} \tag{3.87}$$

are non-conservative forces, including dissipative forces F_i^d and exciting forces F_i^e.

Assume that the dissipative forces are of the form

$$F_i^d = \sum_{j=1}^{n} F_{ij}^d = -\sum_{j=1}^{n} c_{ij}(\boldsymbol{q}, \boldsymbol{p}) \frac{\partial H}{\partial p_j}. \tag{3.88}$$

If c_{ij} is a constant, F_{ij}^d is a linear damping force. If c_{ij} is a function of $\boldsymbol{q}, \boldsymbol{p}$, F_{ij}^d is a nonlinear damping force. Expression (3.88) is quite general since it includes almost all linear and nonlinear damping forces so far used in dynamics.

Assume that the exciting forces are of the form

$$F_i^e = \sum_{k=1}^{m} F_{ik}^e = \sum_{k=1}^{m} f_{ik}(\boldsymbol{q}, \boldsymbol{p}) \xi_k(t), \quad k = 1, 2, \ldots, m, \tag{3.89}$$

where $\xi_k(t)$ may be random processes, such as Gaussian white noises, Poisson white noises, fractional Gaussian noises, wide-band processes, narrow-band processes, etc. Some $\xi_k(t)$ may also be periodic or harmonic functions. If f_{ik} is a constant, F_{ik}^e is an external excitation or additive excitation. If f_{ik} is a function of $\boldsymbol{q}, \boldsymbol{p}$, F_{ik}^e is a parametric excitation or multiplicative excitation.

Let

$$F_i = F_i^d + F_i^e, \quad i = 1, 2, \ldots, n. \tag{3.90}$$

Then equations (3.86) becomes

$$\begin{aligned} \dot{Q}_i &= \frac{\partial H}{\partial P_i}, \\ \dot{P}_i &= -\frac{\partial H}{\partial Q_i} - \sum_{j=1}^{n} c_{ij}(\boldsymbol{Q}, \boldsymbol{P}) \frac{\partial H}{\partial P_j} + \sum_{k=1}^{m} f_{ik}(\boldsymbol{Q}, \boldsymbol{P}) \xi_k(t), \\ i &= 1, 2, \ldots, n. \end{aligned} \tag{3.91}$$

where capital letters $\boldsymbol{Q}, \boldsymbol{P}$ denote random system states. If at least one $\xi_k(t)$ is stochastic process, the system governed by (3.91) is called stochastically excited and dissipated Hamiltonian system.

It is noted that the system described by (3.91) is non-conservative since m random excitations supply energy into the system while n^2 damping forces dissipate energy from the system. When the difference of input energy and dissipated energy are small compared to the system energy itself, the system is called a quasi-Hamiltonian system. It will be shown that the form and dimension of the stochastic average equations of the system will depend on the integrability and resonance of the Hamiltonian system. (see Chaps. 5, 6, 7 of vol. 1 and Chap. 1 of vol. 2)

3.3 The Generalized Hamiltonian System and its Classification

The Hamiltonian system introduced in Sect. 3.2 is a classical Hamiltonian system defined on even dimensional phase space. However, the dimension of the phase space for an actual system may be odd. For example, the Euler equations of rotational motion of a free rigid body have three basic variables of angular momenta, resulting in a 3-dimensional phase space. Therefore, it is necessary to extend classical Hamiltonian systems to generalized Hamiltonian systems (Li et al. 2007; van Moerbeke 1988; Oliver 1986). Below is a brief introduction to the definition, properties, and classification of generalized Hamiltonian systems relevant to this book.

Although the phase space of generalized Hamiltonian systems can be infinite dimensional, this book only considers finite dimensional cases. Generalized Hamiltonian systems can be defined using generalized Poisson bracket, which are Poisson bracket defined in Sect. 3.2.2 without the non-degeneracy condition. Let the local coordinates of the generalized Hamiltonian system be $\boldsymbol{x} = [x_1, x_2, \ldots, x_m]^{\mathrm{T}}$, the generalized Poisson bracket is denoted by

$$[F, G] = \sum_{i,j=1}^{m} J_{ij}(\boldsymbol{x}) \frac{\partial F}{\partial x_i} \frac{\partial G}{\partial x_j}, \tag{3.92}$$

where $\boldsymbol{J}(\boldsymbol{x}) = [J_{ij}(\boldsymbol{x})]_{m\times m}$ is called the structural matrix of the generalized Poisson bracket. $\boldsymbol{J}(\boldsymbol{x})$ is an $(m \times m)$ -order antisymmetric matrix whose elements are functions of the local coordinates x_i. For a given function matrix $\boldsymbol{J}(\boldsymbol{x})$, it is a structural matrix of the generalized Poisson bracket if

$$J_{ij}(\boldsymbol{x}) = -J_{ji}(\boldsymbol{x}), \quad i, j = 1, 2, \ldots, m,$$

$$\sum_{l=1}^{m} \left(J_{il} \frac{\partial J_{jk}}{\partial x_l} + J_{jl} \frac{\partial J_{ki}}{\partial x_l} + J_{kl} \frac{\partial J_{ij}}{\partial x_l} \right) = 0, \quad i, j, k = 1, 2, \ldots, m. \tag{3.93}$$

The second condition of Eq. (3.93) is actually an equivalent representation of the Jacobi identity equation (3.34).

Obviously, any antisymmetric constant matrix satisfies condition (3.93). In particular, when $\boldsymbol{J}(\boldsymbol{x}) = \begin{bmatrix} 0 & \boldsymbol{I}_n \\ -\boldsymbol{I}_n & 0 \end{bmatrix}$ whare $\boldsymbol{I}_n$ is the $(n \times n)$ -order unit matrix, the structural matrix of the generalized Poisson bracket degenerates into the symplectic matrix (3.17). In this case, the generalized Poisson bracket degenerates into the Poisson bracket. Therefore, the classical Hamiltonian systems defined by the Poisson bracket are a subclass of the generalized Hamiltonian systems defined by the generalized Poisson bracket. Based on the generalized Poisson bracket and the structural matrix, the motion equation of the generalized Hamiltonian systems is defined on the local coordinate system $x_1, x_2, \ldots, x_m$ as follows

$$\frac{\mathrm{d}x_i}{\mathrm{d}t} = [x_i, H] = \sum_{j=1}^{m} J_{ij}(\boldsymbol{x}) \frac{\partial H(\boldsymbol{x})}{\partial x_j}, \quad i = 1, 2, \ldots, m, \tag{3.94}$$

where $H(\boldsymbol{x}, t)$ is the generalized Hamiltonian function, and $J_{ij}(\boldsymbol{x})$ are the elements of the structural matrix of the generalized Poisson bracket.

For example, consider the Euler equations of the rotational motion of a rigid body fixed at a point in 3-dimensional space. Define a rectangular coordinate system with the origin at the mass center of the rigid body. The Euler equations of motion of the rigid body are

$$\frac{\mathrm{d}m_1}{\mathrm{d}t} = \frac{I_2 - I_3}{I_2 I_3} m_2 m_3, \quad \frac{\mathrm{d}m_2}{\mathrm{d}t} = \frac{I_3 - I_1}{I_3 I_1} m_3 m_1, \quad \frac{\mathrm{d}m_3}{\mathrm{d}t} = \frac{I_1 - I_2}{I_1 I_2} m_1 m_2, \tag{3.95}$$

where m_i $(i = 1, 2, 3)$ is the angular momentum, and I_i $(i = 1, 2, 3)$ is the principal inertia moment. Letting $\boldsymbol{m} = [m_1, \ m_2, \ m_3]^{\mathrm{T}}$, the Poisson bracket is defined as follows

$$[F, G] = -\boldsymbol{m} \cdot (\nabla_m F \times \nabla_m G), \tag{3.96}$$

where $\nabla_m F = \left[\frac{\partial F}{\partial m_1}, \frac{\partial F}{\partial m_2}, \frac{\partial F}{\partial m_3}\right]^{\mathrm{T}}$, $\nabla_m G = \left[\frac{\partial G}{\partial m_1}, \frac{\partial G}{\partial m_2}, \frac{\partial G}{\partial m_3}\right]^{\mathrm{T}}$, and $\cdot$ and $\times$ denote the vector dot product and cross product respectively. The total energy $H(\boldsymbol{m}) = \sum_{i=1}^{3} \frac{1}{2I_i} m_i^2$ is the generalized Hamiltonian function. Equation (3.95) can be rewritten as

$$\frac{\mathrm{d}m_i}{\mathrm{d}t} = [m_i, H], \quad i = 1, 2, 3. \tag{3.97}$$

It can be seen that the Euler equations of the rotational motion of a rigid body with a fixed point is a 3-dimensional generalized Hamiltonian system.

If a transformation $x_i \to y_i = \varphi_i(\boldsymbol{x})$ can preserve the Poisson structure, it is called a generalized canonical transformation. Similar to classical Hamiltonian systems, the phase flow of generalized Hamiltonian equations is a generalized canonical transformation.

Unlike classical Hamiltonian systems, the generalized Hamiltonian systems defined in this section have a special property, i.e., there exist Casimir functions. Assume that the rank of the structural matrix $\boldsymbol{J}(\boldsymbol{x})$ in Eq. (3.94) at any point is a constant $r = 2n < m$, and the co-rank $M = m - 2n$, then there exist M independent Casimir functions $C_\nu(\boldsymbol{x})$, $\nu = 1, 2, \ldots, M$, which are not constants and satisfy

$$[C_\nu, F] \equiv 0, \quad \nu = 1, 2, \ldots, M, \tag{3.98}$$

for all continuously differentiable functions F. Assume that $\boldsymbol{C}(\boldsymbol{x}) = [C_1, \ldots, C_M]^{\mathrm{T}}$ is a vector composed of M Casimir functions, then any composite function $\Phi(\boldsymbol{C})$ is also a Casimir function. The above conclusion about Casimir functions is generally local. If Casimir functions $C_1(\boldsymbol{x}), \ldots, C_M(\boldsymbol{x})$ hold for all global points, then the conclusion is global. This section only considers the latter case.

Similar to Eqs. (3.36) and (3.38), for any continuously differentiable real-valued function $F(\boldsymbol{x}, t)$, its derivative with respect to t is

$$\frac{\mathrm{d}F}{\mathrm{d}t} = \frac{\partial F}{\partial t} + [F, H]. \tag{3.99}$$

If $\frac{\mathrm{d}F}{\mathrm{d}t} = 0$, then F is called the first integral of the generalized Hamiltonian system. If F doesn't explicitly include time t, then the necessary and sufficient condition for F to be a first integral is

$$[F, H] = 0. \tag{3.100}$$

From the definition (3.98) of Casimir functions, it is known that the M Casimir functions $C_1(\boldsymbol{x}), \ldots, C_M(\boldsymbol{x})$ must be first integrals. From Eq. (3.100), it is known that when the generalized Hamiltonian function $H(\boldsymbol{x})$ does not explicitly include time t, it also must be a first integral.

Similar to classical Hamiltonian systems, generalized Hamiltonian systems can also be classified into integrable and non-integrable. Consider a generalized Hamiltonian system that is completely integrable in the algebraic sense (van Moerbeke 1988). If, in addition to the M Casimir functions, there exist $n = (m - M)/2$ independent, mutual involutory (see Eq. (3.52)) first integrals $H_1, \ldots, H_n$, then the system is called completely integrable. If there exist action vectors $\boldsymbol{I} = [I_1, \ldots, I_n]^{\mathrm{T}}$ and angle vectors $\boldsymbol{\theta} = [\theta_1, \ldots, \theta_n]^{\mathrm{T}}$, the functions $H_1, \ldots, H_n$ depend only on $\boldsymbol{I}$ and $\boldsymbol{C}$, and in the coordinates of $\boldsymbol{I}, \boldsymbol{\theta}, \boldsymbol{C}$, the followings hold

$$\begin{aligned} &\left[I_i, I_j\right] = \left[\theta_i, \theta_j\right] = \left[I_i, C_{\nu_1}\right] = \left[\theta_i, C_{\nu_1}\right] = \left[C_{\nu_2}, C_{\nu_1}\right] = 0, \\ &\left[I_i, \theta_j\right] = -\delta_{ij}, \quad i, j = 1, 2, \ldots, n; \quad \nu_1, \nu_2 = 1, 2, \ldots, M, \end{aligned} \tag{3.101}$$

then corresponding generalized Hamiltonian equations are

$$\begin{aligned} &\frac{\mathrm{d}I_i}{\mathrm{d}t} = \left[I_i, \hat{H}(\boldsymbol{I}, \boldsymbol{C})\right] = 0, \quad \frac{\mathrm{d}\theta_i}{\mathrm{d}t} = \left[\theta_i, \hat{H}(\boldsymbol{I}, \boldsymbol{C})\right] = \frac{\partial \hat{H}}{\partial I_i} = \omega_i(\boldsymbol{I}, \boldsymbol{C}), \\ &\frac{\mathrm{d}C_\nu}{\mathrm{d}t} = \left[C_\nu, \hat{H}(\boldsymbol{I}, \boldsymbol{C})\right] = 0, \quad i = 1, 2, \ldots, n; \quad \nu = 1, 2, \ldots, M, \end{aligned} \tag{3.102}$$

where $\hat{H}(\boldsymbol{I}, \boldsymbol{C})$ is the generalized Hamiltonian function in the new coordinates, and $\omega = [\omega_1, \omega_2, \ldots, \omega_n]^{\mathrm{T}}$ is the angular frequency vector. The transformation from $\boldsymbol{x}$ to $\boldsymbol{I}, \boldsymbol{\theta}, \boldsymbol{C}$ is a generalized canonical transformation.

From Eq. (3.102), it is known that $\boldsymbol{I}$ is a first integral vector. When the frequencies $\omega_1, \omega_2, \ldots, \omega_n$ do not have linear dependence with integer coefficients similar to Eq. (3.61), the generalized Hamiltonian system is called completely integrable and non-resonant. In this case, the system has $n + M$ first integrals $I_1, \ldots, I_n, C_1, \ldots, C_M$.

Assume that $F_1 = F_1(I_1, \ldots, I_n, C_1, \ldots, C_M)$ is any continuously differentiable function, then it is in involution with the generalized Hamiltonian function, i.e.,

$$[F_1, \hat{H}] = \sum_{i=1}^{n} \frac{\partial F_1}{\partial I_i}[I_i, \hat{H}] + \sum_{l=1}^{M} \frac{\partial F_1}{\partial C_l}[C_l, \hat{H}] = 0. \tag{3.103}$$

Conversely, for a completely integrable and non-resonant generalized Hamiltonian system, if a function F_1 satisfies Eq. (3.103), then it is an arbitrary function of the action variables $I_1, \ldots, I_n$ and the Casimir function $C_1, \ldots, C_M$. Similar to the case that the n-DOF completely integrable and non-resonant Hamiltonian systems are ergodic on the n-dimensional torus, it is assumed that the m-dimensional generalized Hamiltonian systems are ergodic on the submanifolds where $I_1, \ldots, I_n, C_1, \ldots, C_M$ are constants.

When the frequencies in Eq. (3.102) satisfy the resonance relation as follows

$$\sum_{i=1}^{n} k_i^u \omega_i = 0, \quad u = 1, 2, \ldots, \alpha, \tag{3.104}$$

the completely integrable generalized Hamiltonian system is resonant. Here, k_i^u are an integers, and for a given u, not all k_i^u are zero, $\sum_{i=1}^{n} \left|k_i^u\right|$ is called the resonance order. When $\alpha = n - 1$, it is called complete resonance; when $1 \leq \alpha < n - 1$, it is called partial resonance. By introducing the angle variable combination

$$\psi_u = \sum_{i=1}^{n} k_i^u \theta_i, \quad u = 1, 2, \ldots, \alpha, \tag{3.105}$$

the motion equation of ψ_u is

$$\frac{\mathrm{d}\psi_u}{\mathrm{d}t} = \sum_{i=1}^{n} k_i^u \frac{\mathrm{d}\theta_i}{\mathrm{d}t} = \sum_{i=1}^{n} k_i^u \omega_i = 0, \quad u = 1, 2, \ldots, \alpha. \tag{3.106}$$

Therefore, $\boldsymbol{\Psi} = [\psi_1, \psi_2, \ldots, \psi_\alpha]^{\mathrm{T}}$ is also a first integral vector. In this case, the completely integrable and resonant generalized Hamiltonian system has $n + \alpha + M$ first integrals $I_1, \ldots, I_n, \psi_1, \ldots, \psi_\alpha, C_1, \ldots, C_M$.

Assuming that $F_2 = F_2(I_1, \ldots, I_n, \psi_1, \ldots, \psi_\alpha, C_1, \ldots, C_M)$ is any continuously differentiable function, then

$$[F_2, \hat{H}] = \sum_{i=1}^{n} \frac{\partial F_2}{\partial I_i}[I_i, \hat{H}] + \sum_{u=1}^{\alpha} \frac{\partial F_2}{\partial \psi_u}[\psi_u, \hat{H}] + \sum_{l=1}^{M} \frac{\partial F_2}{\partial C_l}[C_l, \hat{H}] = 0. \tag{3.107}$$

On the other hand, if a function F_2 of state variables satisfies Eq. (3.107) for a completely integrable and resonant generalized Hamiltonian system, then it must be an arbitrary function of the action variables $I_1, \ldots, I_n$, the angle variable combinations $\psi_1, \ldots, \psi_\alpha$, and Casimir functions $C_1, \ldots, C_M$. Similar to the case that the n-DOF completely integrable and resonant Hamiltonian systems are ergodic on the $(n-\alpha)$–dimensional (α is the number of resonance relations) subtorus, it is also assumed that the m–dimensional generalized Hamiltonian systems are ergodic on the submanifolds where $I_1, \ldots, I_n, \psi_1, \ldots, \psi_\alpha, C_1, \ldots, C_M$ are constants.

Assume that besides the M Casimir functions $C_1, \ldots, C_M$, a generalized Hamiltonian system has only the generalized Hamiltonian function itself as its first integral, it is called completely non-integrable. Let $F_3 = F_3(H, C_1, \ldots, C_M)$ be any continuously differentiable function, then it is in involution with the generalized Hamiltonian function, i.e.,

$$[F_3, H] = \frac{\partial F_3}{\partial H}[H, H] + \sum_{l=1}^{M} \frac{\partial F_3}{\partial C_l}[C_l, H] = 0. \tag{3.108}$$

For a completely non-integrable generalized Hamiltonian system, if a state function F_3 satisfies the condition (3.108), then it is an arbitrary function of the generalized Hamiltonian function H and Casimir function $C_1, \ldots, C_M$. Similar to the non-integrable Hamiltonian systems being ergodic on a subspace, the non-integrable generalized Hamiltonian systems are assumed to be ergodic on submanifolds where $H, C_1, \ldots, C_M$ are constants.

Completely integrable and completely non-integrable generalized Hamiltonian systems are two special cases of generalized Hamiltonian systems. In most cases, generalized Hamiltonian systems are partially integrable, meaning that besides M Casimir functions, there exist r $(1 < r < n)$ independent, mutual involutory first integrals $H_1, H_2, \ldots, H_r$. If the r first integrals satisfy the relations in Eqs. (3.80) and (3.81), then the first $r-1$ integrals together with M Casimir functions constitute an integrable generalized Hamiltonian subsystem, while H_r constitutes a non-integrable Hamiltonian subsystem.

Now consider a special case of separable partially integrable generalized Hamiltonian systems with $\boldsymbol{x} = [\boldsymbol{x}_1, \boldsymbol{x}_2, \boldsymbol{x}_3]$, where $\boldsymbol{x}_1$ is $2n_1$–dimensional vector; $\boldsymbol{x}_2$ is $2n_2$–dimensional vector; $\boldsymbol{x}_3$ is M–dimensional vector; $\overline{H}_1(\boldsymbol{x}_1) = \overline{H}_1(\boldsymbol{q}_1, \boldsymbol{p}_1)$ where $\boldsymbol{q}_1, \boldsymbol{p}_1$ are n_1–dimensional vector respectively; $\overline{H}_2(\boldsymbol{x}_2) = \overline{H}_2(\boldsymbol{q}_2, \boldsymbol{p}_2)$ where $\boldsymbol{q}_2, \boldsymbol{p}_2$ are n_2-dimensional vector respectively; and $\boldsymbol{C} = \boldsymbol{C}(\boldsymbol{x}_3)$. The Hamiltonian subsystem with $\overline{H}_1$ as the Hamiltonian function is completely integrable, and the action vectors $\overline{\boldsymbol{I}} = [I_1(\boldsymbol{x}_1), \ldots, I_{n_1}(\boldsymbol{x}_1)]^{\mathrm{T}}$ and angle vectors $\overline{\boldsymbol{\theta}} = [\theta_1(\boldsymbol{x}_1), \ldots, \theta_{n_1}(\boldsymbol{x}_1)]^T$ are available. $\overline{H}_2$ is the Hamiltonian function of the non-integrable Hamiltonian subsystem. Similar to Eq. (3.102), the integrable generalized Hamiltonian subsystem satisfies the following motion equations

$$\begin{aligned}
&\frac{\mathrm{d}I_i}{\mathrm{d}t} = \left[I_i, \widehat{\overline{H}}_1\right] = 0, \quad \frac{\mathrm{d}\theta_i}{\mathrm{d}t} = \left[\theta_i, \widehat{\overline{H}}_1\right] = \frac{\partial \widehat{\overline{H}}_1}{\partial I_i} = \omega_i(\overline{\boldsymbol{I}}, \overline{\boldsymbol{C}}), \\
&\frac{\mathrm{d}C_v}{\mathrm{d}t} = \left[C_v, \widehat{\overline{H}}_1\right] = 0, \quad i = 1, 2, \ldots, n_1; \quad v = 1, 2, \ldots, M_1,
\end{aligned} \tag{3.109}$$

where $\widehat{\overline{H}}_1\left(\overline{\boldsymbol{I}}, \overline{\boldsymbol{\theta}}\right)$ is the $\overline{H}_1$ after $\boldsymbol{x}_1$ being transformed into $\overline{\boldsymbol{I}}, \overline{\boldsymbol{\theta}}$, and ω_i is the angular frequencies.

When the frequencies ω_i in Eq. (3.109) do not have linear dependencies with integer coefficients, the integrable generalized Hamiltonian subsystem is non-resonant. The original generalized Hamiltonian system is called partially integrable and non-resonant, with $n_1 + 1 + M$ first integrals $I_1, \ldots, I_{n_1}, \overline{H}_2, C_1, \ldots, C_M$. Suppose $F_4 = F_4\left(\overline{\boldsymbol{I}}, \overline{H}_2, \boldsymbol{C}\right)$ is any continuously differentiable function, then F_4 is in involution with the generalized Hamiltonian function H, i.e.,

$$\begin{aligned}[F_4, H] &= \sum_{i=1}^{n_1} \frac{\partial F_4}{\partial I_i}[I_i, H] + \frac{\partial F_4}{\partial \overline{H}_2}\left[\overline{H}_2, H\right] + \sum_{v=1}^{M} \frac{\partial F_4}{\partial C_v}[C_v, H] \\ &= \sum_{i=1}^{n_1} \frac{\partial F_4}{\partial I_i}\left[I_i, \widehat{\overline{H}}_1\right] + \frac{\partial F_4}{\partial \overline{H}_2}\left[\overline{H}_2, \overline{H}_2\right] + \sum_{v=1}^{M} \frac{\partial F_4}{\partial C_v}[C_v, H] = 0. \end{aligned} \tag{3.110}$$

For the partially integrable and non-resonant generalized Hamiltonian systems, if F_4 satisfies condition (3.110), then F_4 is an arbitrary function of $\overline{\boldsymbol{I}}, \overline{H}_2, \boldsymbol{C}$.

For partially integrable and non-resonant generalized Hamiltonian systems, similar to partially integrable and non-resonant Hamiltonian systems, the integrable generalized Hamiltonian subsystems are ergodic on submanifolds where $I_1, \ldots, I_{n_1}, C_1, \ldots, C_M$ are constants, while the non-integrable generalized Hamiltonian subsystem is ergodic on submanifolds where $\overline{H}_2$ is constant.

When the frequencies ω_i in Eq. (3.109) satisfy the resonance relation as follows

$$\sum_{i=1}^{n_1} k_i^u \omega_i = 0, \quad u = 1, 2, \ldots, \alpha_1, \tag{3.111}$$

the partially integrable generalized Hamiltonian system (3.109) is resonant. Here, notation k_i^u is the same as k_i^u in Eq. (3.104). Similar to Eq. (3.106), byintroducing the angle variable combinations

$$\psi_u = \sum_{i=1}^{n_1} k_i^u \theta_i, \quad u = 1, 2, \ldots, \alpha_1, \tag{3.112}$$

ψ_u is also a first integral of this partially integrable and resonant generalized Hamiltonian system. In this case, the partially integrable and resonant generalized Hamiltonian system has $n_1 + \alpha_1 + 1 + M$ first integrals $\overline{\boldsymbol{I}}, \overline{\boldsymbol{\Psi}} = [\psi_1, \ldots, \psi_{\alpha_1}]^{\mathrm{T}}, \overline{H}_2$, and $\boldsymbol{C}$. Assuming that $F_5 = F_5(\overline{\boldsymbol{I}}, \overline{\Psi}, \overline{H}_2, \boldsymbol{C})$ is any continuously differentiable function, it is in involution with the generalized Hamiltonian function H, i.e.,

$$[F_5, H] = \sum_{i=1}^{n_1} \frac{\partial F_5}{\partial I_i}[I_i, H] + \sum_{u=1}^{\alpha_1} \frac{\partial F_5}{\partial \psi_u}[\psi_u, H]$$

$$+ \frac{\partial F_5}{\partial \overline{H}_2}[\overline{H}_2, H] + \sum_{v=1}^{M} \frac{\partial F_5}{\partial C_v}[C_v, H] = 0 \tag{3.113}$$

If a function F_5 satisfies the condition (3.113), then F_5 must be an arbitrary function of $\overline{\boldsymbol{I}}$, $\overline{\boldsymbol{\Psi}}$, $\overline{H}_2$, and $\boldsymbol{C}$.

For partially integrable and resonant generalized Hamiltonian systems, like partially integrable and resonant Hamiltonian systems, the integrable Hamiltonian subsystem is ergodic on $(n_1 - \alpha_1)$–dimensional submanifold where $\overline{\boldsymbol{I}}$, $\overline{\boldsymbol{\Psi}}$, and $\boldsymbol{C}$ are constant, while the non-integrable Hamiltonian subsystem is ergodic on surface with $\overline{H}_2$ being constant.

For more detailed discussion about generalized Hamiltonian systems, refer to (Li et al. 2007).

3.4 Forces with Genetic Effects

In Sect. 3.2.9, the non-conservative forces include excitations and dissipation forces. It is assumed that a dissipation force is a function of system state at the present time instant, for example denoted as $f\ [\boldsymbol{X}(\mathrm{t})]$. However, in practical problems, some nonconservative forces exhibit behaviors that they depend not only on the current motion state but also on the past history of system motion. Here we name these forces as forces with genetic effects. Examples are hysteretic forces, visco-elastic forces, and damping forces with fractional derivatives. By adding such forces in Eq. (3.91), the system is called stochastically excited dissipated Hamiltonian system with genetically related forces. This section introduces three such forces, hysteretic forces, visco-elastic forces, and damping forces with fractional derivatives.

3.4.1 Hysteretic Forces

Mechanical and structural systems may exhibit hysteretic behaviors when subjected to heavy loadings. Intelligent materials used more and more in structural control, such as piezoceramics, shape memory alloys, and electro-/magneto-rheological fluids also demonstrate hysteretic characteristics. The restoring forces from these materials depend not only on the current state of displacement and deformation but also on the past history of the state. Several types of analytical models have been established to describe the hysteretic constitutive law. The popular models are bilinear model, Bouc-Wen model, Duhem model, and Preisach model.

(i) Bilinear hysteretic model

The bilinear model is a simple and intuitive model for a hysteretic force. Figure 3.4 shows the hysteretic force with respect to the displacement. It consists of several straight lines with the following equations

$$f(x,\dot{x}) = \begin{cases} x, & \text{for } a \le 1, \\ \alpha x - (1-\alpha), & \text{for } a > 1,\ \dot{x} < 0,\ -a \le x < a-2, \\ x - (1-\alpha)(a-1), & \text{for } a > 1,\ \dot{x} < 0,\ a-2 \le x \le a, \\ x + (1-\alpha)(a-1), & \text{for } a > 1,\ \dot{x} > 0,\ -a \le x < 2-a, \\ \alpha x + (1-\alpha), & \text{for } a > 1,\ \dot{x} > 0,\ 2-a \le x \le a. \end{cases} \tag{3.114}$$

where a is the amplitude, and α is the slope of a line, as shown in Fig. 3.4. The area within the loop is the energy loss per cycle, denoted by A_r. It is obtained as

$$A_r = \begin{cases} 0, & \text{for } a \le 1, \\ 4(1-\alpha)(a-1), & \text{for } a > 1. \end{cases} \tag{3.115}$$

Schematically, the bilinear model can be represented by a linear spring and a Jenkin element in parallel, and the latter consists of a linear spring in series with a Coulomb damper, as shown in Fig. 3.5.

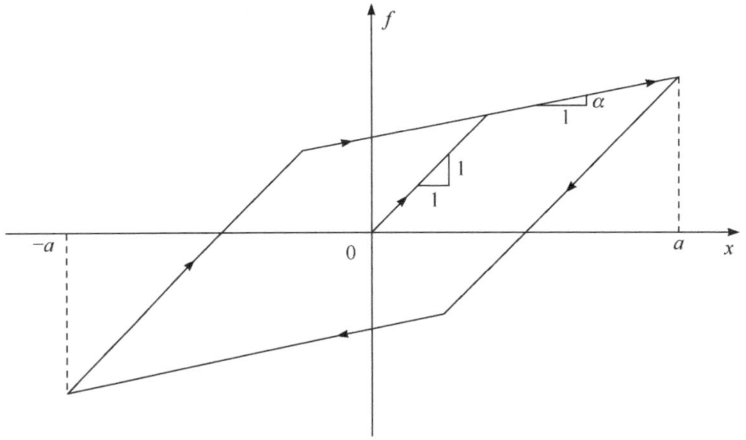

Fig. 3.4 A bilinear hysteresis model

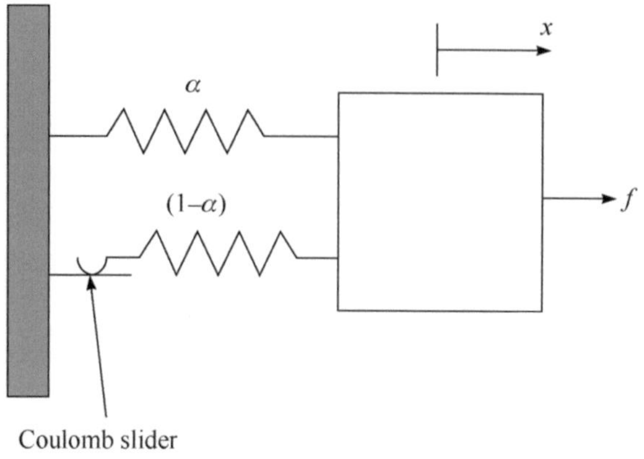

Fig. 3.5 Schematic representation of a bilinear hysteresis model

(ii) Bouc-Wen hysteretic model

It is noted that the abrupt changes of slope in the bilinear model shown in Fig. 3.4 are not realistic. To remedy this problem, a smooth hysteretic model was proposed initially by Bouc and extended by Wen (Bouc 1967; Wen 1976). According to the model, the hysteretic restoring force is given by

$$f(x, \dot{x}) = \alpha x + (1 - \alpha)z(x, \dot{x}), \quad 0 \le \alpha \le 1, \tag{3.116}$$

where α is a constant, αx is the restoring force, $z(x, \dot{x})$ is the hysteretic force which is described by a one-dimensional differential equation

$$\dot{z} = -\gamma|\dot{x}|z|z|^{n-1} - \beta\dot{x}|z|^n + A\dot{x}, \tag{3.117}$$

where γ, β, n, and A are parameters depending on the hysteretic characteristics. Sketch of force f versus x for the Bouc-Wen model is shown in Fig. 3.6. The smoothness of the force-displacement curve is controlled by value n, the general slope of the curve is controlled by value $\gamma + \beta$, and the slimness of the hysteretic loop is controlled by γ. By choosing different sets of parameters γ, β, n, and A, the behaviors of real hysteretic systems can be approximated closely (Baber and Wen 1980).

By integrating Eq. (3.117), the functional relationship between z and x can be obtained (Cai and Lin 1990). For example, for the case of $A = n = 1$ and $\beta \neq \gamma$,

$$z(x) = \begin{cases} \dfrac{\left[e^{(\gamma+\beta)(x-x_0)} - 1\right]}{(\gamma+\beta)}, & \text{for } \dot{x} < 0, \quad -a \le x < x_0, \\ \dfrac{\left[e^{(\gamma-\beta)(x-x_0)} - 1\right]}{(\gamma-\beta)}, & \text{for } \dot{x} < 0, \quad x_0 \le x \le a, \\ \dfrac{\left[1 - e^{-(\gamma-\beta)(x+x_0)}\right]}{(\gamma-\beta)}, & \text{for } \dot{x} \ge 0, \quad -a \le x < -x_0, \\ \dfrac{\left[1 - e^{-(\gamma+\beta)(x+x_0)}\right]}{(\gamma+\beta)}, & \text{for } \dot{x} \ge 0, \quad -x_0 \le x \le a, \end{cases} \tag{3.118}$$

Fig. 3.6 Bouc-Wen hysteretic model

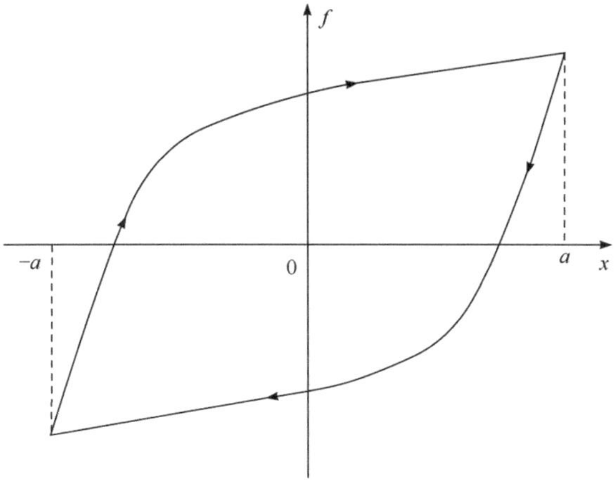

where x_0 can be determined from equation $z(-x_0) = 0$ $(\dot{x} \geq 0)$ for a given amplitude a. The area of the hysteretic loop corresponding to a given amplitude a can be calculated from the integral

$$A_r = 2(1-\alpha)\int_{-a}^{a} z(x)\mathrm{d}x. \tag{3.119}$$

(iii) Duhem hysteretic model

A more flexible and versatile hysteresis model known as Duhem model was proposed (Krasnoselskii and Pokrovskii 1989; Mayergoyz 1991; Visintin 1994). The hysteretic force f is governed by the equation

$$\dot{f} = g\left[x, f, \operatorname{sgn}(\dot{x})\right]\dot{x} = \begin{cases} g_1(x, f)\dot{x}, \dot{x} > 0 \\ g_2(x, f)\dot{x}, \dot{x} < 0 \end{cases}. \tag{3.120}$$

The loop of model (3.120) consists of the ascending segment $f_1(x)$ $(\dot{x} > 0)$ and the descending segment $f_2(x)$ $(\dot{x} < 0)$. The area within the hysteretic loop is

$$A_r = \int_{-a_1}^{a_2} f_1(x)\mathrm{d}x + \int_{a_2}^{-a_1} f_2(x)\mathrm{d}x, \tag{3.121}$$

where a_1 and a_2 are the amplitudes in negative and positive directions, respectively. Model (3.120) allows the ascending segment $f_1(x)$ and the descending segment $f_2(x)$ not symmetric. However, for most existing hysteretic models, the two segments are asymmetric, i.e., $g_2(x, f) = g_1(-x, -f)$. As an example considered in (Ying et al. 2002), a nonlinear elastic and anti-symmetric Duhem model is

$$g_1 = k_1 + 3k_3x^2 + \frac{\gamma}{\beta}\left[\alpha - \beta\left(f - k_1x - k_3x^3\right)\right], \tag{3.122}$$

where k_1 and k_1 are coefficients of the linear and nonlinear stiffness, α, β, and γ are parameters of hysteresis. In this case, the hysteretic restoring force includes two parts, elastic force and hysteretic force. Substituting g_1, g_2 in Eq. (3.120) and solving Eq. (3.120) for f leads to the two branches of the hysteretic restoring force

$$\begin{aligned} f_1 &= k_1x + k_3x^3 + \frac{1}{\beta}\left[1 - e^{-\gamma(x+x_0)}\right], \quad \dot{x} > 0, \\ f_2 &= k_1x + k_3x^3 - \frac{1}{\beta}\left[1 - e^{\gamma(x-x_0)}\right], \quad \dot{x} < 0, \end{aligned} \tag{3.123}$$

where x_0 is determined from the amplitude a as

$$x_0 = -a + \frac{1}{\gamma}\ln\left(\frac{1 + e^{2a\gamma}}{2}\right). \tag{3.124}$$

The area of the hysteretic loop is calculated from (3.121) as

$$A_r = 4\frac{\left[(1 + a\gamma) - e^{\gamma(a-x_0)}\right]}{(\beta\gamma)}. \tag{3.125}$$

Figure 3.7 depict the hysteretic restoring forces for two cases: hardening stiffness $k_3 = 0.03 > 0$ and softening stiffness $k_3 = -0.03 < 0$.

(iv) Preisach hysteretic model

Preisach hysteretic model is a model for more complicated hysteretic restoring forces. It can describe the hysteresis with nonlocal memory. The model can be expressed as the following integral form

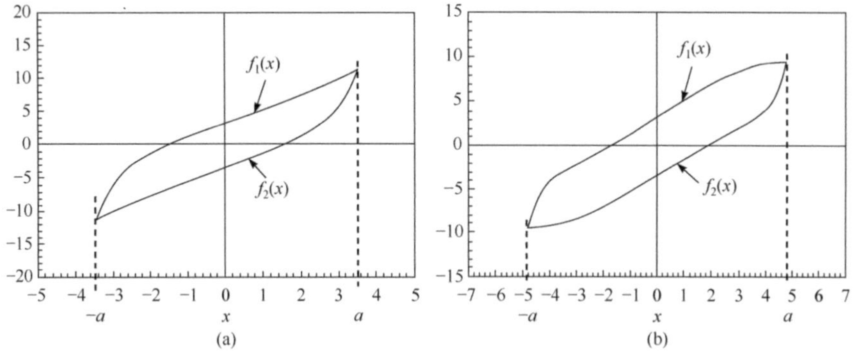

Fig. 3.7 Duhem hysteretic model: **a** hardening stiffness, **b** softening stiffness (From Ying et al. 2002)

$$f(x, \dot{x}) = \iint\limits_{\alpha \geq \beta} \mu(\alpha, \beta)\gamma_{\alpha\beta}(x(t))\mathrm{d}\alpha\mathrm{d}\beta, \tag{3.126}$$

where $\mu(\alpha, \beta)$ is a weighting function, called the Preisach function. It is defined in a triangle D on (α, β)-plane with the line $\alpha = \beta$ as the hypotenuse and point $(\alpha_P, \beta_P = -\alpha_P)$ as a vertex. The triangle on the half-plane $\alpha \geq \beta$ is called the Preisach plane shown in Fig. 3.8. $\mu(\alpha, \beta)$ is equal to zero outside the triangle D. In Eq. (3.126), $\gamma_{\alpha\beta}(x)$ is called the hysteretic relay operator, as shown in Fig. 3.9. It takes two values of $+1$ and -1, corresponding to "up" and "down" position, respectively, as given by

$$\gamma_{\alpha\beta}(x) = \begin{cases} +1, & x > \alpha \text{ or } (x > \beta \text{ and decreasing}), \\ -1, & x < \beta \text{ or } (x < \alpha \text{ and increasing}). \end{cases} \tag{3.127}$$

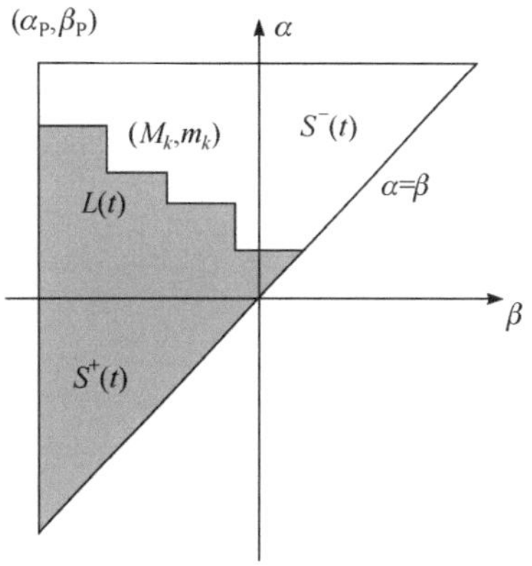

Fig. 3.8 Preisach plane (Wang et al. 2009)

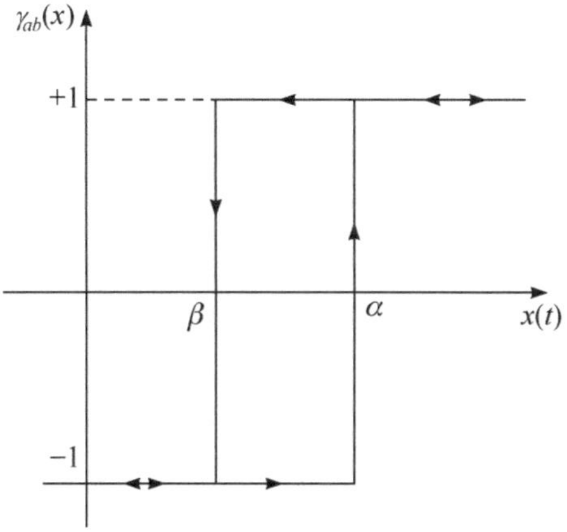

Fig. 3.9 Hysteretic relay operator (Wang et al. 2009)

The Preisach model, expressed as the superposition of a continuous family of elementary rectangular loops, can be interpreted in terms of the spectral decomposition of a complicated hysteretic constitutive law with nonlocal memory into the simplest hysteretic relay operator $\gamma_{\alpha\beta}(x)$ with local memory. The Preisach hysteretic behavior is completely characterized by the weighting function $\alpha_{\alpha\beta}(x)$. For arbitrary displacement input $x(t)$, the Preisach hysteretic force output $f(t)$ can be determined by the weighting function together with a staircase line $L(t)$ (see Fig. 3.8). And the staircase line divides the Preisach plane into two parts: the domain $S^+(t)$ encompassing the set of the relay in the +1 status, and the domain $S^-(t)$ encompassing the set of the relay in the −1 status. Each vertex of the staircase line is associated with dominant maximum M_k or minimum m_k of the displacement $x(t)$. The nonlocal selective memory is then stored in this way. Therefore, the Preisach hysteretic force depends not only on the present displacement but also on the selective memory and nonlocal history of the displacement, namely it has the characteristic of nonlocal memory.

The Preisach hysteretic restoring force $f(t)$ can be equivalently expressed as

$$\begin{aligned} f(t) &= \iint\limits_{S^+(t)} \mu(\alpha,\beta)\mathrm{d}\alpha\mathrm{d}\beta - \iint\limits_{S^-(t)} \mu(\alpha,\beta)\mathrm{d}\alpha\mathrm{d}\beta \\ &= 2\iint\limits_{S^+(t)} \mu(\alpha,\beta)\mathrm{d}\alpha\mathrm{d}\beta - \iint\limits_{D} \mu(\alpha,\beta)\mathrm{d}\alpha\mathrm{d}\beta, \end{aligned} \tag{3.128}$$

where the integration domain is the triangle D in Fig. 3.8.

The Preisach function of a specific hysteretic system can be determined by the first-order transition curves (Lubarda et al. 1993). Applying this procedure, the Preisach function $\alpha_{\alpha\beta}(x)$ is derived in a closed form for classical rheological models. For the Iwan–Jenkins model, it is

$$\begin{aligned} \mu(\alpha,\beta) = & \frac{k_J}{2}\left\{\delta(\alpha-\beta) - \frac{k_J}{2}\frac{1}{f_{y,\max}-f_{y,\min}}\right. \\ & \left.\left[H\left(\alpha-\beta-2\frac{f_{y,\min}}{k_J}\right) - H\left(\alpha-\beta-2\frac{f_{y,\max}}{k_J}\right)\right]\right\}, \end{aligned} \tag{3.129}$$

where $\delta(\cdot)$ and $H(\cdot)$ are Dirac delta function and Heaviside function, respectively. The symbol k_J represents the linear stiffness of the individual Jenkins element. f_y is the yielding force, and $f_{y,\min} \le f_y \le f_{y,\max}$. The weighting function is defined in domain A as shown in Fig. 3.10. Using (3.127), the corresponding Preisach hysteretic force can be obtained as

$$f(t) = \frac{k_J}{2}\left[\int_{-\alpha_P}^{\alpha_P} \gamma_{\alpha\alpha}(x)\mathrm{d}\alpha - \frac{k_J}{2}\frac{1}{f_{y,\max}-f_{y,\min}}\iint\limits_{A} \gamma_{\alpha\beta}(x)\mathrm{d}\alpha\mathrm{d}\beta\right]. \tag{3.130}$$

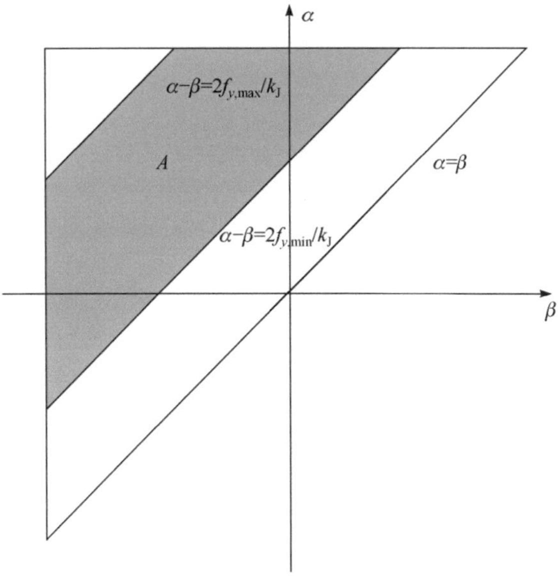

Fig. 3.10 Definition domain of the weighing function for the Iwan–Jenkins model (Wang et al. 2009)

By using Eq. (3.129), the hysteretic force can be rewritten as

$$f(t) = k_J x(t) - \frac{k_J^2}{4} \frac{1}{f_{y,\max} - f_{y,\min}} \left(2 \iint\limits_{S_A^+(t)} \mathrm{d}\alpha \mathrm{d}\beta - \iint\limits_{A} \mathrm{d}\alpha \mathrm{d}\beta \right). \tag{3.131}$$

Let $\alpha_P = -\beta_P = f_{y,\max}/k_J$. Then the shaded area A is a triangle defined by three straight lines, $\alpha = \alpha_P$, $\beta = \beta_P$ and $\alpha - \beta = 2f_{y,\min}/k_J$. Domain A is initially divided into two equal parts to guarantee $f(0) = 0$. Introduce a new function $F(\alpha_i, \beta_i)$

$$F(\alpha_i, \beta_i) = \frac{\left[2f_{y,\min} + k_J\left(\beta_j - \alpha_j\right)\right]^2}{2k_J^2}, \tag{3.132}$$

which represents the area of a triangle surrounded by lines $\alpha = \alpha_i$, $\beta = \beta_i$ and $\alpha - \beta = 2f_{y,min}/k_J$, as shown in Fig. 3.11.

It is assumed that the vibration amplitude varies slowly with respect to time. When the present vibration amplitude $\overline{a} \geq f_{y,min}/k_J$, shown in Fig. 3.12, the hysteretic force for the ascending stage is given by

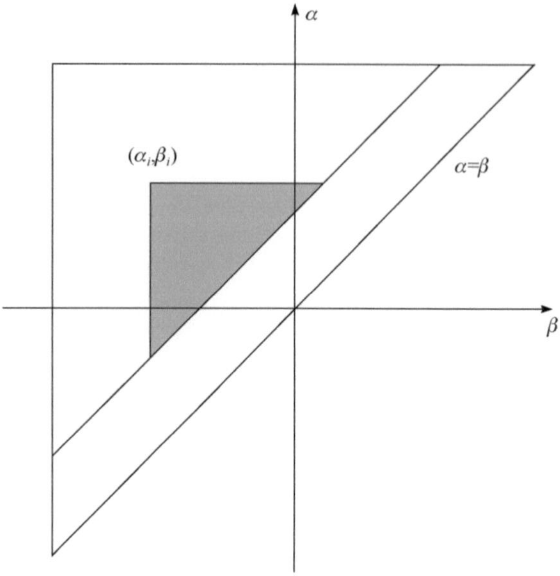

Fig. 3.11 Geometric interpretation of the function F (Wang et al. 2009)

$$
\begin{aligned}
f(t) = k_J x(t) - \frac{k_J^2}{4}\frac{1}{f_{y,\max} - f_{y,\min}}\Bigg[4F\left(\frac{f_{y,\min}}{k_J}, \beta_P\right) - 4F\left(\frac{f_{y,\min}}{k_J}, \beta_P\right) \\
+ 2H\left(x(t) - \beta_{n-1} - \frac{2f_{y,\min}}{k_J}\right)F(x(t), \beta_{n-1}) \\
+2\sum_{j=2}^{n-1}\left(F\left(\alpha_j, \beta_{j-1}\right) - F\left(\alpha_j, \beta_j\right)\right) - F(\alpha_P, \beta_P)\Bigg],
\end{aligned}
\tag{3.133}
$$

and that for the descending stage is

$$
\begin{aligned}
f(t) = k_J x(t) - \frac{k_J^2}{4}\frac{1}{f_{y,\max} - f_{y,\min}}\Bigg[4F\left(\frac{f_{y,\min}}{k_J}, \beta_P\right) - 4F\left(\frac{f_{y,\min}}{k_J}, \beta_1\right) \\
+ 2F(\alpha_n, \beta_{n-1}) - 2H\left(\alpha_n - \frac{2f_{y,\min}}{k_J} - x(t)\right)F(\alpha_n, x(t)) \\
+2\sum_{j=2}^{n-1}\left(F\left(\alpha_j, \beta_{j-1}\right) - F\left(\alpha_j, \beta_j\right)\right) - F(\alpha_P, \beta_P)\Bigg].
\end{aligned}
\tag{3.134}
$$

When the present vibration amplitude $\overline{a} < f_{y,\min}/k_J$, shown in Fig. 3.13, the hysteretic force for both the ascending and descending stages is as follows:

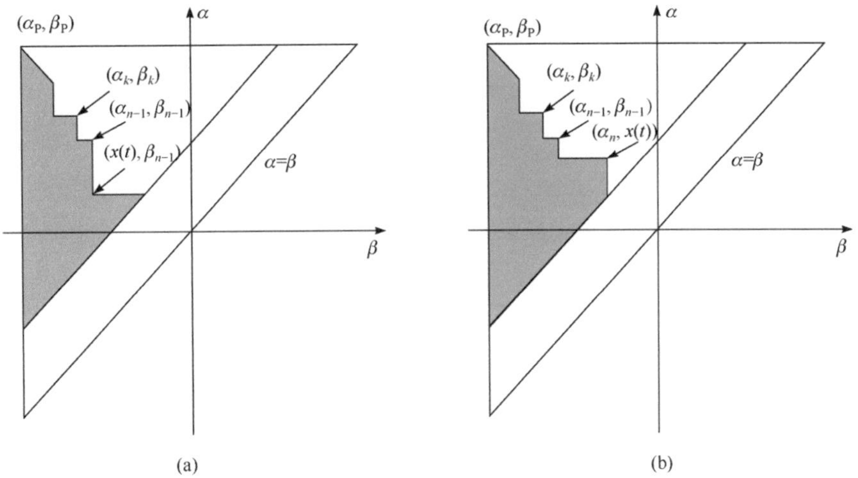

Fig. 3.12 Geometric interpretation of the Preisach hysteretic force when $\overline{a} \geq f_{y,min}/k_J$. **a** For the ascending stages, **b** for the descending stages (Wang et al. 2009)

$$f(t) = k_J x(t) - k_J^2 \frac{1}{f_{y,\max} - f_{y,\min}} \left[F\left(\frac{f_{y,\min}}{k_J}, \beta_P\right) - F\left(\frac{f_{y,\min}}{k_J}, \beta_1\right) \right]$$

$$- \frac{k_J^2}{4} \frac{1}{f_{y,\max} - f_{y,\min}} \left[2 \sum_{j=2}^{n-1} \left(F\left(\alpha_j, \beta_{j-1}\right) - F\left(\alpha_j, \beta_j\right) \right) - F\left(\alpha_P, \beta_P\right) \right]. \quad (3.135)$$

In Eqs. (3.133)–(3.135), α_i and β_i are the local maximum and local minimum of displacement $x(t)$, respectively. It is noticed that the maximum and the minimum,

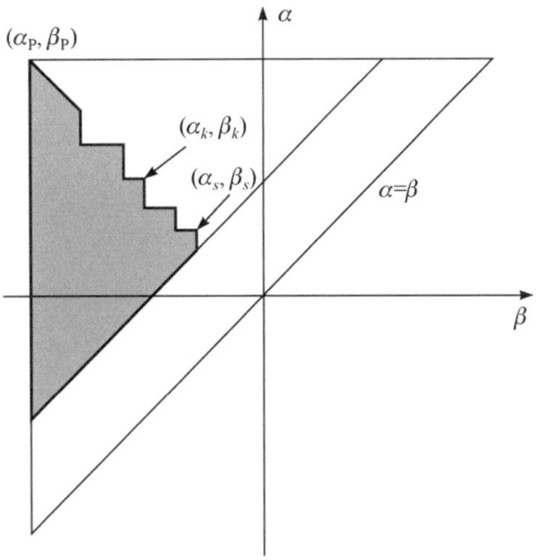

Fig. 3.13 Geometric interpretation of the Preisach hysteretic force when $\overline{a} < f_{y,\min}/k_J$ (Wang et al. 2009)

if less than $f_{y,\min}/k_J$, should not be the local maximum and local minimum. In the present case of $\overline{a} < f_{y,\min}/k_J$, the local minimum of the dominant maxima α_s and local maximum of the dominant minima β_s are approximately equal to $f_{y,\min}/k_J$ and $-f_{y,\min}/k_J$.

3.4.2 *Visco-Elastic Force*

When dealing with a solid material, the classical mechanics theory often treats it as completely elastic. If the loading is below a certain limit, the deformations of the material under an external loading will disappear after the loading is withdrawn. When dealing with a fluid material, the classical mechanics theory often treats it as viscous material. It exhibits flowing characteristic under a shear force, and will not recover the previous original flowing form. However, some materials in practical applications show both elasticity and viscosity, such as plastics and rubbers, concretes and rocks in civil engineering, and bloods and mussels in bioengineering. These materials are called visco-elastic materials, which will have both deformations and flowing phenomena under external loadings. Relaxation and creeping are typical phenomena of visco-elastic materials under loadings. Relaxation is a phenomenon of decreasing stress under a constant strain, while creeping means increasing strain under a constant stress. Figures 3.14 and 3.15 illustrate schematically the relaxation and creeping behaviors of visco-elastic materials.

When investigating mechanical behaviors of materials, a key question is how to describe the mechanical constitutive law, namely the stress-strain relationship. It is $\sigma = E\varepsilon$ for linear elastic materials, is $\sigma = \eta\dot{\varepsilon}$ for Newtonian fluids. Duo to various types of visco-elastic materials, the constitutive law is complicated, and there is no

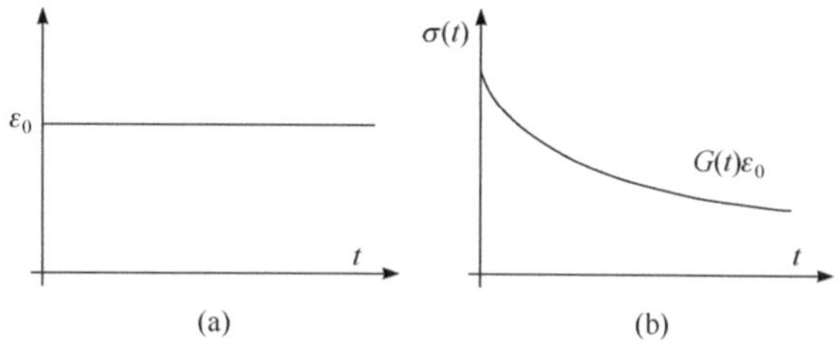

Fig. 3.14 Relaxation phenomenon of visco-elastic materials, **a** constant strain, **b** decreasing stress

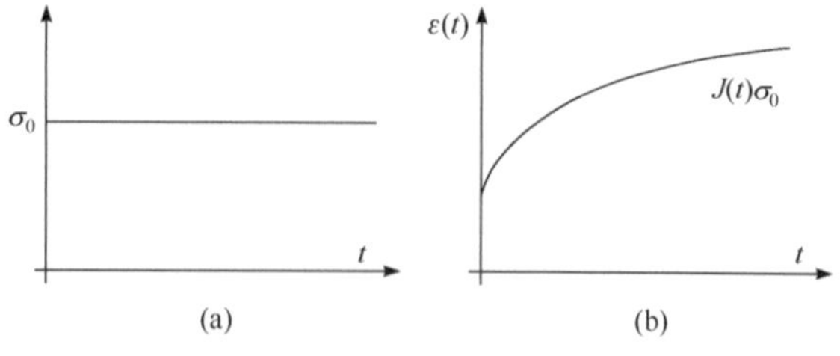

Fig. 3.15 Creeping phenomenon of visco-elastic materials, **a** constant stress, **b** increasing strain

unified theory. Although a general form $\sigma = f(\varepsilon, t)$ can be written for the stress-strain relationship, but it is difficult to specify the functional form for $f(\varepsilon, t)$. One assumption for a type of materials is called linear visco-elasticity with $\sigma = f(t)\varepsilon$. For a linear visco-elastic material with the stress relaxation phenomenon shown in Fig. 3.14, we may write

$$\sigma(t) = G(t)\varepsilon_0, \tag{3.136}$$

where $G(t)$ is called relaxation modulus, a monotonically decreasing function. For the creeping phenomenon shown in Fig. 3.15, we may write

$$\varepsilon(t) = J(t)\sigma_0, \tag{3.137}$$

where $J(t)$ is creep compliance, also a monotonically increasing function. Figures 3.14 and 3.15, as well as Eqs. (3.136) and (3.137) show that the responses of stress and strain are functions of time, and the relaxation modulus $G(t)$ and creep compliance $J(t)$ are important functions describing the characteristics of visco-elastic materials. They reveal visco-elasticity properties of materials from different points of view, and they ought to be related with each other in certain ways. In order to derive $G(t)$ and $J(t)$, theories and equations should be established. Currently, equations to describe constitutive laws of visco-elastic materials can be classified as two major types of differential equations and integral equations (Zhang 1994).

Since visco-elastic materials possess both the elastic and viscous characteristics, simple and intuitive mechanical models are so-called component model using elastic components and viscous components connected in certain ways. Some commonly used component models are introduced as follows.

(i) Kelvin-Voigt model, as shown in Fig. 3.16a. Based on the constitutive law of an elastic component $\sigma = E\varepsilon$ and that of a viscous component $\sigma = \eta\dot{\varepsilon}$, the visco-elastic constitutive law is obtained as

$$\eta\dot{\sigma} + E\sigma = E\eta\dot{\varepsilon}. \tag{3.138}$$

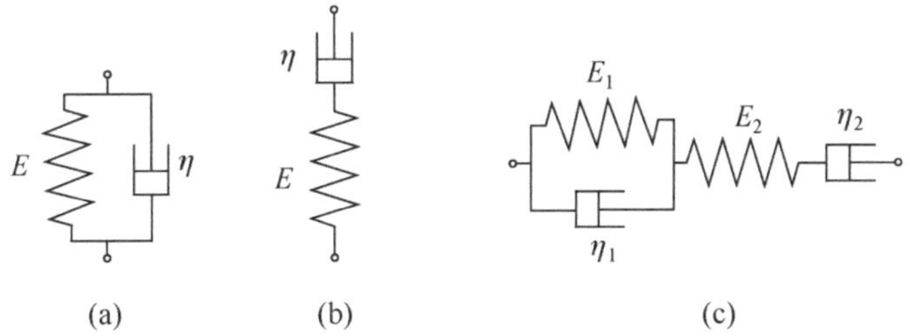

Fig. 3.16 Component models for constitutive laws of visco-elastic materials, **a** Kelvin-Voigt model, **b** Maxwell model, **c** Burgers model

(ii) Maxwell model shown in Fig. 3.16b, from which we have

$$\eta\dot{\varepsilon} + E\varepsilon = \sigma. \tag{3.139}$$

(iii) Burgers model shown in Fig. 3.16c with the constitutive law given as

$$E_1E_2\sigma + (\eta_1E_2 + \eta_1E_1 + \eta_2E_1)\dot{\sigma} + \eta_1\eta_2\ddot{\sigma} = \eta_1E_1E_2\dot{\varepsilon} + \eta_1\eta_2E_1\ddot{\varepsilon}. \tag{3.140}$$

More component models were proposed by using elastic and viscous components connected more complicatedly in series and/or in parallel. It should be pointed out that the component models are not true physical structures of the visco-elastic materials. The constitutive laws derived from component models are verified by experimental data to certain extent. It is noted that different component models may be applied to different materials. The constitutive laws from component models are similar to Eqs. (3.138) through (3.140), expressed as linear combinations of material parameters, stress and strain, and their derivatives of various orders. They are called differential visco-elastic constitutive laws, given by the following general form

$$\sigma + p_1\dot{\sigma} + p_2\ddot{\sigma} + \cdots = q_0\varepsilon + q_1\dot{\varepsilon} + q_2\ddot{\varepsilon} + \cdots. \tag{3.141}$$

Equation (3.141) can be solved directly in time domain, but more simple way is to carry out Laplace transform for (3.141) to result in

$$\overline{\sigma}(s) = \frac{q_0 + q_1s + q_2s^2 + \cdots}{1 + p_1s + p_2s^2 + \cdots}\overline{\varepsilon}(s), \tag{3.142}$$

where $\overline{\sigma}(s)$ and $\overline{\varepsilon}(s)$ are the Laplace transforms of stress $\sigma(t)$ and strain $\varepsilon(t)$, respectively.

Consider to maintain a constant strain $\varepsilon(t) = \varepsilon_0H(t)$ under certain loading, where $H(t)$ is a Heaviside function as shown in (2.140). Substituting its Laplace transform $\overline{\varepsilon}(s) = \frac{\varepsilon_0}{s}$ into Eq. (3.142), and combining (3.143), the Laplace transform of the relaxation modulus $G(t)$ is obtained as

$$\overline{G}(s) = \frac{q_0 + q_1s + q_2s^2 + \cdots}{s\left(1 + p_1s + p_2s^2 + \cdots\right)}. \tag{3.143}$$

The relaxation modulus $G(t)$ can be calculated from $\overline{G}(s)$ by performing the inverse Laplace transform. The result is the same as that from solving the differential equation (3.141). Obviously, the method of Laplace transform is a simpler way to calculate the relaxation modulus. Using the method, for example, the relaxation modulus of Maxwell model is derived as $G(t) = E\exp(-Et/\eta)$.

Combining Eqs. (3.142) and (3.143), the constitutive law in s-domain corresponding to the original constitutive law (3.141) can be obtained in terms of the relaxation modulus as

$$\overline{\sigma}(s) = \overline{G}(s)s\overline{\varepsilon}(s). \tag{3.144}$$

Similarly, the Laplace transform $\overline{J}(s)$ of creep compliance $J(t)$ and the constitutive law in s-domain in terms of creep compliance are

$$\overline{J}(s) = \frac{1 + p_1 s + p_2 s^2 + \cdots}{s\left(q_0 + q_1 s + q_2 s^2 + \cdots\right)}, \quad \overline{\varepsilon}(s) = \overline{J}(s)s\overline{\sigma}(s). \tag{3.145}$$

The creep compliance $J(t)$ can be calculated from $\overline{J}(s)$ using the inverse Laplace transform. For example, the creep compliances of Kelvin-Voigt model and Maxwell model are $J(t) = [1 - \exp(-Et/h)]/E$ and $J(t) = t/\eta + 1/E$, respectively. The relationship between the relaxation modulus and creep compliances can be obtained as (Zhang 1994)

$$\begin{gathered} \overline{J}(s)\overline{G}(s) = \frac{1}{s^2}, \\ \int_0^t J(t-\tau)G(\tau)\mathrm{d}\tau = t \quad \text{or} \quad \int_0^t G(t-\tau)J(\tau)\mathrm{d}\tau = t. \end{gathered} \tag{3.146}$$

Based on the Boltzmann superposition principle, the integral form of the visco-elastic constitutive law can be established. Assume that the strain response is $\varepsilon_1(t)$ when stress σ_1 is applied to a visco-elastic material alone. Also assume that the strain response is $\varepsilon_2(t)$ when stress σ_2 is applied to the same visco-elastic material alone. The Boltzmann superposition principle says that, when stress $\sigma_1 + \sigma_2$ is applied to the same visco-elastic material, the strain response is $\varepsilon_1(t) + \varepsilon_2(t)$. It is clear that the Boltzmann superposition is a linear superposition which neglects the interaction between the two stresses, and the nonlinear terms of higher orders. It approximately agrees with the nature of many non-degrading visco-elastic materials (Drozdov 1998), and brings conveniences for theoretical analysis.

Assume that $\sigma(t)$ is a stress loading process, which can be considered as an accumulation of many times of stress application, and the strain response is a superposition of these stress applications. For example, stress $\Delta\sigma_i$ is applied at time τ_i, then according to (3.137), the strain response due to this stress application is $\Delta\varepsilon_i(t) = J(t - \tau_i)\Delta\sigma_i$, $t > \tau_i$. According to the Boltzmann superposition principle, the total strain response during the entire stress loading is

$$\varepsilon(t) = J(t)\sigma_0 + \sum_{i=1}^{n} J(t - \tau_i)\Delta\sigma_i. \tag{3.147}$$

where σ_0 is the initial stress at $t = 0$. Letting $n \to \infty$ in (3.147), we have the integral form of the visco-elastic constitutive law in terms of the creep compliance $J(t)$

$$\varepsilon(t) = J(t)\sigma_0 + \int_0^t J(t-\tau)\mathrm{d}\sigma(\tau). \tag{3.148}$$

The integral term in (3.148) indicates the memory effect in the stress-strain relationship of visco-elastic materials, namely, the total strain is relates to the history of stress process. Carrying out a integration by part for (3.148), we obtain an equivalent form of (3.148)

$$\varepsilon(t) = J(0)\sigma(t) - \int_0^t \frac{\partial J(t-\tau)}{d\tau}\sigma(\tau)\mathrm{d}\tau. \tag{3.149}$$

Similarly, assume that the strain process is $\varepsilon(t)$ under loading. By using the Boltzmann superposition principle, the integral form of the visco-elastic constitutive law in terms of the relaxation modulus $G(t)$ is obtained as

$$\sigma(t) = G(t)\varepsilon_0 + \int_0^t G(t-\tau)\mathrm{d}\varepsilon(\tau)$$

$$\text{or} \quad \sigma(t) = G(0)\varepsilon(t) - \int_0^t \frac{\partial G(t-\tau)}{\partial\tau}\varepsilon(\tau)\mathrm{d}\tau. \tag{3.150}$$

The integral forms of Eqs. (3.148) through (3.150) show that the creep compliance $J(t)$ and relaxation modulus $G(t)$ are the important functions to describe the visco-elastic behaviors. In practical applications, the two functions must be specified in order to use (3.148) through (3.150). Drozdov proposed (Drozdov 1998) that the relaxation modulus of any non-degrading visco-elastic material can be approached by using the sum of multiple Maxwell models

$$G(t) = \sum_{i=1}^{M}\left(\beta_i e^{-\frac{t}{\lambda_i}}\right), \quad \sum_{i=1}^{M}\beta_i = G(0). \tag{3.151}$$

where λ_i is called the relaxation time of the ith viso-elastic component.

3.4.3 Damping Force with Fractional Derivative

It is noted that the stress-strain relationship of linear elastic materials can be interpreted as $\sigma(t) \sim \frac{\mathrm{d}^0\varepsilon(t)}{\mathrm{d}t^0}$, and that of viscous fluids as $\sigma(t) \sim \frac{\mathrm{d}^1\varepsilon(t)}{\mathrm{d}t^1}$. Thus, some researcher naturally think that stress-strain relationship of visco-elastic materials may

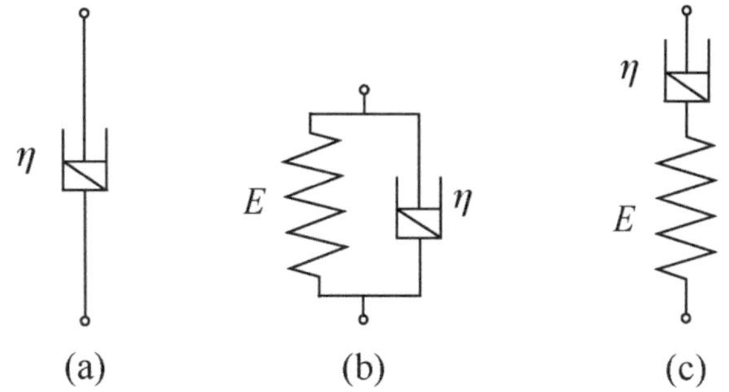

Fig. 3.17 Fractional component models, **a** Abel glue pot, **b** fractional Kelvin-Voigt model, **c** fractional Maxwell model

be expressed as $\sigma(t) \sim \frac{d^{\beta}\varepsilon(t)}{dt^{\beta}}$ $(0 \leq \beta \leq 1)$ since their properties are between linear elastic and viscous materials. It is therefore possible to establish fractional constitutive laws for visco-elastic materials in terms of fractional calculus. Bagley (Bagley and Torvik 1983) carried out experiments to show that the relaxation and creep phenomena of many high-molecule compounds could be described by a constitutive law with fractional derivative as follows

$$\sigma(t) = \eta \frac{d^{\beta}\varepsilon(t)}{dt^{\beta}}, \quad (0 \leq \beta \leq 1). \tag{3.152}$$

Equation (3.122) constitutes so-called Abel glue pot (Kempfle et al. 2002). As shown in Figure 3.17a, Abel glue pot itself is a model of visco-elastic materials without being connected to other components. In order to have the creep compliance $J(t)$ of Abel glue pot, let $\sigma(t) = \sigma_0$ in (3.152) and carry out a $\boldsymbol{\beta}$-order integration. According to the Riemann-Liouville definition of fractional integration given in (2.5.3), the creep compliance of Abel glue pot is obtained as

$$J(t) = \frac{t^{\beta}}{\eta\Gamma(1+\beta)}, \quad (0 \leq \beta \leq 1). \tag{3.153}$$

Similarly, by letting $\varepsilon(t) = \varepsilon_0$, the relaxation modulus of Abel glue pot is derived as

$$G(t) = \frac{t^{-\beta}}{\eta\Gamma(1-\beta)}, \quad (0 \leq \beta \leq 1). \tag{3.154}$$

By connecting the Abel glue pot and other components, fractional component models can be established (Chen and Zhu 2010). For example, the fractional Kelvin-Voigt model and fractional Maxwell model are shown in Fig. 3.17b and c. Fractional component models can remedy some disadvantages that classical component models cannot simulate some visco-elastic materials. Moreover, fractional component models have the benefit of less components and less parameters. By introducing fractional derivatives, the differential constitutive law (3.141) of visco-elastic materials can be modified as

$$\begin{aligned}&\frac{\mathrm{d}^{\alpha_0}\sigma(t)}{\mathrm{d}t^{\alpha_0}}+p_1\frac{\mathrm{d}^{\alpha_1}\sigma(t)}{\mathrm{d}t^{\alpha_1}}+p_2\frac{\mathrm{d}^{\alpha_2}\sigma(t)}{\mathrm{d}t^{\alpha_2}}+\ldots\\&=q_0\frac{\mathrm{d}^{\beta_0}\varepsilon(t)}{\mathrm{d}t^{\beta_0}}+q_1\frac{\mathrm{d}^{\beta_1}\varepsilon(t)}{\mathrm{d}t^{\beta_1}}+q_2\frac{\mathrm{d}^{\beta_2}\varepsilon(t)}{\mathrm{d}t^{\beta_2}}+\ldots.\end{aligned} \tag{3.155}$$

By using similar procedure given in Sect. 3.4.2, the Laplace transforms of relaxation modulus $G(t)$ and creep compliance $J(t)$ can be obtained as

$$\overline{G}(s)=\frac{q_0s^{\beta_0}+q_1s^{\beta_1}+q_2s^{\beta_2}+\ldots}{s(s^{\alpha_0}+p_1s^{\alpha_1}+p_2s^{\alpha_2}+\ldots)},\quad \overline{J}(s)=\frac{s^{\alpha_0}+p_1s^{\alpha_1}+p_2s^{\alpha_2}+\ldots}{s\left(q_0s^{\beta_0}+q_1s^{\beta_1}+q_2s^{\beta_2}+\ldots\right)}. \tag{3.156}$$

The relaxation modulus $G(t)$ and creep compliance $J(t)$ can be calculated from inverse Laplace transform. In terms of relaxation modulus and creep compliance, the fractional constitutive laws are the same as (3.144) and (3.145) in s-domain, and the fractional integral constitutive laws are the same as (3.148)–(3.150).

References

Baber T T, Wen Y K. 1980. Stochastic equivalent linearization for hysteretic degrading, multistory structures. Civil Engineering Studies, Structural Research Series, No. 471, University of Illinois at Urbana-Champaign.

Bagley R L, Torvik P J. 1983. A theoretical basis for the application of fractional calculus to viscoelasticity. Journal of Rheology, 27(3): 201–210.

Bouc R. 1967. Forced vibration of mechanical system with hysteresis. Proceedings of the 4th Conference on Nonlinear Oscillations, (abstract), Prague, Czechoslovakia.

Cai G Q, Lin Y K. 1990. On randomly excited hysteretic structures. ASME Journal of Applied Mechanics, 57: 442–448.

Chen L C, Zhu W Q. 2010. Feedback minimization of reliability of a simple power system under random perturbations. Journal of Dynamics and Control, 8(1): 19–23. (in Chinese)

Drozdov A D. 1998. Viscoelastic Structures: Mechanics of Growth and Aging. San Diego: Academic Press.

Hénon M, Heiles C. 1964. The applicability of the third integral of motion: some numerical experiments. Astronomy Journal, 69: 73–79.

Kempfle S, Schafer I, Beyer H. 2002. Fractional calculus via functional calculus: theory and application. Nonlinear Dynamics, 29(1): 99–127.

Krasnoselskii M A, Pokrovskii A V. 1989. Systems with Hysteresis. Berlin: Springer.

Li J B, Zhao X H, Liu Z R. 2007. Theory of Generalized Hamiltonian Systems and Its Applications. Beijing: Science Press. (in Chinese).

Lubarda V, Sumarac D, Krajcinovic D. 1993. Hysteretic response of ductile materials. European Journal of Mechanics A-Solids, 12(4): 445–470.

Mayergoyz I D. 1991. Mathematical Models of Hysteresis. New York: Springer-Verlag.

Oliver P J. 1986. Application of Lie Groups to Differential Equations. New York: Springer-Verlag.

Spanos P D, Cacciola P, Muscolino G. 2004. Stochastic averaging of Preisach hysteretic systems. ASCE Journal of Engineering Mechanics, 130(11): 1257–1267.

van Moerbeke P. 1988. The geometry of the Painleve analysis, Proceedings of the Workshop on Finite Dimensional Integrable Nonlinear Dynamical Systems. Singapore: Word Scientific, pp. 1–33.

Visintin A. 1994. Differential Models of Hysteresis. Berlin: Springer-Verlag.
Wang Y, Ying Z G, Zhu W Q. 2009. Stochastic averaging of energy envelope of Preisach hysteretic systems. Journal of Sound and Vibration, 321: 976–993.
Wen Y K. 1976. Method for random vibration of hysteretic systems. ASCE Journal of Engineering Mechanics, 103: 249–263.
Ying Z G, Zhu W Q, Ni Y Q, Ko J M. 2002. Stochastic averaging of Duhem hysteretic systems. Journal of Sound and Vibration, 254: 91–104.
Zhang C Y. 1994. Fracture Mechanics of Visoelasticity. Wu Han: Huazhong University of Science and Technology Press. (in Chinese).

Chapter 4
Stochastic Averaging Methods of Single-Degree-Of-Freedom Systems

It is known that Gaussian white noise excitations are required in most cases to acquire exact or approximate solutions of system responses which are Markov diffusion processes (Cai and Zhu 2016). However, the white noise is a mathematical idealization with infinite energy and a correlation function being a delta function. Such a requirement is not met for real stochastic processes. Stochastic averaging method is therefore developed to approximate a non-white process by a white noise under certain conditions, and use the method of diffusion processes to obtain approximate solutions of nonlinear stochastic dynamical systems.

Another advantage of stochastic averaging is to reduce the system dimensions based on different time scales of system responses. For certain systems, different response processes with different time scales can be distinguished as fast-varying and slow-varying response processes. By performing time averaging, the fast-varying response processes disappear and only the slow-varying response are left. By this procedure, the dimension of a system is reduced. However, it is difficult to carry out the time averaging since the system responses are changing randomly. A feasible way is to replace the time averaging by a space averaging based on the ergodicity of system responses over certain manifolds.

In this chapter, the principles and mathematical basis of the stochastic averaging are presented first in Sect. 4.1. The method is then applied to general SDOF stochastic dynamical systems in Sect. 4.2 for the cases of linear restoring force and strongly nonlinear restoring force. Two different schemes of the stochastic averaging method using the amplitude envelope and energy envelope are described in detail. Stochastic averaging procedures of the two schemes are different depending on the nature of excitations. If the excitations are Gaussian white noises, only the time averaging is needed, and the procedure is relatively straightforward, as described in Sect. 4.3. However, if the excitations are of broad band, application of the stochastic averaging is more complicated. Several techniques to treat the broad-band excitations are presented in Sect. 4.4. Although the basic principle and procedures are described in Sects. 4.1–4.4 for SDOF systems, certain flexibility may be needed to treat some

W.-Q. Zhu et al., *Stochastic Averaging*, https://doi.org/10.1007/978-981-96-3829-1_4

special systems. Two such cases are viscoelastic systems and systems with double-well potential. Section 4.5 and 4.6 show how to apply the stochastic averaging method to these two types of systems, respectively. The stochastic averaging method is also applicable to systems subjected to both harmonic and random excitation, to Poisson white noises, and to fractional Gaussian noises, as shown in Sects. 4.7, 4.8, and 4.9, respectively.

4.1 Stochastic Averaging Principles

The stochastic averaging method was first proposed by Stratonovich (1963). Khasminskii (1966) provided its theoretical base in terms of a limit theorem. The theorem is briefly described below.

Consider the following stochastic differential equation

$$\frac{\mathrm{d}}{\mathrm{d}t}\boldsymbol{X} = \varepsilon \boldsymbol{f}(\boldsymbol{X}, t) + \varepsilon^{1/2}\boldsymbol{g}\big[\boldsymbol{X}, \boldsymbol{\xi}(t), t\big], \tag{4.1}$$

where ε is a small parameter, $\boldsymbol{X}$ is a n-dimensional state vector, and $\xi(t)$ is a m-dimensional vector of stochastic excitations. If the stochastic processes in $\xi(t)$ are broad-band processes with zero means, and functions in $\boldsymbol{f}$ and $\boldsymbol{g}$ are bounded and smooth enough, then the system response process $\boldsymbol{X}(t)$ converges weakly to a Markov diffusion process as $\varepsilon \to 0$, governed by the following Itô differential equation

$$\mathrm{d}\boldsymbol{X}(t) = \varepsilon \boldsymbol{m}[\boldsymbol{X}(t)]\mathrm{d}t + \varepsilon^{1/2}\boldsymbol{\sigma}[\boldsymbol{X}(t)]\mathrm{d}\boldsymbol{B}(t). \tag{4.2}$$

The drift vector $\boldsymbol{m}$ and diffusion matrix $\boldsymbol{\sigma}$ are defined as

$$\boldsymbol{m} = \big\langle E[\boldsymbol{f}(\boldsymbol{X}, t)]_t\big\rangle + \left\langle \int_{-\infty}^{0} E\left\{\left[\frac{\partial}{\partial \boldsymbol{X}}\boldsymbol{g}(\boldsymbol{X}_t, \boldsymbol{\xi}(t), t)\right]\boldsymbol{g}(\boldsymbol{X}_{t+\tau}, \boldsymbol{\xi}(t+\tau), t+\tau)\right\}\mathrm{d}\tau\right\rangle_t, \tag{4.3}$$

$$\boldsymbol{\sigma}\boldsymbol{\sigma}^{\mathrm{T}} = \left\langle \int_{-\infty}^{\infty} E\big[\boldsymbol{g}(\boldsymbol{X}_t, \boldsymbol{\xi}(t), t)\boldsymbol{g}^T(\boldsymbol{X}_{t+\tau}, \boldsymbol{\xi}(t+\tau), t+\tau)\big]\mathrm{d}\tau\right\rangle_t, \tag{4.4}$$

where $\boldsymbol{X}_t$ is an abbreviation of $\boldsymbol{X}(t)$, $E[\cdot]$ denotes the ensemble averaging, the superscript T denotes an operation of matrix transpose, and $\langle[\cdot]\rangle_t$ denotes time averaging defined by

$$\langle[\cdot]\rangle_t = \lim_{T\to\infty}\frac{1}{T}\int_{t_0}^{t_0+T}[\cdot]\mathrm{d}t. \tag{4.5}$$

The matrix differentiation in (4.3) is defined as

$$\frac{\partial \boldsymbol{g}}{\partial \boldsymbol{X}} = \begin{bmatrix} \frac{\partial g_1}{\partial X_1} & \frac{\partial g_1}{\partial X_2} & \cdots & \frac{\partial g_1}{\partial X_n} \\ \frac{\partial g_2}{\partial X_1} & \frac{\partial g_2}{\partial X_2} & \cdots & \frac{\partial g_2}{\partial X_n} \\ \cdots & \cdots & \cdots & \cdots \\ \frac{\partial g_n}{\partial X_1} & \frac{\partial g_n}{\partial X_2} & \cdots & \frac{\partial g_n}{\partial X_n} \end{bmatrix}. \tag{4.6}$$

In evaluation (4.3) and (4.4), elements in $\boldsymbol{X}(t)$ are treated as constants since the right- hand-side of Eq. (4.2) are small, and $\boldsymbol{X}(t)$ is slowly varying.

In system Eqs. (4.1) and (4.2), a small parameter $\varepsilon << 1$ is introduced to indicate that the first and second terms on the right-hand-side are of order ε and $\varepsilon^{1/2}$, respectively. As it will be explained later, the contributions from the two terms to the system response are of the same order.

In the above procedure, two conditions are critical. One is that the stochastic excitations are broad-band processes so that they can be approximated as Gaussian white noises and the response is an approximate Markov process. Another is the right-hand-side of Eq. (4.1) is small, indicating that the responses are slow-varying processes. We will discuss the rationales and implications of the two conditions.

Consider the system

$$\frac{\mathrm{d}}{\mathrm{d}t} X_j(t) = f_j(\boldsymbol{X}, t) + \sum_{l=1}^{m} g_{jl}(\boldsymbol{X}, t)\xi_l(t), \quad j = 1, 2, \cdots, n, \tag{4.7}$$

where $\xi_l(t)$ are jointly stationary processes with zero mean and the following correlation functions

$$E[\xi_l(u)\xi_s(v)] = R_{ls}(\tau), \quad \tau = v - u. \tag{4.8}$$

Define the correlation time of $\xi_l(t)$ and $\xi_s(t)$ as (Cai and Zhu 2016)

$$\tau_{ls} = \frac{1}{\sqrt{R_{ll}(0)R_{ss}(0)}} \int_{-\infty}^{0^-} |R_{ls}(\tau)|\mathrm{d}\tau, \tag{4.9}$$

which is a measure of "memory" of the present $\xi_s(t)$ with respect to the past $\xi_l(t\text{-}\tau)$. It is zero for white noises. The shorter the correlation time is, the broader the band of the process is. If all excitations $\xi_l(t)$ can be approximated as white noises, the system response $\boldsymbol{X}(t)$ is a vector Markov diffusion process, governed by the Itô equations

$$\mathrm{d}X_j(t) = m_j(\boldsymbol{X}, t)\mathrm{d}t + \sum_{l=1}^{m} \sigma_{jl}(\boldsymbol{X}, t)\mathrm{d}B_l(t), \quad j = 1, 2, \cdots, n, \tag{4.10}$$

where $B_l(t)$ are independent unit Wiener processes. It is known that the PDF of the system response $\boldsymbol{X}(t)$ satisfies the FPK equation. Thus, the main objective of the approximation is to determine the drift and diffusion coefficients m_j and σ_{jl} in the Itô equations, or the first and second derivate moments in the FPK equation. Denote the time interval as Δt between two consecutive measurements. According to Eq. (2.107), we calculate the increments as follows

$$X_j(t+\Delta t) - X_j(t) = \int_t^{t+\Delta t} f_j(\boldsymbol{X}_u, u)\mathrm{d}u + \sum_{l=1}^{m} \int_t^{t+\Delta t} g_{jl}(\boldsymbol{X}_u, u)\xi_l(u)\mathrm{d}u. \tag{4.11}$$

Expand f_j and g_{jl} in (4.11) at time instant t,

$$\begin{aligned} f_j(\boldsymbol{X}_u, u) =& f_j(\boldsymbol{X}_t, t) + (u-t)\frac{\partial}{\partial t} f_j(\boldsymbol{X}_t, t) \\ &+ \sum_{r=1}^{n} [X_r(u) - X_r(t)] \frac{\partial}{\partial X_r} f_j(\boldsymbol{X}_t, t) + \cdots, \end{aligned} \tag{4.12}$$

$$\begin{aligned} g_{jl}(\boldsymbol{X}_u, u) =& g_{jl}(\boldsymbol{X}_t, t) + (u-t)\frac{\partial}{\partial t} g_{jl}(\boldsymbol{X}_t, t) \\ &+ \sum_{r=1}^{n} [X_r(u) - X_r(t)] \frac{\partial}{\partial X_r} g_{jl}(\boldsymbol{X}_t, t) + \cdots. \end{aligned} \tag{4.13}$$

In (4.12) and (4.13), we again substitute $X_r(u) - X_r(t)$ as

$$X_r(u) - X_r(t) = \int_t^{u} f_r(\boldsymbol{X}_v, v)\mathrm{d}v + \sum_{s=1}^{m} \int_t^{u} g_{rs}(\boldsymbol{X}_v, v)\xi_s(v)\mathrm{d}v. \tag{4.14}$$

Combining (4.11) through (4.14) and keeping the leading terms, we obtain

$$\begin{aligned} X_j(t+\Delta t) - X_j(t) =& \int_t^{t+\Delta t} f_j(\boldsymbol{X}_t, t)\mathrm{d}u + \sum_{l=1}^{m} \int_t^{t+\Delta t} g_{jl}(\boldsymbol{X}_t, t)\xi_l(u)\mathrm{d}u \\ &+ \sum_{l=1}^{m} \int_t^{t+\Delta t} (u-t)\left[\frac{\partial}{\partial t} g_{jl}(\boldsymbol{X}_t, t)\right]\xi_l(u)\mathrm{d}u \\ &+ \sum_{l,s=1}^{m} \int_t^{t+\Delta t} \xi_l(u)\mathrm{d}u \sum_{r=1}^{n} \frac{\partial}{\partial X_r} g_{jl}(\boldsymbol{X}_t, t) \int_t^{u} g_{rs}(\boldsymbol{X}_v, v)\xi_s(v)\mathrm{d}v. \end{aligned} \tag{4.15}$$

Substitution of (4.15) into (2.67) results in

$$a_j(\boldsymbol{x}_t, t) = f_j(\boldsymbol{x}_t, t) + \frac{1}{\Delta t}\sum_{l,s=1}^{m}\sum_{r=1}^{n}\int_{t}^{t+\Delta t} du \int_{t}^{u}\left[\frac{\partial}{\partial x_r} g_{jl}(\boldsymbol{x}_t, t)\right] \times g_{rs}(\boldsymbol{x}_v, v) E[\xi_l(u)\xi_s(v)]dv + O(\Delta t), \tag{4.16}$$

where $O(\Delta t)$ indicates the remaining terms of order Δt. In deriving (4.16), we assume that the variations of f_j and g_{jl} functions are not affected significantly by the randomness in $\boldsymbol{X}(t)$ within the interval Δt since $\boldsymbol{X}(t)$ is slowly varying; thus, they are not included in the ensemble averaging. Substituting (4.8) into (4.16) and change integration variable from v to $\tau = v–u$, we obtain the first derivate moments

$$a_j(\boldsymbol{x}_t, t) = f_j(\boldsymbol{x}_t, t) + \frac{1}{\Delta t}\sum_{l,s=1}^{m}\sum_{r=1}^{n}\int_{t}^{t+\Delta t} du \int_{t-u}^{0}\left[\frac{\partial}{\partial x_r} g_{jl}(\boldsymbol{x}_t, t)\right] \times g_{rs}(\boldsymbol{x}_{u+\tau}, u+\tau) R_{ls}(\tau)d\tau + O(\Delta t), \tag{4.17}$$

If Δt is so adopted that it is much larger than the correlation time τ_{ls}, i.e., $R_{ls}(\tau) \approx 0$ for $\tau > \Delta t$, then by changing the integration order, i.e., integrating on u first, (4.17) can be written as

$$a_j(\boldsymbol{x}_t, t) = f_j(\boldsymbol{x}_t, t) + \sum_{l,s=1}^{m}\sum_{r=1}^{n}\int_{-\Delta t}^{0}\left[\frac{\partial}{\partial x_r} g_{jl}(x_t, t)\right] g_{rs}(x_{t+\tau}, t+\tau) R_{ls}(\tau)d\tau. \tag{4.18}$$

For the second derivate moments, substituting (4.11) into (2.67) and following the similar procedure, we obtain

$$b_{jk}(\boldsymbol{x}_t, t) = \sum_{l,s=1}^{m}\int_{-\Delta t}^{\Delta t} g_{jl}(\boldsymbol{x}_t, t) g_{ks}(\boldsymbol{x}_{t+\tau}, t+\tau) R_{ls}(\tau)d\tau. \tag{4.19}$$

In above derivation, we also assume that functions f_j and g_{jl} are varying slowly within the interval Δt. To meet this condition, we may require Δt not greater than the system relaxation time, which measures the change rate of the system motion without excitations (Cai and Zhu 2016). The relaxation time, denoted by τ_{rel}, has been defined for either an oscillatory or non-oscillatory system. For an oscillatory system, τ_{rel} is the time needed for the amplitude of oscillation to reduce by a factor of e^{-1}, or increase by a factor of e. In the non-oscillatory case, the amplitude is replaced by the motion itself. If Δt is close to or larger than τ_{rel}, functions f_j and g_{jl} may have non-negligible change and too many details may be lost.

From the derivation of (4.18) and (4.19), it is known that the time interval Δt should be much larger than the excitation correlation times and smaller than the system relaxation time. Therefore, the condition for (4.18) and (4.19) to be valid is that the correlation times of all excitations are much shorter than the relaxation times of the system. Under this condition, the system response can be approximated as a Markov diffusion process.

Since Δt is much larger than the excitation correlation times, the correlation functions are nonzero only within a small neighborhood around $\tau = 0$. Thus, the low limit of the integrations in (4.18) and (4.19) may be extended to $-\infty$, and the upper limit in (4.19) may be extended to ∞, i.e.,

$$a_j(\boldsymbol{x}_t, t) = f_j(\boldsymbol{x}_t, t) + \sum_{l,s=1}^{m} \sum_{s=1}^{n} \int_{-\infty}^{0} \left[\frac{\partial}{\partial x_r} g_{jl}(\boldsymbol{x}_t, t) \right] g_{rs}(\boldsymbol{x}_{t+\tau}, t+\tau) R_{ls}(\tau) \mathrm{d}\tau, \quad (4.20)$$

$$b_{jk}(\boldsymbol{x}_t, t) = \sum_{l,s=1}^{m} \int_{-\infty}^{\infty} g_{jl}(\boldsymbol{x}_t, t) g_{ks}(\boldsymbol{x}_{t+\tau}, t+\tau) R_{ls}(\tau) \mathrm{d}\tau. \quad (4.21)$$

These two equations allow us to calculate the approximate first and second derivate moments in the FPK equation, and the drift and diffusion coefficients in the Itô equation according to (2.109), i.e., $m_j(\boldsymbol{X}, t) = a_j(\boldsymbol{x}, t)|_{x \to X}$, $[\boldsymbol{\sigma}(\boldsymbol{X}, t)\boldsymbol{\sigma}^{\mathrm{T}}(\boldsymbol{X}, t)]_{jk} = b_{jk}(\boldsymbol{x}, t)|_{x \to X}$. They will be used for different versions of stochastic averaging to approximate the non-white excitations as white noises, and the system response as a vector Markov diffusion process.

It is noted that Eq. (4.7) is different from (4.1) in that there is no small parameter ε. It indicates that, as long as the bandwidths of the excitations are broad enough, satisfying certain conditions mentioned above. The excitations can be approximated as Gaussian white noises, and the responses as Markov diffusion processes.

If the excitations are Gaussian white noises, i.e.,

$$E[\xi_l(u)\xi_s(v)] = 2\pi K_{ls} \delta(v - u), \quad (4.22)$$

where K_{ls} are constant spectral densities. Equation (4.20) and (4.21) are simplified as

$$a_j(\boldsymbol{x}_t, t) = f_j(\boldsymbol{x}_t, t) + \sum_{l,s=1}^{m} \sum_{r=1}^{n} \pi K_{ls} \left[\frac{\partial}{\partial x_r} g_{jl}(\boldsymbol{x}_t, t) \right] g_{rs}(\boldsymbol{x}_t, t), \quad (4.23)$$

$$b_{jk}(\boldsymbol{x}_t, t) = \sum_{l,s=1}^{m} 2\pi K_{ls} g_{jl}(\boldsymbol{x}_t, t) g_{ks}(\boldsymbol{x}_t, t). \quad (4.24)$$

The second term on the right-hand-side of (4.23) is the Wong-Zakai correction.

As shown in Eq. (4.3) and (4.4), another scheme used in the stochastic averaging method is the time averaging which is applied to the situations in which the system state variables may be divided into two different types: fast varying and slowly varying. Without loss of generality, assume that the first n_1 ($n_1 \le n$) state variables in system (4.7) are slowly varying, and all others are fast varying. It includes the special case of $n_1 = n$, i.e., all are slowly varying variables. Then we may write the first n_1 equations as

$$\frac{\mathrm{d}}{\mathrm{d}t}X_j(t) = \varepsilon f_j(\boldsymbol{X}, t) + \varepsilon^{1/2}\sum_{l=1}^{m} g_{jl}(\boldsymbol{X}, t)\xi_l(t), \quad j = 1, \cdots, n_1. \tag{4.25}$$

Using (4.22) to replace the first n_1 equations in system (4.7) and carrying out the stochastic averaging and the time-averaging, we obtain Itô equations for $X_j(t)$ ($j = 1, 2, \ldots, n_1$) from (4.25) as follows

$$\mathrm{d}X_j(t) = m_j(\tilde{\boldsymbol{X}})\mathrm{d}t + \sum_{l=1}^{m} \sigma_{jl}(\tilde{\boldsymbol{X}})\mathrm{d}B_l(t), \quad j = 1, \cdots, n_1 \tag{4.26}$$

where $B_l(t)$ are independent unit Wiener processes, $\tilde{\boldsymbol{X}} = [X_1, X_2, \cdots X_{n_1}]^{\mathrm{T}}$, and from (4.3) and (4.4)

$$m_j(\tilde{\boldsymbol{X}}) = \varepsilon\left\{\left\langle f_j(\boldsymbol{X}_t, t)\right\rangle_t + \sum_{l,s=1}^{m}\sum_{r=1}^{n}\int_{-\infty}^{0}\left\langle g_{rs}(\boldsymbol{X}_{t+\tau}, t+\tau)\frac{\partial}{\partial X_r}g_{jl}(\boldsymbol{X}_t, t)\right\rangle_t R_{ls}(\tau)\mathrm{d}\tau\right\}, \tag{4.27}$$

$$\left(\boldsymbol{\sigma}\boldsymbol{\sigma}^{\mathrm{T}}\right)_{jk} = \sum_{l=1}^{m}\sigma_{jl}(\tilde{\boldsymbol{X}})\sigma_{kl}(\tilde{\boldsymbol{X}}) = \varepsilon\sum_{l,s=1}^{m}\int_{-\infty}^{\infty}\left\langle g_{jl}(\boldsymbol{X}_t, t)g_{ks}(\boldsymbol{X}_{t+\tau}, t+\tau)\right\rangle_t R_{ls}(\tau)\mathrm{d}\tau. \tag{4.28}$$

In (4.27) and (4.28), $j, k = 1, 2, \ldots, n_1$. When carrying out the time-averaging, the slowly varying variables are considered to be constants, and all fast varying variables are averaged out; therefore, only the slowly varying variables remain, and they also constitute a Markov diffusion process themselves. The first and second derivate moments in the FPK equation can be obtained from (4.27) and (4.28) according to (2.109). The slowly varying processes may reflect the overall trend of the system motion. As a result, the system dimension is reduced to that of the slowly varying processes, and the problem is simplified. The procedure to obtain (4.26) through (4.28) was known as the stochastic averaging theorem of Itô equations with fast and slowly varying variables, proposed by Khasminskii (1968).

In the case of Gaussian white noise excitations, (4.27) and (4.28) are simplified as

$$m_j(\tilde{\boldsymbol{X}}) = \varepsilon f_j(\boldsymbol{X}_t, t)_t + \sum_{l,s=1}^{m}\sum_{r=1}^{n}\pi K_{ls}\left\langle g_{rs}(\boldsymbol{X}_t, t)\frac{\partial}{\partial X_r}g_{jl}(\boldsymbol{X}_t, t)\right\rangle_t, \tag{4.29}$$

$$\left(\sigma\sigma^{\mathrm{T}}\right)_{jk} = \sum_{l=1}^{m} \sigma_{jl}(\tilde{X})\sigma_{kl}(\tilde{X}) = \varepsilon \sum_{l,s=1}^{m} 2\pi K_{ls}\left\langle g_{jl}(X_t, t) g_{ks}(X_t, t)\right\rangle_t. \tag{4.30}$$

If the fast varying variables are periodic functions of time with a period T_0. Then the time averaging can be done in one period as

$$\langle[\cdot]\rangle_t = \lim_{T\to\infty} \frac{1}{T} \int_t^{t+T} [\cdot]\mathrm{d}t = \frac{1}{T_0} \int_0^{T_0} [\cdot]\mathrm{d}t. \tag{4.31}$$

If the fast varying variables are not periodic functions, the time averaging can be replaced by space averaging with respect to the fast varying variables by using the ergodicity characteristic on a certain manifold (see Chap. 5 for details).

It is clear from above analysis that the stochastic averaging may include two procedures. The first one is to approximate broad-band excitation processes as Gaussian white noises and the system response as a Markov diffusion process. The obtained Eqs. (4.10), (4.20) and (4.21), and (4.23) and (4.24) known as the unsmoothed version, in which all state variables remain. The second procedure is to carry out the time averaging, eliminate the fast varying variables, and reduce the system dimensions, resulting in (4.26) through (4.30). It is known as the smoothed version. The stochastic averaging may refer to either the unsmoothed version or the smoothed version.

In (4.27) and (4.28), and (4.29) and (4.30), the averaged drift and diffusion coefficients are smoothed, and no longer depend on the time explicitly. It may happen that only a single response (or a subset of the response vector) constitutes a diffusion process (or vector diffusion process) after the time averaging. In this case, system dimension is reduced. The stochastic averaging will be applied to different situations, as presented in the following sections.

4.2 Stochastic Averaging Methods of SDOF Systems

As shown above, a main advantage of the stochastic averaging is to reduce the system dimension by time averaging when the system equations contain both slowly and fast varying variables. This is usually not the case with the original set of state variables, such as displacements and velocities in oscillatory systems, or generalized displacements and momenta in Hamiltonian systems. However, if certain types of slowly varying variables can be identified, the system equations can be transformed to those including both the slowly and fast varying variables as state variables. The required transformations are different for different situations, as discussed in the following sections for SDOF systems.

4.2.1 Stochastic Averaging of Amplitude Envelope

Consider the following SDOF system with linear stiffness, weak nonlinear damping, and weak excitations of wideband processes,

$$\ddot{X} + \varepsilon h\big(X, \dot{X}\big) + \omega_0^2 X = \varepsilon^{1/2} \sum_{l=1}^{m} g_l(X, \dot{X})\xi_l(t). \tag{4.32}$$

It is clear that both X and $\dot{X}$ are not slowly varying. Let

$$X = A(t)\cos\Phi, \dot{X} = -A(t)\omega_0 \sin\Phi, \Phi = \omega_0 t + \Theta(t). \tag{4.33}$$

where $A(t)$ is the amplitude process. Differentiating the first equation of (4.32), we have

$$\dot{X} = \dot{A}\cos\Phi - A\big(\omega_0 + \dot{\Theta}\big)\sin\Phi. \tag{4.34}$$

Combine (4.34) and the second equation of (4.32) to obtain

$$\dot{A}\cos\Phi - A\dot{\Theta}\sin\Phi = 0. \tag{4.35}$$

Differentiation of the second equation of (4.33) leads to

$$\ddot{X} = -\dot{A}\omega_0 \sin\Phi - A\omega_0\big(\omega_0 + \dot{\Theta}\big)\cos\Phi. \tag{4.36}$$

Substituting $\ddot{X}$ in Eq. (4.32) into (4.36), we obtain

$$\begin{aligned} \dot{A}\sin\Phi + A\dot{\Theta}\cos\Phi = {} & \frac{1}{\omega_0}\varepsilon h(A\cos\Phi, -A\omega_0\sin\Phi) \\ & - \frac{1}{\omega_0}\varepsilon^{\frac{1}{2}} \sum_{l=1}^{m} g_l(A\cos\Phi, -A\omega_0\sin\Phi)\xi_l(t). \end{aligned} \tag{4.37}$$

Solve $\dot{A}$ and $\dot{\Phi}$ from (4.35) and (4.37) to obtain

$$\dot{A} = \frac{\sin\Phi}{\omega_0}\left[\varepsilon h(A\cos\Phi, -A\omega_0\sin\Phi) - \varepsilon^{1/2}\sum_{l=1}^{m} g_l(A\cos\Phi, -A\omega_0\sin\Phi)\xi_l(t)\right], \tag{4.38}$$

$$\dot{\Theta} = \frac{\cos\Phi}{\omega_0 A}\left[\varepsilon h(A\cos\Phi, -A\omega_0\sin\Phi) - \varepsilon^{1/2}\sum_{l=1}^{m} g_l(A\cos\Phi, -A\omega_0\sin\Phi)\xi_l(t)\right]. \tag{4.39}$$

The right-hand sides of (4.38) and (4.39) are small, indicating that both $A(t)$ and $\Theta(t)$ are slowly varying. Physically, they represent the amplitude and phase of a nearly linear oscillator, perturbed by small damping and weak excitations.

Treating $A(t)$ as $X_1(t)$ and $\Theta(t)$ as $X_2(t)$, and comparing (4.38) and (4.39) with the standard form (4.25), we have

$$\begin{aligned} f_1(A,\Phi) &= \frac{\sin\Phi}{\omega_0} h(A\cos\Phi, -A\omega_0\sin\Phi), \\ f_2(A,\Phi) &= \frac{\cos\Phi}{\omega_0 A} h(A\cos\Phi, -A\omega_0\sin\Phi), \end{aligned} \tag{4.40}$$

$$\begin{aligned} g_{1l}(A,\Phi) &= -\frac{\sin\Phi}{\omega_0} g_l(A\cos\Phi, -A\omega_0\sin\Phi), \\ g_{2l}(A,\Phi) &= -\frac{\cos\Phi}{\omega_0 A} g_l(A\cos\Phi, -A\omega_0\sin\Phi). \end{aligned} \tag{4.41}$$

Since the system is nearly linear, the motion is nearly periodic with a "quasi-period" $2\pi/\omega_0$. The time-averaging (4.31) can then be calculated over one quasi-period, i.e.,

$$\langle[\cdot]\rangle_t = \frac{1}{2\pi}\int_0^{2\pi}[\cdot]\mathrm{d}\Phi. \tag{4.42}$$

The drift and diffusion coefficients can be found from Eqs. (4.27) and (4.28), respectively

$$\begin{aligned} m(A) = m_1 &= \varepsilon\langle f_1(A,\Phi)\rangle_t + \varepsilon\sum_{l,s=1}^{m}\int_{-\infty}^{0}\left\langle g_{1l}[A,\Phi(t)]\frac{\partial}{\partial A}g_{1s}[A,\Phi(t+\tau)]\right\rangle_t R_{ls}(\tau)\mathrm{d}\tau \\ &+ \varepsilon\sum_{l,s=1}^{m}\int_{-\infty}^{0}\left\langle g_{2l}[A,\Phi(t)]\frac{\partial}{\partial\theta}g_{1s}[A,\Phi(t+\tau)]\right\rangle_t R_{ls}(\tau)\mathrm{d}\tau, \end{aligned} \tag{4.43}$$

$$\sigma^2(A) = \left(\boldsymbol{\sigma}\boldsymbol{\sigma}^{\mathrm{T}}\right)_{11} = \varepsilon\sum_{l,s=1}^{m}\int_{-\infty}^{\infty}\langle g_{1l}[A,\Phi(t)]g_{1s}[A,\Phi(t+\tau)]\rangle_t R_{ls}(\tau)\mathrm{d}\tau. \tag{4.44}$$

It is found that the averaged equation for the smoothed amplitude process $A(t)$ does not contain $\Phi(t)$, hence $\Theta(t)$, and $A(t)$ itself is a Markov diffusion process governed by an Itô differential equation

$$\mathrm{d}A = m(A)\mathrm{d}t + \sigma(A)\mathrm{d}B(t). \tag{4.45}$$

The drift and diffusion coefficients $m(A)$ and $\sigma(A)$ can be determined when the functions h and g_l are given. Equation (4.45) is a one-dimensional Itô equation, and its analysis becomes much simpler, as shown in the book by Cai and Zhu (2016).

The stationary PDF is obtained as

$$p(a) = \frac{C}{\sigma^2(a)} \exp\left[\int \frac{2m(a)}{\sigma^2(a)} da\right]. \tag{4.46}$$

Although the amplitude process $A(t)$ is an important feature of the motion, the joint stationary probability density, as well as the marginal PDFs of $X(t)$ and $\dot{X}(t)$ may be also needed for analysis, which will be derived below. The joint distribution function $F_{XA}(x, a)$ can be written as

$$\begin{aligned} F_{XA}(x, a) &= \text{Prob}[(X \le x) \cap (A \le a)] \\ &= \text{Prob}\left[(X \le x) \cap \left(\sqrt{X^2 + \dot{X}^2/\omega_0^2} \le a\right)\right] \\ &= \text{Prob}\left[(X \le x) \cap \left(-y \le \dot{X} \le y\right)\right] \\ &= \text{Prob}\left[(X \le x) \cap \left(\dot{X} \le y\right)\right] - \text{Prob}\left[(X \le x) \cap \left(\dot{X} \le -y\right)\right] \\ &= F_{X\dot{X}}(x, y) - F_{X\dot{X}}(x, -y), \end{aligned} \tag{4.47}$$

where

$$y = \omega_0 \sqrt{a^2 - x^2}. \tag{4.48}$$

Thus, we have from (4.57)

$$\begin{aligned} p_{XA}(x, a) &= \frac{\partial^2}{\partial x \partial a} F_{XA}(x, a) \\ &= \left[\frac{\partial^2}{\partial x \partial y} F_{X\dot{X}}(x, y)\right] \frac{\partial y}{\partial a} - \left[\frac{\partial^2}{\partial x \partial y} F_{X\dot{X}}(x, -y)\right] \frac{\partial(-y)}{\partial a} \\ &= p_{X\dot{X}}(x, \dot{x}) \frac{2\omega_0 a}{\sqrt{a^2 - x^2}}. \end{aligned} \tag{4.49}$$

In deriving (4.49), the symmetry of $\dot{X}$ is used, i.e., $p_{X\dot{X}}(x, -\dot{x}) = p_{X\dot{X}}(x, \dot{x})$. By omitting the subscripts, (4.49) can be written as

$$p(x, \dot{x}) = \frac{\sqrt{a^2 - x^2}}{2\omega_0 a} p(x, a) = \frac{\sqrt{a^2 - x^2}}{2\omega_0 a} p(x|a) p(a), \tag{4.50}$$

where $p(x|a)$ is the conditional probability density. For a fixed a, the probability of $X(t)$ in a region near x is inversely proportional to the velocity; therefore, $p(x|a)$ can be expressed as

$$p(x|a) = \frac{C}{|\dot{x}|} = \frac{C}{\omega_0\sqrt{a^2 - x^2}}. \tag{4.51}$$

Integrating both sides of (4.51) with respect to x from $-a$ to a, we obtain

$$C = \left[\int_{-a}^{a} \frac{\mathrm{d}x}{\omega_0\sqrt{a^2 - x^2}}\right]^{-1} = \frac{\omega_0}{\pi}. \tag{4.52}$$

Substituting (4.52) into (4.51), then into (4.50), we obtain

$$p(x, \dot{x}) = \frac{1}{2\pi\omega_0 a} p(a); \quad a = \sqrt{x^2 + \frac{\dot{x}^2}{\omega_0^2}}. \tag{4.53}$$

The marginal PDF $p(x)$ or $p(\dot{x})$ can then be calculated from (4.53) by integrating it with respect to $\dot{x}$ or x.

4.2.2 Stochastic Averaging of Energy Envelope

Consider the system governed by

$$\ddot{X} + \varepsilon h\big(X, \dot{X}\big) + u(X) = \varepsilon^{1/2}\sum_{l=1}^{m} g_l(X, \dot{X})\xi_l(t), \tag{4.54}$$

where $u(X)$ represents a strongly nonlinear restoring force, assumed to be an odd function of X. Now the transformation (4.33) is not appropriate since the frequency of the corresponding undamped free motion is changing with the amplitude.

First, let us study the undamped free vibration of the system

$$\ddot{x} + u(x) = 0. \tag{4.55}$$

Since $\ddot{x} = \frac{d\dot{x}}{dx}\frac{dx}{dt} = \dot{x}\frac{d\dot{x}}{dx}$, Eq. (4.55) can be rewritten as

$$\dot{x}\frac{d\dot{x}}{dx} + u(x) = 0. \tag{4.56}$$

Integrate (4.56) to obtain

$$\frac{1}{2}\dot{x}^2 + U(x) = \Lambda, \tag{4.57}$$

where

$$U(x) = \int u(x)\mathrm{d}x = \int_0^x u(z)\mathrm{d}z. \tag{4.58}$$

It is known that $U(x)$ is the potential energy, and Λ is the total energy, which is constant for an undamped free vibration for given initial conditions. Assuming that $u(x)$ is a symmetric and monotonic function in the range of $x > 0$ or $x < 0$, then the amplitude A of the oscillation is reached when the kinetic energy is zero, and the total energy is converted entirely to the potential energy, i.e.,

$$\Lambda = U(A) \text{ or } A = U^{-1}(\Lambda). \tag{4.59}$$

The period of the free oscillation is obtained as

$$T = 4T_{1/4} = 4\int_0^A \frac{1}{\sqrt{2\Lambda - 2U(x)}}\mathrm{d}x. \tag{4.60}$$

It is clear that the period of the free motion depends on the energy level. Letting

$$\mathrm{sgn}x\sqrt{U(x)} = \sqrt{\Lambda}\cos\theta,$$

$$\dot{x} = -\sqrt{2\Lambda}\sin\theta. \tag{4.61}$$

Treating (4.61) as a transformation from X and $\dot{X}$ to Λ and Φ, we have from (4.55).

$$\dot{\Lambda} = 0,$$

$$\dot{\Phi} = \frac{u}{\sqrt{2\Lambda}\cos\theta}. \tag{4.62}$$

The first equation of (4.62) is a re-statement that the total energy is conserved. If the restoring force u is a linear function, the right-hand side of the second equation reduces to a constant, namely, the natural frequency of free motion. It is, therefore, reasonable to rewrite the equation as

$$\omega(\Lambda, \Phi) = \dot{\Phi} = \frac{u}{\sqrt{2\Lambda}\cos\theta}, \tag{4.63}$$

which is the definition of the instantaneous frequency for the nonlinear system. The phase angle can be written as

$$\Phi(t) = \int_0^t \omega(\Lambda, s)\mathrm{d}s. \tag{4.64}$$

The average $\omega(\Lambda, \Phi)$ in one period, denoted by ω_Λ, is given by

$$\omega_\Lambda = \frac{1}{T}\int_0^T \omega(\Lambda, \Phi)\mathrm{d}t = \frac{1}{T}\int_0^{2\pi} \mathrm{d}\Phi = \frac{2\pi}{T}. \tag{4.65}$$

The above analysis shows that the undamped free motion is periodic with a period T depending on the energy level Λ, and the motion is not harmonic with a time-dependent frequency $\omega(t)$.

Now return to Eq. (4.54). Differentiating (4.57) and using (4.54), we obtain

$$\dot{\Lambda} = -\varepsilon \dot{X} h\big(X, \dot{X}\big) + \varepsilon^{1/2}\dot{X}\sum_{l=1}^{m} g_l(X, \dot{X})\xi_l(t). \tag{4.66}$$

Taking $\Lambda(t)$ as one state variable, the equation of motion of the displacement process $X(t)$ as another state variable is

$$\dot{X} = \pm\sqrt{2\Lambda - 2U(X)}, \tag{4.67}$$

where the positive and negative signs correspond to increasing and decreasing X, respectively. Equations (4.66) and (4.67) now constitute the governing equations for the system, instead of the original equation of motion (4.54). Equation (4.66) shows that the total energy $\Lambda(t)$ varies slowly if the damping and excitations are small. But the displacement process $X(t)$ is fast varying. Treating $\Lambda(t)$ as $X_1(t)$ and $X(t)$ as $X_2(t)$, and comparing (4.66) and (4.67) with the standard form (4.25), we know

$$f_1(\Lambda, X) = -\dot{X} h\big(X, \dot{X}\big),$$

$$f_2(\Lambda, X) = \pm\sqrt{2\Lambda - 2U(X)}, \tag{4.68}$$

$$g_{1l} = \dot{X} g_l\big(X, \dot{X}\big), \quad g_{2l} = 0. \tag{4.69}$$

In (4.68) and (4.69), $\dot{X}$ is treated as function of Λ and X according to (4.67). Equations (4.27) and (4.28) are then specified as

$$m(\Lambda) = m_1 = \varepsilon\left\{\langle f_1(\Lambda, X)\rangle_t + \sum_{l,s=1}^{m}\int_{-\infty}^{0}\left\langle g_{1l}[\Lambda, X(t)]\right.\right.$$
$$\left.\left.\frac{\partial}{\partial\Lambda}g_{1s}[\Lambda, X(t+\tau)]\right\rangle_t R_{ls}(\tau)\mathrm{d}\tau\right\}, \tag{4.70}$$

$$\sigma^2(\Lambda) = \left(\boldsymbol{\sigma}\boldsymbol{\sigma}^{\mathrm{T}}\right)_{11} = \varepsilon\sum_{l,s=1}^{m}\int_{-\infty}^{\infty}\langle g_{1l}[\Lambda, X(t)]g_{1s}[\Lambda, X(t+\tau)]\rangle_t R_{ls}(\tau)\mathrm{d}\tau. \tag{4.71}$$

The time-averaging is then calculated as

$$\langle[\cdot]\rangle_t = \frac{1}{T}\int_0^T[\cdot]\mathrm{d}t = \frac{1}{T}\int_0^T\frac{[\cdot]}{\dot{X}}\mathrm{d}X = \frac{1}{T_{1/4}}\int_0^A\frac{[\cdot]_{\dot{X}=\sqrt{2\Lambda-2U(X)}}}{\sqrt{2\Lambda-2U(X)}}\mathrm{d}X. \tag{4.72}$$

When evaluating the integration, the integration limit A is a function of the energy level Λ according to Eq. (4.59). Given the specific functions of h and g_l, (4.70) and (4.71) can be calculated either analytically or numerically.

The above procedure of stochastic averaging of energy envelope is also known as the quasi-conservative averaging (Landa and Stratonovich 1962; Khasminskii 1964). The stationary PDF of $\Lambda(t)$ is obtained as

$$p(\lambda) = \frac{C}{\sigma^2(\lambda)}\exp\left[\int\frac{2m(\lambda)}{\sigma^2(\lambda)}\mathrm{d}\lambda\right]. \tag{4.73}$$

The joint stationary PDF of $X(t)$ and $\dot{X}(t)$ can be calculated following the similar procedure as in Sect. 4.2.1. For the present situation, Eq. (4.50) is changed to

$$p(x, \dot{x}) = \frac{1}{2}\sqrt{2\lambda - 2U(x)}p(x, \lambda) = \frac{1}{2}\sqrt{2\lambda - 2U(x)}p(x|\lambda)p(\lambda), \tag{4.74}$$

where $p(x|\lambda)$ is the conditional probability density. For a fixed energy level λ, the probability of $X(t)$ in a region near x is inversely proportional to the velocity; therefore, $p(x|\lambda)$ can be expressed as

$$p(x|\lambda) = \frac{C}{|\dot{x}|} = \frac{C}{\sqrt{2\lambda - 2U(x)}}. \tag{4.75}$$

Integrating both sides of (4.75) with respect to x from $-a$ to a, we obtain

$$C(\lambda) = \left[\int_{-a}^{a}\frac{\mathrm{d}x}{\sqrt{2\lambda - 2U(x)}}\right]^{-1} = \frac{2}{T(\lambda)}. \tag{4.76}$$

Substituting (4.76) into (4.75), and then into (4.74), we obtain

$$p(x, \dot{x}) = \frac{p(\lambda)}{T(\lambda)}; \quad \lambda = \frac{1}{2}\dot{x}^2 + U(x). \tag{4.77}$$

4.3 Systems Under Gaussian White Noise Excitations

In the case of Gaussian white noise excitations, the first procedure of the stochastic averaging, i.e., to approximate broad-band excitations as Gaussian white noises, is not needed. Only time averaging is performed with the objective to reduce the system dimension. Note that the correlation functions of Gaussian white noises $W_{gl}(t)$ are delta functions, that is,

$$E\left[W_{gl}(u)W_{gs}(v)\right] = R_{ls}(v-u) = 2\pi K_{ls}\delta(v-u). \tag{4.78}$$

4.3.1 Linear Restoring Force

Consider system (4.32) with excitations $\xi_l(t)$ replaced by Gaussian white noises $W_{gl}(t)$. Substitute (4.78), (4.40) and (4.41) into (4.43) and (4.44) to yield

$$\begin{aligned} m(A) &= m_1 \\ &= \varepsilon\left\{\langle f_1(A,\Phi)\rangle_t + \pi\sum_{l,s=1}^{m} K_{ls}\left\langle g_{1l}(A,\Phi)\frac{\partial}{\partial A}g_{1s}(A,\Phi) + g_{2l}(A,\Phi)\frac{\partial}{\partial \Phi}g_{1s}(A,\Phi)\right\rangle_t\right\}, \end{aligned} \tag{4.79}$$

$$\sigma^2(A) = \varepsilon 2\pi \sum_{l,s=1}^{m} K_{ls}\langle g_{1l}(A,\Phi)g_{1s}(A,\Phi)\rangle_t. \tag{4.80}$$

It is known that the second term on the right-hand side of (4.79) is the well-known Wong-Zakai correction term.

For the present case of Gaussian white noise excitations, the averaged Itô equation for the amplitude $A(t)$ can be calculated using another way. According to the procedure described in Sect. 2.3.5, the equivalent Itô equations from system (4.32) can be derived as

$$
\begin{aligned}
\mathrm{d}X_1 &= X_2 \mathrm{d}t, \\
\mathrm{d}X_2 &= \left\{ -\varepsilon h(X_1, X_2) - \omega_0^2 X_1 + \varepsilon\pi \sum_{l,s=1}^{m} K_{ls} g_l(X_1, X_2) \frac{\partial}{\partial x_2} g_s(X_1, X_2) \right\} \mathrm{d}t \\
&\quad + \varepsilon^{1/2} \sqrt{2\pi \sum_{l,s=1}^{m} K_{ls} g_l(X_1, X_2) g_s(X_1, X_2)}\, \mathrm{d}B(t). \qquad (4.81)
\end{aligned}
$$

$A(t)$ can be expressed as

$$
A(t) = \sqrt{X^2 + \frac{\dot{X}^2}{\omega_0^2}}. \qquad (4.82)
$$

Applying the Itô differential rule (2.111) to (4.82) to obtain an Itô equation for amplitude $A(t)$ as

$$
\mathrm{d}A = \frac{\partial A}{\partial X_1} \mathrm{d}X_1 + \frac{\partial A}{\partial X_2} \mathrm{d}X_2 + \frac{1}{2} \frac{\partial^2 A}{\partial X_2^2} (\mathrm{d}X_2)^2. \qquad (4.83)
$$

The following partial derivatives are obtained from transformation (4.33),

$$
\frac{\partial A}{\partial X_1} = \cos\Phi, \quad \frac{\partial A}{\partial X_2} = -\frac{\sin\Phi}{\omega_0}, \quad \frac{\partial^2 A}{\partial X_2^2} = \frac{\cos^2\Phi}{A\omega_0^2}. \qquad (4.84)
$$

Substituting (4.84) and (4.81) into (4.83), and applying the time-averaging (4.42) for the drift and diffusion coefficients, we can obtain a smoothed version of the Itô equation for $A(t)$.

Example 4.1 Consider the nonlinear stochastic system,

$$
\ddot{X} + \alpha\dot{X} + \beta X^2 \dot{X} + \gamma \dot{X}^3 + \omega_0^2 X = X W_{g1}(t) + \dot{X} W_{g2}(t) + W_{g3}(t), \qquad (4.85)
$$

where $W_{g1}(t)$, $W_{g2}(t)$, and $W_{g3}(t)$ are independent Gaussian white noises with spectral densities K_{ii} $(i = 1, 2, 3)$. It is assumed that the damping and excitations are small. We will use the two procedures described above to derive the smoothed Itô equation for the amplitude $A(t)$.

Using the transformation of (4.33), we obtain the equations of $A(t)$ and $\Phi(t)$ according to (4.38) and (4.39),

$$
\begin{aligned}
\dot{A} &= -\alpha A \sin^2\Phi - \beta A^3 \cos^2\Phi \sin^2\Phi - \gamma\omega_0^2 A^3 \sin^4\Phi \\
&\quad - \frac{1}{\omega_0} A \sin\Phi \cos\Phi W_{g1}(t) + A \sin^2\Phi W_{g2}(t) - \frac{1}{\omega_0} \sin\Phi W_{g3}(t), \qquad (4.86)
\end{aligned}
$$

$$\dot{\Theta} = -\alpha \sin\Phi \cos\Phi - \beta A^2 \cos^3\Phi \sin\Phi - \gamma\omega_0^2 A^2 \sin^3\Phi \cos\Phi$$
$$-\frac{1}{\omega_0}\cos^2\Phi W_{g1}(t) + \sin\Phi\cos\Phi W_{g2}(t) - \frac{1}{\omega_0 A}\cos\Phi W_{g3}(t). \quad (4.87)$$

Also from (4.40) and (4.41) we have

$$\begin{aligned}
f_1 &= -\alpha A \sin^2\Phi - \beta A^3 \cos^2\Phi \sin^2\Phi - \gamma\omega_0^2 A^3 \sin^4\Phi, \\
f_2 &= -\alpha \sin\Phi\cos\Phi - \beta A^2\cos^3\Phi\sin\Phi - \gamma\omega_0^2 A^2 \sin^3\Phi\cos\Phi, \\
g_{11} &= -\frac{1}{\omega_0} A \sin\Phi\cos\Phi, \quad g_{12} = A\sin^2\Phi, \quad g_{13} = -\frac{1}{\omega_0}\sin\Phi, \\
g_{21} &= -\frac{1}{\omega_0}\cos^2\Phi, \quad g_{22} = \sin\Phi\cos\Phi, \quad g_{23} = -\frac{1}{A\omega_0}\cos\Phi.
\end{aligned} \quad (4.88)$$

Substituting (4.88) into (4.79) and (4.80) and carrying out the time-averaging, we obtain

$$m(A) = \left(-\frac{\alpha}{2} + \frac{3\pi}{8\omega_0^2}K_{11} + \frac{5\pi}{8}K_{22}\right)A - \frac{1}{8}(\beta + 3\gamma\omega_0^2)A^3 + \frac{\pi}{2A\omega_0^2}K_{33}, \quad (4.89)$$

$$\sigma^2(A) = \frac{\pi}{4\omega_0^2}(K_{11} + 3\omega_0^2 K_{22})A^2 + \frac{\pi}{\omega_0^2}K_{33}. \quad (4.90)$$

In the above time-averaging over a quasi-period, the amplitude A is considered to be a constant, and the smoothed drift and diffusion coefficients are then functions of the amplitude. Now the amplitude process $A(t)$ itself is a Markov diffusion process, governed by the Itô equation

$$\mathrm{d}A = \left[\left(-\frac{\alpha}{2} + \frac{3\pi}{8\omega_0^2}K_{11} + \frac{5\pi}{8}K_{22}\right)A - \frac{1}{8}(\beta + 3\gamma\omega_0^2)A^3 + \frac{\pi}{2A\omega_0^2}K_{33}\right]\mathrm{d}t$$
$$+\left[\frac{\pi}{4\omega_0^2}(K_{11} + 3\omega_0^2 K_{22})A^2 + \frac{\pi}{\omega_0^2}K_{33}\right]^{1/2}\mathrm{d}B(t). \quad (4.91)$$

To use another procedure, letting $X = X_1$ and $\dot{X} = X_2$, Eq. (4.85) is replaced by two first-order equations

$$\dot{X}_1 = X_2, \quad (4.92)$$

$$\dot{X}_2 = -\omega_0^2 X_1 - \alpha X_2 - \beta X_1^2 X_2 - \gamma X_2^3 + X_1 W_{g1}(t) + X_2 W_{g2}(t) + W_{g3}(t), \quad (4.93)$$

which can be converted to Itô differential equations

$$dX_1 = X_2 dt, \tag{4.94}$$

$$\begin{aligned} dX_2 = &\left[-\omega_0^2 X_1 - \alpha X_2 - \beta X_1^2 X_2 - \gamma X_2^3 + \pi K_{22} X_2\right] dt \\ &+ \left[2\pi\left(K_{11} X_1^2 + K_{22} X_2^2 + K_{33}\right]^{1/2} dB(t). \end{aligned} \tag{4.95}$$

Applying (4.83) to the present case and using (4.84), we obtain an Itô equation for amplitude $A(t)$,

$$dA = m(A, \Phi)dt + \sigma(A, \Phi)dB(t), \tag{4.96}$$

where

$$\begin{aligned} m(A, \Phi) = &-\alpha A \sin^2\Phi - \beta A^3 \cos^2\Phi \sin^2\Phi - \gamma\omega_0^2 A^3 \sin^4\Phi \\ &+ \pi K_{22} A\left(\sin^2\Phi + \sin^2\Phi \cos^2\Phi\right) \\ &+ \frac{1}{\omega_0^2}\pi K_{11} A \cos^4\Phi + \frac{1}{A\omega_0^2}\pi K_{33}\cos^2\Phi, \end{aligned} \tag{4.97}$$

$$\begin{aligned} \sigma^2(A, \Phi) = &\frac{1}{\omega_0^2}\pi K_{11} A \cos^4\Phi + \pi K_{22} A \sin^2\Phi \cos^2\Phi \\ &+ \frac{1}{A\omega_0^2}\pi K_{33}\cos^2\Phi. \end{aligned} \tag{4.98}$$

Equations (4.96) through (4.98) are exact governing equations for system (4.32). Applying the time-averaging to the drift and diffusion coefficients given in (4.97) and (4.98), we obtain a smoothed Itô equation

$$dA = m(A)dt + \sigma(A)dB(t), \tag{4.99}$$

where $m(A)$ and $\sigma(A)$ are the same as those in (4.89) and (4.90). Note that Eq. (4.99) is an approximation of the original system.

The stationary PDF of $A(t)$ is calculated as (Cai and Zhu 2016)

$$\begin{aligned} p(a) = &\frac{C}{\sigma^2(a)}\exp\left[\int \frac{2m(a)}{\sigma^2(a)} da\right] = Ca\left[\left(K_{11} + 3\omega_0^2 K_{22}\right)a^2 + 4K_{33}\right]^{\delta} \times \\ &\exp\left[-\frac{\omega_0^2\left(\beta + 3\omega_0^2\gamma\right)}{2\pi\left(K_{11} + 3\omega_0^2 K_{22}\right)} a^2\right], \end{aligned} \tag{4.100}$$

where

$$\delta = \frac{2\omega_0^2\left[K_{33}\left(\beta + 3\omega_0^2\gamma\right) - \left(\alpha + \pi K_{22}\right)\left(K_{11} + 3\omega_0^2 K_{22}\right)\right]}{\pi\left(K_{11} + 3\omega_0^2 K_{22}\right)^2}. \tag{4.101}$$

It is of interest to note that the nonlinear damping is strong enough to bring the system back when the response is very large ($a \to \infty$). On the other hand, due to the existence of the external excitation, the system motion will be repulsed from the left boundary $a = 0$. Therefore, a nontrivial stationary PDF exists. However, if the external excitation is absent, the linear damping plays the crucial role near $a = 0$. If it is weak enough to meet the condition

$$\alpha < \frac{\pi}{2}\left(\frac{K_{11}}{\omega_0^2} + K_{22}\right), \tag{4.102}$$

it can be shown that a non-trivial stationary PDF $p(a)$ exists based on the integrability of $p(a)$ or the nature of the boundaries (Lin and Cai 1995). If condition (4.102) is not met, a non-trivial stationary PDF does not exist.

The joint PDF of $X_1(t)$ and $X_2(t)$ can be obtained from (4.100) according to (4.53) as

$$\begin{aligned} p(x_1, x_2) = C_1\left[\left(K_{11} + 3\omega_0^2 K_{22}\right)\left(x_1^2 + \frac{x_2^2}{\omega_0^2}\right) + 4K_{33}\right]^{\delta} \times \\ \exp\left[-\frac{\omega_0^2(\beta + 3\omega_0^2\gamma)}{2\pi(K_{11} + 3K_{22})}\left(x_1^2 + \frac{x_2^2}{\omega_0^2}\right)\right]. \end{aligned} \tag{4.103}$$

In this example, the excitations are Gaussian white noises; therefore, the procedure to approximate non-white noises by white noises is not needed. The main task here is to identify the slowly varying variable and carry out the time-averaging. It is noted that an assumption is made for the transformation (4.33) that the system linear restoring force is not changed due to the Wong-Zakai correction. If the Wong-Zakai correction generates an additional restoring force (Cai and Zhu 2016), the ω_0 in the transformation (4.33) needs to be modified accordingly.

4.3.2 *Nonlinear Restoring Force*

Consider system (4.54) with excitations $\xi_l(t)$ replaced by Gaussian white noises $W_{gl}(t)$. Substitute (4.78), (4.68) and (4.69) into (4.70) and (4.71) to yield

$$m(\Lambda) = -\varepsilon\left\langle \dot{X} h(X, \dot{X})\right\rangle_t + \varepsilon\pi\left\langle \sum_{l,s=1}^{m} K_{ls}\left[\dot{X} g_l(X, \dot{X})\right]\frac{\partial}{\partial \Lambda}\left[\dot{X} g_s(X, \dot{X})\right]\right\rangle_t, \tag{4.104}$$

$$\sigma^2(\Lambda) = \varepsilon 2\pi\left\langle \sum_{l,s=1}^{m} K_{ls}\dot{X}^2 g_l(X, \dot{X}) g_s(X, \dot{X})\right\rangle_t. \tag{4.105}$$

The time-averaging is calculated according to (4.72).

Similar to Sect. 4.3.1, another procedure can be used to derive the averaged Itô equation for the energy process by using Itô differential rule. First write corresponding Itô equations for system (4.54) for the case of $\xi_l(t) = W_{gl}(t)$,

$$\begin{aligned} dX_1 &= X_2 dt, \\ dX_2 &= \left\{ -\varepsilon h(X_1, X_2) - u(X_1) + \pi \sum_{l,s=1}^{m} K_{ls} g_l(X_1, X_2) \frac{\partial}{\partial X_2} g_S(X_1, X_2) \right\} dt \\ &\quad + \varepsilon^{1/2} \sqrt{2\pi \sum_{l,s=1}^{m} K_{ls} g_l(X_1, X_2) g_S(X_1, X_2)}\; dB(t). \end{aligned} \tag{4.106}$$

From (4.60), use the Itô differential rule for energy function $\Lambda(X_1, X_2)$ to yield

$$d\Lambda = \frac{\partial \Lambda}{\partial X_1} dX_1 + \frac{\partial \Lambda}{\partial X_2} dX_2 + \frac{1}{2}\frac{\partial^2 \Lambda}{\partial X_2^2}(dX_2)^2. \tag{4.107}$$

Note that the following partial derivatives are obtained from (4.57)

$$\frac{\partial \Lambda}{\partial X_1} = u(X_1), \quad \frac{\partial \Lambda}{\partial X_2} = X_2, \quad \frac{\partial^2 \Lambda}{\partial X_2^2} = 1. \tag{4.108}$$

Substituting (4.60) and (4.62) into (4.61), we have

$$\begin{aligned} d\Lambda &= \varepsilon \left[-X_2 h + \pi X_2 \sum_{l,s=1}^{m} K_{ls} g_l \frac{\partial g_S}{\partial X_2} + \sum_{l,s=1}^{m} \pi K_{ls} g_l g_S \right] dt \\ &\quad + \varepsilon^{1/2} X_2 \sqrt{\sum_{l,s=1}^{m} 2\pi K_{ls} g_l g_S}\, dB(t). \end{aligned} \tag{4.109}$$

Since

$$\frac{\partial (X_2 g_S)}{\partial \Lambda} = \frac{1}{X_2}\frac{\partial (X_2 g_S)}{\partial X_2} = \frac{1}{X_2} g_s + \frac{\partial g_S}{\partial X_2}, \tag{4.110}$$

Equation (4.109) is consistent with Eqs. (4.104) and (4.105). Applying the time-averaging (4.72) for the drift and diffusion coefficients, we can obtain a smoothed version of the Itô equation.

Example 4.2 Consider the nonlinear system

$$\ddot{X} + \alpha \dot{X} + \beta \dot{X}^3 + \delta X^3 = W_g(t), \quad \beta, \delta > 0, \tag{4.111}$$

The equation of the corresponding undamped free vibration is

$$\ddot{x} + \delta x^3 = 0, \tag{4.112}$$

with the potential energy and total energy

$$U = \frac{1}{4}\delta x^4, \quad \Lambda = \frac{1}{2}\dot{x}^2 + \frac{1}{4}\delta x^4, \tag{4.113}$$

and a period

$$T = 4T_{1/4} = 4\int_0^A \frac{\mathrm{d}x}{\sqrt{2\Lambda - \frac{1}{2}\delta x^4}}, \tag{4.114}$$

where the integration limit A is determined from $A = (4\Lambda/\delta)^{1/4}$. Changing the integration variable according to

$$x = \left(\frac{4\Lambda}{\delta}\sin^2\theta\right)^{1/4}, \tag{4.115}$$

we have

$$T_{1/4} = \frac{1}{4}\delta^{-1/4}\Lambda^{-1/4}\mathrm{B}\left(\frac{1}{4}, \frac{1}{2}\right), \tag{4.116}$$

where $\mathrm{B}(\cdot,\cdot)$ is the Beta function. In calculating the integral, the following formula has been used (Gradshteyn and Ryzhik 1980)

$$\int_0^{\pi/2} \sin^{m-1}\theta\cos^{n-1}\theta\mathrm{d}\theta = \frac{1}{2}\mathrm{B}\left(\frac{m}{2}, \frac{n}{2}\right). \tag{4.117}$$

From (4.104) and (4.105), we obtain

$$m(\Lambda) = -\alpha\langle\dot{X}^2\rangle_t - \beta\langle\dot{X}^4\rangle_t + \pi\left\langle K\dot{X}\frac{\partial\dot{X}}{\partial\Lambda}\right\rangle_t, \tag{4.118}$$

$$\sigma^2(\Lambda) = 2\pi K\langle\dot{X}^2\rangle_t. \tag{4.119}$$

Evaluating the time-average terms in (4.118) and (4.119) according to (4.73) and noting that $\partial\dot{X}/\partial\Lambda = 1/\dot{X}$ due to (4.108), we obtain

$$\langle \dot{X}^2 \rangle_t = \frac{1}{T_{1/4}} \int_0^A [2\Lambda - \frac{1}{2}\delta X^4]^{1/2} \mathrm{d}X = \frac{4}{3}\Lambda, \tag{4.120}$$

$$\langle \dot{X}^4 \rangle_t = \frac{1}{T_{1/4}} \int_0^A [2\Lambda - \frac{1}{2}\delta X^4]^{3/2} \mathrm{d}X = \frac{32}{7}\Lambda^2, \tag{4.121}$$

$$\left\langle \dot{X} \frac{\partial \dot{X}}{\partial \Lambda} \right\rangle_t = \frac{1}{T_{1/4}} \int_0^A [2\Lambda - \frac{1}{2}\delta X^4]^{-1/2} \mathrm{d}X = 1. \tag{4.122}$$

Thus the smoothed Itô equation for the energy process $\Lambda(t)$ is

$$\mathrm{d}\Lambda = \left(-\frac{4}{3}\alpha\Lambda - \frac{32}{7}\beta\Lambda^2 + \pi K \right)\mathrm{d}t + \sqrt{\frac{8\pi K}{3}\Lambda}\,\mathrm{d}B(t). \tag{4.123}$$

The stationary PDF of $\Lambda(t)$ can be calculated according to (4.119) as

$$p(\lambda) = C\lambda^{-1/4} \exp\left[-\frac{1}{\pi K}\left(\alpha\lambda + \frac{12}{7}\beta\lambda^2 \right) \right]. \tag{4.124}$$

Using Eq. (4.77), we obtain the stationary PDF for $p(x, \dot{x})$ as

$$p(x, \dot{x}) = \frac{p(\lambda)}{T(\lambda)} = C \exp\left\{ -\frac{1}{2\pi K}\left[\alpha\left(\frac{1}{2}\delta x^4 + \dot{x}^2 \right) + \frac{6\beta}{7}\left(\frac{1}{2}\delta x^4 + \dot{x}^2 \right)^2 \right] \right\}. \tag{4.125}$$

4.4 Systems Under Broad-Band Random Excitations

When excitations are not Gaussian white noises but broad-band processes, the two procedures in the stochastic averaging will be carried out, i.e., to approximate the broad-band processes as Gaussian white noises and to perform the time averaging. The drift and diffusion coefficients in the smoothed Itô Eqs. (4.26) for the slowly varying processes will be determined from Eqs. (4.27) and (4.28), and the calculations are more complicated. If the restoring force of the system is linear, the stochastic averaging of amplitude envelope is the most effective, as described in Sect. 4.2.1. The method is applied to a primary-secondary system in Sect. 4.4.2. However, if the restoring force is nonlinear, the stochastic averaging is not straightforward. Several procedures are proposed for different types of systems (Cai 1995; Dimentberg et al. 1995; Cai and Lin 2001b; Zhu et al. 2001), as described in Sects. 4.4.3–4.4.5 and 8.1.

4.4.1 Linear Restoring Force

If the restoring force is linear, the stochastic averaging of amplitude envelope described in Sect. 4.2.1 can be applied. The procedure is illustrated in the following example.

Example 4.3 Consider a system governed by

$$\ddot{X} + 2\zeta\omega_0\dot{X} + \omega_0^2[1 + \xi_1(t)]X = \xi_2(t), \tag{4.126}$$

where $\xi_1(t)$ and $\xi_2(t)$ are two stationary broadband processes with spectral densities $S_{ij}(\omega)$, $i, j = 1, 2$. Equation (4.126) can be used to describe the first-mode motion of a column subjected to both random axial and transverse loads (Lin and Cai 1995). Using the transformation of (4.33), i.e.,

$$X = A\cos\Phi, \quad \dot{X} = -A\omega_0\sin\Phi, \quad \Phi = \omega_0 t + \Theta, \tag{4.127}$$

we obtain the equations of $A(t)$ and $\Theta(t)$ according to (4.38) and (4.39),

$$\dot{A} = -2\zeta\omega_0 A\sin^2\Phi + \omega_0 A\sin\Phi\cos\Phi\xi_1(t) - \frac{1}{\omega_0}\sin\Phi\xi_2(t), \tag{4.128}$$

$$\dot{\Theta} = -2\zeta\omega_0\sin\Phi\cos\Phi + \omega_0\cos^2\Phi\xi_1(t) - \frac{1}{A\omega_0}\cos\Phi\xi_2(t). \tag{4.129}$$

We assume that the correlation times of $\xi_1(t)$ and $\xi_2(t)$ are much shorter than the system relaxation time which is of the order $(\zeta\omega_0)^{-1}$. Furthermore, we assume that the damping is small and the excitations are weak. Under these assumptions, both $A(t)$ and $\Theta(t)$ are slowly varying, and the stochastic averaging is applicable.

Carrying out the time-averaging in (4.43) and (4.44), we obtain

$$\begin{aligned} m(A) &= -\zeta\omega_0 A + \int_{-\infty}^{0}\left[\frac{3}{8}\omega_0^2 A\cos(2\omega_0\tau)R_{11}(\tau) + \frac{1}{2\omega_0^2 A}\cos(\omega_0\tau)R_{22}(\tau)\right]\mathrm{d}\tau \\ &= -\left[\zeta\omega_0 - \frac{3\pi}{8}\omega_0^2 S_{11}(2\omega_0)\right]A + \frac{\pi}{2\omega_0^2 A}S_{22}(\omega_0), \end{aligned} \tag{4.130}$$

$$\begin{aligned} \sigma^2(A) &= \int_{-\infty}^{\infty}\left[\frac{1}{8}\omega_0^2 A^2\cos(2\omega_0\tau)R_{11}(\tau) + \frac{1}{2\omega_0^2}\cos(\omega_0\tau)R_{22}(\tau)\right]\mathrm{d}\tau \\ &= \frac{\pi}{4}\omega_0^2 S_{11}(2\omega_0)A^2 + \frac{\pi}{\omega_0^2}S_{22}(\omega_0), \end{aligned} \tag{4.131}$$

where use has been made of the relationship between the correlation and spectral density functions, i.e.

$$\int_{-\infty}^{\infty} R_{ij}(\tau)e^{-i\omega\tau}\mathrm{d}\tau = 2\pi S_{ij}(\omega). \tag{4.132}$$

In the above time-averaging over a quasi-period, the amplitude A is considered to be a constant, and the smoothed drift and diffusion coefficients are then functions of the amplitude. Now the amplitude process $A(t)$ itself is a Markov diffusion process, governed by the Itô equation

$$\begin{aligned} \mathrm{d}A = &\left\{\left[-\zeta\omega_0 + \frac{3\pi}{8}\omega_0^2 S_{11}(2\omega_0)\right]A + \frac{\pi}{2\omega_0^2 A}S_{22}(\omega_0)\right\}\mathrm{d}t \\ &+ \left[\frac{\pi}{4}\omega_0^2 S_{11}(2\omega_0)A^2 + \frac{\pi}{\omega_0^2}S_{22}(\omega_0)\right]^{1/2}\mathrm{d}B(t). \end{aligned} \tag{4.133}$$

Equation (4.133) shows that the system behavior depends on the spectrum value of the parametric excitation $\xi_1(t)$ at frequency $2\omega_0$, and the spectrum value of the external excitation $\xi_2(t)$ at frequency ω_0. It is noted that the correlation of $\xi_1(t)$ and $\xi_2(t)$ has no effect on the system response.

The stationary PDF of the amplitude $A(t)$ is obtained as

$$p(a) = \frac{C}{\sigma^2(a)}\exp\left[\int \frac{2m(a)}{\sigma^2(a)}\mathrm{d}a\right] = Ca(a^2 + D)^{-\delta}, \tag{4.134}$$

where

$$D = \frac{4S_{22}(\omega_0)}{\omega_0^4 S_{11}(2\omega_0)}, \quad \delta = \frac{4\zeta}{\pi\omega_0 S_{11}(2\omega_0)}, \quad C = 2(\delta - 1)D^{\delta-1}. \tag{4.135}$$

The integrability of the PDF $p(a)$ requires $\delta > 1$, namely,

$$\zeta > \frac{\pi}{4}\omega_0 S_{11}(2\omega_0). \tag{4.136}$$

The joint stationary PDF of $X(t)$ and $\dot{X}(t)$ can be obtained according to Eq. (4.53).

It is noticed that the system response only depends on the spectral density of the external excitation $\xi_2(t)$ at the system natural frequency ω_0, and that of the parametric excitation $\xi_1(t)$ at $2\omega_0$. This can be understood according to the nature of the stochastic averaging. In the procedure, the correlation times of the excitations are required to be much shorter than the system relaxation time, and the spectra to be much broader than ω_0. Thus, the spectrum value remains almost unchanged near ω_0

or $2\omega_0$, and $S_{11}(2\omega_0)$ and $S_{22}(\omega_0)$ represent the spectrum values near $2\omega_0$ and ω_0, respectively. The significance of $S_{22}(\omega_0)$ is obvious since the system response function has a sharp peak near ω_0 due to small damping. That $S_{11}(2\omega_0)$ plays an important role also agrees with the deterministic analysis if $\xi_1(t)$ were to be replaced by a sinusoidal excitation, system equation would become a damped Mathieu equation, and $2\omega_0$ would be the primary resonant frequency (Lin and Cai 1995).

We digress to remark that if the random parametric excitation were to be multiplied to the velocity $\dot{X}$ rather than the displacement X, the spectrum value at $\omega = 0$ is also involved (Cai and Zhu 2016).

If the parametric excitation is absent, the stationary PDFs are obtained as

$$p(a) = \frac{2\zeta\omega_0^3 a}{\pi S_{22}(\omega_0)} \exp\left[-\frac{\zeta\omega_0^3}{\pi S_{22}(\omega_0)} a^2\right], \tag{4.137}$$

$$p(x, \dot{x}) = \frac{\zeta\omega_0^2}{\pi^2 S_{22}(\omega_0)} \exp\left[-\frac{\zeta\omega_0^3}{\pi S_{22}(\omega_0)}\left(x^2 + \frac{\dot{x}^2}{\omega_0^2}\right)\right]. \tag{4.138}$$

Equations (4.137) and (4.138) indicate that, for a linear oscillator under an external broad-band excitation, the amplitude $A(t)$ is Rayleigh-distributed, and the displacement $X(t)$ and velocity $\dot{X}$ are jointly and marginally Gaussian-distributed, which are the well-known results.

If the external excitation $\xi_2(t)$ is absent, i.e. $D = 0$, it is known from the non-integrability of the PDF (4.134) that a non-trivial stationary PDF does not exist. In this case, the response will be attracted to zero if the damping is strong, while it will grow boundless if the damping is not strong enough.

4.4.2 A Primary-Secondary System

The procedure in Sect. 4.4.1 will be applied to a particular system, referred as the primary-secondary system. The primary system is the main system subjected to a broad-band stochastic excitation. Another system connected the primary system but not directly under stochastic excitation, will be referred as the secondary system. Figure 4.1 show such a primary-secondary system. It indicates that only the damping of the primary system is nonlinear. The governing equations are

$$m_1\ddot{X} + g\big(X, \dot{X}\big) + c_2\dot{X} + (k_1 + k_2)X - c_2\dot{Y} - k_2Y = F(t), \tag{4.139}$$

$$m_2\ddot{Y} + c_2\dot{Y} + k_2Y - c_2\dot{X} - k_2X = 0, \tag{4.140}$$

where $g\big(X, \dot{X}\big)$ is a nonlinear damping force, and $F(t)$ is a broad-band excitation. Letting

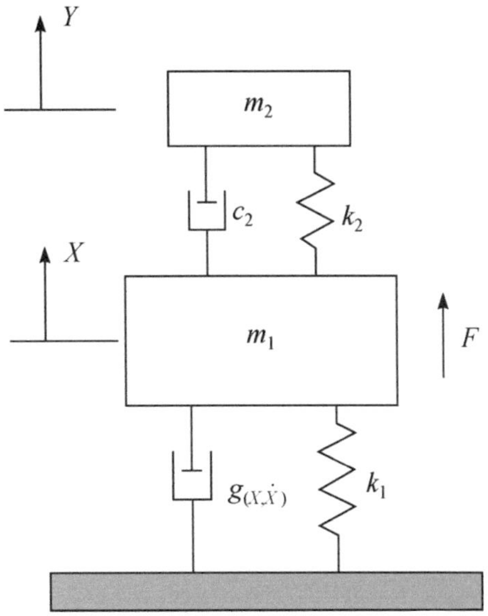

Fig. 4.1 A primary-secondary system

$$\begin{aligned} &\omega_1^2 = \frac{k_1}{m_1}, \quad \omega_2^2 = \frac{k_2}{m_2}, \quad h_1(X, \dot{X}) = \frac{g(X, \dot{X})}{m_1}, \\ &\zeta_2 = \frac{c_2}{2m_2\omega_2}, \quad \eta = \frac{m_2}{m_1}, \quad \xi_1(t) = \frac{F(t)}{m_1}. \end{aligned} \tag{4.141}$$

Equations (4.139) and (4.140) are recast into

$$\ddot{X} + h_1(X, \dot{X}) + 2\eta\zeta_2\omega_2\dot{X} + (\omega_1^2 + \eta\omega_2^2)X - 2\eta\zeta_2\omega_2\dot{Y} - \eta\omega_2^2 Y = \xi_1(t), \tag{4.142}$$

$$\ddot{Y} + 2\zeta_2\omega_2\dot{Y} + \omega_2^2 Y - 2\zeta_2\omega_2\dot{X} - \omega_2^2 X = 0. \tag{4.143}$$

Note that ω_1 and ω_2 would be the natural frequencies of the primary and secondary systems respectively, if they were uncoupled. It is expected that, in a good design, ω_1 and ω_2 are not near each other to avoid resonance. In such a case, the motion of the primary system is still dominated by a single frequency, although it may be very much different from ω_1 due to the effect of the secondary system.

Equation (4.142) may be re-written as follows, using Eq. (4.143)

$$\ddot{X} + h_1(X, \dot{X}) + \omega_1^2 X + \eta\ddot{Y} = \xi_1(t). \tag{4.144}$$

The effects of the secondary system are lumped in one term $\eta\ddot{Y}$ in (4.144), which may contribute to modify mass, damping, and stiffness of the primary system. To take these effects separately into consideration, Eq. (4.144) is recast again into the form

$$\ddot{X} + h\big(X, \dot{X}\big) + \omega_0^2 X = \xi(t), \tag{4.145}$$

where $h\big(X, \dot{X}\big)$ is the total damping, ω_0 is the modified frequency, and $\xi(t)$ is the adjusted excitation due to the mass effect. It is implied in (4.145) that the effect on the stiffness is linear. Now we assume that the damping and excitation are weak, and the stochastic averaging is applicable. Substituting the transformation (4.127) into (4.143) yields

$$\ddot{Y} + 2\zeta_2\omega_2\dot{Y} + \omega_2^2 Y = -2\zeta_2\omega_2 A\omega_0 \sin\Phi + \omega_2^2 A\cos\Phi. \tag{4.146}$$

Treating the terms of the right-hand-side as harmonic forces and neglecting the time-variation of A, the steady-state solution of (4.146) is obtained as

$$Y = \frac{1}{D}A\omega_2^2\big(\omega_2^2 - \omega_0^2 + 4\zeta_2^2\omega_0^2\big)\cos\Phi + \frac{2}{D}A\zeta_2\omega_2\omega_0^3\sin\Phi, \tag{4.147}$$

where

$$D = (\omega_2^2 - \omega_0^2)^2 + 4\zeta_2^2\omega_2^2\omega_0^2. \tag{4.148}$$

Twice differentiation of Eq. (4.147) leads to

$$\ddot{Y} = -\frac{1}{D}A\omega_2^2\omega_0^2\big(\omega_2^2 - \omega_0^2 + 4\zeta_2^2\omega_0^2\big)\cos\Phi - \frac{2}{D}A\zeta_2\omega_2\omega_0^5\sin\Phi. \tag{4.149}$$

The term associated with $\sin\Phi$ on the right-hand side of Eq. (4.149) accounts for the damping effect on the primary system, whereas the term associated with $\cos\Phi$ involves both the mass and stiffness effects. To separate the mass effect from the stiffness effect, two extreme cases are considered. One is the case of a very large ω_2, indicating that the two masses are connected rigidly. In this case, the stiffness and damping effects disappear; thus, $\ddot{Y} = -A\omega_0^2\cos\Phi = \ddot{X}$. Another is the case of very large ω_1, hence very large ω_0, associated with a weak connection between the two masses. Thus the mass effect from the secondary system is negligible, namely, $\ddot{Y} = \omega_2^2 A\cos\Phi + 2\zeta_2\omega_2(-A\omega_0\sin\Phi) = \omega_2^2 X + 2\zeta_2\omega_2\dot{X}$. In order for these two extreme situations be included as special cases, Eq. (4.149) is recast as

$$\begin{aligned}\ddot{Y} &= \frac{1}{D}\omega_2^2\big(\omega_2^2 + 4\zeta_2^2\omega_0^2\big)\big(-A\omega_0^2\cos\Phi\big) + \frac{1}{D}\omega_2^2\omega_0^4(A\cos\Phi)\\ &\quad + \frac{2}{D}\zeta_2\omega_2\omega_0^4(-A\omega_0\sin\Phi)\\ &= \frac{1}{D}\omega_2^2\big(\omega_2^2 + 4\zeta_2^2\omega_0^2\big)\ddot{X} + \frac{1}{D}\omega_2^2\omega_0^4 X + \frac{2}{D}\zeta_2\omega_2\omega_0^4\dot{X}.\end{aligned} \tag{4.150}$$

Substituting (4.150) into (4.144), and comparing with (4.145), we obtain

$$\omega_0^2 = \frac{\omega_1^2 + \frac{1}{D}\eta\omega_2^2\omega_0^2}{1 + \frac{1}{D}\eta\omega_2^2\left(\omega_2^2 + 4\zeta_2^2\omega_0^2\right)}, \tag{4.151}$$

$$h\left(X, \dot{X}\right) = \frac{h_1\left(X, \dot{X}\right) + \frac{2}{D}\eta\zeta_2\omega_2\omega_0^4}{1 + \frac{1}{D}\eta\omega_2^2\left(\omega_2^2 + 4\zeta_2^2\omega_0^2\right)}\dot{X}, \tag{4.152}$$

$$\xi(t) = \frac{1}{1 + \frac{1}{D}\eta\omega_2^2\left(\omega_2^2 + 4\zeta_2^2\omega_0^2\right)}\xi_1(t). \tag{4.153}$$

Equation (4.151) is a nonlinear equation for ω_0, (4.152) gives the total damping, and (4.153) is the excitation adjusted for the mass effect of the secondary system. In the two limiting cases, we have the exact results

$$h\left(X, \dot{X}\right) = \frac{h_1\left(X, \dot{X}\right)}{1+\eta}, \quad \omega_0^2 = \frac{\omega_1^2}{1+\eta}, \quad \xi(t) = \frac{\xi_1(t)}{1+\eta}, \quad \text{as} \quad \omega_2 \to \infty, \tag{4.154}$$

$$h\left(X, \dot{X}\right) = h_1\left(X, \dot{X}\right) + 2\eta\zeta_2\omega_2\dot{X}, \quad \omega_0^2 = \omega_1^2 + \eta\omega_2^2, \quad \xi(t) = \xi_1(t), \quad \text{as} \quad \omega_1 \to \infty. \tag{4.155}$$

Now the stochastic averaging is applied to system (4.145) using the transformation (4.127). The drift and diffusion coefficients for the amplitude process $A(t)$ are obtained as

$$m(A) = \frac{1}{\omega_0}q(A) + \frac{\pi}{2\omega_0^2 A}S(\omega_0), \tag{4.156}$$

$$\sigma^2(A) = \frac{\pi}{\omega_0^2}S(\omega_0), \tag{4.157}$$

where $S(\omega_0)$ is the value of the spectral density of $\xi(t)$ at frequency ω_0, and

$$q(A) = \frac{1}{2\pi}\int_0^{2\pi} h(A\cos\Phi, -A\omega_0\sin\Phi)\sin\Phi \mathrm{d}\Phi. \tag{4.158}$$

The stationary PDF of $A(t)$ is given by

$$p(a) = Ca\exp\left[\frac{2\omega_0}{S(\omega_0)}\int q(a)\mathrm{d}a\right], \tag{4.159}$$

and the statistical moments of $A(t)$ are governed by

$$\frac{\mathrm{d}E[A^n]}{\mathrm{d}t} = \frac{n}{\omega_0}E\left[A^{n-1}q(A)\right] + \frac{\pi n^2}{2\omega_0^2}S(\omega_0)E\left[A^{n-2}\right], \quad n \geq 2. \tag{4.160}$$

To verify the accuracy of the procedure, numerical calculations are carried out for the primary-secondary system. First we assume a linear damping, i.e., $h_1(X, \dot{X}) = 2\zeta_1\omega_1\dot{X}$ in Eq. (4.144), and a Gaussian white noise for $\xi_1(t)$ with a spectral density K, so that both the original system governed by (4.139) and (4.140) and the simplified system (4.145) can be solved exactly. The system parameter are $\eta = 0.1$, $\zeta_1 = \zeta_2 = 0.05$, $\omega_1 = 6$, and $K = 0.5$. Figures 4.2 and 4.3 depict the transient mean-square values of X and $\dot{X}$ for two different secondary systems with $\omega_2 = 3$ and 9, respectively. In the figures, the exact solutions were obtained from the original primary-secondary system, while the approximate solutions from the simplified system (4.145) using $E[X^2] = E[A^2]/2$ and $E[\dot{X}^2] = \omega_0^2 E[A^2]/2$ according to Eq. (4.127). It is shown that the approximate results are quite accurate.

Numerical calculations are also carried out for the case of nonlinear damping $h_1(X, \dot{X}) = 2\zeta_1\omega_1\dot{X} + \beta\dot{X}^3$ in Eq. (4.144). Stationary mean-square values of X are shown in Fig. 4.4 as functions of the nonlinear parameter β. Also depicted in the figures are Monte Carlo simulation results for both the original system given by

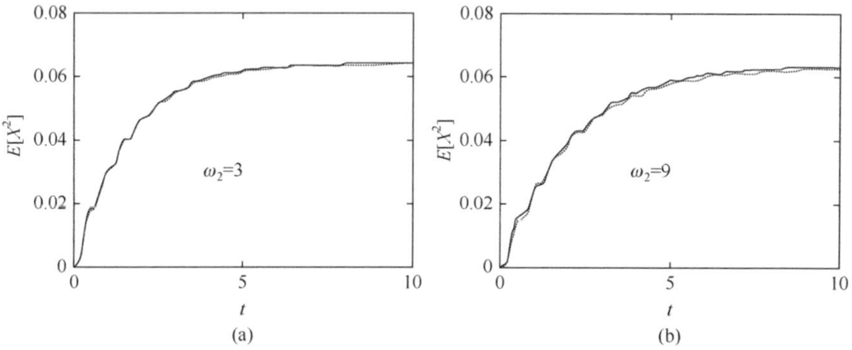

Fig. 4.2 Transient mean-square values of X of linear primary system. Solid lines: exact solutions; Dotted lines: approximate solutions (Cai and Lin 2001a)

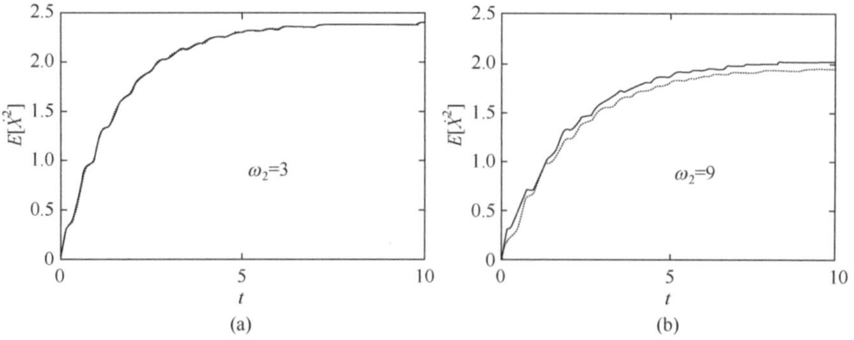

Fig. 4.3 Transient mean-square values of $\dot{X}$ of linear primary system. Solid lines: exact solutions; Dotted lines: approximate solutions (Cai and Lin 2001a)

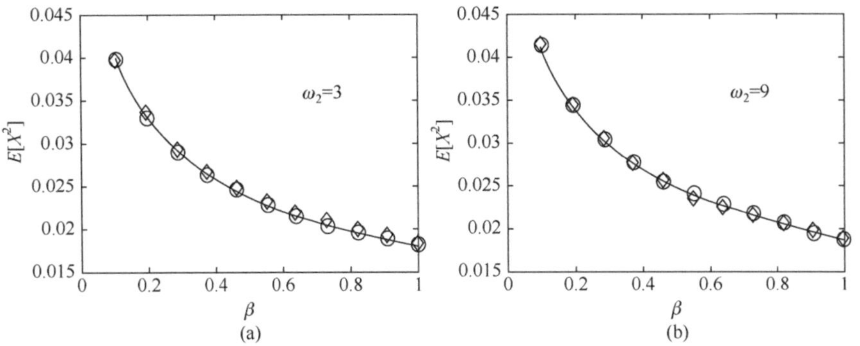

Fig. 4.4 Stationary mean-square values of X of nonlinear primary system. Solid lines: approximate solutions; Diamonds: simulation results for the original system; Circles: simulation results for the simplified system (Cai and Lin 2001a)

(4.139) and (4.140), and the simplified system (4.145), respectively. It is shown that the results from the approximation procedure are accurate even in the case of high nonlinearity.

4.4.3 Energy-Dependent White-Noise Approximation

It is seen that, for a system with linear restoring force $\omega_0^2 X$, the frequency of the undamped free motion is ω_0, which is also the dominant frequency of the motion of the system with weak damping and excitation. Thus the transformation (4.127) can be used, and the stochastic averaging of amplitude envelope can be applied, as shown in Sects. 4.4.1 and 4.4.2. However, if the restoring force is strongly nonlinear, the frequency of the undamped free motion is not fixed, but changes with energy level. Therefore, the stochastic averaging of energy envelope should be applied. Consider the case in which only an external random excitation is present, given by

$$\ddot{X} + h\big(X, \dot{X}\big) + u(X) = \xi(t), \tag{4.161}$$

where $\xi(t)$ is a stationary broad-band process with a correlation function $R(\tau)$. Assume that the damping and excitation are weak, the total energy is a slowly varying process, and the stochastic averaging is applicable. The energy process $\Lambda(t)$ defined by

$$\Lambda = \frac{1}{2}\dot{X}^2 + U(X),\ U(X) = \int_0^X u(z)\mathrm{d}z, \tag{4.162}$$

is approximately a Markov process governed by the Itô equation

$$\mathrm{d}\Lambda = m(\Lambda)\mathrm{d}t + \sigma(\Lambda)\mathrm{d}B(t), \tag{4.163}$$

where $m(\Lambda)$ and $\sigma(\Lambda)$ can be derived from (4.70) and (4.71) as follows

$$m(\Lambda) = \langle \dot{X} h(X, \dot{X}) \rangle_t + \int_{-\infty}^{0} \left\langle \dot{X}(t) \frac{\partial}{\partial \Lambda} \dot{X}(t+\tau) \right\rangle_t R(\tau)\mathrm{d}\tau, \tag{4.164}$$

$$\sigma^2(\Lambda) = \int_{-\infty}^{\infty} \langle \dot{X}(t)\dot{X}(t+\tau) \rangle_t R(\tau)\mathrm{d}\tau. \tag{4.165}$$

Since $\xi(t)$ is a broad-band process, it can be replaced approximately by a white noise with an energy-dependent intensity as follows

$$\xi(t) = \sqrt{2\pi S(\omega_\Lambda)} W_g(t), \tag{4.166}$$

where $S(\omega)$ is the spectral density of $\xi(t)$, and ω_Λ is the average circular frequency corresponding to energy level Λ given by (4.65), and $W_g(t)$ is a unit white noise with

$$E\left[W_g(t)W_g(t+\tau)\right] = \delta(\tau). \tag{4.167}$$

The correlation function of $\xi(t)$ is approximated as

$$R(\tau) = 2\pi S(\omega_\Lambda)\delta(t) \tag{4.168}$$

Equation (4.166) shows that the excitation is a white noise, but with an intensity changing with the motion frequency, hence with time. Thus, it is called the energy-dependent white noise. This procedure was proposed by Dimentberg et al. (1995)

Substitute (4.168) into (4.164) and (4.165), and apply the time averaging in (4.72) to result in

$$m(\Lambda) = -\frac{1}{T_{\frac{1}{4}}} \int_0^A h\left[X, \sqrt{2\Lambda - 2U(X)}\right]\mathrm{d}X + \pi S(\omega_\Lambda), \tag{4.169}$$

$$\sigma^2(\Lambda) = \frac{1}{T_{\frac{1}{4}}} 2\pi S(\omega_\Lambda) \int_0^A \sqrt{2\Lambda - 2U(X)}\mathrm{d}X, \tag{4.170}$$

where A is the amplitude determined from (4.59), i.e., $\Lambda = U(A)$, and $T_{1/4}$ is a quarter of the period calculated from (4.60). Since A, ω_Λ, and $T_{1/4}$ are dependent on the energy level Λ, the right-hand sides of (4.169) and (4.170) are functions of energy Λ.

The stationary PDF of the energy process $\Lambda(t)$ is obtained as

$$\begin{aligned} p(\lambda) &= \frac{C}{\sigma^2(\lambda)} \exp\left[\int \frac{2m(\lambda)}{\sigma^2(\lambda)} \mathrm{d}\lambda\right] \\ &= C_1 \frac{T_{1/4}}{S(\omega_\Lambda)} \exp\left\{-\int \frac{\int_0^a h\left[x, \sqrt{2\lambda - 2U(x)}\right] dx}{\pi S(\omega_\Lambda) \int_0^a \sqrt{2\lambda - 2U(x)} dx} \mathrm{d}\lambda\right\}. \end{aligned} \tag{4.171}$$

The approximation of replacing a broad-band excitation by an energy-dependent white noise implies that the dominant energy input from the excitation to the system is derived from the spectrum near frequency ω_Λ. This approximation is valid only if the excitation is external.

4.4.4 Fourier-Expansion Scheme

Now consider a more general system with strongly nonlinear restoring force under both external and parametric excitations, governed by

$$\ddot{X} + h\left(X, \dot{X}\right) + u(X) = \sum_{l=1}^{m} g_l(X, \dot{X}) \xi_l(t). \tag{4.172}$$

We also assume that the damping and excitation are weak. Using (4.61), i.e.,

$$\begin{aligned} \operatorname{sgn} X \sqrt{U(X)} &= \sqrt{\Lambda} \cos\Phi, \\ \dot{X} &= -\sqrt{2\Lambda} \sin\Phi, \end{aligned} \tag{4.173}$$

system (4.172) is transformed to two first-order differential equations of the energy process $\Lambda(t)$ and the phase process $\Phi(t)$ as follows

$$\dot{\Lambda} = \sqrt{2\Lambda} h(\Lambda, \Phi) \sin\Phi + \sum_{l=1}^{m} g_{1l}(\Lambda, \Phi) \xi_l(t), \tag{4.174}$$

$$\dot{\Phi} = \frac{1}{\sqrt{2\Lambda}} \left[h(\Lambda, \Phi) \cos\Phi + \frac{u(\Lambda, \Phi)}{\cos\Phi}\right] + \sum_{l=1}^{m} g_{2l}(\Lambda, \Phi) \xi_l(t), \tag{4.175}$$

where

$$\begin{aligned} g_{1l}(\Lambda, \Phi) &= -\sqrt{2\Lambda} g_l(\Lambda, \Phi) \sin\Phi, \\ g_{2l}(\Lambda, \Phi) &= -\frac{1}{\sqrt{2\Lambda}} g_l(\Lambda, \Phi) \cos\Phi. \end{aligned} \tag{4.176}$$

In (4.174)–(4.176), $h(\Lambda, \Phi)$, $u(\Lambda, \Phi)$, and $g_l(\Lambda, \Phi)$ are obtained from $h(X, \dot{X})$, $u(X)$, and $g_l(X, \dot{X})$, respectively, by expressing X and $\dot{X}$ in terms of Λ and Φ according to the transformation (4.173). Equations (4.174) and (4.175) show that the energy process $\Lambda(t)$ is slowly varying, but the phase process $\Phi(t)$ is not. By applying the time averaging, the energy process itself can be approximate as a Markov diffusion process with the drift and diffusion coefficients calculated from

$$m(\Lambda) = \left\langle \sqrt{2\Lambda} h(\Lambda, \Phi) \sin \Phi \right\rangle_t + \sum_{l,s=1}^{m} \int_{-\infty}^{0} \left\langle g_{1l}(t+\tau) \frac{\partial}{\partial \Lambda} g_{1s}(t) \right\rangle_t R_{ls}(\tau) \mathrm{d}\tau$$
$$+ \sum_{l,s=1}^{m} \int_{-\infty}^{0} \left\langle g_{2l}(t+\tau) \frac{\partial}{\partial \Phi} g_{1s}(t) \right\rangle_t R_{ls}(\tau) \mathrm{d}\tau, \tag{4.177}$$

$$\sigma^2(\Lambda) = \sum_{l,s=1}^{m} \int_{-\infty}^{\infty} \langle g_{1l}(t+\tau) g_{1s}(t) \rangle_t R_{ls}(\tau) \mathrm{d}\tau, \tag{4.178}$$

where $R_{ls}(\tau)$ is the correlation function of $\xi_l(t)$ and $\xi_s(t)$, i.e.,

$$R_{ls}(\tau) = E[\xi_l(t)\xi_s(t+\tau)]. \tag{4.179}$$

The first term of on the right-hand side of (4.177), representing the energy dissipated by the damping force per unit time, is that same as the first term of the right-hand side of (4.169), and can be calculated directly according to the time averaging (4.72). However, calculations of the second and third terms in (4.177) and the term in (4.178) are not straightforward. A scheme based on Fourier series expansion is proposed below (Cai and Lin 2001b).

It is known that the undamped free motion for a given Λ is not harmonic but periodic with an average frequency ω_Λ. Since the time averaging is based on the undamped free motion, functions $g_{ls}(t)$ and their derivatives in (4.177) and (4.178) can be expanded into Fourier series with a fundamental frequency ω_Λ, and the coefficients of the Fourier series can be calculated based on the undamped free motion. The time averaging can then be carried out to eliminate the fast varying terms, namely the terms involving $\sin(n\omega_\Lambda t)$ and $\cos(n\omega_\Lambda t)$. In the calculations, the following approximations are often made

$$\int_{-\infty}^{0} R_{ls}(\tau) \sin(n\omega_\Lambda \tau) \mathrm{d}\tau \approx 0, \quad \int_{-\infty}^{\infty} R_{ls}(\tau) \sin(n\omega_\Lambda \tau) \mathrm{d}\tau \approx 0. \tag{4.180}$$

These approximations are justified on the basis of short correlation times of the broad-band excitations.

The two methods of energy-dependent white noise approximation and Fourier expansion scheme will be illustrated in the following example.

Example 4.4 (Cai and Lin 2001b) As an example, consider the system

$$\ddot{X} + 2\zeta\omega_0\dot{X} + \omega_0^2\left(1 + \alpha X^2\right)X = \xi(t). \tag{4.181}$$

Since only an external excitation is present, both the methods of energy-dependent white noise approximation and Fourier-expansion scheme are applicable.

Compared (4.181) with the general forms (4.172), we have

$$h\left(X, \dot{X}\right) = 2\zeta\omega_0\dot{X}, \quad u(X) = \omega_0^2\left(1 + \alpha X^2\right)X, \quad g_1\left(X, \dot{X}\right) = 1. \tag{4.182}$$

Carrying out the transformation (4.173), Eq. (4.176) is specified as

$$g_{11}(\Lambda, \Phi) = -\sqrt{2\Lambda}\sin\Phi, \quad g_{21}(\Lambda, \Phi) = -\frac{1}{\sqrt{2\Lambda}}. \tag{4.183}$$

For the energy-dependent white noise approximation, (4.171) leads to

$$p(\lambda) = C_1\frac{T_{1/4}}{S(\omega_\Lambda)}\exp\left\{-\frac{2\zeta\omega_0\lambda}{\pi S(\omega_\Lambda)}\right\}. \tag{4.184}$$

For the Fourier-expansion scheme, substitution of (4.183) into (4.177) and (4.178) leads to

$$m(\Lambda) = 2\zeta\omega_0\left\langle\dot{X}^2\right\rangle_t + \int\limits_{-\infty}^{0}\langle\sin\Phi(t+\tau)\sin\Phi(t) + \cos\Phi(t+\tau)\cos\Phi(t)\rangle_t R(\tau)\mathrm{d}\tau, \tag{4.185}$$

$$\sigma^2(\Lambda) = 2\Lambda\int\limits_{-\infty}^{\infty}\langle\sin\Phi(t+\tau)\sin\Phi(t)\rangle_t R(\tau)\mathrm{d}\tau. \tag{4.186}$$

For a given Λ, X and $\dot{X}$ are treated as periodic functions of t obtained from the undamped free motion. Thus, $\sin\Phi$ and $\cos\Phi$, being functions of X and $\dot{X}$, can be expanded into Fourier series as

$$\sin\Phi(t) = \sum_{n=1}^{\infty} a_n\sin(n\omega_\Lambda t), \cos\Phi(t) = \sum_{n=1}^{\infty} b_n\cos(n\omega_\Lambda t), \tag{4.187}$$

where

$$a_n = \frac{2}{T}\int\limits_0^T \sin\Phi(t)\sin(n\omega_\Lambda t)\mathrm{d}t = \frac{2}{\sqrt{2\Lambda}T_{1/4}}\int\limits_0^{T_{1/4}}\dot{X}(t)\sin(n\omega_\Lambda t)\mathrm{d}t, \tag{4.188}$$

$$b_n = \frac{2}{T}\int_0^T \cos\Phi(t)\cos(n\omega_\Lambda t)\mathrm{d}t = \frac{2}{\sqrt{\Lambda}T_{1/4}}\int_0^{T_{1/4}} \sqrt{U[X(t)]}\cos(n\omega_\Lambda t)\mathrm{d}t. \quad (4.189)$$

Substituting (4.187) into (4.185) and (4.186), and performing the time averaging, we obtain

$$m(\Lambda) = \frac{2\zeta\omega_0}{T_{1/4}}\int_0^{T_{1/4}} \dot{X}^2(t)\mathrm{d}t + \frac{\pi}{2}\sum_{n=1}^{\infty}(a_n^2 + b_n^2)S(n\omega_\Lambda), \quad (4.190)$$

$$\sigma^2(\Lambda) = 2\pi\Lambda\sum_{n=1}^{\infty} a_n^2 S(n\omega_\Lambda). \quad (4.191)$$

To determine $m(\Lambda)$ and $\sigma(\Lambda)$, numerical calculations need to be carried out in general. To proceed, functions $X(t)$ and $\dot{X}(t)$ within a quarter period are obtained first for a given Λ based on the undamped free motion. Then calculate the Fourier expansion coefficients a_n and b_n from (4.188) and (4.189), and $m(\Lambda)$ and $\sigma(\Lambda)$ from (4.190) and (4.191). With the knowledge of $m(\Lambda)$ and $\sigma^2(\Lambda)$, the stationary PDF of $\Lambda(t)$ can be calculated. Furthermore, the joint stationary PDF of $X(t)$ and $\dot{X}(t)$, as well as the marginal stationary density of $X(t)$ or $\dot{X}(t)$, can be calculated following the procedure in Sect. 4.2.2.

Numerical calculation was carried out for system (4.181) with $\xi(t)$ being a filtered noise from a second-order linear system (see Sect. 2.6.1). The spectral density of $\xi(t)$ is given by

$$S(\omega) = \frac{K}{(\omega^2 - \omega_1^2)^2 + 4\zeta_1^2\omega_1^2\omega^2}. \quad (4.192)$$

Two different sets of parameters are considered: (i) $\zeta_1 = 0.5$, $\omega_1 = 5$, $K = 30$, and (ii) $\zeta_1 = 0.3$, $\omega_1 = 2$, $K = 0.5$. The spectral density functions for the two cases are plotted in Fig. 4.5. In the first case, the excitation has a rather broad spectrum in the frequency range $0 < \omega < 5$, while in the second case, the spectrum is not of broad band, but has a peak near $\omega = 2$.

The stationary PDF $p(\lambda)$ of the energy process $\Lambda(t)$ has been calculated for the system with $\zeta = 0.02$, $\omega_0 = 1$, two levels of nonlinearity, $\alpha = 0.1$ and 1, and the above two sets of excitation parameters. The two methods of energy-dependent white noise approximation and the Fourier-expansion scheme are used. The results obtained for the first set of excitation parameters $\zeta_1 = 0.5$, $\omega_1 = 5$, $K = 30$ are shown in Fig. 4.6, and those obtained for the second set of excitation parameters $\zeta_1 = 0.3$, $\omega_1 = 2$, $K = 0.5$ are shown in Fig. 4.7. Also shown are results from Monte Carlo simulations for comparison. As expected, the two methods result in nearly the same accurate stationary PDFs of $\Lambda(t)$ for case (i) when the excitation has slowly varying spectrum

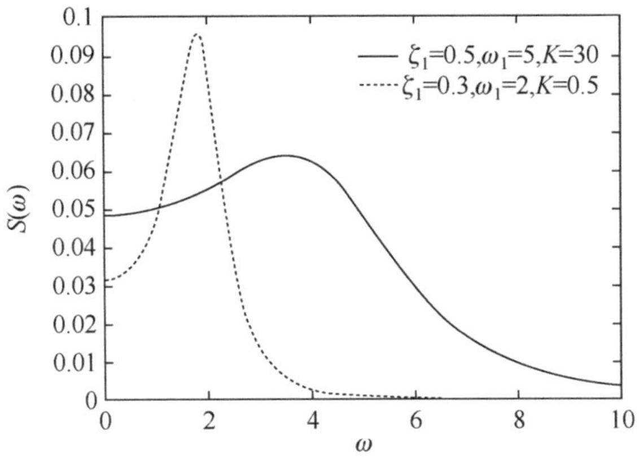

Fig. 4.5 Spectral densities of excitation

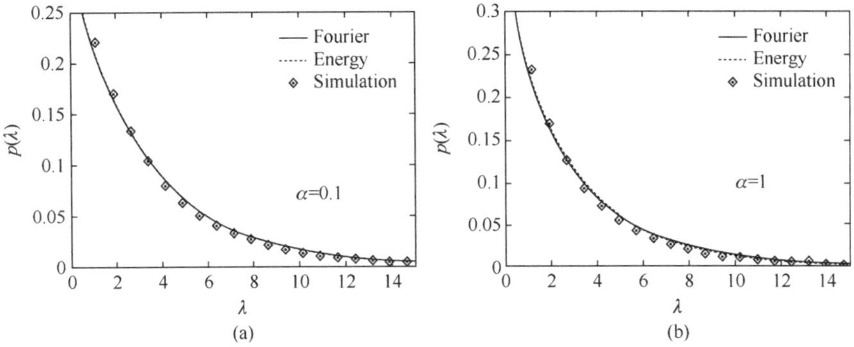

Fig. 4.6 Stationary PDFs of $\Lambda(t)$, $\zeta_1 = 0.5$, $\omega_1 = 5$, $K = 30$ (Cai and Lin 2001b)

in the range of $\omega < 5$. The two methods yield slightly different results in case (ii) when the excitation spectrum is not strictly of broad band.

It was found during the calculations that the coefficients a_n and b_n decrease rapidly with an increasing n, suggesting that only two or three terms need to be retained in the Fourier expansions even for a strongly nonlinear restoring force.

4.4.5 Residual Phase Procedure

For the undamped free motion corresponding to system (4.172), the frequency is time varying, as given by (4.63), and the phase angle in (4.62) is expressed as in (4.64). We repeat these equations as follows

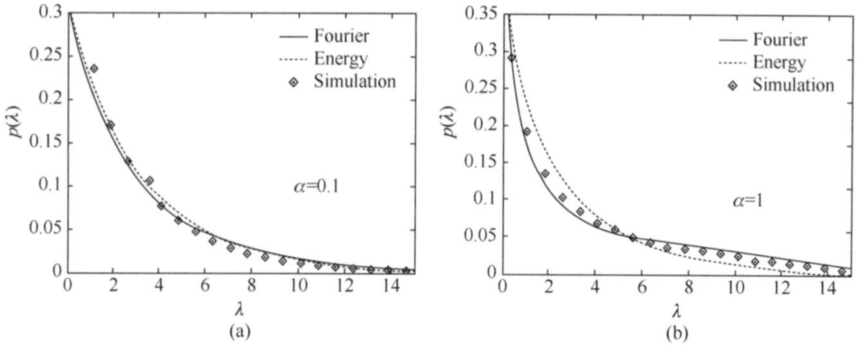

Fig. 4.7 Stationary PDFs of $\Lambda(t)$, $\zeta_1 = 0.3$, $\omega_1 = 2$, $K = 0.5$ (Cai and Lin 2001b)

$$\omega(t) = \dot{\Phi}(t) = \frac{u}{\sqrt{2\Lambda}\cos\Phi}, \quad \Phi(t) = \int_0^t \omega(s)\mathrm{d}s. \tag{4.193}$$

Under weak damping and excitations, the phase angle can be modified as

$$\Phi(t) = \int_0^t \omega(s)\mathrm{d}s + \Theta(t). \tag{4.194}$$

In what follows, $\Phi(t)$ and $\Theta(t)$ are referred as the total phase and residual phase, respectively. With $\Phi(t)$ and $\Theta(t)$ given by (4.194), let the transformation for the stochastic system (4.172) be

$$\mathrm{sgn}X\sqrt{U(X)} = \sqrt{\Lambda}\cos\Phi, \quad \dot{X} = -\sqrt{2\Lambda}\sin\Phi. \tag{4.195}$$

Equation (4.172) is transformed to

$$\dot{\Lambda} = \sqrt{2\Lambda}h(\Lambda, \Phi)\sin\Phi + \sum_{l=1}^{m} g_{1l}(\Lambda, \Phi)\xi_l(t), \tag{4.196}$$

$$\dot{\Theta} = \frac{1}{\sqrt{2\Lambda}}h(\Lambda, \Phi)\cos\Phi + \sum_{l=1}^{m} g_{2l}(\Lambda, \Phi)\xi_l(t), \tag{4.197}$$

where g_{1l} and g_{2l} are given by (4.176). Now, both $\Lambda(t)$ and $\Theta(t)$ are slowly varying. Upon caring the stochastic averaging, $\Lambda(t)$ is approximated as a Markov diffusion process with its drift and diffusion coefficients determined by (4.177) and (4.178), respectively. However, making use of the fact that the residual phase $\Theta(t)$ is slowly varying, we have the following approximations

$$\Theta(t+\tau)=\int_0^{t+\tau}\omega(s)\mathrm{d}s=\int_0^{t}\omega(s)\mathrm{d}s+\int_t^{t+\tau}\omega(s)\mathrm{d}s\approx\Phi_t+\omega_\Lambda\tau, \tag{4.198}$$

$$\begin{aligned}X(t+\tau)&=X(t)+\int_t^{t+\tau}\dot{X}(s)\mathrm{d}s\approx X(t)+\frac{\sqrt{2\Lambda}}{\omega_\Lambda}\{\cos[\Phi(t)+\omega_\Lambda\tau]-\cos\Phi(t)\}\\&=X_t+\frac{\mathrm{sgn}(X_t)\sqrt{2U_t}}{\omega_\Lambda}(\cos\omega_\Lambda\tau-1)+\frac{\dot{X}_t}{\omega_\Lambda}\sin\omega_\Lambda\tau,\end{aligned} \tag{4.199}$$

$$\begin{aligned}\dot{X}(t+\tau)&=-\sqrt{2\Lambda}\sin\Phi(t+\tau)\approx-\sqrt{2\Lambda}\sin[\Phi(t)+\omega_\Lambda\tau]\\&=\dot{X}_t\cos\omega_\Lambda\tau-\mathrm{sgn}(X_t)\sqrt{2U_t}\sin\omega_\Lambda\tau,\end{aligned} \tag{4.200}$$

where the subscript t associated with a variable indicates that it is evaluated at time t. Using approximations in Eqs. (4.198) through (4.200), the terms in (4.177) and (4.178) can be calculated. The method is illustrated in the following example.

Example 4.5 Consider the system (Cai and Lin 2001b).

$$\ddot{X}+2\zeta\omega_0\dot{X}+\omega_0^2X+\gamma X^3=X\xi_1(t)+\xi_2(t), \tag{4.201}$$

where the damping force and the excitations are assumed to be small. The corresponding undamped free system is

$$\ddot{x}+\omega_0^2x+\gamma x^3=0. \tag{4.202}$$

The restoring force and potential energy are

$$u(x)=\omega_0^2x+\gamma x^3,\quad U(x)=\frac{1}{2}\omega_0^2x^2+\frac{1}{4}\gamma x^4. \tag{4.203}$$

The total energy is

$$\Lambda=\frac{1}{2}\dot{x}^2+\frac{1}{2}\omega_0^2x^2+\frac{1}{4}\gamma x^4=\frac{1}{2}\omega_0^2A^2+\frac{1}{4}\gamma A^4, \tag{4.204}$$

and the period of free motion is

$$T=4T_{1/4}=4\int_0^A\frac{\mathrm{d}x}{\sqrt{2\Lambda-\left(\omega_0^2x^2+\gamma x^4/2\right)}}. \tag{4.205}$$

Now consider the stochastic system (4.201). First, the Fourier-expansion scheme will be applied. Comparing (4.201) with the standard forms (4.172), we have

$$h(X,\dot{X}) = 2\zeta\omega_0\dot{X}, \quad g_1(X,\dot{X}) = X, \quad g_2(X,\dot{X}) = 1. \tag{4.206}$$

From (4.176),

$$\begin{aligned} g_{11}(\Lambda,\Phi) &= -\sqrt{2\Lambda}X\sin\Phi, \quad g_{12}(\Lambda,\Phi) = -\sqrt{2\Lambda}\sin\Phi, \\ g_{21}(\Lambda,\Phi) &= -\frac{1}{\sqrt{2\Lambda}}X\cos\Phi, \quad g_{22}(\Lambda,\Phi) = -\frac{1}{\sqrt{2\Lambda}}\cos\Phi, \end{aligned} \tag{4.207}$$

and the derivatives

$$\begin{aligned} \frac{\partial g_{11}}{\partial\Lambda} &= -\frac{1}{\sqrt{2\Lambda}}X\sin\Phi - \sqrt{2\Lambda}\sin\Phi\frac{\partial X}{\partial\Lambda}, \quad \frac{\partial g_{12}}{\partial\Lambda} = -\frac{1}{\sqrt{2\Lambda}}\sin\Phi, \\ \frac{\partial g_{11}}{\partial\Phi} &= -\sqrt{2\Lambda}X\cos\Phi - \sqrt{2\Lambda}\sin\Phi\frac{\partial X}{\partial\Phi}, \quad \frac{\partial g_{12}}{\partial\Lambda} = -\sqrt{2\Lambda}\cos\Phi. \end{aligned} \tag{4.208}$$

By substituting (4.207) and (4.208) into Eqs. (4.177) and (4.178), and taking into consideration the property of each term in (4.177) and (4.178), it is found that the following Fourier expansions are needed

$$\sin\Phi(t) = \sum_{n=1}^{\infty} a_n \sin(n\omega_\Lambda t), \tag{4.209}$$

$$\cos\Phi(t) = \sum_{n=1}^{\infty} b_n \cos(n\omega_\Lambda t), \tag{4.210}$$

$$X(t)\sin\Phi(t) = \sum_{n=1}^{\infty} c_n \sin(n\omega_\Lambda t), \tag{4.211}$$

$$X(t)\cos\Phi(t) = \sum_{n=1}^{\infty} d_n \cos(n\omega_\Lambda t), \tag{4.212}$$

$$2\Lambda\sin\Phi(t)\frac{\partial X(t)}{\partial\Lambda} = \sum_{n=1}^{\infty} e_n \sin(n\omega_\Lambda t), \tag{4.213}$$

$$\sin\Phi(t)\frac{\partial X(t)}{\partial\Phi} = \sum_{n=1}^{\infty} f_n \cos(n\omega_\Lambda t), \tag{4.214}$$

where the coefficients a_n through a_n are calculated from

$$a_n = \frac{2}{T}\int_0^T \sin\Phi(t)\sin(n_\Lambda t)\mathrm{d}t = \frac{2}{\sqrt{2\Lambda}T_{1/4}}\int_0^{T_{1/4}} \dot{X}(t)\sin(n\omega_\Lambda t)\mathrm{d}t, \tag{4.215}$$

$$b_n = \frac{2}{T}\int_0^T \cos\Phi(t)\cos(n\omega_\Lambda t)\mathrm{d}t$$

$$= \frac{1}{\sqrt{\Lambda}T_{1/4}}\int_0^{T_{1/4}} X(t)\sqrt{2\omega_0^2 + \gamma X^2(t)}\cos(n\omega_\Lambda t)\mathrm{d}t, \tag{4.216}$$

$$c_n = \frac{2}{T}\int_0^T X(t)\sin\Phi(t)\sin(n\omega_\Lambda t)\mathrm{d}t$$

$$= -\frac{2}{\sqrt{2\Lambda}T_{1/4}}\int_0^{T_{1/4}} X(t)\dot{X}(t)\sin(n\omega_\Lambda t)\mathrm{d}t, \tag{4.217}$$

$$d_n = \frac{2}{T}\int_0^T X(t)\cos\Phi(t)\cos(n\omega_\Lambda t)\mathrm{d}t$$

$$= \frac{1}{\sqrt{\Lambda}T_{1/4}}\int_0^{T_{1/4}} X^2(t)\sqrt{2\omega_0^2 + \gamma X^2(t)]}\cos(n\omega_\Lambda t)\mathrm{d}t, \tag{4.218}$$

$$e_n = \frac{2}{T}\int_0^T 2\Lambda\sin\Phi(t)\frac{\partial X(t)}{\partial\Lambda}\sin(n\omega_\Lambda t)\mathrm{d}t$$

$$= -\frac{1}{\sqrt{2\Lambda}T_{\frac{1}{4}}}\int_0^{T_{\frac{1}{4}}} \frac{X(t)\dot{X}(t)\left[2\omega_0^2 + \gamma X^2(t)\right]}{\omega_0^2 + \gamma X^2(t)}\sin(n\omega_\Lambda t)\mathrm{d}t, \tag{4.219}$$

$$f_n = \frac{2}{T}\int_0^T \sin\Phi(t)\frac{\partial X(t)}{\partial\Phi}\cos(n\omega_\Lambda t)\mathrm{d}t$$

$$= -\frac{1}{\sqrt{\Lambda}T_{1/4}}\int_0^{T_{1/4}} \frac{\dot{X}^2(t)\sqrt{2\omega_0^2 + \gamma X^2(t)}}{\omega_0^2 + \gamma X^2(t)}\cos(n\omega_\Lambda t)\mathrm{d}t. \tag{4.220}$$

Now perform the time averaging in (4.177) and (4.178), and using (4.209)–(4.214), we obtain

$$m(\Lambda) = \frac{2\zeta\omega_0}{T_{1/4}}\int_0^{T_{1/4}} \dot{X}^2(t)\mathrm{d}t + \frac{\pi}{8}\left(d_0^2 + d_0 f_0\right)S_{11}(0)$$

$$+\frac{\pi}{2}\sum_{n=1}^{\infty}\left(c_n^2+c_ne_n+d_n^2+d_nf_n\right)S_{11}(n\omega_\Lambda)$$
$$+\frac{\pi}{2}\sum_{n=1}^{\infty}\left(a_n^2+b_n^2\right)S_{22}(n\omega_\Lambda), \tag{4.221}$$

$$\sigma^2(\Lambda)=2\pi\Lambda\sum_{n=1}^{\infty}\left[c_n^2S_{11}(n\omega_\Lambda)+a_n^2S_{22}(n\omega_\Lambda)\right]. \tag{4.222}$$

As mentioned in Example 4.4, the Fourier expansion coefficients decrease rapidly with an increasing n, and only first several terms need to be retained even for a strongly nonlinear restoring force.

Now we use the residual phase procedure to solve the problem. Using the approximations in Eqs. (4.198) through (4.200), the following terms in (4.177) and (4.178) are calculated,

$$\begin{aligned}&\left\langle\left[\dot{X}g_1(X,\dot{X})\right]_{t+\tau}\frac{\partial}{\partial\Lambda}\left[\dot{X}g_1(X,\dot{X})\right]_t\right\rangle_t\\&=\left\langle X^2-\frac{X\,\mathrm{sgn}X\sqrt{2U(X)}}{\omega_\Lambda}\right\rangle_t\cos\omega_\Lambda\tau+\left\langle\frac{X\,\mathrm{sgn}X\sqrt{2U(X)}}{\omega_\Lambda}\right\rangle_t\cos2\omega_\Lambda\tau,\end{aligned} \tag{4.223}$$

$$\left\langle\left[\dot{X}g_2(X,\dot{X})\right]_{t+\tau}\frac{\partial}{\partial\Lambda}\left[\dot{X}g_2(X,\dot{X})\right]_t\right\rangle_t=\cos\omega_\Lambda\tau, \tag{4.224}$$

$$\left\langle\left[\dot{X}g_1(X,\dot{X})\right]_{t+\tau}\frac{\partial}{\partial\Lambda}\left[\dot{X}g_2(X,\dot{X})\right]_t\right\rangle_t=\left\langle\dot{X}\left[g_2(X,\dot{X})\right]_{t+\tau}\frac{\partial}{\partial\Lambda}\left[\dot{X}g_1(X,\dot{X})\right]_t\right\rangle_t=0, \tag{4.225}$$

$$\begin{aligned}&\left\langle\left[\dot{X}g_1(X,\dot{X})\right]_{t+\tau}\left[\dot{X}g_1(X,\dot{X})\right]_t\right\rangle_t\\&=\left\langle X^2\dot{X}^2-\frac{\dot{X}^2X\,\mathrm{sgn}X\sqrt{2U(X)}}{\omega_\Lambda}\right\rangle_t\cos\omega_\Lambda\tau+\left\langle\frac{\dot{X}^2X\,\mathrm{sgn}X\sqrt{2U(X)}}{\omega_\Lambda}\right\rangle_t\cos\omega_\Lambda\tau,\end{aligned} \tag{4.226}$$

$$\left\langle\left[\dot{X}g_2(X,\dot{X})\right]_{t+\tau}\left[\dot{X}g_2(X,\dot{X})\right]_t\right\rangle_t=\left\langle\dot{X}^2\right\rangle_t\cos\omega_\Lambda\tau, \tag{4.227}$$

$$\left\langle\left[\dot{X}g_1(X,\dot{X})\right]_{t+\tau}\left[\dot{X}g_2(X,\dot{X})\right]_t\right\rangle_t=\left\langle\left[\dot{X}g_2(X,\dot{X})\right]_{t+\tau}\left[\dot{X}g_1(X,\dot{X})\right]_t\right\rangle_t=0. \tag{4.228}$$

Substituting (4.223) through (4.225) into (4.177) and substituting (4.226) through (4.228) into (4.178), we obtain

$$m(\Lambda)=-2\zeta\omega_0u_2+\pi(u_1-u_3)S_{11}(\omega_\Lambda)+\pi u_3S_{11}(2\omega_\Lambda)+\pi S_{22}(\omega_\Lambda), \tag{4.229}$$

$$\sigma^2(\Lambda) = 2\pi(u_4 - u_5)S(\omega_\Lambda) + 2\pi u_5 S_{11}(2\omega_\Lambda) + 2\pi u_2 S_{22}(\omega_\Lambda), \tag{4.230}$$

where

$$u_1 = \langle X^2 \rangle_t = \frac{1}{T_{1/4}} \int_0^A \frac{X^2}{\sqrt{2\Lambda - 2U(X)}} \mathrm{d}X, \tag{4.231}$$

$$u_2 = \langle \dot{X}^2 \rangle_t = \frac{1}{T_{1/4}} \int_0^A \sqrt{2\Lambda - 2U(X)} \mathrm{d}X, \tag{4.232}$$

$$u_3 = \left\langle \frac{X \operatorname{sgn}(X) \sqrt{2U(X)}}{\omega_\Lambda} \right\rangle_t = \frac{2}{\pi} \int_0^A \frac{X\sqrt{2U(X)}}{\sqrt{2\Lambda - 2U(X)}} \mathrm{d}X, \tag{4.233}$$

$$u_4 = \langle X^2 \dot{X}^2 \rangle_t = \frac{1}{T_{1/4}} \int_0^A X^2 \sqrt{2\Lambda - 2U(X)} \mathrm{d}X, \tag{4.234}$$

$$u_5 = \left\langle \frac{X \dot{X}^2 \operatorname{sgn}(X) \sqrt{2U(X)}}{\omega_\Lambda} \right\rangle_t = \frac{4}{\pi} \int_0^A X\sqrt{U(X)[\Lambda - U(X)]} \mathrm{d}X. \tag{4.235}$$

In deriving (4.229) and (4.230), use has been made of

$$\begin{gathered} 2\int_{-\infty}^{0} \cos(\omega\tau) R_{ii}(\tau) \mathrm{d}\tau = \int_{-\infty}^{\infty} \cos(\omega\tau) R_{ii}(\tau) \mathrm{d}\tau = 2\pi S_{ii}(\omega), \\ \int_{-\infty}^{0} \sin(\omega\tau) R_{ii}(\tau) \mathrm{d}\tau \approx 0, \quad \int_{-\infty}^{\infty} \sin(\omega\tau) R_{ii}(\tau) \mathrm{d}\tau \approx 0. \end{gathered} \tag{4.236}$$

Two special cases are noted below. First consider the case in which both $\xi_1(t)$ and $\xi_2(t)$ are Gaussian white noises. By letting $S_{11}(\omega_\Lambda) = S_{11}(2\omega_\Lambda) = K_{11}$ and $S_{22}(\omega_\Lambda) = K_{22}$, Eqs. (4.221) and (4.222) of Fourier expansion scheme are simplified to

$$\begin{aligned} m(\Lambda) = & \frac{2\zeta\omega_0}{T_{1/4}} \int_0^{T_{1/4}} \dot{X}^2(t) \mathrm{d}t + \frac{\pi}{2} K_{11} \left[\frac{1}{4}(d_0^2 + d_0 f_0) + \right. \\ & \left. \sum_{n=1}^{\infty} (c_n^2 + c_n e_n + d_n^2 + d_n f_n) \right] + \frac{\pi}{2} K_{22} \sum_{n=1}^{\infty} (a_n^2 + b_n^2), \end{aligned} \tag{4.237}$$

$$\sigma^2(\Lambda) = 2\pi\Lambda\sum_{n=1}^{\infty}\left(c_n^2K_{11} + a_n^2K_{22}\right). \tag{4.238}$$

On the other hand, Eqs. (4.229) and (4.230) of the residual phase procedure reduce to

$$m(\Lambda) = -2\zeta\omega_0\left\langle\dot{X}^2\right\rangle_t + \pi K_{11}\left\langle X^2\right\rangle_t + \pi K_{22}, \tag{4.239}$$

$$\sigma^2(\Lambda) = 2\pi K_{11}\left\langle X^2\dot{X}^2\right\rangle_t + 2\pi K_{22}. \tag{4.240}$$

It can be shown that

$$\begin{gathered}\sum_{n=1}^{\infty}a_n^2 = \frac{u_2}{\Lambda},\quad \sum_{n=1}^{\infty}c_n^2 = \frac{u_5}{\Lambda},\quad \sum_{n=1}^{\infty}\left(a_n^2 + b_n^2\right) = 2,\\ \frac{1}{4}d_0^2 + \sum_{n=1}^{\infty}\left(c_n^2 + d_n^2\right) = 2u_1,\quad \frac{1}{4}d_0f_0 + \sum_{n=1}^{\infty}\left(c_ne_n + d_nf_n\right) = 0.\end{gathered} \tag{4.241}$$

Therefore, Eqs. (4.237) and (4.238) are the same as Eqs. (4.239) and (4.240) respectively, and the method of Fourier expansion and the residual phase procedure yield the same results. It can also verify that the method in Sect. 4.3.2 would also lead to the same results.

Another case is the system with a linear stiffness ($\gamma = 0$). In this case, $\omega_A = \omega_0$, $X = \sqrt{2\Lambda}\cos\Phi/\omega_0$, and

$$a_1 = b_1 = 1,\quad c_1 = d_0 = d_1 = e_1 = f_1 = \frac{\sqrt{\Lambda}}{\sqrt{2}\omega_0},\quad f_0 = -\frac{\sqrt{\Lambda}}{\sqrt{2}\omega_0}. \tag{4.242}$$

Other $a_n = b_n = c_n = d_n = e_n = f_n = 0$, and

$$u_1 = u_3 = u_4 = \frac{\Lambda}{\omega_0^2},\quad u_2 = \Lambda,\quad u_5 = u_6 = \frac{\Lambda}{2\omega_0^2}. \tag{4.243}$$

Thus, both the Fourier expansion scheme and the residual phase procedure lead to the same drift and diffusion coefficients as follows

$$\begin{aligned} m(\Lambda) &= -2\zeta\omega_0u_2 + \frac{\pi}{\omega_0^2}S_{11}(2\omega_0)\Lambda + \pi S_{22}(\omega_0),\\ \sigma^2(\Lambda) &= \frac{\pi}{\omega_0^2}S_{11}(2\omega_0)\Lambda^2 + 2\pi S_{22}(\omega_0)\Lambda.\end{aligned} \tag{4.244}$$

Using the Itô differential rule, it can be proven that, in this case, the Itô equation of the energy $\Lambda(t)$ with drift and diffusion coefficients given by (4.244) is equivalent to Itô equation (4.133) of the amplitude $A(t)$.

Numerical calculations are carried out for system (4.201) with $\zeta = 0.05$ and $\omega_0 = 1$ by using the two methods of Fourier expansion scheme and the residual phase procedure. The spectral densities of $\xi_1(t)$ and $\xi_2(t)$ are assumed to be of the following low-pass type

$$S_{ii}(\omega) = \frac{K_{ii}}{\alpha_i^2 + \omega^2}, \quad i = 1, 2, \tag{4.245}$$

where α_i and K_{ii} are the band width and magnitude parameters, respectively. A higher K_{ii} corresponds to a stronger excitation, while a larger α_i indicates a broader band, or equivalently, a shorter correlation time. Each $\xi_i(t)$ can be generated from

$$\dot{\xi}_i + \alpha_i \xi_i = W_{gi}(t), \quad i = 1, 2, \tag{4.246}$$

where $W_{gi}(t)$ is a white noise with a spectral density K_{ii}.

The procedure of numerical calculations is (i) to integrate Eq. (4.202) of the undamped free motion to obtain functions X(t) and $\dot{X}(t)$ for a given Λ within a quarter period, (ii) to calculate coefficients a_n through f_n from Eqs. (4.215)–(4.220) for the Fourier expansion scheme, or parameter u_1 through u_5 from Eqs. (4.231)–(4.235) for the residual phase procedure, (iii) to use Eqs. (4.221) and (4.222) or (4.229) and (4.230) to obtained $m(\Lambda)$ and $\sigma(\Lambda)$ for the two methods respectively, and (iv) to calculate the stationary PDF of energy process $\Lambda(t)$ according to Eq. (4.189), i.e.

$$p(\lambda) = \frac{C}{\sigma^2(\lambda)} \exp\left[\int \frac{2m(\lambda)}{\sigma^2(\lambda)} \mathrm{d}\lambda\right]. \tag{4.247}$$

Shown in Figs. 4.8 and 4.9 are plots of the stationary PDFs of the energy process $\Lambda(t)$ for two different set of parameters: in Fig. 4.8, $\alpha_1 = \alpha_2 = \alpha = 1.0$, $K_{11} = K_{22} = K = 0.1$; in Fig. 4.9, $\alpha_1 = \alpha_2 = \alpha = 1.5$, $K_{11} = K_{22} = K = 0.1$. In each figure, two different nonlinearity levels $\gamma = 0.1$ and 1.0 are indicated. Numerical results are obtained using the Fourier expansion scheme and the residual phase procedure, and they are compared with those from Monte Carlo simulations. The results from both analytical methods agree very well with the simulation results, even at very low PDF level. As expected, the analytical results are more accurate in the cases of weaker nonlinearity (smaller γ) and/or shorter excitation correlation time (larger α).

4.5 Viscoelastic Systems Under Broad-Band Excitations

The phenomenon of viscoelasticity occurs for certain types of materials, such as metals and composites. It describes the capability of a material, not only to store the potential energy due to its elasticity, but also to dissipate energy as a damping. It is known that the viscoelastic force depends on the current state as well as the

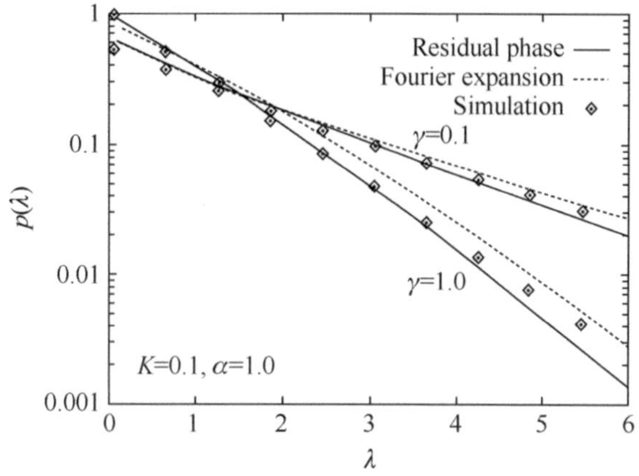

Fig. 4.8 Stationary PDFs of $\Lambda(t)$ for the case of $\alpha = 1.0$, $K = 0.1$ (Cai and Lin 2001b)

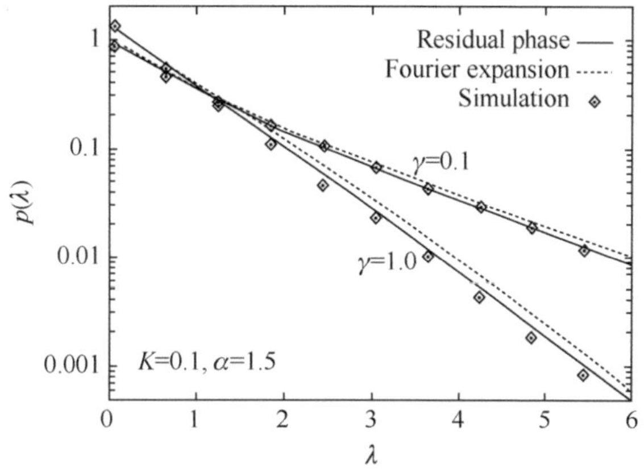

Fig. 4.9 Stationary PDFs of $\Lambda(t)$ for the case of $\alpha = 1.5$, $K = 0.1$ (Cai and Lin 2001b)

past history of the state. Viscoelastic systems under random excitations have been of interest among researchers (Ariaratnam 1993; Cai et al. 1998; Huang and Xie 2008; Lin et al. 2011; Zhu and Cai 2011).

4.5.1 Linear Restoring Force

First consider the following viscoelastic system of linear restoring force (Ariaratnam 1993)

$$\ddot{X} + f(X, \dot{X}) + \omega_0^2(X) + \int_0^t h(t-\tau)\dot{X}(\tau)\mathrm{d}\tau = \sum_{l=1}^{m} g_l(X, \dot{X})\xi_l(t), \tag{4.248}$$

where $f(X, \dot{X})$ is a damping force, the integral term represents a viscoelastic force, $h(t)$ is known as the relaxation function, and $\xi_j(t)$ are random excitations with zero mean and broad-band spectra. It is assumed that the damping force f, the viscoelastic force Z, and the excitations $\xi_j(t)$ are weak so that the stochastic averaging is applicable.

A common choice for the relaxation function $h(t)$ is the Maxwell model (Christensen 1982; Drozdov 1998)

$$h(t) = \sum_i \beta_i \exp(-\alpha_i t), \quad \alpha_i > 0. \tag{4.249}$$

Equations (4.248) and (4.249) indicate that the total viscoelastic force may be composed of a number of viscoelastic components. In (4.249), $1/\alpha_i$ is known as the relaxation time of each viscoelastic component, and β_i specifies its magnitude which can be either positive or negative. Both parameters are to be determined according to the specific problems. In terms of (4.249), Eq. (4.248) can be replaced by the following set of equations

$$\ddot{X} + f(X, \dot{X}) + \omega_0^2 X + \sum_i Z_i = \sum_{l=1}^{m} g_l(X, \dot{X}) \xi_l(t), \tag{4.250}$$

$$\dot{Z}_i = -\alpha_i Z_i + \beta_i \dot{X}. \tag{4.251}$$

To apply the stochastic averaging of amplitude envelope, use the transformation (4.33), i.e.,

$$X = A_0(t) \cos \Phi_0, \quad \dot{X} = -A_0(t)\omega_0 \sin \Phi_0, \quad \Phi_0 = \omega_0 t + \Theta_0(t). \tag{4.252}$$

Substitute (4.252) into (4.251), and solve for the steady-state solutions of Z_i to yield

$$\dot{Z}_i = \frac{\beta_i}{\alpha_i^2 + \omega_0^2}\left(\omega_0^2 A_0 \cos \Phi_0 - \alpha_i \omega_0 A_0 \sin \Phi_0\right) = \frac{\beta_i}{\alpha_i^2 + \omega_0^2}\left(\omega_0^2 X + \alpha_i \dot{X}\right), \tag{4.253}$$

where terms involving $\dot{A}$ and $\dot{\Phi}$ are neglected since A and Φ are slowly varying, and $\dot{A}$ and $\dot{\Phi}$ are small. Combing (4.250) and (4.253), we have

$$\ddot{X} + \left[f(X, \dot{X}) + \sum_i \frac{\beta_i \alpha_i}{\alpha_i^2 +} \dot{X}\right] + \left[\omega_0^2 X + \sum_i \frac{\beta_i \omega_0^2}{\alpha_i^2 + \omega_0^2} X\right] = \sum_{l=1}^{m} g_l(X, \dot{X}) \xi_l(t). \tag{4.254}$$

Equation (4.254) shows that the viscoelastic force changes the system restoring force from $\omega_0^2 X$ to $\omega_1^2 X$, where

$$\omega_1^2 = \omega_0^2 + \sum_i \frac{\beta_i \omega_0^2}{\alpha_i^2 + \omega_0^2}, \tag{4.255}$$

and adds an additional damping to results in the total damping as

$$f(X, \dot{X}) + \sum_i \frac{\beta_i \alpha_i}{\alpha_i^2 + \omega_0^2} \dot{X}. \tag{4.256}$$

In the relaxation function (4.249), α_i must be positive, but β_i could be positive or negative mathematically. However, a viscoelastic material will add extra damping to the system; thus, β_i should also be positive in (4.256).

It is noted that (4.256) is the steady-state solution of Eq. (4.251) to a harmonic excitation $\sin\Phi_0$, and the decaying transient term is neglected. Therefore, (4.255) and (4.256) are only meaningful for the stationary solution of system (4.248).

Taking into account the fact that the linear restoring force is changed, the transformation of (4.252) needs to be modified as

$$X = A_1(t)\cos\Phi_1, \quad \dot{X} = -A_1(t)\omega_1 \sin\Phi_1, \quad \Phi_1 = \omega_1 t + \Theta_1(t), \tag{4.257}$$

and Eqs. (4.255) and (4.256) are also modified to

$$\omega_1^2 = \omega_0^2 + \sum_i \frac{\beta_i \omega_1^2}{\alpha_i^2 + \omega_1^2}, \tag{4.258}$$

$$f(X, \dot{X}) + \sum_i \frac{\beta_i \alpha_i}{\alpha_i^2 + \omega_0^2} \dot{X}. \tag{4.259}$$

The modified ω_1 can be determined from (4.258). Using Eqs. (4.38) and (4.39), we obtain from (4.254)

$$\begin{aligned}\dot{A}_1 = &\frac{\sin\Phi_1}{\omega_0}\Bigg[f(A_1\cos\Phi_1, -A_1\omega_1\sin\Phi_1) - \\ &\sum_i \frac{A_1\beta_i\alpha_i\omega_1\sin\Phi_1}{\alpha_i^2+\omega_1^2} - \sum_{l=1}^{m} g_l(A_1\cos\Phi_1, -A_1\omega_1\sin\Phi_1)\xi_l(t)\Bigg],\end{aligned} \tag{4.260}$$

$$\begin{aligned}\dot{\Theta}_1 = &\frac{\cos\Phi_1}{\omega_0 A_1}\Bigg[f(A_1\cos\Phi_1, -A_1\omega_1\sin\Phi_1) \\ &-\sum_i \frac{A_1\beta_i\alpha_i\omega_1\sin\Phi_1}{\alpha_i^2+\omega_1^2} - \sum_{l=1}^{m} g_l(A_1\cos\Phi_1, -A_1\omega_1\sin\Phi_1)\xi_l(t)\Bigg].\end{aligned} \tag{4.261}$$

Given the specific forms of functions f and g_l, as well as the spectral densities of $\xi_j(t)$, the stochastic averaging can be applied to (4.260) and (4.261).

Example 4.6 Consider the system (Cai et al. 1998).

$$\ddot{X} + \omega_0^2 X + Z(t) = X W_{g1}(t) + W_{g2}(t), \tag{4.262}$$

$$\dot{Z} = -\alpha Z + \beta \dot{X}, \tag{4.263}$$

where $W_{g1}(t)$ and $W_{g2}(t)$ are two Gaussian white noises. Following the above procedure, we obtain the drift and diffusion coefficients for the amplitude process as

$$m(A_1) = \left[-\frac{\beta\alpha}{2(\alpha^2 + \omega_1^2)} + \frac{3\pi}{8\omega_1^2} K_1 \right] A_1 + \frac{\pi}{2\omega_1^2 A_1} K_2, \tag{4.264}$$

$$\sigma^2(A_1) = \frac{\pi}{4\omega_1^2} K_1 A_1^2 + \frac{\pi}{\omega_1^2} K_2, \tag{4.265}$$

where K_1 and K_2 are spectral densities of $W_{g1}(t)$ and $W_{g2}(t)$, respectively, and ω_1 is determined from (4.258) as

$$\omega_1^2 = \frac{1}{2}\left[\omega_0^2 + \beta - \alpha^2 + \sqrt{\left(\omega_0^2 + \beta - \alpha^2\right)^2 + 4\alpha^2\omega_0^2} \right]. \tag{4.266}$$

The condition of asymptotic stability in probability of the system can be obtained as (Cai and Zhu 2016)

$$\frac{\alpha\beta\omega_1^2}{\alpha^2 + \omega_1^2} \geq \frac{1}{2}\pi K_1. \tag{4.267}$$

If the original transformation (4.252) is used, the condition is

$$\frac{\alpha\beta\omega_0^2}{\alpha^2 + \omega_0^2} \geq \frac{1}{2}\pi K_1. \tag{4.268}$$

Numerical calculations were carried out for the system with $\omega_0 = 1$. The stability boundaries are calculated from: (i) (4.268) of the original transformation, (ii) (4.267) of the improved transformation, and (iii) Monte Carlo simulations. The results are shown in Figs. 4.10 and 4.11 for two values of $\beta = 1$ and 2, respectively. It is seen that the results obtained with the improved transformation agree very well with the simulation results. The results obtained from the original transformation are too conservative since the hardening stiffness effect due to the viscoelastic force is not taken into account.

The approximate mean square values of the amplitude are determined from the two versions, the original and improved transformations, as follows

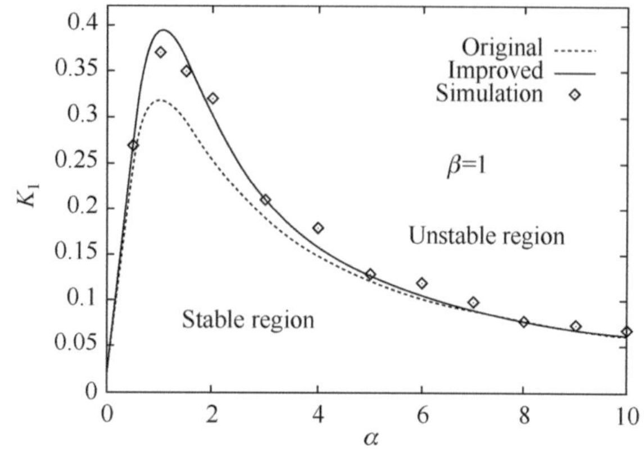

Fig. 4.10 Stability boundaries for $\beta = 1$ (Cai et al. 1998)

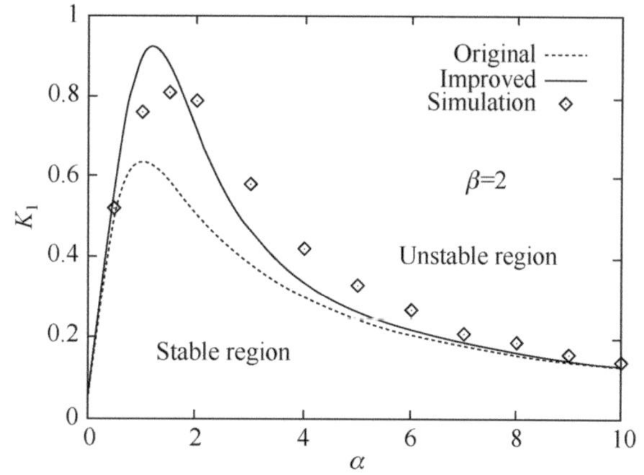

Fig. 4.11 Stability boundaries for $\beta = 2$ (Cai et al. 1998)

$$E\left[A_0^2\right] = \frac{2\pi K_2\left(\omega_0^2 + \alpha^2\right)}{\alpha\beta\omega_0^2 - \pi K_1\left(\alpha^2 + \omega_0^2\right)}, \tag{4.269}$$

$$E\left[A_1^2\right] = \frac{2\pi K_2\left(\omega_1^2 + \alpha^2\right)}{\alpha\beta\omega_1^2 - \pi K_1\left(\alpha^2 + \omega_1^2\right)}. \tag{4.270}$$

Since the system governed by (4.262) and (4.263) is linear, exact mean square values for X, $\dot{X}$, and Z can be obtained. Then the exact mean square values for A_0 and A_1 can be calculated from (4.252) and (4.257), respectively, as follows

$$E\left[A_0^2\right] = E\left[X^2\right] + \frac{E\left[\dot{X}^2\right]}{\omega_0^2} = \frac{\pi K_2\left(2\omega_0^2 + 2\alpha^2 + \beta\right)}{\alpha\beta\omega_0^2 - \pi K_1\left(\alpha^2 + \omega_0^2\right)}, \tag{4.271}$$

$$E\left[A_1^2\right] = E\left[X^2\right] + \frac{E\left[\dot{X}^2\right]}{\omega_1^2} = \frac{\pi K_2[\alpha^2 + \omega_0^2 + \frac{\omega_0^2}{\omega_1^2}\left(\omega_0^2 + \alpha^2 + \beta\right)}{\alpha\beta\omega_0^2 - \pi K_1\left(\alpha^2 + \omega_0^2\right)}. \tag{4.272}$$

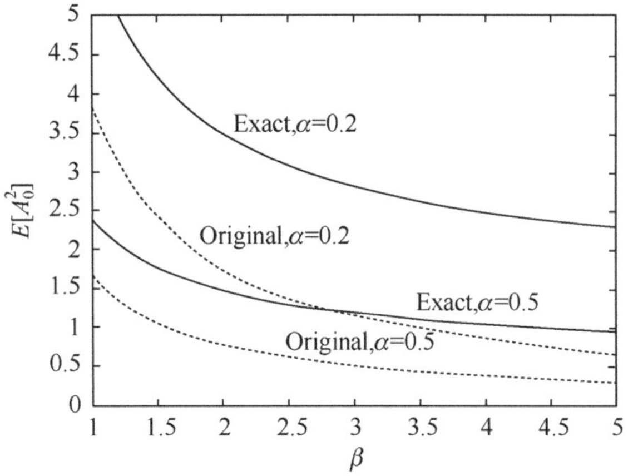

Fig. 4.12 Mean square values of amplitude $A_0(t)$ using the original transformation (4.252) (Cai et al. 1998)

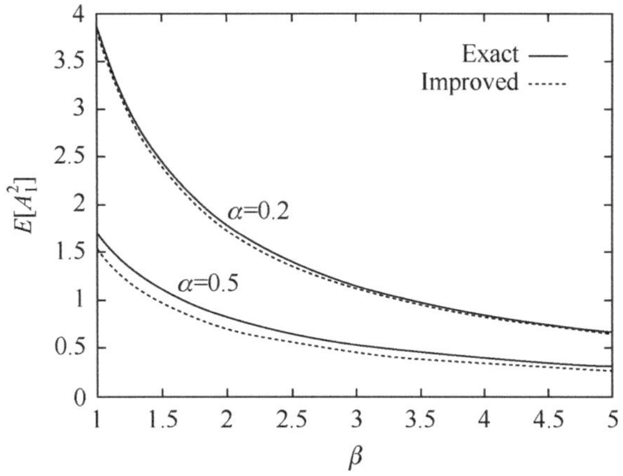

Fig. 4.13 Mean square values of amplitude $A_1(t)$ using the improved transformation (4.257) (Cai et al. 1998)

Figure 4.12 depicts the approximate values of $E[A_0^2]$ calculated from (4.269) of the traditional transformation with $\omega_0 = 1$, $K_1 = 0.01$, and $K_2 = 0.1$. These results are very much in error as compared to the exact values obtained from (4.271). In contrast, with the improved version of transformation, more accurate results are obtained from (4.270) compared with those from (4.272), as shown in Fig. 4.13.

4.5.2 Nonlinear Restoring Force

If the restoring force is strongly nonlinear, the amplitude envelope method is not applicable. Consider the system

$$\ddot{X} + f(X, \dot{X}) + u(X) + \sum_i Z_i = \sum_j g_j(X, \dot{X})\xi_j(t), \tag{4.273}$$

$$Z_i = \int_0^t h_i(t-\tau)X(\tau)\mathrm{d}\tau, \tag{4.274}$$

$$h_i(t) = \sum_i \beta_i \exp(-t/\lambda_i), \quad \lambda_i > 0. \tag{4.275}$$

Note that the model for the viscoelastic force (4.274) is different from that in (4.248). The viscoelastic phenomenon in (4.274) is associated with the displacement X, while it is related to the velocity $\dot{X}$ in (4.248). But the relaxation function $h(t)$ remains the same. It is noted that Eq. (4.274) is similar to the right-hand side of Eq. (3.4.37).

It is shown in Sect. 4.5.1 that the viscoelastic force contributes to both the stiffness and the damping. Taking this into consideration, system (4.273) is replaced by

$$\ddot{X} + f_1(X, \dot{X}) + u_1(X) = \sum_j g_j(X, \dot{X})\xi_j(t), \tag{4.276}$$

where the effects of the viscoelastic force have been incorporated in $f_1(X, \dot{X})$ and $u_1(X)$, respectively. For system (4.276), the stochastic averaging of energy envelope can be applied. The potential energy $U(X)$ and the total energy Λ of the system are, respectively,

$$U(X) = \int_0^X u_1(z)\mathrm{d}z,\ \Lambda = \frac{1}{2}\dot{X}^2 + U(X). \tag{4.277}$$

Invoking the transformation (4.173), i.e.,

$$\begin{aligned} \mathrm{sgn}X\sqrt{U(X)} &= \sqrt{\Lambda}\cos\Phi, \\ \dot{X} &= -\sqrt{2\Lambda}\sin\Phi, \end{aligned} \tag{4.278}$$

where

$$\Phi(t) = \int_0^t \omega(s)\mathrm{d}s + \Theta(t); \quad \omega(t) = \frac{u_1(X)}{\sqrt{2\Lambda}\cos\Phi}. \tag{4.279}$$

It is known that $\Phi(t)$, $\Theta(t)$, and $\omega(t)$ are the total phase, residual phase, and instantaneous frequency, respectively. The average frequency ω_Λ is obtained in Eq. (4.65).

Each component in the viscoelastic force given in (4.274) and (4.275) can be expressed as

$$Z_i = \int_0^t \beta_i e^{-(t-\tau)/\lambda_i} X(\tau)\mathrm{d}\tau. \tag{4.280}$$

Carrying out the integration, using transformation (4.278) with $\Phi(t) \approx \omega_\Lambda t + \Theta$, and neglecting the decaying transient part, we obtain

$$Z_i = \frac{\beta_i \lambda_i}{1 + \omega_\Lambda^2 \lambda_i^2}\left(X - \lambda_i \dot{X}\right). \tag{4.281}$$

Taking (4.281) into consideration, functions $f_1(X, \dot{X})$ and $u_1(X)$ in (4.276) are derived as

$$f_1(X, \dot{X}) = f(X, \dot{X}) - \sum_i \frac{\beta_i \lambda_i^2}{1 + \omega_\Lambda^2 \lambda_i^2}\dot{X},$$

$$u_1(X) = u(X) + \sum_i \frac{\beta_i \lambda_i}{1 + \omega_\Lambda^2 \lambda_i^2} X. \tag{4.282}$$

To determine the average frequency ω_Λ, combine (4.65) and (4.60) to obtain

$$\frac{\pi}{\omega_\Lambda} = 2\int_0^A \frac{\mathrm{d}X}{\sqrt{2\Lambda - 2U(X)}}. \tag{4.283}$$

Since $U(X)$ is a function of ω_Λ according to (4.277) and (4.282), (4.283) is a nonlinear algebraic equation for ω_Λ, and can be solved numerically for a given Λ as a parameter.

The new additional terms in the stiffness and damping in (4.282) represent the contributions from the viscoelastic force. It can be proved that (4.281) and (4.282) are exact if the system is linear and the excitation is external and harmonic, in which case the response is also harmonic. It is noted that (4.281) is derived by neglecting the decaying transient terms, indicating the replacement is only applicable for the system in stationary state.

The effects of each viscoelastic component on the stiffness and damping depends on parameter β_i and λ_i. While the relaxation time λ_i must be positive, β_i could be either positive or negative. For a negative β_i, the viscoelastic component increases the damping and reduces the stiffness. This is the case in general for a viscoelastic material since its elastic modulus is less than that of a pure elastic material, but a damping mechanism is introduced. Note that the role of parameter β_i is different from the β_i in Sect. 4.5.1 due to the different viscoelastic models of (4.248) and (4.274).

Equation of motion (4.276) will be used as an equivalent one to the original system (4.273), and both the Fourier-expansion scheme and residual phase procedure will be applied to solve the problem.

Example 4.7 Consider the following Duffing oscillator with a viscoelastic force (Zhu and Cai 2011).

$$\ddot{X} + 2\zeta\omega_0\dot{X} + \omega_0^2 X + \gamma X^3 + Z = X\xi_1(t) + \xi_2(t),$$
$$Z = \beta_1 \int_0^t e^{-\frac{t-\tau}{\lambda_1}} X(\tau)\mathrm{d}\tau, \tag{4.284}$$

where the damping force, viscoelastic force, and $\xi_1(t)$ and $\xi_2(t)$ are assumed to be weak. Approximating the viscoelastic force Z according to (4.281), the equivalent system of (4.284) has the form

$$\ddot{X} + f_1(\dot{X}) + u_1(X) = X\xi_1(t) + \xi_2(t),$$
$$f_1(\dot{X}) = \left[2\xi\omega_0 - \frac{\beta_1\lambda_1^2}{1+\omega_\Lambda^2\lambda_1^2}\right]\dot{X}, \quad u_1(X) = \left[\omega_0^2 + \frac{\beta_1\lambda_1}{1+\omega_\Lambda^2\lambda_1^2}\right]X + \gamma X^3. \tag{4.285}$$

The average frequency ω_Λ is derived from (4.283) for the present case as

$$\frac{\pi}{\omega_\Lambda} = 2\int_0^{\pi/2} \frac{\mathrm{d}\phi}{\sqrt{\omega_0^2 + \frac{\beta_1\lambda_1}{1+\omega_\Lambda^2\lambda_1^2} + \frac{1}{2}\gamma A^2(1+\sin^2\phi)}}. \tag{4.286}$$

Equation (4.286) can be solved numerically for a given energy Λ or amplitude A. Considering the case of $A = 0$ or $\gamma = 0$, we obtain the modified frequency for the linear stiffness

$$\omega_\Lambda = \frac{1}{2\lambda_1^2}\left[\sqrt{(1+\omega_0^2\lambda_1^2)^2 + 4\beta_1\lambda_1^3} - \left(1-\omega_0^2\lambda_1^2\right)\right]. \tag{4.287}$$

Now the residual phase procedure of the stochastic averaging method can be applied, namely, to evaluate (4.177) and (4.178). In carrying the time averaging and stochastic averaging in (4.177) and (4.178), the approximations in (4.198) through (4.200) need to be used. For the present case, we obtain

$$m(\Lambda) = -\left[2\xi\omega_0 - \frac{\beta_1\lambda_1^2}{1+\omega_\Lambda^2\lambda_1^2}\right]u_2 + \pi(u_1 - u_3)S_{11}(\omega_\Lambda)$$
$$+ \pi u_3 S_{11}(2\omega_\Lambda) + \pi S_{22}(\omega_\Lambda), \tag{4.288}$$

$$\sigma^2(\Lambda) = 2\pi(u_4 - u_5)S_{11}(\omega_\Lambda) + 2\pi u_5 S_{11}(2\omega_\Lambda) + 2\pi u_2 S_{22}(\omega_\Lambda). \tag{4.289}$$

where u_k's (k = 1, 2, ..., 5) are functions of Λ given by Eqs. (4.231)–(4.235) with $u(X)$ replaced by $u_1(X)$ in (4.285).

Numerical calculations are carried out for system (4.284) with $\zeta = 0.05$ and $\omega_0 = 1$. Different value of the coefficient γ will be used to show the applicability of the method for the case of strongly nonlinear stiffness. The spectral densities of random excitations $\xi_1(t)$ and $\xi_2(t)$ are assumed to be of the following low-pass type

$$S_{ii}(\omega) = \frac{K_{ii}}{\alpha_i^2 + \omega^2}, \quad i = 1, 2. \tag{4.290}$$

In the calculations, $K_{11} = K_{22} = 0.01$, and $\alpha_1 = \alpha_2 = 2$. For the viscoelastic force, different values of the relaxation time λ_1 and the magnitude parameter β_1 will be selected to illustrate their effects on the system responses.

Monte Carlo simulations are performed to assess the accuracy of the proposed method. Instead of the integration in (4.284), the viscoelastic force Z is generated from the first-order differential equation for computational convenience

$$\dot{Z} + \frac{1}{\lambda_1} Z = \beta_1 X. \tag{4.291}$$

With the spectral density given by (4.290), each random excitation $\xi_i(t)$ can be generated from a first-order filter

$$\dot{\xi}_i + \alpha_i \xi_i = W_{gi}(t), \quad i = 1, 2, \tag{4.292}$$

where $W_{gi}(t)$ is a Gaussian white noise with a spectral density K_{ii}.

Shown in Fig. 4.14 are plots of the stationary PDFs of the energy process $\Lambda(t)$ for $\beta_1 = -0.1$, $\gamma = 0.2$ and three different values of relaxation time λ_1. To show a broad range of the probability density, especially the small probability, the vertical coordinate adopts the log scale. According to (4.282), a negative β_1 value causes an increase of damping and a decrease of stiffness. These two effects on the system response are opposite; i.e., an increase of damping reduces the response, and a decrease of stiffness enhances the response. A larger λ_1 corresponds to a higher damping and a lower stiffness. Figure 4.14 shows that the system response is weaker with a large λ_1, indicating that the damping effect is predominant over the stiffness effect. In fact, the system has a strong nonlinear stiffness, and the influence of the viscoelastic force on the stiffness is indeed insignificant. Figure 4.15 depicts the PDFs $p(\lambda)$ for $\lambda_1 = 1$, $\gamma = 0.2$, and three different β_1 values. Since β_1 is the magnitude of the viscoelastic force, a larger magnitude of β_1 results in a higher damping and a weaker stiffness. As in the situation of Fig. 4.14, the damping effect is more significant, and the system response is weaker with a larger β_1 magnitude. To examine the effect of the strongly nonlinear stiffness, cases of different stiffness coefficient γ are considered. Figure 4.16 depict the calculated results of $p(\lambda)$ for $\lambda_1 = 0.1, \beta_1 = -0.1$, and three different γ values. It is shown that the method is accurate even for strong nonlinear stiffness.

Using the results of $p(\lambda)$, the stationary PDFs $p(x)$ of the state process $X(t)$ can be calculated from Eq. (4.77). For different sets of λ_1, β_1 and γ values, the calculated results are plotted in Figs. 4.17, 4.18, and 4.19, in which $p(x)$ functions are only plotted in the positive x range since they are symmetrical in the negative x range. As expected, increasing of either λ_1, $|\beta_1|$ or γ has similar effects to reduce the system response.

Fig. 4.14 PDFs of energy $\Lambda(t)$ for $\gamma = 0.2, \beta_1 = -0.1$ and different λ_1 values (Zhu and Cai 2011)

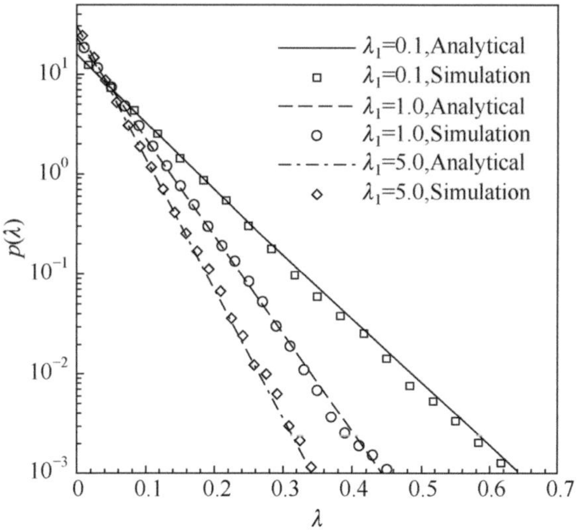

Fig. 4.15 PDFs of energy $\Lambda(t)$ for $\gamma = 0.2, \lambda_1 = 1$ and different β_1 values (Zhu and Cai 2011)

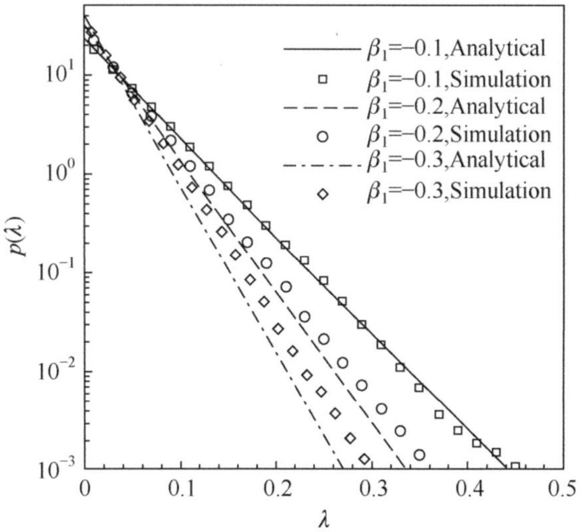

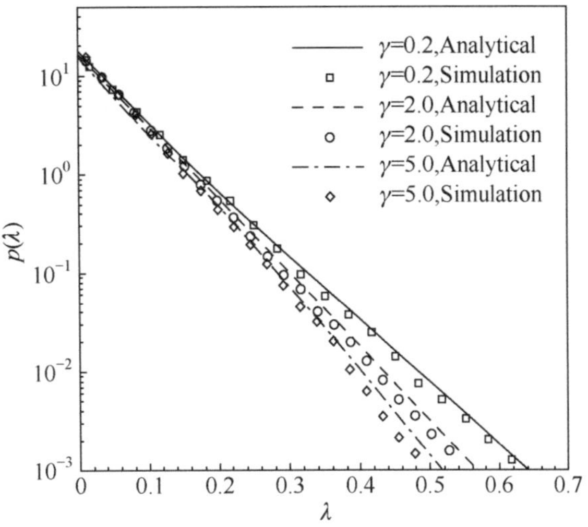

Fig. 4.16 PDFs of energy $\Lambda(t)$ for $\lambda_1 = 0.1, \beta_1 = -0.1$ and different γ values (Zhu and Cai 2011)

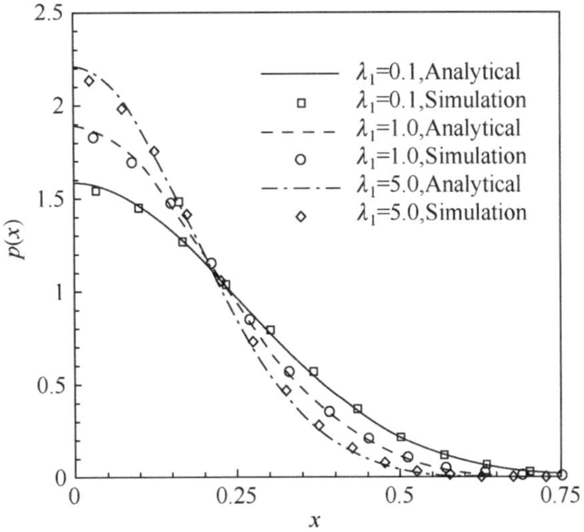

Fig. 4.17 PDFs of state variable $X(t)$ for $\gamma = 0.2, \beta_1 = -0.1$ and different λ_1 values (Zhu and Cai 2011)

In all figures, results obtained from the proposed analytical method agree well with those from Monte Carlo simulations. Errors seem significant in Fig. 4.15 for cases of $\beta_1 = -0.2$ and -0.3. It is expected that the accuracy of the analytical results deteriorates with increasing magnitude of β_1 since the proposed procedure requires a weak viscoelastic force. Note that the log scale is used for the vertical coordinate of $p(\lambda)$, and the discrepancy occurs in regions of small probability. In regions of significant probability, the analytical results are quite accurate, as shown in all figures. Similar situation also takes place for cases of $\gamma = 2$ and 5 in Fig. 4.16 since very strong nonlinear stiffness forces affect the accuracy of the approximation.

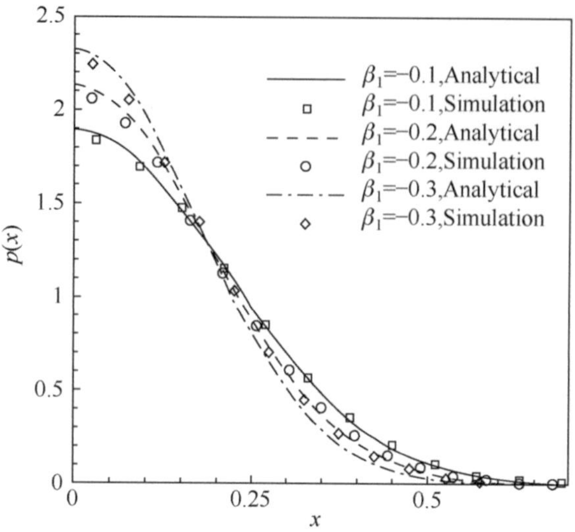

Fig. 4.18 PDFs of state variable $X(t)$ for $\gamma = 0.2$, $\lambda_1 = 1$ and different β_1 values (Zhu and Cai 2011)

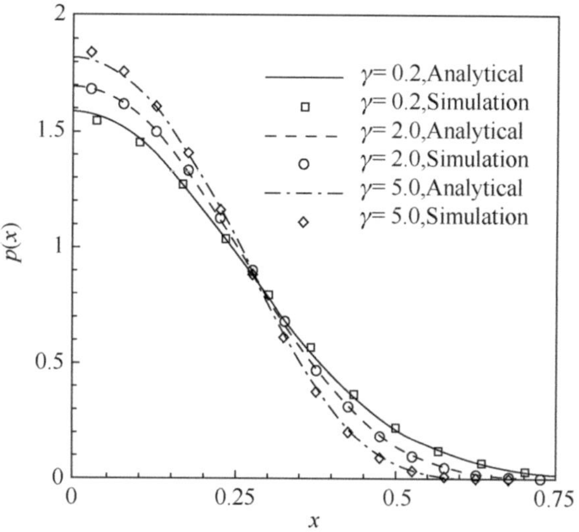

Fig. 4.19 PDFs of state variable $X(t)$ for $\lambda_1 = 0.1, \beta_1 = -0.1$ and different γ values (Zhu and Cai 2011)

4.6 A System with Double-Well Potential

The strongly nonlinear restoring force $u(X)$ considered in the previous sections is assumed to be a monotonic function, namely, its corresponding potential energy has a single-well shape. If the potential has a double-well shape and the restoring force is not monotonic, then the system motion is more complicated. It may move in one well, transit from one well to another, or move all over two wells. Thus, the above developed schemes of the stochastic averaging are no longer applicable, and a new procedure is required to apply the stochastic averaging method appropriately.

4.6.1 Deterministic System with Double-Well Potential

A typical conservative dynamical system with a double-well potential is given by

$$\ddot{x} - \alpha x + \beta x^3 = 0, \tag{4.293}$$

where α and β are two positive constants. The potential energy and total energy of the system are, respectively,

$$U(x) = -\frac{1}{2}\alpha x^2 + \frac{1}{4}\beta x^4 + \frac{\alpha^2}{4\beta}, \tag{4.294}$$

$$\lambda(x, \dot{x}) = \frac{1}{2}\dot{x}^2 - \frac{1}{2}\alpha x^2 + \frac{1}{4}\beta x^4 + \frac{\alpha^2}{4\beta}, \tag{4.295}$$

where the constant $\alpha^2/(4\beta)$ is added so that both the potential energy and total energy are nonnegative. Figure 4.20 shows the double-well potential energy of the system schematically.

Letting $x_1 = x$ and $x_2 = \dot{x}$, system (4.293) can be written in the state space as follows

$$\begin{aligned} \dot{x}_1 &= x_2, \\ \dot{x}_2 &= \alpha x_1 - \beta x_1^3. \end{aligned} \tag{4.296}$$

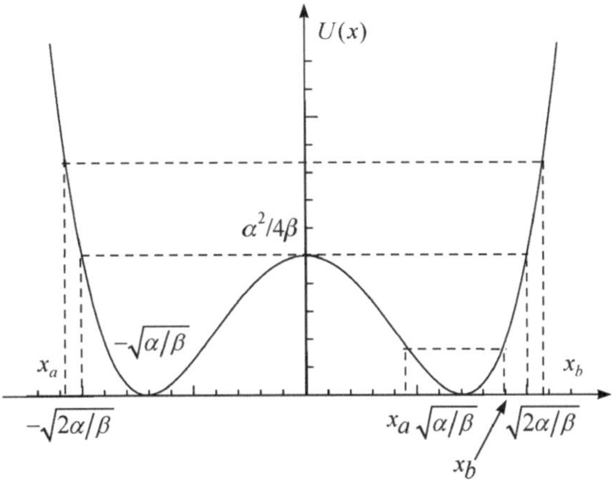

Fig. 4.20 Double-well potential energy of system (4.293)

The system has three equilibrium points: saddle point $(0, 0)$, centers $(-\sqrt{\alpha/\beta},\ 0)$, and $(\sqrt{\alpha/\beta},\ 0)$. For initial conditions other than at these three points, the system motion is periodic. There are two homoclinic orbits which divide the phase plane (x_1, x_2) into three regions, as shown in Fig. 4.21. Depending on the initial condition, the periodic motion may be in one of the potential wells, or pass through both wells. Figure 4.21 shows these periodic motions schematically. If the total energy is less than $\alpha^2/(4\beta)$, there are two possible motions, located on either side of the phase plane. For a given initial state of (x_{10}, x_{20}) in this case, the periodic trajectory is restricted on one side of the phase plane depending on the sign of x_{10}. The lower of the total energy is, the closer of the trajectory is to one of the stable equilibriums. When the total energy exceeds $\alpha^2/(4\beta)$, the system moves across the entire phase plane, and only one periodic trajectory corresponds to a given energy level. When the total energy equals $\alpha^2/(4\beta)$, the system motion is on the homoclinic orbits, which are not periodic. With time approached infinity, the system state approaches a saddle point. It is noted that (i) the periodic motion in either case is far from harmonic except for a very low energy level, (ii) the concept of amplitude with respect to the origin is no longer meaningful in the case of $\lambda < \alpha^2/(4\beta)$.

For a given energy level $\lambda < \alpha^2/(4\beta)$, the natural period of the motion can be calculated from

$$T(\lambda) = \oint dt = \oint \frac{\mathrm{d}x}{\dot{x}} = 2\int_{x_a}^{x_b} \frac{\mathrm{d}x}{\sqrt{2\lambda - \frac{\alpha^2}{2\beta} + \alpha x^2 - \frac{\beta}{2}x^4}}, \tag{4.297}$$

where x_a and x_b are the smallest and largest values of x respectively (see x_a and x_b for right potential well in Fig. 4.20), calculated from

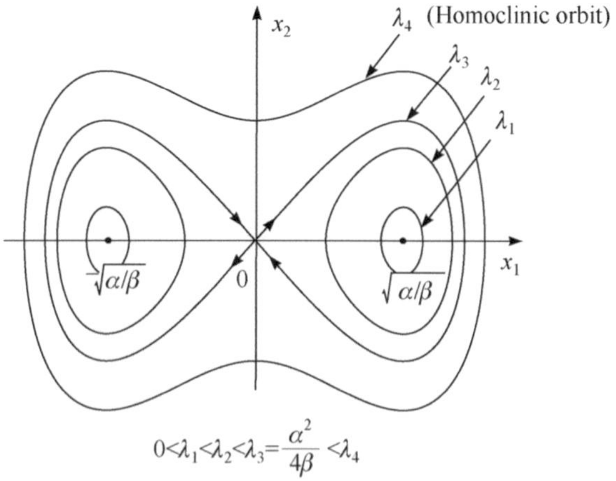

Fig. 4.21 Periodic motions of system (4.293) corresponding to different energy levels

$$x_a = \sqrt{\frac{1}{\beta}\left(\alpha - \sqrt{4\beta\lambda}\right)}, \quad x_b = \sqrt{\frac{1}{\beta}\left(\alpha + \sqrt{4\beta\lambda}\right)}. \tag{4.298}$$

In deriving (4.297), it is assumed that the motion is on the right side of the phase plane, i.e., x is always positive. Due to the symmetry, the period is the same if the motion is on the other side. Neither x_a nor x_b has the meaning of amplitude in this case.

If the energy level $\lambda > \alpha^2/(4\beta)$, the natural period is obtained from

$$T(\lambda) = \oint dt = \oint \frac{\mathrm{dx}}{\dot{\mathrm{x}}} = 4\int_0^{x_b} \frac{\mathrm{d}x}{\sqrt{2\lambda - \frac{\alpha^2}{2\beta} + \alpha x^2 - \frac{\beta}{2}x^4}}, \tag{4.299}$$

where x_b is also given in (4.298) (see right x_b in Fig. 4.20), known as the amplitude of the periodic motion in the case. Figure 4.22 shows the natural period and circular frequency ($\omega = 2\pi/T$) versus the energy level for the case of $\alpha = 2$ and $\beta = 1$. At $\lambda = \alpha^2/(4\beta) = 1$, the period has a jump of twice value. This is because the motion jumps from a small trajectory on one side of the phase plane to a twice large trajectory on the entire plane. When $\lambda < \alpha^2/(4\beta) = 1$, the term $-\alpha x$ plays the dominant role, the system stiffness decreases with an increasing energy level, leading to an increasing period and decreasing frequency. In the energy level of $0 < \lambda < 1$, the natural frequency is in the range of $1 < \omega < 2$. On the other hand, when $\lambda > \alpha^2/(4\beta) = 1$, the term βx^3 is dominant, indicating a hardening stiffness. Thus, the larger the energy level, the larger the natural frequency is and the shorter the period is. However, the natural frequency remains in the range of $1 < \omega < 2$ even up to quite high energy level of $\lambda = 12$. Figure 4.22 also shows that the behavior of the system natural frequency and period is very much different from those of a system with a single potential well.

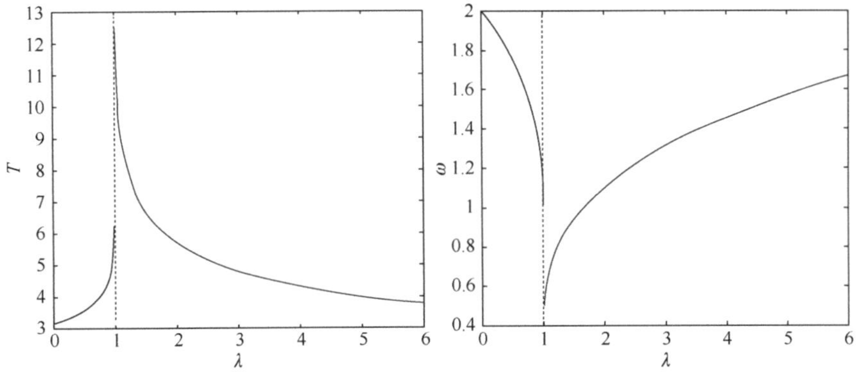

Fig. 4.22 Natural period and frequency of system (4.293) with respect to energy level λ (Zhu et al. 2013)

For the limiting case of $\lambda \to 0$, x approaches either $\sqrt{\alpha/\beta}$ or $-\sqrt{\alpha/\beta}$. Without loss of generality, assume that it approaches $\sqrt{\alpha/\beta}$. Denoting

$$x_e = x - \sqrt{\frac{\alpha}{\beta}}, \tag{4.300}$$

and neglecting higher-order terms, we have

$$\ddot{x}_e + 2\alpha x_e = 0. \tag{4.301}$$

Equation (4.301) indicates that the system can be approximated as a linear oscillator around the equilibrium point ($\sqrt{\alpha/\beta}$, 0), and with a limiting period,

$$\lim_{\lambda \to 0} T = \frac{\sqrt{2}\pi}{\sqrt{\alpha}}. \tag{4.302}$$

4.6.2 Stochastic Averaging

Consider the following stochastic system with a double-well potential

$$\ddot{X} + h(X, \dot{X}) - \alpha X + \beta X^3 = \sum_{i=1}^{n} g_i(X, \dot{X})\xi_i(t), \tag{4.303}$$

where $h(X, \dot{X})$ represents a damping force, $\xi_i(t)$ are stationary broad-band excitations with zero means and the correlation functions $R_{ij}(\tau)$. It is assumed that damping and excitations are weak so that the stochastic averaging is applicable.

In terms of $X(t)$ and $\Lambda(t)$, the system equation of motion, Eq. (4.303), can be replaced by a set of two first-order equations

$$\begin{aligned} \dot{X} &= \pm\sqrt{2\Lambda - \frac{\alpha^2}{2\beta} + \alpha X^2 - \frac{1}{2}\beta X^4}, \\ \dot{\Lambda} &= -\dot{X}h(X, \dot{X}) + \sum_{i=1}^{n} \dot{X} g_i(X, \dot{X})\xi_i(t), \end{aligned} \tag{4.304}$$

where $\dot{X}$ in the second equation is treated as a function of X and Λ according to (4.295). The second equation in (4.304) shows that the energy process $\Lambda(t)$ is slowly varying, and is approximately a Markov diffusion process, governed by an Itô stochastic differential equation

$$\mathrm{d}\Lambda = m(\Lambda)\mathrm{d}t + \sigma(\Lambda)\mathrm{d}B(t), \tag{4.305}$$

where $m(\Lambda)$ and $\sigma(\Lambda)$ can be calculated according to (4.177) and (4.178). It is noticed that the time averaging (4.72) is only applicable for the case in which the restoring force is monotonic. For the present case, the time averaging in one period needs to be reformulated for different ranges of the energy Λ. For $\Lambda < \alpha^2/(4\beta)$, we have

$$\langle[\cdot]\rangle_t = \frac{1}{T}\int_0^T [\cdot]\mathrm{d}t = \frac{2}{T}\int_{x_a}^{x_b} \frac{[\cdot]_{\dot{X}=\sqrt{2\Lambda-2U(X)}}}{\sqrt{2\Lambda - 2U(X)}}\mathrm{d}X, \tag{4.306}$$

where x_a and x_b are given in (4.298). If the energy level $\Lambda > \alpha^2/(4\beta)$,

$$\langle[\cdot]\rangle_t = \frac{1}{T}\int_0^T [\cdot]\mathrm{d}t = \frac{1}{T_{1/4}}\int_0^{x_b} \frac{[\cdot]_{\dot{X}=\sqrt{2\Lambda-2U(X)}}}{\sqrt{2\Lambda - 2U(X)}}\mathrm{d}X, \tag{4.307}$$

where x_b is also given in (4.298). With correlation functions $R_{ij}(\tau)$ given, $m(\Lambda)$ and $\sigma(\Lambda)$ can be calculated numerically.

Example 4.8 As an example, consider the following oscillator (Zhu et al. 2013).

$$\ddot{X} + \gamma\dot{X} - \alpha X + \beta X^3 = X\xi_1(t) + \xi_2(t). \tag{4.308}$$

The drift and diffusion coefficients of the energy process $\Lambda(t)$ are obtained from (4.177) and (4.178) as follows

$$\begin{aligned} m(\Lambda) = & -\gamma\dot{X}_t^2 + \int_{-\infty}^{0} \left\langle \frac{X(t)X(t+\tau)\dot{X}(t+\tau)}{\dot{X}(t)} \right\rangle_t R_{11}(\tau)\mathrm{d}\tau \\ & + \int_{-\infty}^{0} \left\langle \frac{\dot{X}(t+\tau)}{\dot{X}(t)} \right\rangle_t R_{22}(\tau)\mathrm{d}\tau, \end{aligned} \tag{4.309}$$

$$\begin{aligned} \sigma^2(\Lambda) = & \int_{-\infty}^{\infty} \left\langle X(t)\dot{X}(t)X(t+\tau)\dot{X}(t+\tau)\right\rangle_t R_{11}(\tau)\mathrm{d}\tau \\ & + \int_{-\infty}^{\infty} \left\langle \dot{X}(t)\dot{X}(t+\tau)\right\rangle_t R_{22}(\tau)\mathrm{d}\tau. \end{aligned} \tag{4.310}$$

Governed by the Itô stochastic differential equation (4.305) with the drift and diffusion coefficients given by (4.309) and (4.310), the two boundaries at $\Lambda = 0$ and $\Lambda = \infty$ can be classified based on a theory described by Lin and Cai (1995).

As the system approaches the left boundary $\Lambda = 0$, $\dot{X}$ approaches zero and X approaches either $\sqrt{\alpha/\beta}$ or $-\sqrt{\alpha/\beta}$. Assume that the motion is in the well on the right side, and define

$$X_e = X - \sqrt{\alpha/\beta}. \tag{4.311}$$

Substituting (4.311) into (4.293), we obtain the equation of motion for X_e as

$$\ddot{X}_e + \gamma\dot{X}_e + 2\alpha X_e + 3\beta\sqrt{\frac{\alpha}{\beta}}X_e^2 + \beta X_e^3 = X_e\xi_1(t) + \sqrt{\frac{\alpha}{\beta}}\xi_1(t) + \xi_2(t). \tag{4.312}$$

It can then be shown that

$$m(\Lambda) \to \pi\frac{\alpha}{\beta}S_{11}\left(\sqrt{2\alpha}\right) + \pi S_{22}\left(\sqrt{2\alpha}\right), \quad \text{as} \quad \Lambda \to 0, \tag{4.313}$$

$$\sigma^2(\Lambda) \to 2\pi\Lambda\left[\frac{\alpha}{\beta}S_{11}\left(\sqrt{2\alpha}\right) + S_{22}\left(\sqrt{2\alpha}\right)\right], \quad \text{as} \quad \Lambda \to 0. \tag{4.314}$$

According to (Lin and Cai 1995), the left boundary $\Lambda = 0$ is singular of the first kind, and it is an entrance. As the probability flow approaches this boundary, the repulsive force becomes larger, and it forces the system motion back to its defining range.

As $\Lambda(t)$ approaches the right boundary at infinity, the analytical cxprcssions for $m(\Lambda)$ and $\sigma^2(\Lambda)$ are difficult to obtain. But their orders of magnitude can be accessed as follows

$$m(\Lambda) \sim \mathrm{O}(-\Lambda), \quad \sigma^2(\Lambda) \sim \mathrm{O}\left(\Lambda^{3/2}\right), \quad \text{as} \quad \Lambda \to \infty, \tag{4.315}$$

where $\mathrm{O}(\cdot)$ denotes the order of magnitude. Thus, the right boundary $\Lambda = \infty$ is singular of the second kind, and is repulsively natural (Lin and Cai 1995), similar to, but weaker than an entrance.

The behaviors of sample functions of process $\Lambda(t)$ near the two boundaries are represented schematically in Fig. 4.23. It is concluded that the stationary PDF of $\Lambda(t)$ exists.

Consider the case of low-pass random processes for the random excitations $\xi_1(t)$ and $\xi_2(t)$. The correlation functions are

$$R_{ii}(\tau) = D_i e^{-\alpha_i|\tau|}, \quad i = 1,\ 2, \tag{4.316}$$

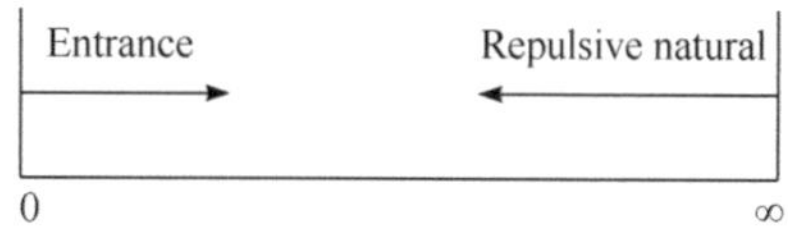

Fig. 4.23 Boundary behaviors of sample functions of process $\Lambda(t)$

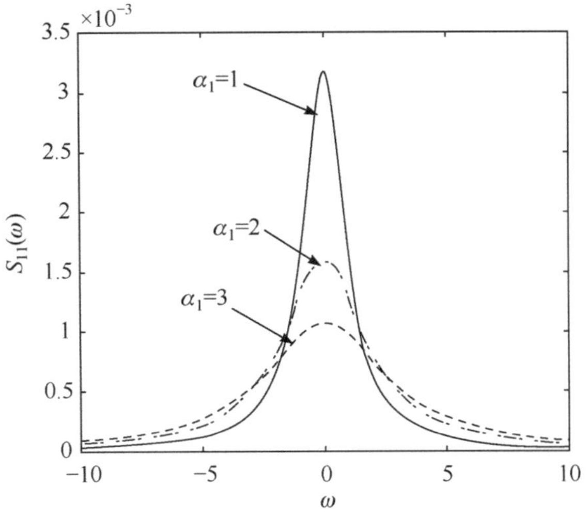

Fig. 4.24 Power spectral densities of the low-pass process with different band-width parameter α_i

and the power spectral densities are

$$S_{ii}(\omega) = \frac{D_i \alpha_i}{\pi \left(\omega^2 + \alpha_i^2\right)}, \tag{4.317}$$

where α_i and D_i are the band width and intensity parameters, respectively. Figure 4.24 depicts the power spectral densities of the low-pass process with $D_1 = 0.01$ and three different α_1 values. The case of $\alpha_1 = 1$ corresponds to a narrow band width, while the process is of broad band if $\alpha_1 = 3$.

Following the proposed procedure, we calculate $m(\lambda)$ and $\sigma^2(\lambda)$ from Eqs. (4.308) and (4.309), and then $p(\lambda)$ from (4.73), $p(x, \dot{x})$ from (4.77), and finally $p(x)$. The numerical results of the stationary PDF $p(x)$ in the range of $x \geq 0$ are depicted in Fig. 4.25 for system parameters $\alpha = 2$, $\beta = 1$, $\gamma = 0.015$, and excitation intensities $D_1 = D_2 = 0.01$. Three different values of the band-width parameter $\alpha_1 = \alpha_2$ are adopted. $p(x)$ in the range of $x < 0$ is symmetric.

Monte Carlo simulations are performed to assess the accuracy of the proposed method. For computational convenience, the excitations $\xi_i(t)$ are generated from the first-order differential equations

$$\dot{\xi}_i + \alpha_i \xi_i = W_{gi}(t), \quad i = 1, 2, \tag{4.318}$$

where each $W_{gi}(t)$ is a Gaussian white noise with a spectral density $K_i = \frac{D_i \alpha_i}{\pi}$. Results from the Monte Carlo simulation are also depicted in Fig. 4.25 to substantiate the accuracy of the analytical results.

Assume that the system is in one well initially. After the random excitations are applied, it begins to oscillate randomly in the well. When the energy exceeds the critical value $\lambda_c = \alpha^2/(4\beta)$, it will jump into another well. The average transition

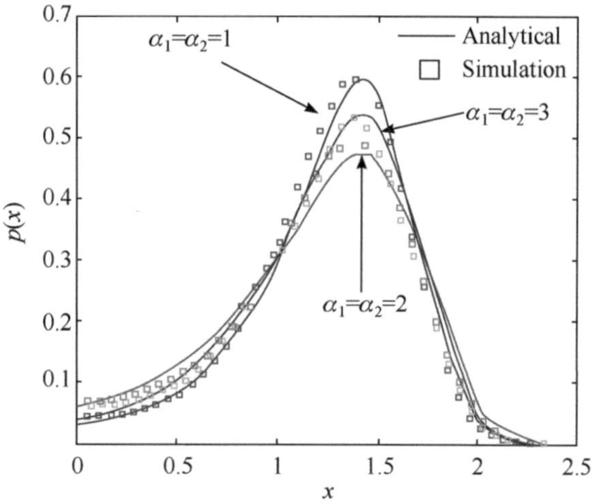

Fig. 4.25 Stationary PDFs of system response subjected to low-pass excitations with different band-width parameters α_1 and α_2 (Zhu et al. 2013)

time beginning from an initial energy level λ_0, denoted by $\mu(\lambda_0)$, is governed by the well-known Pontryagin equation (Andrnov et al. 1933)

$$1 + m(\lambda_0)\frac{\mathrm{d}\mu}{\mathrm{d}\lambda_0} + \frac{1}{2}\sigma^2(\lambda_0)\frac{\mathrm{d}^2\mu}{\mathrm{d}\lambda_0^2} = 0, \tag{4.319}$$

where $m(\lambda_0)$ and $\sigma^2(\lambda_0)$ are given by Eqs. (4.308) and (4.309) with Λ replaced by λ_0. The boundary conditions for Eq. (4.318) are

$$\mu(\lambda_c) = 0, \quad \left.\frac{\mathrm{d}\mu}{\mathrm{d}\lambda_0}\right|_{\lambda_0=0} = -\frac{1}{m(0)}. \tag{4.320}$$

The second condition can be derived directly from Eq. (4.318) since $\sigma^2(0) = 0$ according to (4.311). The solution of (4.318) satisfying the two boundary conditions are derived as (Cai and Zhu 2016)

$$\mu(\lambda_0) = -\int_{\lambda_0}^{\lambda_c} f(z)\mathrm{d}z, \tag{4.321}$$

$$f(z) = \exp\left[-\int_0^z \frac{2m(u)}{\sigma^2(u)}\mathrm{d}u\right]\left\{-\int_0^z \frac{2}{\sigma^2(u)}\exp\left[\int_0^u \frac{2m(v)}{\sigma^2(v)}\mathrm{d}v\right]\mathrm{d}u - \frac{1}{m(0)}\right\}. \tag{4.322}$$

The average transition time can be calculated numerically from (4.321) and (4.322).

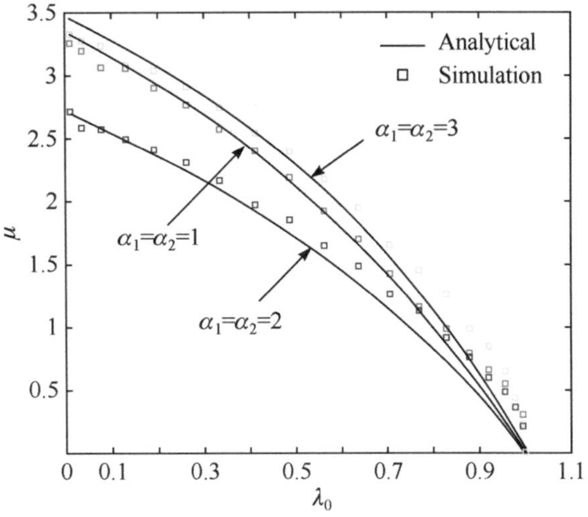

Fig. 4.26 Mean transition time of system subjected to low-pass excitations with different band-width parameter α_i (Zhu et al. 2013)

Assume that, before exposed to the random excitations, the system energy is λ_0, which is less than the critical value $\lambda_c = \alpha^2/(4\beta)$. Upon imposing the excitations, jump to the other well is a random event. Figure 4.26 depicts the calculated results of the mean transition time from one well to another, as well as the simulation ones, with the same system parameters as in Fig. 4.25.

Figures 4.25 and 4.26 show that the band widths of the excitations have significant effects on the system response. It is noticed that, in Fig. 4.25, the curve of the case $\alpha_1 = \alpha_2 = 3$ is between the curses of $\alpha_1 = \alpha_2 = 1$ and $\alpha_1 = \alpha_2 = 2$, indicating the effect of the band width does not follow a one-way trend. Similar phenomenon is also observed in Fig. 4.26. Therefore, the effect of the band width also depends on the system properties, more specifically, the range of the system natural frequency.

Finally, It needs to be clarified that the stochastic averaging method based on Khasminskii theory (Khasminskii 1966, 1968) is not applicable since the motion of the conservative system (4.296) is not periodic at the saddle point and on homoclinic orbits. Therefore, the averaging is performed differently in the different ranges, and neglects the homoclinic orbits. The effect of the saddle point and homoclinic orbits on the system probability distribution is very small, therefore, the results obtained in this section are satisfactory. For more perfection mathematically, Freidlin and other researchers established theories of graphical stochastic averaging for stochastically perturbated Hamiltonian systems with saddle point and homoclinic orbits, and pointed out that certain glueing condition needs to be met at saddle point and homoclinic orbits. The method has been promoted and improved in recent years. Readers of interest can refer to relative literatures (Brin and Freidlin 2000; Choi and Sri Namachchivaya 2006; Freidlin 2001; Freidlin and Sheu 2000; Freidlin and Weber 1998, 1999; Freidlin and Wentzell 1994; 2012; Skorokhod et al. 2002; Sowers 2003, 2005, 2007, 2001; Sowers and Sri Namachchivaya 2002).

4.7 Systems Under Combined Random and Harmonic Excitations

Consider the following SDOF system

$$\ddot{X} + \varepsilon h(X, \dot{X}) + \omega_0^2 X = \varepsilon f(X, \dot{X}) \sin \Omega t + \varepsilon^{1/2} g(X, \dot{X}) \xi(t), \tag{4.323}$$

where $\xi(t)$ is a broad-band process. In system (4.323), the restoring force is linear, the damping and excitations are assumed to be weak. With the small parameter ε appearing in the equation, the contributions from the damping, the harmonic and random excitations are commensurable. Under these assumptions, the stochastic averaging method is applicable.

Consider a simplest case of an external harmonic excitation, i.e., $f(X, \dot{X}) = D$ where D is a constant. Applying the stochastic averaging of amplitude envelop, i.e. letting

$$X = A(t) \cos \Phi, \quad \dot{X} = -A(t)\omega_0 \sin \Phi, \quad \Phi = \omega_0 t + \Theta(t), \tag{4.324}$$

we obtain the equation for the amplitude $A(t)$

$$\begin{aligned} \dot{A} = \frac{\sin \Phi}{\omega_0} [\varepsilon h(A \cos \Phi, -A\omega_0 \sin \Phi) - \varepsilon D \sin \Omega t \\ -\varepsilon^{1/2} g(A \cos \Phi, -A\omega_0 \sin \Phi) \xi(t) \big]. \end{aligned} \tag{4.325}$$

The time averaging for the harmonic excitation term is $\langle \sin(\omega_0 t + \Theta) \sin \Omega t \rangle_t$. If Ω is not close to ω_0, the time averaging is zero, and the effect of the harmonic excitation can be neglected. If Ω is close to ω_0, say the extreme case of exact tuning $\Omega = \omega_0$, we have

$$\langle \sin(\omega_0 t + \Theta) \sin \Omega t \rangle_t = \langle \sin(\omega_0 t + \Theta) \sin \omega_0 t \rangle_t = \cos \Theta. \tag{4.326}$$

Thus, the averaged equations for the two slow variables $A(t)$ and $\Theta(t)$ are still coupled, and the amplitude $A(t)$ cannot be solved simply as a one-dimensional diffusion process. From the above observation, we have the following statement.

Under both harmonic and random excitations, the resonance and non-resonance cases must be treated differently. For the non-resonance case, the harmonic excitation can be ignored, and the stochastic averaging described in Sects. 4.2–4.4 can be applied. If Ω is near the resonance frequency, the harmonic excitation in (4.323) would have a significant effect on the system response. The location of the resonance frequency depends on the function $f(X, \dot{X})$ accompanying the harmonic excitation $\sin\Omega t$. For example, if $f(X, \dot{X}) = \text{constant}$, the resonance frequency is ω_0, and if $f(X, \dot{X}) = aX$, it is $2\omega_0$. In this book, only resonance cases are investigated for nonlinear systems without restriction of exact tuning. For illustration, only an external random excitation is conserved and Eq. (4.323) is changed to the following form

$$\ddot{X} + 2\zeta\omega_0\dot{X} + h(X, \dot{X}) + \omega_0^2 X = f(X, \dot{X})\sin(j\nu t) + \xi(t), \tag{4.327}$$

where ν is a constant close to ω_0. For convenience, the small parameter ε is dropped with the understanding that the terms $2\zeta\omega_0\dot{X}$, $h(X, \dot{X})$, and $f(X, \dot{X})$ are of order ε, and $\xi(t)$ is of order $\varepsilon^{1/2}$

It is shown that the following transformation is more convenient for carrying the stochastic averaging than using the amplitude and phase,

$$X = X_c\cos(\nu t) + X_s\sin(\nu t),$$

$$\dot{X} = \nu - X_c[\sin(\nu t) + X_s\cos(\nu t)]. \tag{4.328}$$

The transformation was used by Ariaratnum and Tam (1976, 1977), Dimentberg (1988), and Zhu and Huang (1984) for stability analysis of linear systems. With transformation (4.328), Eq. (4.327) is transformed to two first-order equations for the new variables X_c and X_s,

$$\begin{aligned}\dot{X}_c = {} & 2\zeta\omega_0\sin(\nu t)[-X_c\sin(\nu t) + X_s\cos(\nu t)] + \frac{1}{\nu}h(X_c, X_s)\sin(\nu t)\\ & - 2\gamma\sin(\nu t)[X_c\cos(\nu t) + X_s\sin(\nu t)]\\ & - \frac{1}{\nu}f(X_c, X_s)\sin(\nu t)\sin(j\nu t) - \frac{1}{\nu}\sin(\nu t)\xi(t),\end{aligned} \tag{4.329}$$

$$\begin{aligned}\dot{X}_s = {} & -2\zeta\omega_0\cos(\nu t)[-X_c\sin(\nu t) + X_s\cos(\nu t)] - \frac{1}{\nu}h(X_c, X_s)\cos(\nu t)\\ & +2\gamma\cos(\nu t)[X_c\cos(\nu t) + X_s\sin(\nu t)]\\ & +\frac{1}{\nu}f(X_c, X_s)\cos(\nu t)\sin(j\nu t) + \frac{1}{\nu}\cos(\nu t)\xi(t),\end{aligned} \tag{4.330}$$

where $h(X_c, X_s)$ and $f(X_c, X_s)$ are obtained from $h(X, \dot{X})$ and $f(X, \dot{X})$, respectively, by changing their arguments from X and $\dot{X}$ to X_c and X_s, according to (4.328), and

$$\gamma = \frac{\nu^2 - \omega_0^2}{2\nu}. \tag{4.331}$$

For $\nu \approx \omega_0$,

$$\gamma = \frac{\nu^2 - \omega_0^2}{2\nu} = \frac{(\nu + \omega_0)(\nu - \omega_0)}{2\nu} \approx \nu - \omega_0. \tag{4.332}$$

Therefore, γ is approximately the difference between ν and ω_0, and it will be referred as the detuning parameter. In this case, the detuning parameter γ is small, the right-hand sides of Eqs. (4.329) and (4.330) are small, both X_c and X_s are slowly varying, and the stochastic averaging can be applied. The broad-band excitation $\xi(t)$

is replaced by a white noise and the fast varying terms in Eqs. (4.329) and (4.330) are eliminated, resulting in the following averaged Itô differential equations

$$\mathrm{d}X_c = \left[-\zeta\omega_0 X_c - \gamma X_s + h_c(X_c, X_s)\right]\mathrm{d}t + \sqrt{2\pi K}\mathrm{d}B_1(t), \tag{4.333}$$

$$\mathrm{d}X_s = \left[\gamma X_c - \zeta\omega_0 X_s + h_s(X_c, X_s)\right]\mathrm{d}t + \sqrt{2\pi K}\mathrm{d}B_2(t), \tag{4.334}$$

where $B_1(t)$ and $B_2(t)$ are independent Wiener processes, functions $h_c(X_c, X_s)$ and $h_s(X_c, X_s)$ will be determined according to the specific $h(X, \dot{X})$ and $f(X, \dot{X})$ functions for the investigated problem, and

$$K = \frac{S_{\xi\xi}(\nu)}{2\nu^2}. \tag{4.335}$$

Equations (4.329) and (4.330) show that whether the harmonic excitation $\sin(j\nu t)$ has a significant contribution after time averaging depends on the nature of its accompanying function $f(X, \dot{X})$. The system is said to be in resonance if the harmonic excitation contributes significantly to the system response.

Equations (4.333) and (4.334) indicate that processes $X_c(t)$ and $X_s(t)$ will approach stationary states if the system is stable. The reduced FPK equation for the stationary PDF $p(x_c, x_s)$ is

$$\begin{aligned} &\frac{\partial}{\partial x_c}\left\{\left[-\zeta\omega_0 x_c - \gamma x_s + h_c(x_c, x_s)\right]p\right\} \\ &+ \frac{\partial}{\partial x_s}\left\{\left[\gamma x_c - \zeta\omega_0 x_s + h_s(x_c, x_s)\right]p\right\} \\ &- \pi K\frac{\partial^2 p}{\partial x_c^2} - \pi K\frac{\partial^2 p}{\partial x_s^2} = 0. \end{aligned} \tag{4.336}$$

The two-dimensional partial differential Eq. (4.336) can be solved analytically or numerically. Two examples are presented below for illustration.

Example 4.9 Consider a linear system under an external random excitation and a parametric harmonic excitation (Lin and Cai 1995). The system equation of motion is

$$\ddot{X} + 2\zeta\omega_0\dot{X} + \omega_0^2 X[1 + \lambda\sin(2\nu t)] = \xi(t), \tag{4.337}$$

where $\xi(t)$ is a broad-band process with a spectral density $\Phi(\omega)$. Comparing (4.337) and (4.327), we have $h = 0, f = \lambda\omega_0^2 X$, and it is found that $h_c = \zeta\omega_0 r X_c$, and $h_s = -\zeta\omega_0 r X_s$. Equations (4.333) and (4.334) are then specified to

$$\mathrm{d}X_c = \left[-\zeta\omega_0(1 - r)X_c - \gamma X_s\right]\mathrm{d}t + \sqrt{2\pi K}\mathrm{d}B_1(t), \tag{4.338}$$

$$dX_s = \left[\gamma X_c - \zeta\omega_0(1+r)X_s\right]dt + \sqrt{2\pi K}dB_2(t), \tag{4.339}$$

where

$$\gamma = \frac{\nu^2 - \omega_0^2}{2\nu}, \quad r = \frac{\omega_0\lambda}{4\zeta\nu}, \quad K = \frac{S(\nu)}{2\nu^2}. \tag{4.340}$$

It is noted that the stochastic averaging used to obtain (4.338) and (4.339) is only applicable for the case of resonance, i.e., $\nu \approx \omega_0$, and the detuning parameter γ is small.

In the following, we try to solve the following reduced FPK equation corresponding to the system given by (4.338) and (4.339). The reduced FPK equation may be replaces by the following sufficient conditions

$$\left[-\zeta\omega_0(1-r)x_c - \gamma x_s\right]p - \pi K\frac{\partial p}{\partial x_c} - D\frac{\partial p}{\partial x_s} = 0, \tag{4.341}$$

$$\left[\gamma x_c - \zeta\omega_0(1+r)x_s\right]p + D\frac{\partial p}{\partial x_c} - \pi K\frac{\partial p}{\partial x_s} = 0. \tag{4.342}$$

Letting

$$p(x_c, x_s) = C\exp[-\phi(x_c, x_s)], \tag{4.343}$$

where $\phi(x_c, x_s)$ is known as the probability potential (Lin and Cai 1995). Equations (4.341) and (4.342) are transformed to

$$\pi K\frac{\partial\phi}{\partial x_c} + D\frac{\partial\phi}{\partial x_s} = \zeta\omega_0(1-r)x_c + \gamma x_s, \tag{4.344}$$

$$-D\frac{\partial\phi}{\partial x_c} + \pi K\frac{\partial\phi}{\partial x_s} = -\gamma x_c + \zeta\omega_0(1+r)x_s. \tag{4.345}$$

From (4.344) and (4.345), $\partial\phi/\partial x_c$ and $\partial\phi/\partial x_s$ can be solved as

$$\frac{\partial\phi}{\partial x_c} = \frac{1}{\pi^2K^2 + D^2}\left\{\left[\pi K\zeta\omega_0(1+r) + \gamma D\right]x_c + \left[\pi K\gamma - D\zeta\omega_0(1+r)\right]x_s\right\}, \tag{4.346}$$

$$\frac{\partial\phi}{\partial x_s} = \frac{1}{\pi^2K^2 + D^2}\left\{\left[D\zeta\omega_0(1-r) - \pi K\gamma\right]x_c + \left[D\gamma + \pi K\zeta\omega_0(1+r)\right]x_s\right\}. \tag{4.347}$$

In order for function $\phi(x_c, x_s)$ to exist, the following compatibility must be satisfied

$$\frac{\partial}{\partial x_s}\left(\frac{\partial \phi}{\partial x_c}\right) = \frac{\partial}{\partial x_c}\left(\frac{\partial \phi}{\partial x_s}\right). \tag{4.348}$$

Substitution of Eqs. (4.346) and (4.347) into (4.348) leads to

$$D = \pi K \eta, \quad \eta = \frac{\gamma}{\zeta \omega_0}. \tag{4.349}$$

With D given by (4.349), the probability potential function is solved from (4.346) and (4.347) as

$$\phi(x_c, x_s) = \frac{\zeta \omega_0}{2\pi K\left(1+\eta^2\right)}\left[(1 - r + \eta^2)x_c^2 - 2r\eta x_c x_s + \left(1 + r + \eta^2\right)x_s^2\right], \tag{4.350}$$

and the stationary PDF is

$$p(x_c, x_s) = \frac{\zeta \omega_0 \sqrt{1-\mu^2}}{2\pi^2 K} \times \exp\left\{-\frac{\zeta \omega_0}{2\pi K\left(1+\eta^2\right)}\left[(1 - r + \eta^2)x_c^2 - 2r\eta x_c x_s + \left(1 + r + \eta^2\right)x_s^2\right]\right\}, \tag{4.351}$$

where $\mu = \frac{r}{\sqrt{1+\eta^2}}$. The condition for (4.351) to be a valid PDF is

$$\mu = \frac{r}{\sqrt{1+\eta^2}} < 1. \tag{4.352}$$

The PDF in (4.351) is the exact solution for the averaged system represented by the Itô Eqs. (4.338) and (4.339), and the averaged system belong to the class of stationary potential (Lin and Cai 1995). However, the solution (4.351) is an approximate solution of system (4.337) due to the stochastic averaging. Solution (4.351) shows that processes $X_c(t)$ and $X_s(t)$ are Gaussian distributed.

The stability boundary is defined by $\mu = 1$. In term of the system original parameters it is

$$\lambda^2 = 4\left(\frac{\nu}{\omega_0}\right)^4 - 8\left(1 - 2\zeta^2\right)\left(\frac{\nu}{\omega_0}\right)^2 + 4. \tag{4.353}$$

Equation (4.353) confirms the known fact that the stability of linear system (4.337) only depends on the parametric excitation. Figure 4.27 shows the stability region in the parameter space (λ, ν/ω_0) for different damping ratio ζ. Such a system can be more stable by reducing the intensity λ of the harmonic excitation, by increasing the detuning parameter γ, or by increasing the damping ratio ζ.

The PDF of $X(t)$ and $\dot{X}(t)$ can be determined from (4.351) according to the transformation (4.328). It is clear that, although $X_c(t)$ and $X_s(t)$ are stationary, $X(t)$

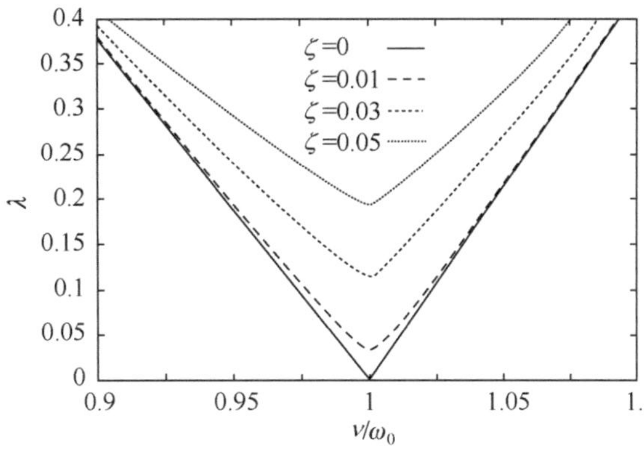

Fig. 4.27 Stability boundary for system (4.337)

and $\dot{X}(t)$ are non-stationary except for the case $\lambda = 0$, i.e., without the harmonic excitation.

Since $X_c(t)$ and $X_s(t)$ defined by (4.328) do not have explicit physical meaning, we transform them into another pair of variables

$$X_c = A\cos\Theta, \quad X_c = A\sin\Theta, \tag{4.354}$$

which is equivalent to

$$X = A\cos(\nu t - \Theta), \quad \dot{X} = -\nu A\sin(\nu t - \Theta). \tag{4.355}$$

The new processes $A(t)$ and $\Theta(t)$ are the amplitude and phase for the narrow-band process $X(t)$, and they are also slowly varying variables. The joint PDF of A and Θ can be obtained from (4.351) by change of variables using the equation (Cai and Zhu 2016)

$$p(a,\theta) = p[x_c(a,\theta), x_s(a,\theta)] \begin{vmatrix} \frac{\partial x_c}{\partial a} & \frac{\partial x_c}{\partial \theta} \\ \frac{\partial x_s}{\partial a} & \frac{\partial x_s}{\partial \theta} \end{vmatrix}. \tag{4.356}$$

The result is

$$p(a,\theta) = \frac{\zeta\omega_0\sqrt{1-\mu^2}}{2\pi^2 K} a \exp\left\{-\frac{\zeta\omega_0 a^2}{2\pi K\left(1+\eta^2\right)}[1-\mu\cos(2\theta-2\psi)]\right\}, \tag{4.357}$$

where η and μ are given in (4.349) and (4.352), respectively, and ψ is determined from

$$\cos(2\psi) = \frac{1}{\sqrt{1+\eta^2}}, \quad \sin(2\psi) = \frac{\eta}{\sqrt{1+\eta^2}}. \tag{4.358}$$

The marginal PDF for A and Θ can be obtained by integrating (4.357) with respect to θ and a respectively as follows

$$p(a) = \frac{\zeta\omega_0\sqrt{1-\mu^2}}{\pi K} a\mathrm{I}_0\left(\frac{\mu\zeta\omega_0 a^2}{2\pi K}\right)\exp\left(-\frac{\zeta\omega_0 a^2}{2\pi K}\right), \tag{4.359}$$

$$p(\theta) = \frac{\sqrt{1-\mu^2}}{2\pi[1-\mu\cos(2\theta-2\psi)]}, \tag{4.360}$$

where $\mathrm{I}_0(\cdot)$ is the modified Bessel function of the zeroth order. As expected for the case without the harmonic excitation, i.e., $\lambda = 0$, hence $\mu = 0$, the amplitude A is Rayleigh distributed and the phase angle Θ is uniformly distributed.

Stationary PDF $p(a)$ was calculated for system (4.337) with $\xi = 0.05$, $\omega_0 = 10$, $\nu = 11$, and $S(\nu) = 10$. The numerical results obtained from the analytical solution (4.359) and from simulations are shown in Fig. 4.28 for $\lambda = 0.05$ and 0.2. Figure 4.29 depicts the mean value of A versus the frequency ν for different intensities of the harmonic excitation. The analytical results agree very well with the simulation results. As mentioned before, the small errors are from the stochastic averaging.

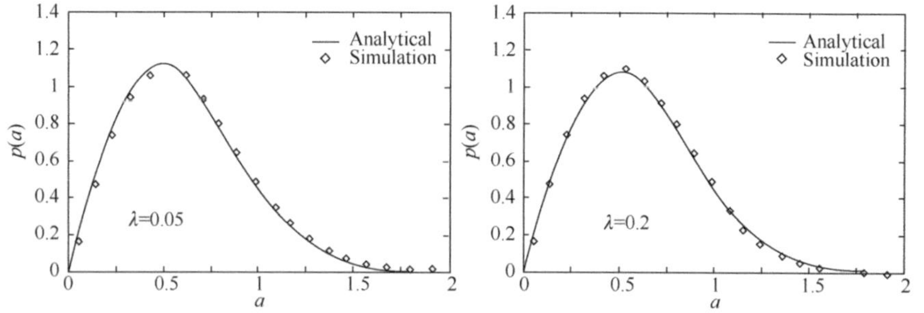

Fig. 4.28 Stationary PDF of response amplitude A for system (4.337) (Lin and Cai 1995)

Fig. 4.29 Mean value of response amplitude A for system (4.337) (Lin and Cai 1995)

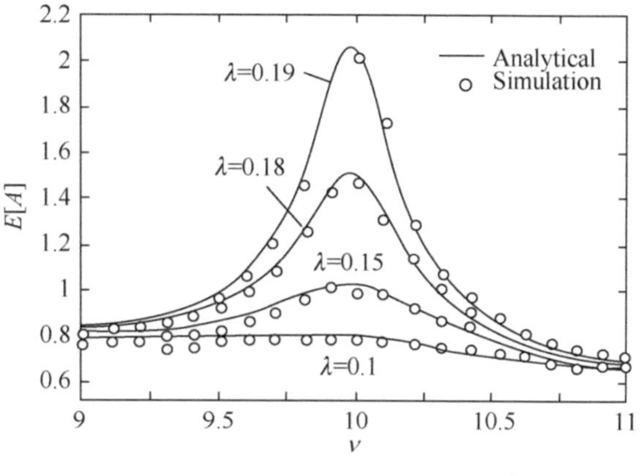

Example 4.10 A nonlinearly damped system is under an external random excitation and a parametric harmonic excitation (Lin and Cai 1995). The system equation of motion is

$$\ddot{X} + 2\zeta\omega_0\dot{X} + \beta X^2\dot{X} + \omega_0^2 X[1 + \lambda\sin(2\nu t)] = \xi(t), \tag{4.361}$$

where $\xi(t)$ is a broad-band process with a spectral density $S(\omega)$. Comparing (4.361) and (4.327), we have $h = \beta X^2\dot{X}, f = \lambda\omega_0^2 X$, and it is found that h_c and h_s functions in (4.333) and (4.334) are

$$\begin{aligned} h_c &= \zeta\omega_0 r X_c - \frac{\beta}{8}X_c\left(X_c^2 + X_s^2\right), \\ h_s &= -\zeta\omega_0 r X_s - \frac{\beta}{8}X_s\left(X_c^2 + X_s^2\right). \end{aligned} \tag{4.362}$$

For the resonance case, i.e., $\nu \approx \omega_0$, the FPK equation is (4.335) which may be replaces by the following sufficient conditions

$$\left[-\zeta\omega_0 x_c - \gamma x_s + h_c(x_c, x_s)\right]p - \pi K\frac{\partial p}{\partial x_c} - D\frac{\partial p}{\partial x_s} = 0, \tag{4.363}$$

$$\left[\gamma x_c - \zeta\omega_0 x_s + h_s(x_c, x_s)\right]p + D\frac{\partial p}{\partial x_c} - \pi K\frac{\partial p}{\partial x_s} = 0, \tag{4.364}$$

where γ, r, and K are given in (4.340). In terms of the probability potential $\phi(x_c, x_s)$ defined in (4.343), Eqs. (4.363) and (4.364) are transformed to

$$\pi K\frac{\partial\phi}{\partial x_c} + D\frac{\partial\phi}{\partial x_s} = \zeta\omega_0 x_c + \gamma x_s - h_c(x_c, x_s), \tag{4.365}$$

$$-D\frac{\partial\phi}{\partial x_c} + \pi K\frac{\partial\phi}{\partial x_s} = -\gamma x_c + \zeta\omega_0 x_s - h_s(x_c, x_s). \tag{4.366}$$

Solving $\partial\phi/\partial x_c$ and $\partial\phi/\partial x_s$ from (4.365) and (4.366), and applying the compatibility condition (4.348), we obtain

$$2\pi K\gamma - 2\zeta\omega_0 D - \frac{\beta}{2}D\left(x_c^2 + x_s^2\right) = 0. \tag{4.367}$$

Since x_c and x_s are variables, condition (4.367) can be met only if $\gamma = 0$ and $D = 0$, which is the case of exact tuning, i.e. $\nu = \omega_0$. In this case, we have

$$\phi(x_c, x_s) = \frac{1}{2\pi K}\left[\zeta\omega_0(1 - r)x_c^2 + \zeta\omega_0(1 + r)x_s^2 + \frac{\beta}{16}(x_c^2 + x_s^2)^2\right]. \tag{4.368}$$

By coordinate transformation from X_c and X_s to A and Θ according to Eq. (4.354), the joint PDF of the amplitude $A(t)$ and the phase $\Theta(t)$ is obtained as

$$p(a,\theta) = Ca\exp\left\{-\frac{\zeta\omega_0 a^2}{2\pi K}[1-\mu\cos(2\theta)] - \frac{\beta}{32\pi K}a^4\right\}. \tag{4.369}$$

The stationary PDF of $A(t)$ is determined by integrating (4.369) with respect to θ to yield

$$p(a) = CaI_0\left(\frac{\mu\zeta\omega_0 a^2}{2\pi K}\right)\exp\left(-\frac{\zeta\omega_0 a^2}{2\pi K} - \frac{\beta a^4}{32\pi K}\right). \tag{4.370}$$

The PDFs in (4.369) and (4.370) are valid ones as long as $\beta > 0$, which guarantees a stable system (4.361).

In the case of not exact tuning, i.e., $\gamma \neq 0$, Eq. (4.367) cannot be satisfied, an approximate solution procedure is required. The idea is to replace functions h_c and h_s in (4.362) appropriately so that the averaged system of (4.333) and (4.334) possesses an exact solution, which can serve as approximate solution for the original system with h_c and h_s functions. Assume that the replacement systems is represented by

$$\mathrm{d}X_c = \left[-\zeta\omega_0 X_c - \gamma X_s + H_c(X_c, X_s)\right]\mathrm{d}t + \sqrt{2\pi K}\mathrm{d}B_1(t), \tag{4.371}$$

$$\mathrm{d}X_s = \left[\gamma X_c - \zeta\omega_0 X_s + H_s(X_c, X_s)\right]\mathrm{d}t + \sqrt{2\pi K}\mathrm{d}B_2(t), \tag{4.372}$$

and system of (4.371) and (4.372) has an exact PDF and potential functions

$$\begin{aligned} p(x_c, x_s) &= C\exp[-\phi(x_c, x_s)],\\ \phi(x_c, x_s) &= \phi_l(x_c, x_s) + c_1 x_c^4 + c_2 x_c^2 x_s^2 + c_3 x_s^4 u, \end{aligned} \tag{4.373}$$

where $\phi_l(x_c, x_s)$ is the potential function for the linear system if $\beta = 0$, given by Eq. (4.350), i.e.,

$$\phi_l(x_c, x_s) = \frac{\zeta\omega_0}{2\pi K\left(1+\eta^2\right)}\left[\left(1 - r + \eta^2\right)x_c^2 - 2r\eta x_c x_s + \left(1 + r + \eta^2\right)x_s^2\right]. \tag{4.374}$$

In (4.373), the fourth-order power functions are added due to the nonlinear damping term in the system (4.361). Using Eq. (4.365) and (4.366), we find that functions H_c and H_s should have the following forms

$$H_c = \zeta\omega_0 r X_c - 4\pi K c_1 x_c^3 - 2Dc_2 x_c^2 x_s - 2\pi K c_2 x_c x_s^2 - 4Dc_3 x_s^3, \tag{4.375}$$

$$H_s = -\zeta\omega_0 r X_s + 4Dc_1 x_c^3 - 2\pi K c_2 x_c^2 x_s + 2Dc_2 x_c x_s^2 - 4\pi K c_3 x_s^3. \tag{4.376}$$

The FPK equation for the replacement system (4.371) and (4.372) is

$$\begin{aligned}&\frac{\partial}{\partial x_c}\{[-\zeta\omega_0 x_c-\gamma x_s+H_c(x_c,x_s)]p\}\\&+\frac{\partial}{\partial x_s}\{[\gamma x_c-\zeta\omega_0 x_s+H_s(x_c,x_s)]p\}\\&-\pi K\frac{\partial^2 p}{\partial x_c^2}-\pi K\frac{\partial^2 p}{\partial x_s^2}=0.\end{aligned}\tag{4.377}$$

Since $p(x_c, x_s)$ is not exact solution for the original system, we have the residue error

$$\begin{aligned}\delta=&\frac{\partial}{\partial x_c}\{[-\zeta\omega_0 x_c-\gamma x_s+h_c(x_c,x_s)]p\}\\&+\frac{\partial}{\partial x_s}\{[\gamma x_c-\zeta\omega_0 x_s+h_s(x_c,x_s)]p\}\\&-\pi K\frac{\partial^2 p}{\partial x_c^2}-\pi K\frac{\partial^2 p}{\partial x_s^2}.\end{aligned}\tag{4.378}$$

Subtraction of (4.377) from (4.378) leads to

$$\delta=\frac{\partial}{\partial x_c}[h_c-H_c)p]+\frac{\partial}{\partial x_s}[h_s-H_s)p].\tag{4.379}$$

This residue error should be as small as possible according to a certain criterion. Here we apply the method of weighted residuals (Lin and Cai 1995), and obtain

$$\Delta_M=\int M(x_c,x_s)\delta dx_c dx_s=E\left[(h_c-H_c)\frac{\partial M}{\partial x_c}+(h_s-H_s)\frac{\partial M}{\partial x_s}\right]=0,\tag{4.380}$$

where M is a selected weighting function. Letting $M = x_c^2, x_c x_s$, and x_s^2, and substituting h_c and h_s from (4.362) and H_c and H_s from (4.375) and (4.376), we obtain

$$2\pi K m_{40}c_1+(Dm_{31}+\pi K m_{22})c_2+2Dm_{13}c_3=\frac{\beta}{16}(m_{40}+m_{22}),\tag{4.381}$$

$$\begin{aligned}&2(-Dm_{40}+\pi K m_{31})c_1+\pi K(m_{31}+m_{13})c_2\\&+2(\pi K m_{13}+Dm_{04})c_3=\frac{\beta}{8}(m_{31}+m_{13}),\end{aligned}\tag{4.382}$$

$$-2Dm_{31}c_1+(\pi K m_{22}-Dm_{13})c_2+2\pi K m_{04}c_3=\frac{\beta}{16}(m_{22}+m_{04}),\tag{4.383}$$

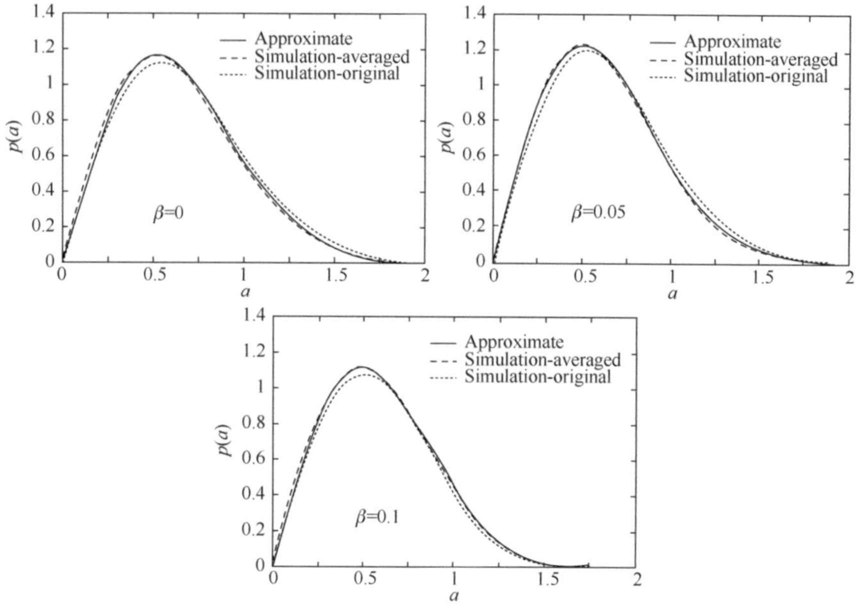

Fig. 4.30 Stationary PDF of response amplitude A for system (4.361) (Lin and Cai 1995)

where $m_{ij} = E[x_c^i x_s^j]$. Equations (4.381)–(4.383) are a set of linear algebraic equations for c_1, c_2, and c_3, and they can be solved numerically in conjunction with Eqs. (4.373) and (4.374).

Numerical calculations were carried out for system (4.361) with $\xi = 0.05$, $\omega_0 = 10$, $v = 11$, $\lambda = 0.05$, $\Phi(v) = 10$, and different values of nonlinear parameter β. Monte Carlo simulations were also performed for both the original system (4.361) and the averaged system given by (4.333), (4.334), and (4.362). It is seen that the analytical approximation results are indistinguishable from the simulation results based on the averaged system. The errors are from the stochastic averaging procedure Fig. 4.30.

4.8 Systems Under Poisson White Noise Excitations

Consider the following n-dimensional stochastic system

$$\dot{X}_i = -f_i(\boldsymbol{X}) + h_i(\boldsymbol{X}) W_{pi}(t), \quad i = 1, 2, \cdots, n, \tag{4.384}$$

where $\boldsymbol{X} = [X_1, X_2, \ldots, X_n]^{\mathrm{T}}$, and $W_{pi}(t)$, $i = 1, 2, \ldots, n$ are independent Poison white noises. In terms of compound Poisson processes (2.174), (4.384) is equivalent to

$$\mathrm{d}X_i = -f_i\mathrm{d}t + h_i\mathrm{d}C_i(t) + \frac{1}{2}h_i\frac{\partial h_i}{\partial X_i}[\mathrm{d}C_i(t)]]^2 + \frac{1}{3!}\left[h_i^2\frac{\partial^2 h_i}{\partial X_i^2} + h_i\frac{\partial h_i}{\partial X_i})^2\right][\mathrm{d}C_i(t)]^3$$
$$+\frac{1}{4!}\left[h_i^3\frac{\partial^3 h_i}{\partial X_i^3} + 4h_i^2\frac{\partial h_i}{\partial X_i}\frac{\partial^2 h_i}{\partial X_i^2} + h_i\left(\frac{\partial h_i}{\partial X_i}\right)^3\right][\mathrm{d}C_i(t)]^4 + \cdots\cdots. \quad (4.385)$$

For function $F(\boldsymbol{X}, t)$, we have according to (2.171)

$$\mathrm{d}F(\boldsymbol{X}, t) = \frac{\partial F}{\partial t}\mathrm{d}t + \sum_{j=1}^{n}\frac{\partial F}{\partial X_i}\mathrm{d}X_i + \frac{1}{2!}\sum_{i,j=1}^{n}\frac{\partial^2 F}{\partial X_i\partial X_j}\mathrm{d}X_i\mathrm{d}X_j$$
$$+\frac{1}{3!}\sum_{i,j,k=1}^{n}\frac{\partial^3 F}{\partial X_i\partial X_j\partial X_k}\mathrm{d}X_i\mathrm{d}X_j\mathrm{d}X_k + \cdots\cdots. \quad (4.386)$$

4.8.1 Amplitude Envelope

The above theories are now applied to a SDOF system with linear stiffness, weak damping, and weak excitations,

$$\ddot{X} + f\left(X, \dot{X}\right) + \omega_0^2 X = h\left(X, \dot{X}\right)W_p(t). \quad (4.387)$$

Letting $X_1 = X, X_2 = \dot{X}$, in terms of compound Poisson processes, (4.387) is equivalent to

$$\mathrm{d}X_1 = X_2\mathrm{d}t,$$
$$\mathrm{d}X_2 = \left(-f - \omega_0^2 X_1\right)\mathrm{d}t + h\mathrm{d}C(t) + \frac{1}{2}h\frac{\partial h}{\partial X_2}[\mathrm{d}C(t)]^2$$
$$+\frac{1}{3!}\left[h^2\frac{\partial^2 h}{\partial X_2^2} + h\left(\frac{\partial h}{\partial X_2}\right)\right]^2[\mathrm{d}C(t)]^3$$
$$+\frac{1}{4!}\left[h^3\frac{\partial^3 h}{\partial X_2^3} + 4h^2\frac{\partial h}{\partial X_2}\frac{\partial^2 h}{\partial X_2^2} + h\left(\frac{\partial h}{\partial X_2}\right)^3\right][\mathrm{d}C(t)]^4 + \cdots\cdots. \quad (4.388)$$

Applying the transformation (4.33), the amplitude process is

$$A = \left(X_1^2 + \frac{1}{\omega_0^2}X_2^2\right)^{1/2}. \quad (4.389)$$

From (4.33) and (4.388), we obtain

$$\frac{\partial A}{\partial X_1} = \cos\Phi, \frac{\partial A}{\partial X_2} = -\frac{\sin\Phi}{\omega_0}, \frac{\partial^2 A}{\partial X_2^2} = \frac{\cos^2\Phi}{A\omega_0^2},$$

$$\frac{\partial^3 A}{\partial X_2^3} = \frac{3\cos^2\Phi\sin\Phi}{A^2\omega_0^3}, \frac{\partial^4 A}{\partial X_2^4} = -\frac{3\cos^2\Phi\left(1-5\sin^2\Phi\right)}{A^3\omega_0^4}. \tag{4.390}$$

Application of (4.386) for the amplitude process $A(t)$, we obtain

$$\begin{aligned} \mathrm{d}A = & -\omega_0 A\sin\Phi\cos\Phi \mathrm{d}t - \frac{\sin\Phi}{\omega_0}\mathrm{d}X_2 + \frac{1}{2}\frac{\cos^2\Phi}{A\omega_0^2}(\mathrm{d}X_2)^2 \\ & + \frac{1}{2}\frac{\cos^2\Phi\sin\Phi}{A^2\omega_0^3}(\mathrm{d}X_2)^3 - \frac{1}{8}\frac{\cos^2\Phi\left(1-5\sin^2\Phi\right)}{A^3\omega_0^4}(\mathrm{d}X_2)^4 + \cdots. \end{aligned} \tag{4.391}$$

Given functions $f\left(X,\dot{X}\right)$ and $h\left(X,\dot{X}\right)$ in Eq. (4.387), Eq. (4.391) can be formulated specifically. The derivate moments in FPK equation can be calculated from (4.391). After carrying out the time averaging with respect to Φ as shown in (4.42), the derivate moments can be calculated are function of the amplitude only. Since the FPK equation corresponding to (4.384) has infinite number of terms, approximation procedures are needed to solve for PDF function.

Example 4.11 Consider a linear stochastic system governed by

$$\ddot{X} + 2\zeta\omega_0\dot{X} + \omega_0^2 X - W_p(t), \tag{4.392}$$

where $W_p(t)$ is a Poisson white noise. Note that there is only an external excitation. Letting $X_1 = X,\ X_2 = \dot{X}$, (4.392) can be written in terms of compounded Poisson process as

$$\begin{aligned} \mathrm{d}X_1 &= X_2 \mathrm{d}t, \\ \mathrm{d}X_2 &= \left(-2\xi\omega_0 X_2 - \omega_0^2 X_1\right)\mathrm{d}t + \mathrm{d}C(t). \end{aligned} \tag{4.393}$$

From (4.391) we have

$$\begin{aligned} \mathrm{d}A = & -2\xi\omega_0 A\sin^2\Phi dt - \frac{\sin\Phi}{\omega_0}\mathrm{d}C(t) + \frac{1}{2}\frac{\cos^2\Phi}{A\omega_0^2}[\mathrm{d}C(t)]^2 \\ & + \frac{1}{2}\frac{\cos^2\Phi\sin\Phi}{A^2\omega_0^3}[\mathrm{d}C(t)]^3 - \frac{1}{8}\frac{\cos^2\Phi\left(1-5\sin^2\Phi\right)}{A^3\omega_0^4}[\mathrm{d}C(t)]^4 + \cdots. \end{aligned} \tag{4.394}$$

Carrying out time averaging, assuming $E[C^k(t)] = 0$ for $k =$ odd, and neglecting D_n for $n > 4$ in (2.145) and (2.146), the first four derivate moments are

$$a_1(a) = -\xi\omega_0 a + \frac{D_2}{4a\omega_0^2} + \frac{D_4}{64a^3\omega_0^4}, \quad b_{11}(a) = \frac{D_2}{2\omega_0^2} - \frac{D_4}{32a^2\omega_0^4},$$
$$c_{111}(a) = \frac{3D_4}{16a\omega_0^4}, \quad d_{1111}(a) = \frac{3D_4}{8\omega_0^4}. \tag{4.395}$$

The reduced FPK equation for the amplitude process is then

$$-\frac{\mathrm{d}}{da}(a_1 p) + \frac{1}{2!}\frac{\mathrm{d}^2}{da^2}(b_{11}p) - \frac{1}{3!}\frac{\mathrm{d}^3}{da^3}(c_{111}p) + \frac{1}{4!}\frac{\mathrm{d}^4}{da^4}(d_{1111}p) - \cdots\cdots = 0. \tag{4.396}$$

Equation (4.396) is a partial differential equation with infinite number of terms, cannot be solved exactly, and approximate procedure must be applied. Here a perturbation scheme is proposed to solve the problems approximately.

Noting the fact that the Poisson white noise tends to Gaussian white noise when the average arrival rate $\lambda \to \infty$, we denote a perturbation parameter $\varepsilon = \lambda^{-1/2}$. As described in Sect. 2.4, the spectral density of the Poisson white noise must be finite, namely, $\lambda E[Y^2]$ is finite. When λ approaches infinite, $E[Y^2]$ is of order of ε^2, and $E[Y^k]$ is at most of order $\lambda^{-k/2}$, i.e. ε^k. Define parameter I_k as

$$\varepsilon^k I_k = \lambda E\left[Y^{k+2}\right] = D_{k+2}, \quad k = 0, 1, 2, \ldots. \tag{4.397}$$

With Eq. (4.397), the first four derivate moments given in (4.395) can be written as

$$a_1(a) = -\xi\omega_0 a + \frac{I_0}{4a\omega_0^2} + \varepsilon^2\frac{I_2}{64a^3\omega_0^4}, \quad b_{11}(a) = \frac{I_0}{2\omega_0^2} - \varepsilon^2\frac{I_2}{32a^2\omega_0^4},$$
$$c_{111}(a) = \varepsilon^2\frac{3I_2}{16a\omega_0^4}, \quad d_{1111}(a) = \varepsilon^2\frac{3I_2}{8\omega_0^4}. \tag{4.398}$$

The solution of Eq. (4.396) is written as

$$p(a) = p_0(a) + \varepsilon p_1(a) + \varepsilon^2 p_2(a) + \cdots. \tag{4.399}$$

Substituting (4.398) and (4.399) into (4.396), and grouping the terms of the same order of ε, we obtain a sequence of ordinary differential equations. The first two equations are

$$-\frac{\mathrm{d}}{da}\left[\left(-\xi\omega_0 a + \frac{I_0}{4a\omega_0^2}\right)p_0\right] + \frac{I_0}{4\omega_0^2}\frac{\mathrm{d}^2 p_0}{da^2} = 0, \tag{4.400}$$

$$-\frac{\mathrm{d}}{da}\left[\left(-\xi\omega_0 a + \frac{I_0}{4a\omega_0^2}\right)p_2\right] + \frac{I_0}{4\omega_0^2}\frac{\mathrm{d}^2 p_2}{da^2}$$

$$= \frac{\mathrm{d}}{\mathrm{d}a}\left(\frac{I_2}{64a^3\omega_0^4}p_0\right) + \frac{1}{2!}\frac{\mathrm{d}^2}{\mathrm{d}a^2}\left(\frac{I_2}{32a^2\omega_0^4}p_0\right)$$
$$+ \frac{1}{3!}\frac{\mathrm{d}^3}{\mathrm{d}a^3}\left(\frac{3I_2}{16a\omega_0^4}p_0\right) - \frac{1}{4!}\frac{\mathrm{d}^4}{\mathrm{d}a^4}\left(\frac{3I_2}{8\omega_0^4}p_0\right). \tag{4.401}$$

Equation (4.400) is the FPK equation for the case of Gaussian white noise excitation. Its solution is the well-known Rayleigh distribution

$$p_0(a) = Ca\exp\left(-\frac{C}{2}a^2\right), \quad C = \frac{4\xi\omega_0^3}{I_0}. \tag{4.402}$$

Substituting $p0(a)$ into the right-hand side of (4.401), $p2(a)$ can be solved either analytically or numerically.

Example 4.12 Consider a system governed by

$$\ddot{X} + 2\zeta\omega_0\dot{X} + \omega_0^2X = XW_p(t), \tag{4.403}$$

where $W_p(t)$ is a Poisson white noise. Letting $X_1 = X,\ X_2 = \dot{X}$, (4.403) can be written in terms of compounded Poisson process as

$$\begin{aligned} \mathrm{d}X_1 &= X_2\mathrm{d}t, \\ \mathrm{d}X_2 &= \left(-2\xi\omega_0X_2 - \omega_0^2X_1\right)\mathrm{d}t + X_1\mathrm{d}C(t). \end{aligned} \tag{4.404}$$

From (4.389) and (4.390) we have

$$\mathrm{d}A = -2\xi\omega_0A\sin^2\Phi\mathrm{d}t - \frac{\sin\Phi\cos\Phi}{\omega_0}A\mathrm{d}C(t) + \frac{1}{2}\frac{\cos^4\Phi}{\omega_0^2}A[\mathrm{d}C(t)]^2$$
$$+ \frac{1}{2}\frac{\cos^5\Phi\sin\Phi}{\omega_0^3}A[\mathrm{d}C(t)]^3 - \frac{1}{8}\frac{\cos^6\Phi\left(1 - 5\sin^2\Phi\right)}{\omega_0^4}A[\mathrm{d}C(t)]^4 + \cdots. \tag{4.405}$$

Carrying out time averaging, assuming $E[C^k(t)] = 0$ for $k =$ odd, and neglecting D_n for $n > 4$, the first four derivate moments are

$$a_1(a) = \left(-\xi\omega_0 + \frac{3D_2}{16\omega_0^2} - \frac{15D_4}{1024\omega_0^4}\right)a, \quad b_{11}(a) = \left(\frac{D_2}{8\omega_0^2} + \frac{15D_4}{512\omega_0^4}\right)a^2,$$
$$c_{111}(a) = \frac{15D_4}{256\omega_0^4}a^3, \quad d_{1111}(a) = \frac{3D_4}{128\omega_0^4}a^4. \tag{4.406}$$

The same perturbation procedure in Example 4.11 can be used to solve the reduced FPK equation for this problem.

Example 4.13 Consider a Rayleigh oscillator excited by Poisson white noise

$$\ddot{X} + \varepsilon(\beta X^2 - 1)\dot{X} + \omega_0^2 X = W_p(t). \tag{4.407}$$

Letting $X_1 = X,\ X_2 = \dot{X}$, system (4.385) can be described by

$$\begin{aligned} \mathrm{d}X_1 &= X_2 \mathrm{d}t, \\ \mathrm{d}X_2 &= \left[\varepsilon\left(1 - \beta X_1^2\right)X_2 - \omega_0^2 X_1\right]\mathrm{d}t + \mathrm{d}C(t). \end{aligned} \tag{4.408}$$

Using (4.389) and (4.390), we obtain

$$\begin{aligned} \mathrm{d}A &= \varepsilon\left(1 - A^2\beta\cos^2\Phi\right)A\sin^2\Phi dt - \frac{\sin\Phi}{\omega_0}\mathrm{d}C(t) + \frac{1}{2}\frac{\cos^2\Phi}{A\omega_0^2}[\mathrm{d}C(t)]^2 \\ &+ \frac{1}{2}\frac{\cos^2\Phi\sin\Phi}{A^2\omega_0^3}[\mathrm{d}C(t)]^3 - \frac{1}{8}\frac{\cos^2\Phi\left(1 - 5\sin^2\Phi\right)}{A^3\omega_0^4}[\mathrm{d}C(t)]^4 + \cdots. \end{aligned} \tag{4.409}$$

Carrying out time averaging, assuming $E[C^k(t)] = 0$ for $k =$ odd, and neglecting D_n for $n > 4$, the first four derivate moments are

$$\begin{gathered} a_1(a) = \frac{1}{2}\varepsilon a - \frac{1}{8}\varepsilon\beta a^3 + \frac{D_2}{4a\omega_0^2} + \frac{D_4}{64a^3\omega_0^4}, \quad b_{11}(a) = \frac{D_2}{2\omega_0^2} - \frac{D_4}{32a^2\omega_0^4}, \\ c_{111}(a) = \frac{3D_4}{16a\omega_0^4}, \quad d_{1111}(a) = \frac{3D_4}{8\omega_0^4}. \end{gathered} \tag{4.410}$$

The corresponding FPK equation can be solved using the same perturbation procedure in Example 4.11.

Example 4.14 Consider a van der Pol oscillator subjected to a Poisson white noise

$$\ddot{X} + \varepsilon(\dot{X}^2 - 1)\dot{X} + \omega_0^2 X = W_p(t). \tag{4.411}$$

Letting $X_1 = X,\ X_2 = \dot{X}$, Eq. (4.385) is equivalent to

$$\begin{aligned} \mathrm{d}X_1 &= X_2 \mathrm{d}t, \\ \mathrm{d}X_2 &= \left[\varepsilon\left(1 - X_2^2\right)X_2 - \omega_0^2 X_1\right]\mathrm{d}t + \mathrm{d}C(t). \end{aligned} \tag{4.412}$$

Following the same procedure as in Example 4.11, we have the following results and solve the problem.

$$\begin{aligned} \mathrm{d}A &= \varepsilon\left(A\sin^2\Phi - A^2\omega_0^2\sin^4\Phi\right)\mathrm{d}t - \frac{\sin\Phi}{\omega_0}\mathrm{d}C(t) + \frac{1}{2}\frac{\cos^2\Phi}{A\omega_0^2}[\mathrm{d}C(t)]^2 \\ &+ \frac{1}{2}\frac{\cos^2\Phi\sin\Phi}{A^2\omega_0^3}[\mathrm{d}C(t)]^3 - \frac{1}{8}\frac{\cos^2\Phi\left(1 - 5\sin^2\Phi\right)}{A^3\omega_0^4}[\mathrm{d}C(t)]^4 + \cdots, \end{aligned} \tag{4.413}$$

and

$$a_1(a) = \frac{1}{2}\varepsilon a - \frac{3}{8}\varepsilon\omega_0^2 a^3 + \frac{D_2}{4a\omega_0^2} + \frac{D_4}{64a^3\omega_0^4}, \quad b_{11}(a) = \frac{D_2}{2\omega_0^2} - \frac{D_4}{32a^2\omega_0^4},$$
$$c_{111}(a) = \frac{3D_4}{16a\omega_0^4}, \quad d_{1111}(a) = \frac{3D_4}{8\omega_0^4}. \tag{4.414}$$

Example 4.15 Consider a system with an energy-dependent damping and subjected to a Poisson white noise

$$\ddot{X} + \varepsilon\left[g(E) - 1\right]\dot{X} + \omega_0^2 X = W_p(t), \tag{4.415}$$

where $E = \left(\dot{X}^2 + \omega_0^2 X^2\right)/2$, is the system energy. As an example, we assume that $g(E) = 2E$. Using the same procedure as in the previous examples, we have

$$\begin{aligned} \mathrm{d}X_1 &= X_2 \mathrm{d}t, \\ \mathrm{d}X_2 &= \left[\varepsilon(1 - 2E)X_2 - \omega_0^2 X_1\right]\mathrm{d}t + \mathrm{d}C(t), \end{aligned} \tag{4.416}$$

$$\begin{aligned} \mathrm{d}A = {} & \varepsilon\left(1 - \omega_0^2 A^2\right)A\sin^2\Phi\big)\mathrm{d}t - \frac{\sin\Phi}{\omega_0}\mathrm{d}C(t) + \frac{1}{2}\frac{\cos^2\Phi}{A\omega_0^2}[\mathrm{d}C(t)]^2 \\ & + \frac{1}{2}\frac{\cos^2\Phi\sin\Phi}{A^2\omega_0^3}[\mathrm{d}C(t)]^3 - \frac{1}{8}\frac{\cos^2\Phi\left(1 - 5\sin^2\Phi\right)}{A^3\omega_0^4}[\mathrm{d}C(t)]^4 + \cdots, \end{aligned} \tag{4.417}$$

and

$$a_1(a) = \frac{1}{2}\varepsilon a - \frac{1}{2}\varepsilon\omega_0^2 a^3 + \frac{D_2}{4a\omega_0^2} + \frac{D_4}{64a^3\omega_0^4}, \quad b_{11}(a) = \frac{D_2}{2\omega_0^2} - \frac{D_4}{32a^2\omega_0^4},$$
$$c_{111}(a) = \frac{3D_4}{16a\omega_0^4}, \quad d_{1111}(a) = \frac{3D_4}{8\omega_0^4}. \tag{4.418}$$

Example 4.16 Consider a system with a damping of power-order and subjected to a Poisson white noise

$$\ddot{X} + \varepsilon\left|\dot{X}\right|^{\alpha-1}\dot{X} + \omega_0^2 X = W_p(t). \tag{4.419}$$

With changing of parameter α, the damping can describe different material internal friction from small motion ($\alpha = 1.3$) to large motion ($\alpha = 7$). Following the same procedure as in previous examples, we have

$$\begin{aligned} \mathrm{d}X_1 &= X_2 \mathrm{d}t, \\ \mathrm{d}X_2 &= \left[\varepsilon(1 - 2E)X_2 - \omega_0^2 X_1\right]\mathrm{d}t + \mathrm{d}C(t), \end{aligned} \tag{4.420}$$

$$
\begin{aligned}
\mathrm{d}A = & -\varepsilon|A\omega_0 \sin\Phi|^{\alpha-1}A\sin^2\Phi dt - \frac{\sin\Phi}{\omega_0}\mathrm{d}C(t) + \frac{1}{2}\frac{\cos^2\Phi}{A\omega_0^2}[\mathrm{d}C(t)]^2 \\
& + \frac{1}{2}\frac{\cos^2\Phi\sin\Phi}{A^2\omega_0^3}[\mathrm{d}C(t)]^3 - \frac{1}{8}\frac{\cos^2\Phi\left(1-5\sin^2\Phi\right)}{A^3\omega_0^4}[\mathrm{d}C(t)]^4 + \cdots,
\end{aligned} \tag{4.421}
$$

and

$$
\begin{aligned}
a_1(a) = \frac{-\varepsilon\alpha\omega_0^{\alpha-1}\Gamma(\alpha/2)}{2\sqrt{\pi}\Gamma(3/2+\alpha/2)} + \frac{D_2}{4a\omega_0^2} + \frac{D_4}{64a^3\omega_0^4}, \quad b_{11}(a) = \frac{D_2}{2\omega_0^2} - \frac{D_4}{32a^2\omega_0^4}, \\
c_{111}(a) = \frac{3D_4}{16a\omega_0^4}, \quad d_{1111}(a) = \frac{3D_4}{8\omega_0^4}.
\end{aligned} \tag{4.422}
$$

4.8.2 Energy Envelope

For a system with nonlinear restoring force and excited by a Poisson white noise

$$
\ddot{X} + f\left(X, \dot{X}\right) + u(X) = h\left(X, \dot{X}\right)W_p(t). \tag{4.423}
$$

we assume weak damping and weak excitation. Letting $X_1 = X$, $X_2 = \dot{X}$, in terms of compound Poisson processes, (4.423) is equivalent to

$$
\begin{aligned}
\mathrm{d}X_1 &= X_2\mathrm{d}t, \\
\mathrm{d}X_2 &= (-f - u)\mathrm{d}t + h\mathrm{d}C(t) + \frac{1}{2}h\frac{\partial h}{\partial X_2}[\mathrm{d}C(t)]^2 \\
&\quad + \frac{1}{3!}h\left[h\frac{\partial^2 h}{\partial X_2^2} + \left(\frac{\partial h}{\partial X_2}\right)2\right][\mathrm{d}C(t)]^3 \\
&\quad + \frac{1}{4!}h\left[h^2\frac{\partial^3 h}{\partial X_2^3} + 4h\frac{\partial h}{\partial X_2}\frac{\partial^2 h}{\partial X_2^2} + \left(\frac{\partial h}{\partial X_2}\right)^3\right][\mathrm{d}C(t)]^4
\end{aligned} \tag{4.424}
$$

The energy process is

$$
\Lambda = \frac{1}{2}X_2^2 + U(X_1), \tag{4.425}
$$

where $U(X_1)$ is the potential energy. The following partial derivatives can be obtained from (4.425)

$$
\frac{\partial\Lambda}{\partial X_1} = u(X_1), \quad \frac{\partial\Lambda}{\partial X_2} = X_2, \quad \frac{\partial^2\Lambda}{\partial X_2^2} = 1, \quad \frac{\partial^{(i)}\Lambda}{\partial X_2^{(i)}} = 0 \quad \text{for} \quad i > 2. \tag{4.426}
$$

Application of (4.386) for the energy process $\Lambda(t)$, we obtain

$$\mathrm{d}\Lambda = -X_2 f\,\mathrm{d}t + X_2 h \mathrm{d}C(t) + \frac{1}{2}\left[X_2 h \frac{\partial h}{\partial X_2} + h^2\right][\mathrm{d}C(t)]^2$$
$$+ \left\{\frac{1}{2}h^2\frac{\partial h}{\partial X_2} + \frac{1}{6}X_2 h\left[h\frac{\partial^2 h}{\partial X_2^2} + \left(\frac{\partial h}{\partial X_2}\right)^2\right]\right\}[\mathrm{d}C(t)]^3$$
$$+ \left\{\frac{7}{24}h^2\left(\frac{\partial h}{\partial X_2}\right)^2 + \frac{1}{6}h^3\frac{\partial^2 h}{\partial X_2^2} + \frac{1}{4!}X_2 h\left[h^2\frac{\partial^3 h}{\partial X_2^3} + 4g\frac{\partial h}{\partial X_2}\frac{\partial^2 h}{\partial X_2^2} + \left(\frac{\partial h}{\partial X_2}\right)^3\right]\right\}\times$$
$$[\mathrm{d}C(t)]^4 + \cdots\cdots, \tag{4.427}$$

where X_2 is treated as a function of Λ and X_1 as follows

$$X_2 = \pm\sqrt{2\Lambda - 2U(X_1)}. \tag{4.428}$$

Assuming $E[C^k(t)] = 0$ for $k = \text{odd}$, and neglecting D_n for $n > 4$, the first four derivate moments can be obtained from (4.427)

$$a_1 = \langle -x_2 f\rangle_t + \frac{1}{2}D_2\left\langle x_2 h\frac{\partial h}{\partial x_2} + h^2\right\rangle_t + D_4\left\langle \frac{7}{24}h^2\left(\frac{\partial h}{\partial x_2}\right)^2 + \frac{1}{6}h^3\frac{\partial^2 h}{\partial x_2^2}\right\rangle_t$$
$$+ \frac{1}{4!}D_4\left\langle x_2\left[h^3\frac{\partial^3 h}{\partial x_2^3} + 4h^2\frac{\partial h}{\partial x_2}\frac{\partial^2 h}{\partial x_2^2} + h\left(\frac{\partial h}{\partial x_2}\right)^3\right]\right\rangle_t, \tag{4.429}$$

$$b_{11} = D_2\langle x_2^2 h^2\rangle_t + \frac{1}{4}D_4\left\langle h^2\left(x_2\frac{\partial h}{\partial x_2} + h\right)^2\right\rangle_t$$
$$+ D_4\left\langle x_2 h^3\frac{\partial h}{\partial X_2} + \frac{1}{3}x_2^2 h^2\left[h\frac{\partial^2 h}{\partial x_2^2} + \left(\frac{\partial h}{\partial x_2}\right)^2\right]\right\rangle_t, \tag{4.430}$$

$$c_{111} = \frac{3}{2}D_4\left\langle x_2^2 h^3\left(x_2\frac{\partial h}{\partial x_2} + h\right)\right\rangle_t, \tag{4.431}$$

$$d_{1111} = D_4\langle x_2^4 h^4\rangle_t, \tag{4.432}$$

where time averaging is carried out according to (4.72), i.e.,

$$\langle[\cdot]\rangle_t = \frac{1}{T_{1/4}}\int_0^A \frac{[\cdot]_{x_2=\sqrt{2\lambda-2U(x_1)}}}{\sqrt{2\lambda - 2U(x_1)}}\mathrm{d}x_1. \tag{4.433}$$

In (4.433), $T_{1/4}$ is calculated from (4.60), and A is determined from $U(\lambda) = A$. The reduced FPK equation is

$$-\frac{\mathrm{d}}{\mathrm{d}\lambda}(a_1 p) + \frac{1}{2!}\frac{\mathrm{d}^2}{\mathrm{d}\lambda^2}(b_{11}p) - \frac{1}{3!}\frac{\mathrm{d}^3}{\mathrm{d}\lambda^3}(c_{111}p) + \frac{1}{4!}\frac{\mathrm{d}^4}{\mathrm{d}\lambda^4}(d_{1111}p) - \cdots\cdots = 0. \tag{4.434}$$

By applying the same perturbation procedure, an approximate solution can be obtained.

Example 4.17 Consider the nonlinear system

$$\ddot{X} + \alpha\dot{X} + \beta\dot{X}^3 + \delta X^3 = W_p(t), \quad \beta, \delta > 0. \tag{4.435}$$

The equation for the energy $\Lambda(t)$ can be derived from (4.429), and the derivate moments from (4.429)–(4.432) as follows

$$a_1 = \left\langle -\alpha x_2^2 - \beta x_2^4 \right\rangle_t + \frac{1}{2}D_2 = -\frac{4}{3}\alpha\lambda - \frac{32}{7}\beta\lambda^2 + \frac{1}{2}D_2,$$

$$b_{11} = D_2\left\langle x_2^2 \right\rangle_t + \frac{1}{4}D_4 = \frac{4}{3}D_2\lambda + \frac{1}{4}D_4,$$

$$c_{111} = 2D_4\lambda, \quad d_{1111} = \frac{32}{7}D_4\lambda^2. \tag{4.436}$$

Applying the perturbation procedure described in Example 4.11, write the PDF of $\Lambda(t)$ as

$$p(\lambda) = p_0(\lambda) + \varepsilon p_1(\lambda) + \varepsilon^2 p_2(\lambda) + \cdots, \tag{4.437}$$

and the first derivate moments in (4.436) can be obtained as

$$a_1(\lambda) = -\frac{4}{3}\alpha\lambda - \frac{32}{7}\beta\lambda^2 + \frac{I_0}{2}, \quad b_{11}(\lambda) = \frac{4I_0}{3}\lambda + \varepsilon^2\frac{I_2}{4},$$

$$c_{111}(\lambda) = \varepsilon^2 2I_2\lambda, \quad d_{1111}(\lambda) = \varepsilon^2\frac{32I_2}{7}\lambda^2. \tag{4.438}$$

Substituting (4.437) and (4.438) into the FPK Eq. (4.434), and grouping equations of the same order of ε, we obtain

$$-\frac{\mathrm{d}}{\mathrm{d}\lambda}\left[\left(-\frac{3}{4}\alpha\lambda - \frac{32}{7}\beta\lambda^2 + \frac{I_0}{2}\right)p_0\right] + \frac{\mathrm{d}^2}{\mathrm{d}\lambda^2}\left(\frac{4I_0}{3}\lambda p_0\right) = 0, \tag{4.439}$$

$$
\begin{aligned}
&-\frac{\mathrm{d}}{\mathrm{d}\lambda}\left[\left(-\frac{3}{4}\alpha\lambda-\frac{32}{7}\beta\lambda^2+\frac{I_0}{2}\right)p_2\right]+\frac{1}{2}\frac{\mathrm{d}^2}{\mathrm{d}\lambda^2}\left(\frac{4I_0}{3}\lambda p_2\right)\\
&=\frac{\mathrm{d}}{\mathrm{d}\lambda}\left[\left(-\frac{3}{4}\alpha\lambda-\frac{32}{7}\beta\lambda^2+\frac{I_0}{2}\right)p_0\right]-\frac{1}{2}\frac{\mathrm{d}^2}{\mathrm{d}\lambda^2}\left(\frac{I_2}{4}p_0\right)\\
&+\frac{1}{3!}\frac{\mathrm{d}^3}{\mathrm{d}\lambda^3}(2I_2\lambda p_0)-\frac{1}{4!}\frac{\mathrm{d}^4}{\mathrm{d}\lambda^4}\left(\frac{32I_2}{7}\lambda^2 p_0\right).
\end{aligned}
\tag{4.440}
$$

Equation (4.439) is the FPK equation for the case of Gaussian white noise excitation. Its solution is given in Example 4.2, i.e., Eq. (4.124),

$$
p_0(\lambda)=C\lambda^{-\frac{1}{4}}\exp\left[-\frac{1}{\pi K}\left(\alpha\lambda+\frac{12}{7}\beta\lambda^2\right)\right].
\tag{4.441}
$$

Substituting $p0(\lambda)$ into the right-hand side of (4.440), $p2(\lambda)$ can be solved either analytically or numerically. The stationary PDF for $p(x,\dot{x})$ can be calculated according to Eq. (4.77).

4.9 Systems Excited by Fractional Gaussian Noises

For nonlinear SDOF systems under fractional Gaussian noises, stochastic averaging methods of amplitude envelope and energy envelope can be applied for linear and nonlinear restoring forces, respectively. This section only describes the stochastic averaging method of energy envelope.

Consider a strongly nonlinear SDOF systems excited by fractional Gaussian noises

$$
\ddot{X}+\varepsilon^{2\mathcal{H}}h(X,\dot{X})+u(X)=\varepsilon^{\mathcal{H}}\sum_{l=1}^{m}g_l(X,\dot{X})W_{Hl}(t),
\tag{4.442}
$$

where $u(X)$ is a strongly nonlinear restoring force with $u(0)=0$, ε is a small positive parameter, $\varepsilon^{2\mathcal{H}}h(X,\dot{X})$ represents a weak linear or nonlinear damping, $W_{Hl}(t)$ (l = 1, 2, ···, m) are fractional Gaussian noises with zero-mean and Hurst indexes $1/2<\mathcal{H}<1$. The auto-correlation function $R_l(\tau)$ and power spectral density $S_l(\omega)$ are given in (2.217) and (2.219). The energy functions of system (4.442) are

$$
\Lambda(X,\dot{X})=\frac{\dot{X}^2}{2}+U(X),\quad U(X)=\int_0^X u(x)\mathrm{d}x.
\tag{4.443}
$$

Similarly to the derivation of (4.60), the motion period of system (4.442) in the absence of damping and excitations is

$$T(\Lambda) = 2\int_{x'}^{x''} \frac{1}{\sqrt{2\Lambda - 2U(x)}} \mathrm{d}x, \tag{4.444}$$

where x', x'' are the negative and positive roots of $\Lambda = U(x)$.

Letting $X_1 = X$, $X_2 = \dot{X}$, (4.442) can be modeled as a fractional stochastic differential equation defined based on symmetric integral, and then transformed to one defined as forward integral,

$$\begin{aligned} \mathrm{d}X_1 &= X_2 \mathrm{d}t, \\ \mathrm{d}X_2 &= \left[-u(X_1) - \varepsilon^{2\mathcal{H}} h(X_1, X_2)\right]\mathrm{d}t + \varepsilon^{\mathcal{H}} \sum_{l=1}^{m} g_l\left(X, \dot{X}\right)\mathrm{d}B_{Hl}(t). \end{aligned} \tag{4.445}$$

As shown in (2.222) through (2.224), there is no Wong-Zakai correction term if $\mathcal{H}>1/2$. According to the stochastic differential rule for the fractional Brownian motion, a fractional stochastic differential equation governing the energy process $\Lambda(t)$ can be obtained and replaces the second equation in (4.445),

$$\begin{aligned} \mathrm{d}X_1 &= \pm\sqrt{2\Lambda - 2U(X_1)}\mathrm{d}t, \\ \mathrm{d}\Lambda &= \left[\mp\varepsilon^{2\mathcal{H}}\sqrt{2\Lambda - 2U(X_1)}h\left(X_1, \pm\sqrt{2\Lambda - 2U(X_1)}\right)\right]\mathrm{d}t \\ &\quad + \varepsilon^{\mathcal{H}}\sqrt{2\Lambda - 2U(X_1)}\sum_{l=1}^{m} g_l\left(X, \pm\sqrt{2\Lambda - 2U(X_1)}\right)\mathrm{d}B_{Hl}(t). \end{aligned} \tag{4.446}$$

In the case of weak damping and weak excitations, the energy process$\Lambda(t)$ in (4.446) is slowly varying while the displace process $X_1(t)$ is fast varying. According to the averaging principle about the fractional stochastic differential equation (Xu et al. 2014a, b), an averaged fractional stochastic differential equation governing $\Lambda(t)$ is obtained

$$\mathrm{d}\Lambda = m(\Lambda)\mathrm{d}t + \sigma(\Lambda)\mathrm{d}B_H(t), \tag{4.447}$$

where $m(\Lambda)$ and $\sigma(\Lambda)$ are determined from the time averaging

$$\begin{aligned} &\lim_{T\to\infty}\frac{1}{T}\int_0^T \left|\mp\varepsilon^{2\mathcal{H}}\sqrt{2\Lambda - 2U(X_1)}h\left(X_1, \pm\sqrt{2\Lambda - 2U(X_1)}\right) - m(\Lambda)\right|\mathrm{d}t = 0, \\ &\lim_{T\to\infty}\frac{1}{T}\int_0^T |[(2\Lambda - 2U(X_1))\varepsilon^{2\mathcal{H}}\sum_{l,k=1}^{m} g_l g_k]^{\frac{1}{2}} - \sigma(\Lambda)|^2\mathrm{d}t = 0. \end{aligned} \tag{4.448}$$

Using the first equation in (4.446), the time averaging in (4.448) can be replaced by space averaging with respect to the fast variable X_1, resulting in

$$m(\Lambda) = \frac{2\varepsilon^{2\mathcal{H}}}{T(\Lambda)} \int_{X_{1'}}^{X_1''} g\Big(X_1, \pm\sqrt{2\Lambda - 2U(X_1)}\Big) \mathrm{d}X_1,$$

$$\sigma^2(\Lambda) = \frac{2\varepsilon^{2\mathcal{H}}}{T(\Lambda)} \int_{X_1'}^{X_1'} \sqrt{2\Lambda - 2U(X_1)} \sum_{l,k=1}^{m} g_l g_k \mathrm{d}X_1, \tag{4.449}$$

where $T(\Lambda)$ is given in (4.444).

The response of system (4.447) under the fractional Gaussian process, i.e. the energy process $\Lambda(t)$, is not a Markov process. Simulations must be carried out to solve the system response.

Example 4.18 Consider a Duffing oscillator subjected to a fractional Gaussian noise (Deng and Zhu 2016),

$$\ddot{X} + \gamma\dot{X} + \omega^2 X + kX^3 = \sqrt{2D} W_H(t), \tag{4.450}$$

where $W_H(t)$ is a unit fractional Gaussian noise with a Hurst index $1/2 < \mathcal{H} < 1$, $2D$ is the excitation intensity, γ and k are positive parameters. It is assumed that D and γ are of order $\varepsilon^{2\mathcal{H}}$.

The total energy and potential energy of system (4.450) are

$$\Lambda = \frac{1}{2}\dot{X}^2 + U(X), \quad U(X) = \frac{1}{2}\omega^2 X^2 + \frac{1}{4}kX^4. \tag{4.451}$$

Figure 4.31a depicts the distribution of 10^5 sample points on the phase plane $\big(X(t), \dot{X}(t)\big)$ at $t = 80$. Figure 4.31b shows the corresponding contours of the PDF and three equi-energy curves. It is seen from Fig. 4.31b that the equi-energy curves and the contours are almost coincident, indicating the rational of replacing time averaging by space averaging.

Letting $X_1 = X$, $X_2 = \dot{X}$, (4.450) can be modeled as a fractional stochastic differential equation defined based on symmetric integral, and then transformed to one defined as forward integral,

$$\begin{aligned} \mathrm{d}X_1 &= X_2 \mathrm{d}t, \\ \mathrm{d}X_2 &= \big(-\omega^2 X_1 - kX_1^3 - \gamma X_2\big)\mathrm{d}t + \sqrt{2D}\mathrm{d}B_H(t). \end{aligned} \tag{4.452}$$

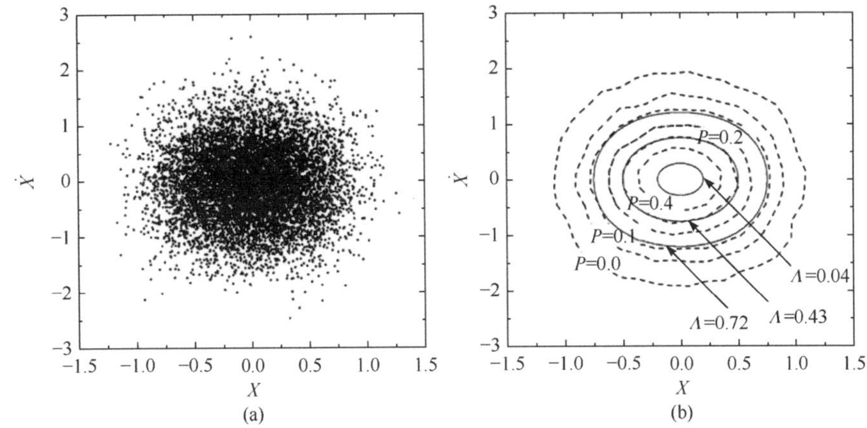

Fig. 4.31 **a** Sample points on the phase plane at $t = 80$. **b** Contours of PDF and three equi-energy curves. Pd represents values of probability density. The parameters are $\omega = 1.414$, $\gamma = 0.01$, $k = 2$, $D = 0.01$, $\mathcal{H} = 0.75$. (Deng and Zhu 2016)

Using (4.446) and (4.447), we obtain a fractional stochastic differential equation governing the energy process $\Lambda(t)$

$$\mathrm{d}\Lambda = -\gamma X_2^2 \mathrm{d}t + \sqrt{2D} X_2 \mathrm{d}B_H(t). \tag{4.453}$$

After the time averaging, an averaged fractional stochastic differential equation of the form of (4.447) is obtained. The two coefficients $m(\Lambda)$ and $\sigma(\Lambda)$ are determined from (4.449) and (4.453) as follows

$$\begin{aligned} &m(\Lambda) = -\gamma\rho, \quad \sigma^2(\Lambda) = 2D\rho, \\ &\rho = 2\Lambda\left(\frac{1+\mu}{3\mu}\frac{\mathrm{E}(\mu)}{\mathrm{K}(\mu)} + \frac{\mu - 1}{3\mu}\right), \quad \mu = \eta\left(\sqrt{\eta^2 + 2} - \eta\right) - 1, \quad \eta = \frac{\omega^2}{\sqrt{2k\Lambda}}, \end{aligned} \tag{4.454}$$

where $\mathrm{E}(x) = \int_0^{\frac{\pi}{2}} (1 - x\sin^2\theta)^{\frac{1}{2}} \mathrm{d}\theta$ and $\mathrm{K}(x) = \int_0^{\frac{\pi}{2}} (1 - x\sin^2\theta)^{-\frac{1}{2}} \mathrm{d}\theta$ are the complete elliptical integrals of the second kind and the first kind, respectively.

Numerical simulations were carried out for the original system (4.450) and the averaged system (4.447) and (4.454). The results are shown in Figs. 4.32 and 4.33 for the stationary probability density $p(\lambda)$, and mean $E[\Lambda]$ and mean square $E[\Lambda^2]$. The results agree very well.

According to (4.77), the approximate joint stationary PDF $p(x, \dot{x})$ and the marginal PDFs $p(x)$ and $p(\dot{x})$ of the displacement X and velocity $\dot{X}$ of system (4.450) can be calculated as

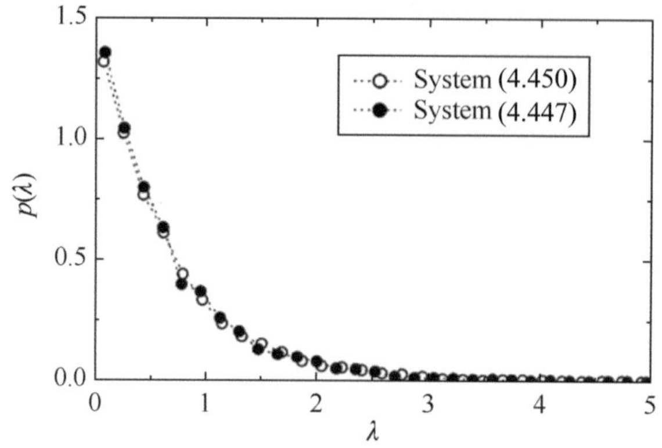

Fig. 4.32 Stationary probability density $p(\lambda)$. Parameters are the same as those in Fig. 4.31 (Deng and Zhu 2016)

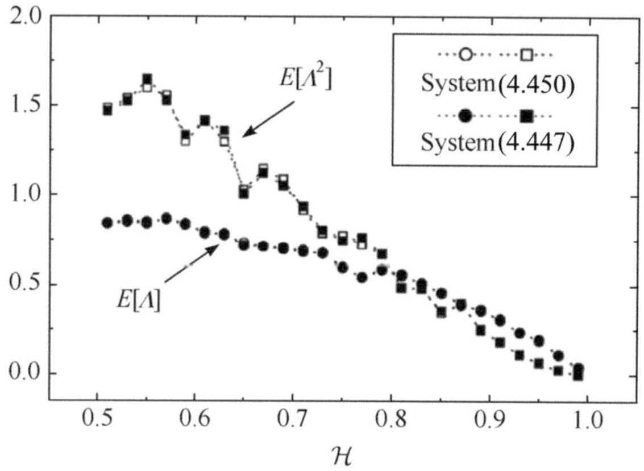

Fig. 4.33 Mean $E[\Lambda]$ and mean square $E[\Lambda^2]$. Parameters are the same as those in Fig. 4.31 (Deng and Zhu 2016)

$$p(x, \dot{x}) = \left.\frac{p(\lambda)}{T(\lambda)}\right|_{\lambda=\frac{\dot{x}^2}{2}+U(x)},$$

$$p(x) = \int_{-\infty}^{\infty} p(x, \dot{x})\mathrm{d}\dot{x}, \quad p(\dot{x}) = \int_{-\infty}^{\infty} p(x, \dot{x})\mathrm{d}x,$$

$$E\left[X^2\right] = \int_{-\infty}^{\infty} x^2 p(x)\mathrm{d}x, \quad E\left[\dot{X}^2\right] = \int_{-\infty}^{\infty} \dot{x}^2 p(\dot{x})\mathrm{d}\dot{x}. \tag{4.455}$$

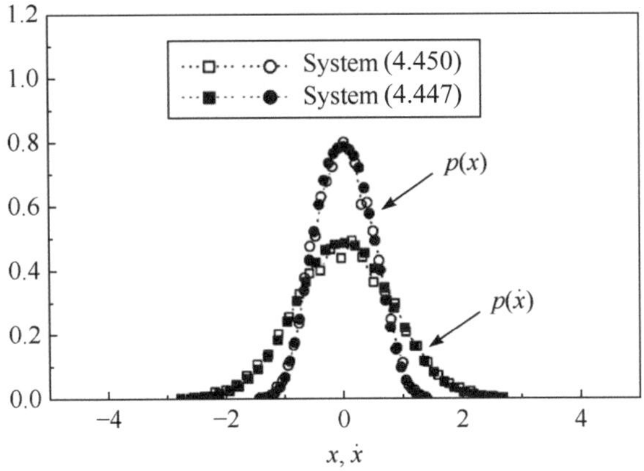

Fig. 4.34 Stationary PDFs $p(x)$ and $p(\dot{x})$. Parameters are the same as those in Fig. 4.31 (Deng and Zhu 2016)

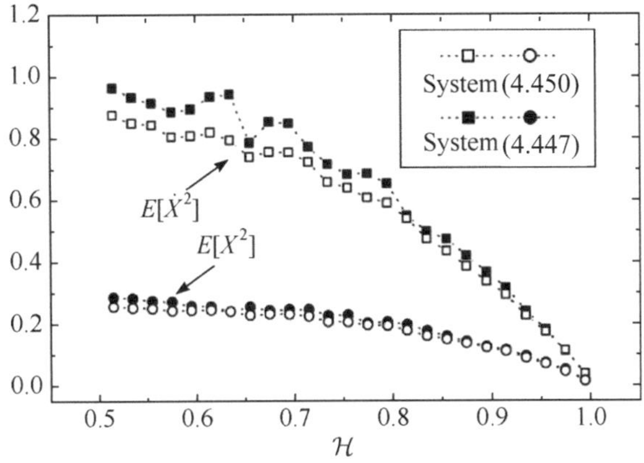

Fig. 4.35 $E[X^2]$ and $E[\dot{X}^2]$ with respect to Hurst index. Parameters are the same as those in Fig. 4.31. (Deng and Zhu 2016)

While carrying out the numerical simulations to get the results for Figs. 4.32 and 4.33, the stationary PDFs $p(x)$ and $p(\dot{x})$, and mean square values $E[X^2]$ and $E[\dot{X}^2]$ were also obtained. The results are shown in Figs. 4.34 and 4.35. Two results agree quite well, except the discrepancies of mean square velocity occur in the range of $0.5 < \mathcal{H} < 0.8$. Both Figs. 4.33 and 4.35 show the tendency of decreasing system responses with an increasing Hurst index. This is because the total power is decreasing with an increasing Hurst index, which can be clearly seen in Fig. 2.5.2 from the area under the power spectral density curves. Figure 4.36 gives the comparison of the PDF $p(x)$ for different nonlinearity levels of k values. When $k > 0$, $p(x)$ deviates from Gaussian distribution corresponding to the linear case of $k = 0$.

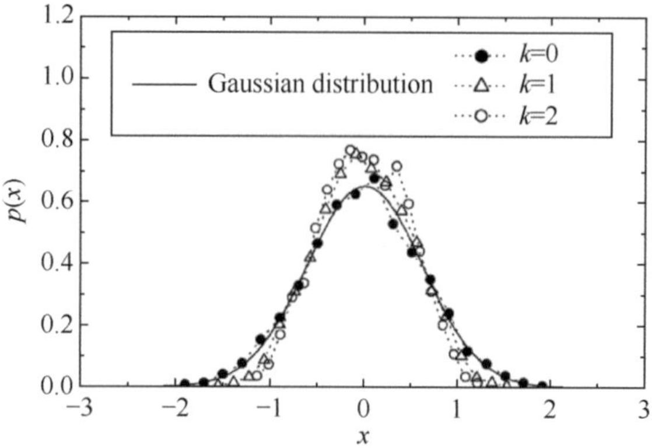

Fig. 4.36 Stationary PDF $p(x)$ for different values of k. Parameters: $\omega = 1.414$, $\gamma = 0.5\omega$, $D = 0.5$, $\mathcal{H} = 0.75$ (Deng and Zhu 2016)

References

Andrnov A, Pontryagin L, Witt A. 1933. On the statistical investigation of dynamical systems. Zh. Eksp. Teor. Fiz., 3: 165–180 (in Russian).

Ariaratnam S T. 1993. Stochastic stability of linear viscoelastic systems. Probabilistic Engineering Mechanics, 8: 153–155.

Ariaratnam S T, Tam D S F. 1976. Parametric random excitation of a damped Mathieu oscillator. Zeitschrift Fur Angewandte Mathematik Und Mechanik, 56: 449–452.

Ariaratnam S T, Tam D S F. 1977. Moment stability of coupled linear systems under combined harmonic and stochastic excitation. In Stochastic Problems in Dynamics, ed. B. L. Clarkson, Pitman, London, 90–103.

Brin M, Freidlin M I. 2000. On stochastic behavior of perturbed Hamiltonian systems, Ergodic Theory and Dynamical Systems, 20: 55–76.

Cai G Q. 1995. Random vibration of nonlinear system under nonwhite excitations. ASCE Journal of Engineering Mechanics, 121(5): 633–639.

Cai G Q, Lin Y K. 2001a, Effect of secondary systems on response of nonlinear primary system to random excitation. in Structural Safety and Reliability, Proceedings of ICOSSAR'01, the 8th International Conference on Structural Safety and Reliability, New Port Beach, CA, June 17–21, 2001, CD-ROM.

Cai G Q, Lin Y K. 2001b. Random vibration of strongly nonlinear systems. Nonlinear Dynamics, 24: 3–15.

Cai G Q, Zhu W Q. 2016. Elements of Stochastic Dynamics. Singapore: World Scientific Publisher.

Cai G Q, Lin Y K, Xu W. 1998. Visco-elastic systems under both additive and multiplicative random excitations, in Computational Stochastic Mechanics. Proceedings of the Third International Conference on Computational Stochastic Mechanics, Balkema, Rottendam, Netherlands, 411–416.

Choi S, Sri Namachchivaya N. 2006. Stochastic dynamics of a nonlinear aeroelastic system. AIAA Journal, 44(9): 1921–1931.

Christensen R M. 1982. Theory of Viscoelasticity, An Introduction. New York: Academic Press.

Deng M L, Zhu W Q. 2016. Stochastic averaging of quasi-non-integrable Hamiltonian systems under fractional Gaussian noise excitation. Nonlinear Dynamics, 83(1-2): 1015–1027.

Dimentberg M F. 1988. Statistical Dynamics of Nonlinear and Time-Varying Systems. Taunton: Research Studies Press.

Dimentberg M F, Cai G Q, Lin Y K. 1995. Application of quasi-conservative averaging to a non-linear system under non-white excitations. International Journal of Non-Linear Mechanics, 30(5): 677–685.

Drozdov A D. 1998. Viscoelastic Structures: Mechanics of Growth and Aging. San Diego: Academic Press, CA.
Freidlin M I. 2001. On stable oscillations and equilibriums induced by small noise. Journal of Statistical Physics, 103(1-2): 283–300.
Freidlin M I, Sheu S J. 2000. Diffusion processes on graphs: stochastic differential equations, large deviation principle, Probability Theory and Related Fields, 116(2): 181–220.
Freidlin M I, Weber M. 1998. Random perturbations of nonlinear oscillators, Annals of Probability, 26(3): 925–967.
Freidlin M I, Weber M. 1999. A remark on random perturbations of the nonlinear pendulum, Annals of Applied Probability, 9(3): 611–628.
Freidlin M I, Wentzell A D. 1994. Random perturbations of Hamiltonian Systems. In Memoirs of American Mathematical Society, Vol.109, No.523.
Freidlin M I, Wentzell A D. 2012. Random Perturbations of Dynamical Systems, 3rd edition. Springer.
Gradshteyn I S, Ryzhik I M. 1980. Table of Integrals, Series, and Products. New York: Academic Press.
Huang Q, Xie, W C. 2008. Stability of SDOF linear viscoelastic system under the excitation of wideband noise. ASME Journal of Applied Mechanics, 75: 12–21.
Khasminskii R Z. 1964. On the behavior of a conservative system with small friction and small random noise. Prikladnaya Mathematika (Applied Mathematics and Mechanics), 28(5): 1126–1130. (in Russian)
Khasminskii R Z. 1966. A limit theorem for the solution of differential equations with random right hand sides. Theory of Probability and Application, 11(3): 390–405.
Khasminskii R Z. 1968. On the averaging principle for Itô stochastic differential equations. Kibernetika, 3(4): 260–279. (in Russian)
Landa P S, Stratonovoch R L. 1962. Theory of stochastic transition in various systems between different states. Vestnik MGU (Proceedings of Moscow University), Series III(1), 33–45. (in Russian)
Lin Q, Jin X L, Huang Z L. 2011. Response and stability of SDOF viscoelastic system under wideband noise excitations. Journal of the Franklin Institute-Engineering and Applied Mathematics, 348: 2026–2043.
Lin Y K, Cai G Q. 1995. Probabilistic Structural Dynamics: Advanced Theory and Applications. New York: McGraw-Hill.
Skorokhod A V, Hoppensteadt F C, Salehi H. 2002. Random Perturbation Methods with Applications in Science and Engineering. New York: Springer.
Sowers R B. 2003. Stochastic averaging near a homoclinic orbit with multiplicative noise. Stochastics and Dynamics, 3: 299–391.
Sowers R B. 2005. A boundary-layer theory for diffusively perturbed transport around a heteroclinic cycle. Communications on Pure and Applied Mathematics, 58: 30–84.
Sowers R B. 2007. Stochastic averaging near long heteroclinic orbits. Stochastics and Dynamics, 7: 187–228.
Sowers R B, Sri Namachchivaya N. 2002. Rigorous stochastic averaging at a center with additive noise. Meccanica, 6: 85–114.
Sowers R B, Sri Namachchivaya N, Vedula L. 2001. Nonstandard reduction of noisy Duffing-van der Pol equation. Dynamical Systems: an International Journal, 16: 223–245.
Stratonovich R L. 1963. Topics in the Theory of Random Noise, Vol. 1, New York: Gordon and Breach.
Xu Y, Guo R, Xu W. 2014a. A limit theorem for the solutions of slow-fast systems with fractional Brownian motion. Theoretical and Applied Mechanics Letters, 4: 013003.
Xu Y, Guo R, Liu D, Zhang H Q, Duan J Q. 2014b. Stochastic averaging principle for dynamical systems with fractional Brownian motion. AIMS Discrete and Continuous Dynamical Systems-Series B, 19: 1197–1212.

Zhu W Q, Cai G Q. 2011. Random vibration of viscoelastic system under broad-band excitations. International Journal of Non-Linear Mechanics, 46(5): 720–726.

Zhu W Q, Cai G Q, Hu R C. 2013. Stochastic analysis of dynamical systems with double-well potential. International Journal of Dynamics and Control, 1(1): 12–19.

Zhu W Q, Huang T C. 1984. Dynamic stability of a liquid-free surface in a container with an elastic bottom under combined harmonic and stochastic longitudinal excitation. In Random Vibrations, ASME Winter Annual Meeting, AMD Vol. 65, eds. Huang, T.C., and Spanos, P.D., ASME New York.

Zhu W Q, Huang Z L, Suzuki Y. 2001. Response and stability of strongly non-linear oscillators under wide-band random excitation. International Journal of Nonlinear Mechanics, 36(8): 1235–1250.

Chapter 5
Stochastic Averaging Methods of Quasi-Hamiltonian Systems Under Gaussian White Noise Excitations

In Chap. 4, the mathematical principle and physical interpretation of the stochastic averaging methods along with the derivation and solution of the averaged FPK equation for single degree-of-freedom (SDOF) nonlinear systems under various random excitations are elucidated. Starting from this chapter, the stochastic averaging methods for multiple degrees-of-freedom (MDOF) nonlinear systems, particularly strongly nonlinear systems under various random excitations will be presented. It is well known that the purpose of predicting the response of a stochastic system is to obtain the probability distribution of the response in the entire phase space, which is a global solution. When performing stochastic averaging over a MDOF nonlinear random systems, it is necessary to know the global relationship between degrees-of-freedom. For a MDOF quasi-linear random system, the degenerate system is linear and can be separated through linear transformation. However, for a MDOF strongly nonlinear random system, the degenerate system remains strongly nonlinear. In this case, the third mathematical expression in Sect. 3.1, i.e., the Hamiltonian formulation, can be used to describe clearly the global relationship between degrees-of-freedom. Specifically, this is achieved by classifying the system into five classes: i.e., non-integrable, integrable and non-resonant, integrable and resonant, partially integrable and non-resonant, and partially integrable and resonant. As the most basic and common situation, in this chapter the stochastic averaging methods for five classes of quasi-Hamiltonian systems under Gaussian white noise excitations will be presented. Also, how to select different stochastic averaging methods for a system with different parameters will be illustrated using an example of two-DOF vibration-impact system. Finally, the stochastic averaging method of quasi-non-integrable Hamiltonian systems will be extended to the systems with Markov jump parameters.

W.-Q. Zhu et al., *Stochastic Averaging*, https://doi.org/10.1007/978-981-96-3829-1_5

5.1 Quasi-Non-Integrable Hamiltonian Systems

Consider the following n–DOF stochastically excited and dissipated Hamiltonian system under random excitations (Zhu and Yang 1997)

$$\dot{Q}_i = \frac{\partial H'}{\partial P_i},$$
$$\dot{P}_i = -\frac{\partial H'}{\partial Q_i} - \varepsilon \sum_{j=1}^{n} c_{ij}(Q, P)\frac{\partial H'}{\partial P_j} + \varepsilon^{1/2} \sum_{l=1}^{m} g_{il}(Q, P) W_{gl}(t),$$
$$i = 1, 2, \ldots, n, \tag{5.1}$$

where $\boldsymbol{Q} = [Q_1, Q_2, \ldots, Q_n]$ is a vector of generalized displacements; $\boldsymbol{P} = [P_1, P_2, \ldots, P_n]$ is a vector of generalized momenta; $H'(\boldsymbol{Q}, \boldsymbol{P})$ is Hamiltonian function; $c_{ij}(\boldsymbol{Q}, \boldsymbol{P})$ are damping coefficients; $g_{il}(\boldsymbol{Q}, \boldsymbol{P})$ are amplitudes of excitation; ε is a small parameter; $W_{gl}(t)$ are the Gaussian white noises with correlation functions

$$E[W_{gl}(t)W_{gs}(t+\tau)] = 2\pi K_{ls}\delta(\tau), \quad l, s = 1,\ 2, \ldots, m. \tag{5.2}$$

Due to the light damping forces and weak excitations, system (5.1) is named as quasi-Hamiltonian system.

System (5.1) can be modeled as Stratonovich stochastic differential equations and then transformed into the following Itô stochastic differential equations after adding Wong-Zakai correction terms

$$\mathrm{d}Q_i = \frac{\partial H'}{\partial P_i}\mathrm{d}t,$$
$$\mathrm{d}P_i = \left[-\frac{\partial H'}{\partial Q_i} - \varepsilon \sum_{j=1}^{n} c_{ij}\frac{\partial H'}{\partial P_j} + \varepsilon\pi \sum_{j=1}^{n}\sum_{l,s=1}^{m} K_{ls} g_{js}\frac{\partial g_{il}}{\partial P_j} \right] \times$$
$$\mathrm{d}t + \varepsilon^{1/2} \sum_{l=1}^{m} g_{il}\mathrm{d}B_l'(t),$$
$$i = 1, 2, \ldots, n, \tag{5.3}$$

where $B_l'(t)$ are Wiener processes or Brownian motions. The correlation functions of their increments are

$$E[\mathrm{d}B_l'(t)\mathrm{d}B_s'(t+\tau)] = 2\pi K_{ls}\delta(\tau)\mathrm{d}t, \quad l, s = 1,\ 2, \ldots, m. \tag{5.4}$$

The two-fold summation terms in Eq. (5.3) are known as the Wong-Zakai correction terms, which can be separated into two terms, one affects the restoring forces and the other affects damping forces. By combining these two terms with the restoring forces and damping forces and changing the original Hamiltonian function $H'(\boldsymbol{Q}, \boldsymbol{P})$

and the damping coefficients $c_{ij}(\boldsymbol{Q}, \boldsymbol{P})$ to $H(\boldsymbol{Q}, \boldsymbol{P})$ and $m_{ij}(\boldsymbol{Q}, \boldsymbol{P})$, respectively, system (5.3) is updated as

$$
\begin{aligned}
\mathrm{d}Q_i &= \frac{\partial H}{\partial P_i}\mathrm{d}t, \\
\mathrm{d}P_i &= -\left(\frac{\partial H}{\partial Q_i} + \varepsilon\sum_{j=1}^{n} m_{ij}\frac{\partial H}{\partial P_j}\right)\mathrm{d}t + \varepsilon^{1/2}\sum_{l=1}^{m} g_{il}\mathrm{d}B_l'(t), \\
i &= 1, 2, \ldots, n.
\end{aligned} \tag{5.5}
$$

Assuming that the Hamiltonian system with Hamiltonian function H possesses only one conservative quantity, i.e., Hamiltonian H, , system (5.5) is called quasi-non-integrable Hamiltonian system. The quasi-non-integrable Hamiltonian system possesses only one slowly-varying process $H(t)$, which is function of $\boldsymbol{Q}(t)$ and $\boldsymbol{P}(t)$. Using Itô stochastic differential rule, the Itô stochastic differential equation for $H(t)$ is derived as follows

$$
\begin{aligned}
\mathrm{d}H &= \sum_{j=1}^{n}\frac{\partial H}{\partial Q_j}\mathrm{d}Q_j + \sum_{j=1}^{n}\frac{\partial H}{\partial P_j}\mathrm{d}P_j + \frac{1}{2}\sum_{j,k=1}^{n}\frac{\partial^2 H}{\partial P_j \partial P_k}(\mathrm{d}P_j)(\mathrm{d}P_k) \\
&= \varepsilon\left(-\sum_{j,k=1}^{n} m_{jk}\frac{\partial H}{\partial P_j}\frac{\partial H}{\partial P_k} + \sum_{j,k=1}^{n}\sum_{l,s=1}^{m}\pi K_{ls}g_{jl}g_{ks}\frac{\partial^2 H}{\partial P_j \partial P_k}\right)\mathrm{d}t \\
&\quad + \varepsilon^{1/2}\sum_{j=1}^{n}\sum_{l=1}^{m} g_{jl}\frac{\partial H}{\partial P_j}\mathrm{d}B_l'(t).
\end{aligned} \tag{5.6}
$$

Note that the righ side of Eq. (5.6) is small, indicating $H(t)$ is a slowly-varying process. Replacing P_1 by H, and keeping all other Q_i, P_i unchanged, the system equations consist of Eqs. (5.6) and (5.5) except $\mathrm{d}P_1$. In this new system, $H(t)$ is slowly-varying process while $Q_1(t), \ldots, Q_n(t), P_2(t), \ldots, P_n(t)$ are fast-varying processes. According to the Khasminskii stochastic averaging theorem (Khasminskii 1968), when $\varepsilon \to 0$, $H(t)$ weakly converges to a Markov diffusion process. This process is governed by the following one-dimensional averaged Itô stochastic differential equation

$$
\mathrm{d}H = m(H)\mathrm{d}t + \sigma(H)\mathrm{d}B(t), \tag{5.7}
$$

where $B(t)$ is a unit Wiener process. According to Eqs. (4.29) and (4.30), the drift coefficient and diffusion coefficient in Eq. (5.7) are obtained by performing time averaging over Eq. (5.6) as follow

$$
m(H) = \varepsilon\left\langle -\sum_{j,k=1}^{n} m_{jk}\frac{\partial H}{\partial P_j}\frac{\partial H}{\partial P_k} + \sum_{j,k=1}^{n}\sum_{l,s=1}^{m}\pi K_{ls}g_{jl}g_{ks}\frac{\partial^2 H}{\partial P_j \partial P_k}\right\rangle_t, \tag{5.8}
$$

$$\sigma^2(H) = \varepsilon \left\langle \sum_{j,k=1}^{n} \sum_{l,s=1}^{m} 2\pi K_{ls} g_{jl} g_{ks} \frac{\partial H}{\partial P_j} \frac{\partial H}{\partial P_k} \right\rangle_t. \tag{5.9}$$

Since all the variables in the right side of Eqs. (5.8) and (5.9) are random processes, it's difficult to complete the time averaging. However, since a n-DOF non-integrable Hamiltonian system is ergodic on the $(2n-1)$-dimensional isoenergetic surface, the time averaging in Eqs. (5.8) and (5.9) can be replaced by spatial averaging over the isoenergetic surface. According to the first equation in Eq. (5.5), $\mathrm{d}t$ is replaced by $(\partial H/\partial p_1)^{-1}\mathrm{d}q_1$. Then Eqs. (5.8) and (5.9) can be replaced by

$$m(H) = \frac{\varepsilon}{T(H)} \int_{\Omega} \left[\left(\frac{\partial H}{\partial p_1} \right)^{-1} \left(-\sum_{j,k=1}^{n} m_{jk} \frac{\partial H}{\partial p_j} \frac{\partial H}{\partial p_k} + \sum_{j,k=1}^{n} \sum_{l,s=1}^{m} \pi K_{ls} g_{jl} g_{ks} \frac{\partial^2 H}{\partial p_j \partial p_k} \right) \right] \times \mathrm{d}q_1 \ldots \mathrm{d}q_n \mathrm{d}p_2 \ldots \mathrm{d}p_n, \tag{5.10}$$

$$\sigma^2(H) = \frac{\varepsilon}{T(H)} \int_{\Omega} \left[\left(\frac{\partial H}{\partial p_1} \right)^{-1} \sum_{j,k=1}^{n} \sum_{l,s=1}^{m} 2\pi K_{ls} g_{jl} g_{ks} \frac{\partial H}{\partial p_j} \frac{\partial H}{\partial p_k} \right] \mathrm{d}q_1 \ldots \mathrm{d}q_n \mathrm{d}p_2 \ldots \mathrm{d}p_n, \tag{5.11}$$

where

$$T(H) = \int_{\Omega} \left(\frac{\partial H}{\partial p_1} \right)^{-1} \mathrm{d}q_1 \ldots \mathrm{d}q_n \mathrm{d}p_2 \ldots \mathrm{d}p_n, \tag{5.12}$$

$$\Omega = \{(q_1, \ldots, q_n, p_2, \ldots, p_n) | H(q_1, \ldots, q_n, 0, p_2, \ldots, p_n) \le H\}. \tag{5.13}$$

Substituting Eqs. (5.10) and (5.11) into Eq. (5.7), taking account of boundary conditions (i) p is finite when $h = 0$ and (ii) $p = 0$ when $h \to \infty$, and solving the reduced averaged FPK equation associated with averaged Itô Eq. (5.7) yield the following stationary PDF of Hamiltonian

$$p(h) = \frac{C}{\sigma^2(h)} \exp\left[\int \frac{2m(h)}{\sigma^2(h)} \mathrm{d}h \right]. \tag{5.14}$$

The approximate stationary PDF of generalized displacement $\boldsymbol{Q}$ and generalized momenta $\boldsymbol{P}$ in Eq. (5.5) can be calculated from Eq. (5.14) as follows

$$p(\boldsymbol{q}, \boldsymbol{p}) = p(\boldsymbol{q}, p_2, \ldots, p_n, h) \left| \frac{\partial h}{\partial p_1} \right| = p(\boldsymbol{q}, p_2, \ldots, p_n | h) p(h) \left| \frac{\partial h}{\partial p_1} \right| \tag{5.15}$$

For a fixed Hamiltonian $H(\boldsymbol{q}, \boldsymbol{p}) = h$, the conditional PDF $p(\boldsymbol{q}, \boldsymbol{p}_2, \ldots, p_n | H(\boldsymbol{q}, \boldsymbol{p}) = h)$ is inversely proportional to $\dot{q}_1 = \left| \partial h/\partial p_1 \right|$. Thus,

$$p(\boldsymbol{q}, p_2, \ldots, p_n | H(\boldsymbol{q}, \boldsymbol{p}) = h) = C_1 \left| \frac{\partial h}{\partial p_1} \right|^{-1}. \tag{5.16}$$

Integrating both sides of Eq. (5.16), and using Eq. (5.12) yield

$$C_1 = [T(h)]^{-1}. \tag{5.17}$$

Substituting Eq. (5.17) into Eq. (5.16), and then substituting the obtained result into Eq. (5.15) lead to

$$p(\boldsymbol{q}, \boldsymbol{p}) = \left[\frac{p(h)}{T(h)} \right]_{h=H(\boldsymbol{q}, \boldsymbol{p})} \tag{5.18}$$

It is known from Sect. 3.2.6 that the motion of a non-integrable Hamiltonian system with large values of Hamiltonian is chaotic. After adding Gaussian white noise excitation, the motion of quasi-Hamiltonian system should be more complicated. However, for any finite n–DOF quasi-non-integrable Hamiltonian system, the stationary PDF of Hamiltonian is given by Eq. (5.14) and the stationary joint PDF of generalized displacements and momenta is given by Eq. (5.18). For different systems, only $m(H)$ and $\sigma^2(H)$ are different. It is wonderfully concise.

Example 5.1 Consider the following 2-DOF nonlinear stochastic system

$$\begin{aligned} &\ddot{X} - \alpha_1 \dot{X} + \beta_1 X^2 \dot{X} + \omega_1^2 X + aY + b(X - Y)^3 = c_1 X W_{g1}(t), \\ &\ddot{Y} - (\alpha_1 - \alpha_2)\dot{Y} + \beta_2 Y^2 \dot{Y} + \omega_2^2 Y + aX + b(Y - X)^3 = c_2 Y W_{g2}(t), \end{aligned} \tag{5.19}$$

where $W_{g1}(H)$ and $W_{g2}(H)$ are independent Gaussian white noises with spectral densities K_1 and K_2, respectively. Let $\boldsymbol{Q} = [X, Y]^{\mathrm{T}}$, $\boldsymbol{P} = [\dot{X}, \dot{Y}]^{\mathrm{T}}$. Note that the Wong-Zakai correction term for system (5.19) vanish. Equation (5.19) can be rewritten as the form of Eq. (5.5). The Hamiltonian, i.e., the total energy of system is

$$H = \frac{1}{2}\dot{X}^2 + \frac{1}{2}\dot{Y}^2 + U(X, Y), \tag{5.20}$$

where

$$U(X, Y) = \frac{1}{2}\omega_1^2 X^2 + \frac{1}{2}\omega_2^2 Y^2 + aXY + \frac{1}{4}b(X - Y)^4. \tag{5.21}$$

is the potential energy of system. For most parameter domain, the Hamiltonian system with Hamiltonian in Eq. (5.20) is non-integrable. Assume that $\alpha_i, \beta_i, c_i^2 K_i \ \ (i = 1, 2)$ are small quantities of same order. Then system (5.19) is a quasi-non-integrable Hamiltonian system and the stochastic averaging method of quasi-non-integrable Hamiltonian systems can be applied.

The averaged Itô equation is of the form of Eq. (5.7). The averaged drift coefficient and diffusion coefficients can be obtained by using Eqs. (5.10) to (5.13) as follows

$$m(H) = \frac{1}{T(H)} \int\limits_{\Omega} \left[(\alpha_1 - \beta_1 x^2)\dot{x} + (\alpha_1 - \alpha_2 - \beta_2 y^2)\frac{\dot{y}^2}{\dot{x}} + \pi K_1 c_1^2 \frac{x^2}{\dot{x}} + \pi K_2 c_2^2 \frac{y^2}{\dot{x}} \right] \mathrm{d}x\mathrm{d}y\mathrm{d}\dot{y}, \tag{5.22}$$

$$\sigma^2(H) = \frac{2\pi}{T(H)} \int\limits_{\Omega} \left(K_1 c_1^2 x^2 \dot{x} + K_2 c_2^2 \frac{y^2 \dot{y}^2}{\dot{x}} \right) \mathrm{d}x\mathrm{d}y\mathrm{d}\dot{y}, \tag{5.23}$$

where

$$T(H) = \int\limits_{\Omega} \frac{1}{\dot{x}} \mathrm{d}x\mathrm{d}y\mathrm{d}\dot{y}. \tag{5.24}$$

The integral domain Ω is determined by $H(x, y, 0, \dot{y}) = \frac{1}{2}\dot{y}^2 + U(x, y) \le H$.

The multiple integrals in Eqs. (5.22)–(5.24) can be simplified to a single integral with respect to θ by using the transformation $x = r\cos\theta$ and $y = r\sin\theta$. For system (5.19) with parameters $\alpha_1 = \alpha_2 = 0.01$, $\beta_1 = 0.01$, $\beta_2 = 0.02$, $\omega_1 = 1$, $\omega_2 = 2$, $a = 0.01$, $b = 1$, $c_1 = c_2 = 0.2$ and $K_1 = K_2 = 0.3$, calculations were performed for the stationary PDF of $H(t)$, the joint stationary PDFs of displacement and velocity, and their mean-square values (Zhu and Yang 1997). In Fig. 5.1, the solid line represents the stationary PDF of $H(t)$ obtained by using the stochastic averaging method, while black dots represent the results from Monte Carlo simulation of system (5.19). In Fig. 5.2, the solid line represents $E[X^2(t)]$ obtained by using the stochastic averaging method, black dots represent results from Monte Carlo simulation, and the dashed line represents the results obtained by using the equivalent nonlinear system method. The system parameters for Fig. 5.2 are $\alpha_1 = \alpha_2 = 0.03$, $\beta_1 = 0.03$, $\beta_2 = 0.04$, $\omega_1 = 1$, $\omega_2 = 2$, $a = 0.01$, $b = 1$, $c_1 = c_2 = 0.2$. It can be observed that the stochastic averaging method yields pretty accurate results.

The stochastic averaging method of quasi-non-integrable Hamiltonian systems has been applied to several 2-DOF strongly nonlinear stochastic dynamical systems (Gan and Zhu 2001; Zhu and Deng 2004; Zhu and Huang 2004). However, applying the method to three or more DOF strongly nonlinear stochastic dynamical systems, difficulty arises in completing the multiple domain integrals in evaluating the averaged drift and diffusion coefficients in Eqs. (5.10) to (5.12). To overcome this difficulty, Sun et al. (2021) proposed the following two-step generalized elliptic coordinate transformation to convert the $(2n - 1)$-dimensional domain integrals into more manageable n-fold integrals.

Suppose that the Hamiltonian function in Eqs. (5.10) to (5.12) is of the form

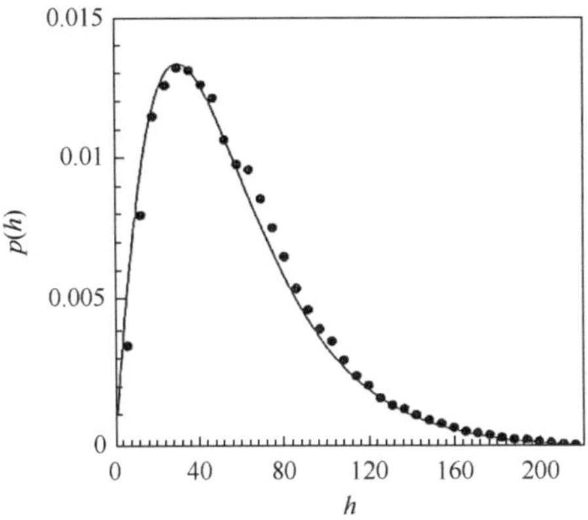

Fig. 5.1 Stationary PDF of Hamiltonian function $H(t)$ of system (5.19). ——, result from the stochastic averaging method; ●, result from Monte Carlo simulation (Zhu and Yang 1997)

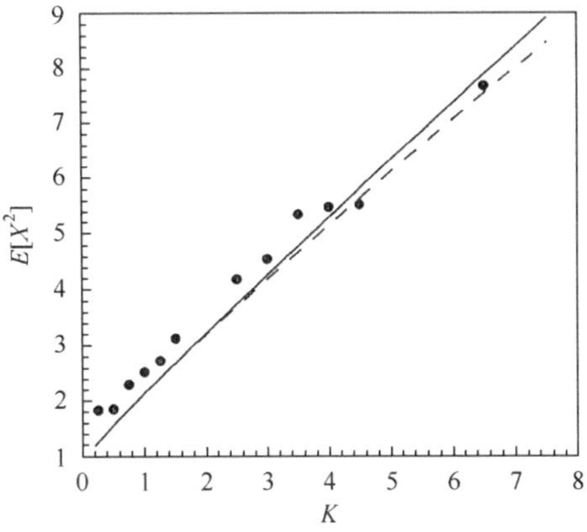

Fig. 5.2 Stationary mean-square value of displacement $X(t)$ of system (5.19). ——, result from the stochastic averaging method; -----, result from equivalent nonlinear system method; ●, result from Monte Carlo simulation (Zhu and Yang 1997)

$$H(\boldsymbol{q}, \boldsymbol{p}) = \frac{1}{2}\sum_{i=1}^{n} p_i^2 + U(\boldsymbol{q}) \tag{5.25}$$

where $U(\boldsymbol{q})$ is potential energy. Express the integral domain in Eqs. (5.10) to (5.12) as the Cartesian product of two subdomains Ω_1 and Ω_2, i.e.

$$\Omega = \Omega_1 \times \Omega_2 = \{(q_1, \ldots, q_n, p_2, \ldots, p_n) | (q_1, \ldots, q_n) \in \Omega_1, (p_2, \ldots, p_n) \in \Omega_2\} \tag{5.26}$$

where $\times$ represents the Cartesian product and

$$\Omega_1 = \{(q_1, \ldots, q_n) | U(\boldsymbol{q}) \leq H\} \tag{5.27}$$

$$\Omega_2 = \left\{ (p_2, \ldots, p_n) | \sum_{i=2}^{n} p_i^2 \leq 2[H - U(\boldsymbol{q})] \right\} \tag{5.28}$$

Thus, the expressions of averaged drift and diffusion coefficients in Eqs. (5.10) to (5.12) become

$$
\begin{aligned}
m(H) =& \frac{\varepsilon}{T(H)} \int_{\Omega_1} \left\{ \int_{\Omega_2} \left[\left(\frac{\partial H}{\partial p_1} \right)^{-1} \left[-\sum_{j,k=1}^{n} m_{jk} \frac{\partial H}{\partial p_j} \frac{\partial H}{\partial p_k} \right. \right. \right. \\
&\left. \left. \left. + \sum_{j,k=1}^{n} \sum_{l,s=1}^{m} \pi K_{ls} g_{jl} g_{ks} \frac{\partial^2 H}{\partial p_j \partial p_k} \right] \right] \times \mathrm{d}p_2 \ldots \mathrm{d}p_n \right\} \mathrm{d}q_1 \ldots \mathrm{d}q_n,
\end{aligned} \tag{5.29}
$$

$$
\begin{aligned}
\sigma^2(H) = & \frac{\varepsilon}{T(H)} \int_{\Omega_1} \left\{ \int_{\Omega_2} \left[\left(\frac{\partial H}{\partial p_1} \right)^{-1} \sum_{j,k=1}^{n} \sum_{l,s=1}^{m} 2\pi K_{ls} g_{jl} g_{ks} \frac{\partial H}{\partial p_j} \frac{\partial H}{\partial p_k} \right] \right. \\
& \left. \mathrm{d}p_2 \ldots \mathrm{d}p_n \right\} \mathrm{d}q_1 \ldots \mathrm{d}q_n,
\end{aligned} \tag{5.30}
$$

$$
T(H) = \int_{\Omega_1} \left\{ \int_{\Omega_2} \left(\frac{\partial H}{\partial p_1} \right)^{-1} \mathrm{d}p_2 \ldots \mathrm{d}p_n \right\} \mathrm{d}q_1 \ldots \mathrm{d}q_n. \tag{5.31}
$$

In the first step, treat q_i and H as constant and integrate with respect to p_i. For this purpose, introduce the following generalized elliptic coordinate transformation (Blumenson 1960)

$$
\begin{aligned}
& p_1 = \sqrt{2H - 2U} \cos\theta_1, \\
& p_2 = \sqrt{2H - 2U} \sin\theta_1 \cos\theta_2, \\
& p_3 = \sqrt{2H - 2U} \sin\theta_1 \sin\theta_2 \cos\theta_3, \\
& \vdots \\
& p_{n-1} = \sqrt{2H - 2U} \sin\theta_1 \ldots \sin\theta_{n-2} \cos\theta_{n-1}, \\
& p_n = \sqrt{2H - 2U} \sin\theta_1 \ldots \sin\theta_{n-1}, \\
& 0 \le \theta_i \le \pi, \quad i = 1, 2, \ldots, n-2, \\
& 0 \le \theta_{n-1} \le 2\pi.
\end{aligned} \tag{5.32}
$$

The Jacobian determinant for this transformation is

$$
Jp_n = \frac{\partial(p_2, p_3, \ldots, p_{n-1}, p_n)}{\partial(\theta_1, \theta_2, \ldots, \theta_{n-2}, \theta_{n-1})} = \begin{vmatrix}
\frac{\partial p_2}{\partial \theta_1} & \frac{\partial p_2}{\partial \theta_2} & \cdots & \frac{\partial p_2}{\partial \theta_{n-2}} & \frac{\partial p_2}{\partial \theta_{n-1}} \\
\frac{\partial p_3}{\partial \theta_1} & \frac{\partial p_3}{\partial \theta_2} & \cdots & \frac{\partial p_3}{\partial \theta_{n-2}} & \frac{\partial p_3}{\partial \theta_{n-1}} \\
\vdots & \vdots & \ddots & \vdots & \vdots \\
\frac{\partial p_{n-1}}{\partial \theta_1} & \frac{\partial p_{n-1}}{\partial \theta_2} & \cdots & \frac{\partial p_{n-1}}{\partial \theta_{n-2}} & \frac{\partial p_{n-1}}{\partial \theta_{n-1}} \\
\frac{\partial p_n}{\partial \theta_1} & \frac{\partial p_n}{\partial \theta_2} & \cdots & \frac{\partial p_n}{\partial \theta_{n-2}} & \frac{\partial p_n}{\partial \theta_{n-1}}
\end{vmatrix} \tag{5.33}
$$

Expanding the last expression in Eq. (5.33) along the last column yields

$$
Jp_n = \frac{-\partial p_{n-1}}{\partial \theta_{n-1}} \begin{vmatrix} \frac{\partial p_2}{\partial \theta_1} & \frac{\partial p_2}{\partial \theta_2} & \cdots & 0 & 0 \\ \frac{\partial p_3}{\partial \theta_1} & \frac{\partial p_3}{\partial \theta_2} & \cdots & 0 & 0 \\ \vdots & \vdots & \ddots & \vdots & \vdots \\ \frac{\partial p_{n-2}}{\partial \theta_1} & \frac{\partial p_{n-2}}{\partial \theta_2} & \cdots & \frac{\partial p_{n-2}}{\partial \theta_{n-3}} & \frac{\partial p_{n-2}}{\partial \theta_{n-2}} \\ \frac{\partial p_n}{\partial \theta_1} & \frac{\partial p_n}{\partial \theta_2} & \cdots & \frac{\partial p_n}{\partial \theta_{n-3}} & \frac{\partial p_n}{\partial \theta_{n-2}} \end{vmatrix} + \frac{\partial p_n}{\partial \theta_{n-1}} \begin{vmatrix} \frac{\partial p_2}{\partial \theta_1} & \frac{\partial p_2}{\partial \theta_2} & \cdots & 0 & 0 \\ \frac{\partial p_3}{\partial \theta_1} & \frac{\partial p_3}{\partial \theta_2} & \cdots & 0 & 0 \\ \vdots & \vdots & \ddots & \vdots & \vdots \\ \frac{\partial p_{n-2}}{\partial \theta_1} & \frac{\partial p_{n-2}}{\partial \theta_2} & \cdots & \frac{\partial p_{n-2}}{\partial \theta_{n-3}} & \frac{\partial p_{n-2}}{\partial \theta_{n-2}} \\ \frac{\partial p_{n-1}}{\partial \theta_1} & \frac{\partial p_{n-1}}{\partial \theta_2} & \cdots & \frac{\partial p_{n-1}}{\partial \theta_{n-3}} & \frac{\partial p_{n-1}}{\partial \theta_{n-2}} \end{vmatrix} \tag{5.34}
$$

Let p_i' $(i = 1, 2, \ldots, n-1)$ be the last generalized momentum of the i th DOF. Introduce a transformation similar to Eq. (5.32),

$$
p_i' = \sqrt{2H - 2U} \sin\theta_1 \ldots \sin\theta_{i-2} \sin\theta_{i-1} \tag{5.35}
$$

Thus,

$$
p_{n-1} = p_{n-1}' \cos\theta_{n-1}, \; p_n = p_{n-1}' \sin\theta_{n-1}. \tag{5.36}
$$

Substituting Eqs. (5.35) and (5.36) into Eq. (5.34) and further derivating yield

$$
\begin{aligned} Jp_n &= p_{n-1}' Jp_{n-1} = p_{n-1}' p_{n-2}' \cdots p_2' Jp_2 \\ &= (2H - 2U)^{(n-1)/2} \cos\theta_1 \sin^{n-2}\theta_1 \ldots \sin\theta_{n-2}, \end{aligned} \tag{5.37}
$$

where

$$
Jp_{n-1} = \begin{vmatrix} \frac{\partial p_2}{\partial \theta_1} & \frac{\partial p_2}{\partial \theta_2} & \cdots & \frac{\partial p_2}{\partial \theta_{n-3}} & \frac{\partial p_2}{\partial \theta_{n-2}} \\ \frac{\partial p_3}{\partial \theta_1} & \frac{\partial p_3}{\partial \theta_2} & \cdots & \frac{\partial p_3}{\partial \theta_{n-3}} & \frac{\partial p_3}{\partial \theta_{n-2}} \\ \vdots & \vdots & \ddots & \vdots & \vdots \\ \frac{\partial p_{n-2}}{\partial \theta_1} & \frac{\partial p_{n-2}}{\partial \theta_2} & \cdots & \frac{\partial p_{n-2}}{\partial \theta_{n-3}} & \frac{\partial p_{n-2}}{\partial \theta_{n-2}} \\ \frac{\partial p_{n-1}'}{\partial \theta_1} & \frac{\partial p_{n-1}'}{\partial \theta_2} & \cdots & \frac{\partial p_{n-1}'}{\partial \theta_{n-3}} & \frac{\partial p_{n-1}'}{\partial \theta_{n-2}} \end{vmatrix}. \tag{5.38}
$$

$$
Jp_2 = \sqrt{2H - 2U} \cos\theta_1 \tag{5.39}
$$

For convenience, denote the integrands in Eqs. (5.29) and (5.30) as $f_1(\boldsymbol{q}, \boldsymbol{p})$ and $f_2(\boldsymbol{q}, \boldsymbol{p})$, respectively, i.e.,

$$f_1(q,p)=\left(\frac{\partial H}{\partial p_1}\right)^{-1}\left(-\sum_{j,k=1}^{n} m_{jk}\frac{\partial H}{\partial p_j}\frac{\partial H}{\partial p_k}+\sum_{j,k=1}^{n}\sum_{l,s=1}^{m}\pi K_{ls}g_{jl}g_{ks}\frac{\partial^2 H}{\partial p_j\partial p_k}\right),\tag{5.40}$$

$$f_2(q,p)=\left(\frac{\partial H}{\partial p_1}\right)^{-1}\sum_{j,k=1}^{n}\sum_{l,s=1}^{m}2\pi K_{ls}g_{jl}g_{ks}\frac{\partial H}{\partial p_j}\frac{\partial H}{\partial p_k}.\tag{5.41}$$

After the transformation (5.32), the Ω_2 domain integrals in Eqs. (5.29) to (5.31) become

$$\tilde{m}(q,H)=\int_0^{2\pi}\int_0^{\pi}\cdots\int_0^{\pi} f_1(q,\theta_1,\ldots,\theta_{n-1})Jp_n\mathrm{d}\theta_1\mathrm{d}\theta_2\ldots\mathrm{d}\theta_{n-1},\tag{5.42}$$

$$\tilde{\sigma}^2(q,H)=\int_0^{2\pi}\int_0^{\pi}\cdots\int_0^{\pi} f_2(q,\theta_1,\ldots,\theta_{n-1})Jp_n\mathrm{d}\theta_1\mathrm{d}\theta_2\ldots\mathrm{d}\theta_{n-1},\tag{5.43}$$

$$\begin{aligned}\tilde{T}(q,H)&=\int_0^{2\pi}\int_0^{\pi}\cdots\int_0^{\pi}\frac{1}{\sqrt{2H-2U\cos\theta_1}}Jp_n\mathrm{d}\theta_1\mathrm{d}\theta_2\ldots\mathrm{d}\theta_{n-1}\\&=\begin{cases}\frac{\Gamma[(n-1)/2]}{\Gamma(n-1)}(4\pi)^{\frac{n-1}{2}}(2H-2U)^{\frac{n-2}{2}}, & n \text{ is odd number}\\ \frac{1}{\Gamma[(n-1)/2]}(2\pi)^{\frac{n}{2}}(H-U)^{\frac{n-2}{2}}, & n \text{ is even number,}\end{cases}\end{aligned}\tag{5.44}$$

where $f_1(\boldsymbol{q},\theta_1,\ldots,\theta_{n-1})$ and $f_2(\boldsymbol{q},\theta_1,\ldots,\theta_{n-1})$ are the functions obtained from Eqs. (5.40) and (5.41) by replacing p_i with θ_i according to the transformation (5.32). If m_{jk} and g_{jk} in Eqs. (5.10) to (5.12) are polynomials of the generalized momentum p_j, the integrals in Eqs. (5.42) and (5.43) can be completed similarly to Eq. (5.44). Thus, Eqs. (5.29) to (5.31) become

$$m(H)=\frac{\varepsilon}{T(H)}\int_{\Omega_1}\tilde{m}(q,H)\mathrm{d}q_1\mathrm{d}q_2\ldots\mathrm{d}q_n,\tag{5.45}$$

$$\sigma^2(H)=\frac{2\pi\varepsilon}{T(H)}\int_{\Omega_1}\tilde{\sigma}^2(q,H)\mathrm{d}q_1\mathrm{d}q_2\ldots\mathrm{d}q_n,\tag{5.46}$$

$$T(H)=\int_{\Omega_1}\tilde{T}(q,H)\mathrm{d}q_1\mathrm{d}q_2\ldots\mathrm{d}q_n.\tag{5.47}$$

In the second step, introduce the following generalized elliptic coordinate transformation

$$
\begin{aligned}
q_1 &= \frac{r}{\omega_1}\cos\varphi_1, \\
q_2 &= \frac{r}{\omega_2}\sin\varphi_1\cos\varphi_2, \\
&\vdots \\
q_{n-1} &= \frac{r}{\omega_{n-1}}\sin\varphi_1\ldots\sin\varphi_{n-2}\cos\varphi_{n-1}, \\
q_n &= \frac{r}{\omega_n}\sin\varphi_1\ldots\sin\varphi_{n-2}\sin\varphi_{n-1}, \\
0 &\le \varphi_i \le \pi, \quad i = 1, 2, \ldots, n-2; \quad 0 \le \varphi_{n-1} \le 2\pi; \quad 0 \le r \le R,
\end{aligned}
\tag{5.48}
$$

where ω_i is the linearized frequency of the i th oscillator, R is the maximum value of r, and it is the positive root of the following algebraic equation

$$
U(R, \varphi_1, \varphi_2, \ldots, \varphi_{n-1}) - H = 0, \tag{5.49}
$$

where $U(R, \varphi_1, \varphi_2, \ldots, \varphi_{n-1})$ is the transformed expression of the potential energy $U(\boldsymbol{q})$ in Eq. (5.25).

Similar to the derivation from Eqs. (5.33) to (5.39), the Jacobian determinant for the transformation (5.48) is

$$
Jq_n = \frac{r^{n-1}}{\omega_1\omega_2\ldots\omega_n}\sin^{n-2}\varphi_1\ldots\sin\varphi_{n-2}. \tag{5.50}
$$

Therefore, after the transformation (5.48), Eqs. (5.45)–(5.47) become

$$
m(H) = \frac{\varepsilon}{T(H)}\int_0^{2\pi}\int_0^{\pi}\cdots\int_0^{\pi}\int_0^{R}\tilde{m}(\Phi, H)Jq_n \mathrm{d}r\mathrm{d}\varphi_1\ldots\mathrm{d}\varphi_{n-1}, \tag{5.51}
$$

$$
\sigma^2(H) = \frac{2\pi\varepsilon}{T(H)}\int_0^{2\pi}\int_0^{\pi}\cdots\int_0^{\pi}\int_0^{R}\tilde{\sigma}^2(\Phi, H)Jq_n \mathrm{d}r\mathrm{d}\varphi_1\ldots\mathrm{d}\varphi_{n-1}, \tag{5.52}
$$

$$
T(H) = \int_0^{2\pi}\int_0^{\pi}\cdots\int_0^{\pi}\int_0^{R}\tilde{T}(\Phi, H)Jq_n \mathrm{d}r\mathrm{d}\varphi_1\ldots\mathrm{d}\varphi_{n-1}, \tag{5.53}
$$

where

$$
\Phi = [\varphi_1, \varphi_2, \ldots, \varphi_{n-1}]^T. \tag{5.54}
$$

Thus, the $(2n-1)$-dimensional integrals over domain Ω in Eqs. (5.10) to (5.12) become the n-fold integrals in Eqs. (5.51) to (5.53), which are much easier to complete.

Example 5.2 Consider the following 5-DOF strongly nonlinear stochastic system

$$\ddot{X}_i + (\beta_i X_i^2 + c_i)\dot{X}_i + g_i(X) = W_{gi}(t), \quad i = 1, 2, \ldots, 5 \tag{5.55}$$

where

$$\begin{aligned}
g_1(\boldsymbol{X}) &= \omega_1^2 X_1 + a_{12}(X_1 - X_2) + b_{12}(X_1 - X_2)^3, \\
g_2(\boldsymbol{X}) &= \omega_2^2 X_2 + a_{12}(X_2 - X_1) + b_{12}(X_2 - X_1)^3 \\
&\quad + a_{23}(X_2 - X_3) + b_{23}(X_2 - X_3)^3, \\
g_3(\boldsymbol{X}) &= \omega_3^2 X_3 + a_{23}(X_3 - X_2) + b_{23}(X_3 - X_2)^3 \\
&\quad + a_{34}(X_3 - X_4) + b_{34}(X_3 - X_4)^3, \\
g_4(\boldsymbol{X}) &= \omega_4^2 X_4 + a_{34}(X_4 - X_3) + b_{34}(X_4 - X_3)^3 \\
&\quad + a_{45}(X_4 - X_5) + b_{45}(X_4 - X_5)^3, \\
g_5(\boldsymbol{X}) &= \omega_5^2 X_5 + a_{45}(X_5 - X_4) + b_{45}(X_5 - X_4)^3,
\end{aligned} \tag{5.56}$$

$W_{gi}(t)$ are independent Gaussian white noises with the correlation functions

$$E[W_{gi}(t)W_{gi}(t+\tau)] = 2\pi K_i \delta(\tau), \quad i = 1, 2, \ldots, 5. \tag{5.57}$$

System (5.55) has only external excitations and thus the Wong-Zakai correction terms vanish. Letting $Q_i = X_i,\ P_i = \dot{X}_i,\ i = 1, 2, \ldots, 5,$, system (5.55) can be rewritten as the following stochastically excited and dissipated Hamiltonian system

$$\begin{aligned}
&\dot{Q}_i = P_i, \\
&\dot{P}_i = -(\beta_i Q_i^2 + c_i)P_i - g_i(\boldsymbol{Q}) + W_{gi}(t), \\
&i = 1, 2, \ldots, 5,
\end{aligned} \tag{5.58}$$

where $g_i(\boldsymbol{Q}) = g_i(\boldsymbol{X})|_{X_i = Q_i}$. The corresponding Hamiltonian function is

$$H(\boldsymbol{Q}, \boldsymbol{P}) = \frac{1}{2}\sum_{i=1}^{5} P_i^2 + U(\boldsymbol{Q}), \tag{5.59}$$

where

$$\begin{aligned}U(\boldsymbol{Q}) =& \sum_{i=1}^{5} \frac{1}{2}\omega_i^2 Q_i^2 + \frac{1}{2}\big[a_{12}(Q_1 - Q_2)^2 + a_{23}(Q_2 - Q_3)^2 \\ &+ a_{34}(Q_3 - Q_4)^2 + a_{45}(Q_4 - Q_5)^2\big] \\ &+ \frac{1}{4}\big[b_{12}(Q_1 - Q_2)^4 + b_{23}(Q_2 - Q_3)^4 \\ &+ b_{34}(Q_3 - Q_4)^4 + b_{45}(Q_4 - Q_5)^4\big].\end{aligned} \tag{5.60}$$

System (5.55) consists of 5 oscillators connected in series with nonlinear springs. It can be viewed as a complex oscillator network coupled with nonlinear springs. Due to the nonlinear coupling between oscillators, the Hamiltonian system associated with system (5.58) is non-integrable. Suppose that β_i, c_i, K_i are small quantities of same order. System (5.58) is a quasi-non-integrable Hamiltonian system and the stochastic averaging method described in Sect. 5.1 can be applied. The averaged drift and diffusion coefficients can be obtained according to Eqs. (5.10) to (5.12) as follows

$$\begin{aligned}m(H) = & \frac{1}{T(H)} \int_{\Omega} \sum_{i=1}^{5} \big[-(\beta_i q_i^2 + c_i) p_i^2 + \pi K_i\big] \times \\ & \frac{1}{p_1} \mathrm{d}q_1 \ldots \mathrm{d}q_5 \mathrm{d}p_2 \ldots \mathrm{d}p_5\end{aligned} \tag{5.61}$$

$$\sigma^2(H) = \frac{1}{T(H)} \int_{\Omega} \sum_{i=1}^{5} (2\pi K_i p_i^2) \frac{1}{p_1} \mathrm{d}q_1 \ldots \mathrm{d}q_5 \mathrm{d}p_2 \ldots \mathrm{d}p_5, \tag{5.62}$$

where

$$T(H) = \int_{\Omega} \frac{1}{p_1} \mathrm{d}q_1 \ldots \mathrm{d}q_5 \mathrm{d}p_2 \ldots \mathrm{d}p_5, \tag{5.63}$$

$$\Omega = \{(q_1, \ldots, q_5, p_2, \ldots, p_5) | H(q_1, \ldots, q_5, 0, p_2, \ldots, p_5) \leq H\}. \tag{5.64}$$

To complete the above integrals, represent the integration domain Ω as the Cartesian product of subdomains Ω_1 and Ω_2, i.e.,

$$\begin{aligned}\Omega = \Omega_1 \times \Omega_2 = \{ & (q_1, \ldots, q_5, p_2, \ldots, p_5) | \\ & (q_1, \ldots, q_5) \in \Omega_1, (p_2, \ldots, p_5) \in \Omega_2\}\end{aligned} \tag{5.65}$$

where

$$\Omega_1 = \{(q_1, \ldots, q_5) | U(\boldsymbol{q}) \leq H\}, \tag{5.66}$$

$$\Omega_2 = \left\{ (p_2, \ldots, p_5) | \sum_{i=2}^{5} p_i^2 \leq 2[H - U(\boldsymbol{q})] \right\}. \tag{5.67}$$

Equations (5.61)–(5.63) become

$$\begin{aligned} m(H) = \frac{1}{T(H)} \int\limits_{\Omega_1} \Bigg\{ \int\limits_{\Omega_2} \sum_{i=1}^{5} [-(\beta_i q_i^2 + c_i) p_i^2 + \pi K_i] \\ \frac{1}{p_1} \mathrm{d}p_2 \ldots \mathrm{d}p_5 \Bigg\} \mathrm{d}q_1 \ldots \mathrm{d}q_5, \end{aligned} \tag{5.68}$$

$$\sigma^2(H) = \frac{1}{T(H)} \int\limits_{\Omega_1} \left\{ \int\limits_{\Omega_2} \sum_{i=1}^{5} (2\pi K_i) \frac{1}{p_1} \mathrm{d}p_2 \ldots \mathrm{d}p_5 \right\} \mathrm{d}q_1 \ldots \mathrm{d}q_5, \tag{5.69}$$

$$T(H) = \int\limits_{\Omega_1} \left\{ \int\limits_{\Omega_2} \frac{1}{p_1} \mathrm{d}p_2 \ldots \mathrm{d}p_5 \right\} \mathrm{d}q_1 \ldots \mathrm{d}q_5. \tag{5.70}$$

Successively applying the generalized elliptic coordinate transformations (5.32) and (5.48) to render the integrals in $m(H)$ and $\sigma^2(H)$ more manageable five-fold integrals. Sun et al. (2021) used Eq. (5.14) to calculate the stationary probability density, mean-value and mean-square value for $H(t)$ of system (5.55) with parameters $\omega_i = 1$, $\beta_i = 0.01$, $c_i = 0.02$ $(i = 1, 2, \ldots, 5)$, $a_{12} = a_{23} = a_{34} = a_{45} = 4$, $b_{12} = b_{23} = b_{34} = b_{45} = 4$, and compared the results with results from Monte Carlo simulation, see Figs. 5.3 and 5.4. Relative errors of mean-value and mean-square value of H are obtained and shown in Fig. 5.5. When $\pi K_i = \pi K$ $(i = 1, 2, \ldots, 5)$ and β_i, c_i are small quantities of same order, these figures showed that the results from stochastic averaging method and those from simulation are in good agreement.

5.2 Quasi-Integrable Hamiltonian Systems

Now return to quasi-Hamiltonian system (5.5) and assume that the Hamiltonian system associated with system (5.5) is integrable. Suppose that the action-angle vectors $\boldsymbol{I}$ and $\boldsymbol{\Theta}$ for the Hamiltonian system are available. Introduce the transformation (Zhu et al. 1997)

$$I_r = I_r(\boldsymbol{Q}, \boldsymbol{P}) \tag{5.71}$$

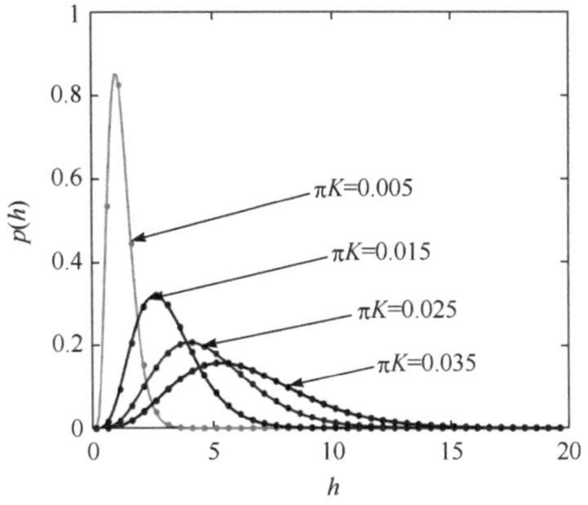

Fig. 5.3 Stationary PDF of the Hamiltonian function H. ——, results from the stochastic averaging method; ●, results from Monte Carlo simulation (Sun et al. 2021)

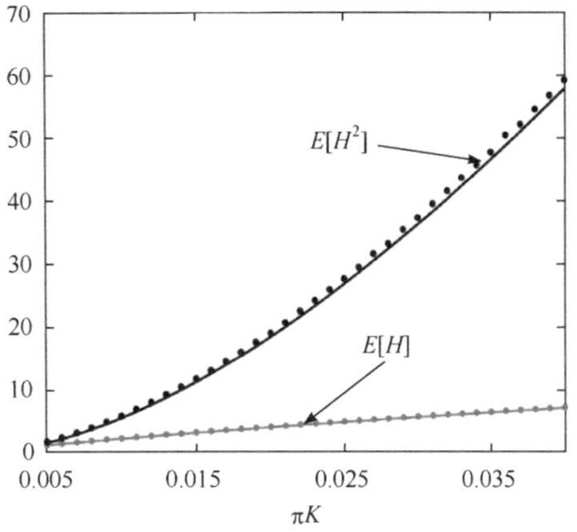

Fig. 5.4 Mean-value and mean-square value of Hamiltonian function H. ——, results from the stochastic averaging method; ●, results from Monte Carlo simulation (Sun et al., 2021)

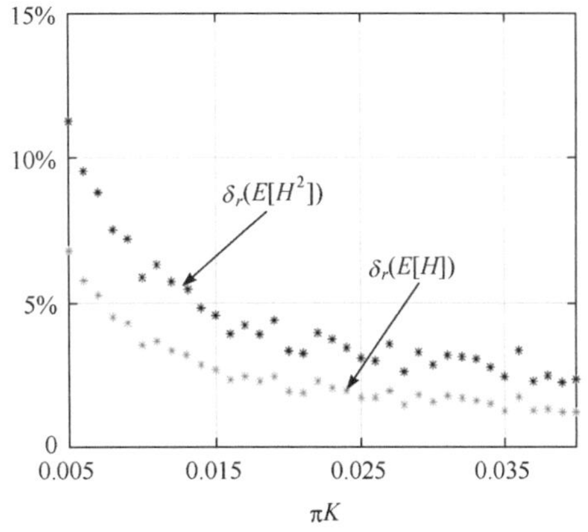

Fig. 5.5 Relative errors of the mean-value and mean-square value of Hamiltonian function (Sun et al. 2021)

$$\Theta_r = \Theta_r(\boldsymbol{Q}, \boldsymbol{P}), \quad r = 1, 2, \ldots, n. \tag{5.72}$$

By applying Itô differential rule, the following Itô stochastic differential equations governing I_r and Θ_r can be obtained from Eq. (5.5)

$$\begin{aligned} \mathrm{d}I_r = {} & \varepsilon \sum_{j,k=1}^{n} \left[-m_{jk} \frac{\partial I_r}{\partial P_j} \frac{\partial H}{\partial P_k} + \sum_{l,s=1}^{m} \left(\pi K_{ls} g_{jl} g_{ks} \frac{\partial^2 I_r}{\partial P_j \partial P_k} \right) \right] \mathrm{d}t \\ & + \varepsilon^{1/2} \sum_{j=1}^{n} \sum_{l=1}^{m} \frac{\partial I_r}{\partial P_j} g_{jl} \mathrm{d}B'_l(t), \end{aligned} \tag{5.73}$$

$$\begin{aligned} \mathrm{d}\Theta_r = {} & \left\{ \omega_r - \varepsilon \sum_{j,k=1}^{n} \left[m_{jk} \frac{\partial \Theta_r}{\partial P_j} \frac{\partial H}{\partial P_k} - \sum_{l,s=1}^{m} \left(\pi K_{ls} g_{jl} g_{ks} \frac{\partial^2 \Theta_r}{\partial P_j \partial P_k} \right) \right] \right\} \\ & \mathrm{d}t + \varepsilon^{1/2} \sum_{j=1}^{n} \sum_{l=1}^{m} \frac{\partial \Theta_r}{\partial P_j} g_{jl} \mathrm{d}B'_l(t), \\ & r = 1, 2, \ldots, n. \end{aligned} \tag{5.74}$$

The dimension and form of the averaged equations will depend on the resonant characteristics of the Hamiltonian system associated with Eq. (5.5). In the following, the non-internal resonant case and the internal resonant case will be considered separately.

5.2.1 Non-Internal Resonant Case

Assume that the integrable Hamiltonian system associated with system (5.5) is non-internal resonant, i.e., there does not exist the following weak internal resonant relationship

$$\sum_{r=1}^{n} k_r \omega_r = O(\varepsilon), \tag{5.75}$$

where k_r are integers; $O(\varepsilon)$ is a small quantity of ε order. Then $I_r(t)$ in Eq. (5.73) are slowly-varying processes and $\Theta_r(t)$ in Eq. (5.74) are fast-varying processes. According to the Khasminskii theorem (Khasminskii 1968), when $\varepsilon \to 0$, $\boldsymbol{I} = [I_1, I_2, \ldots, I_n]^T$ weakly converges to a n-dimensional vector Markov diffusion process. The averaged Itô stochastic differential equations are of the form

$$dI_r = m_r(I)dt + \sum_{l=1}^{m} \sigma_{rl}(I)dB_l(t), \quad r = 1, 2, \ldots, n, \tag{5.76}$$

where $B_l(t)$ $(l = 1, 2, \ldots, m)$ are independent unit Wiener processes. According to Eqs. (4.29) and (4.30), the averaged drift coefficients and diffusion coefficients are obtained from Eq. (5.73) by time averaging, i.e.,

$$m_r(I) = \varepsilon \left\langle \sum_{j,k=1}^{n} \left[-m_{jk} \frac{\partial I_r}{\partial P_j} \frac{\partial H}{\partial P_k} + \sum_{l,s=1}^{m} \left(\pi K_{ls} g_{jl} g_{ks} \frac{\partial^2 I_r}{\partial P_j \partial P_k} \right) \right] \right\rangle_t, \tag{5.77}$$

$$\sum_{l=1}^{m} \sigma_{rl}\sigma_{il}(I) = \varepsilon \left\langle \sum_{j,k=1}^{n} \sum_{l,s=1}^{m} \left(2\pi K_{ls} g_{jl} g_{ks} \frac{\partial I_r}{\partial P_j} \frac{\partial I_i}{\partial P_k} \right) \right\rangle_t, \quad r, i = 1, 2, \ldots, n. \tag{5.78}$$

where $\langle [\cdot] \rangle_t$ denotes the time averaging. In Eqs. (5.77) and (5.78), all $\boldsymbol{Q}$, $\boldsymbol{P}$ should be replaced by $\boldsymbol{I}$, $\boldsymbol{\Theta}$ according to Eqs. (5.71) and (5.72). $\boldsymbol{I}$ is regarded as constant vector. It was pointed out in Sect. 3.2.7 that the integrable and non-resonant Hamiltonian system is ergodic on the n–dimensional torus with I_r $(r = 1, 2, \ldots, n)$ being constants. The time averaging in Eqs. (5.77) and (5.78) can be replaced by spatial averaging with respect to phase angles θ_i $(i = 1, 2, \ldots, n)$. Thus,

$$m_r(I) = \frac{\varepsilon}{(2\pi)^n} \int_0^{2\pi} \cdots \int_0^{2\pi} \sum_{j,k=1}^{n} \left[-m_{jk} \frac{\partial I_r}{\partial p_j} \frac{\partial H}{\partial p_k} + \sum_{l,s=1}^{m} \left(\pi K_{ls} g_{jl} g_{ks} \frac{\partial^2 I_r}{\partial p_j \partial p_k} \right) \right] d\theta_1 \ldots d\theta_n, \tag{5.79}$$

$$\sum_{l=1}^{m} \sigma_{rl}\sigma_{il}(I) = \frac{\varepsilon}{(2\pi)^n} \int_0^{2\pi} \cdots \int_0^{2\pi} \sum_{j,k=1}^{n} \sum_{l,s=1}^{m} \left(2\pi K_{ls} g_{jl} g_{ks} \frac{\partial I_r}{\partial p_j} \frac{\partial I_i}{\partial p_k} \right) d\theta_1 \ldots d\theta_n, \quad r, i = 1, 2, \ldots, n. \tag{5.80}$$

The averaged FPK equation associated with Eq. (5.76) is

$$\frac{\partial p}{\partial t} = -\sum_{r=1}^{n} \frac{\partial}{\partial I_r} [a_r(\boldsymbol{I})p] + \frac{1}{2} \sum_{r,i=1}^{n} \frac{\partial^2}{\partial I_r \partial I_i} [b_{ri}(\boldsymbol{I})p], \tag{5.81}$$

where the first and second derivative moments are

$$a_r(\boldsymbol{I}) = m_r(\boldsymbol{I}), \tag{5.82}$$

$$b_{ri}(\boldsymbol{I}) = \sum_{l=1}^{n} \sigma_{rl}\sigma_{il}(\boldsymbol{I}). \tag{5.83}$$

In Eq. (5.81), $p = p(\boldsymbol{I}, t|\boldsymbol{I}_0)$ is the transition PDF of $\boldsymbol{I}(t)$ with initial condition

$$p(\boldsymbol{I}, 0|\boldsymbol{I}_0) = \delta(\boldsymbol{I} - \boldsymbol{I}_0), \tag{5.84}$$

or p is the PDF $p(\boldsymbol{I}, t)$ of $\boldsymbol{I}(t)$ with initial condition

$$p(\boldsymbol{I}, 0) = p(\boldsymbol{I}_0). \tag{5.85}$$

The boundary conditions of FPK Eq. (5.81) depends on the property of corresponding Hamiltonian system and the constraints on the system. If $\boldsymbol{I}$ can take any value in the first quadrant of R^n, then the boundary conditions are

$$p \text{ is finite}, \quad I_r = 0, \quad r = 1, 2, \ldots, n. \tag{5.86}$$

$$p, \ \partial p/\partial I_r \to 0, \ |\boldsymbol{I}| \to \infty, \quad r = 1, 2, \ldots, n. \tag{5.87}$$

Equation (5.86) indicates that $I_r = 0$ are reflecting boundaries. Equation (5.86) is only qualitative and the quantitative boundary condition can be obtained from Eq. (5.81). Equation (5.87) means that $|\boldsymbol{I}| \to \infty$ is an absorbing boundary. In addition, p has to satisfy the normalization condition. Solving Eq. (5.81) under initial condition, boundary conditions and normalization condition yields the transition PDF or PDF of $\boldsymbol{I}(t)$. The stationary solution $p(\boldsymbol{I})$ of Eq. (5.81) can be obtained by letting $\partial p / \partial t = 0$ and using appropriate boundary conditions.

Suppose that the Hamiltonian system associated with system (5.5) is integrable and non-resonant. The action-angle variables are unavailable and only n independent and involutive first integrals (i.e., conservative quantities) H_r $(r = 1, 2, \ldots, n)$ are available. Introduce transformation

$$H_r = H_r(\boldsymbol{Q}, \boldsymbol{P}), \quad r = 1, 2, \ldots, n \tag{5.88}$$

The following Itô stochastic differential equations governing $H_r(t)$ can be derived from Eq. (5.5) by using Itô differential rule,

$$\mathrm{d}H_r = \varepsilon \sum_{j,k=1}^{n} \left[-m_{jk} \frac{\partial H_r}{\partial P_j} \frac{\partial H}{\partial P_k} + \sum_{l,s=1}^{m} \left(\pi K_{ls} g_{jl} g_{ks} \frac{\partial^2 H_r}{\partial P_j \partial P_k} \right) \right] \mathrm{d}t$$
$$+ \varepsilon^{1/2} \sum_{j=1}^{n} \sum_{l=1}^{m} \frac{\partial H_r}{\partial P_j} g_{jl} \mathrm{d}B_l'(t),$$
$$r = 1, 2, \ldots, n. \tag{5.89}$$

Replace the n equations of P_j in Eq. (5.5) by the n equations of H_r in Eq. (5.89) to constitute a new system, in which $Q_j(t)$ are fast-varying processes while $H_r(t)$ are slowly-varying processes. According to the Khasminskii theorem (Khasminskii 1968), when $\varepsilon \to 0$, $\boldsymbol{H}(t) = [H_1(t), H_2(t), \ldots, H_n(t)]^{\mathrm{T}}$ weakly converges to a n-dimensional vector Markov diffusion process. The averaged Itô stochastic differential equations are of the form

$$\mathrm{d}H_r = \overline{m}_r(\boldsymbol{H})\mathrm{d}t + \sum_{l=1}^{m} \overline{\sigma}_{rl}(\boldsymbol{H})\mathrm{d}B_l(t), \quad r = 1, 2, \ldots, n, \tag{5.90}$$

where $B_l(t)$ $(l = 1, 2, \ldots, m)$ are independent unit Wiener processes. Using Eqs. (4.29) and (4.30), the averaged drift coefficients and diffusion coefficients are obtained from time averaging of the coefficients of Eq. (5.89) as follows

$$\overline{m}_r(\boldsymbol{H}) = \varepsilon \left\langle \sum_{j,k=1}^{n} \left[-m_{jk} \frac{\partial H_r}{\partial P_j} \frac{\partial H}{\partial P_k} + \sum_{l,s=1}^{m} \left(\pi K_{ls} g_{jl} g_{ks} \frac{\partial^2 H_r}{\partial P_j \partial P_k} \right) \right] \right\rangle_t, \tag{5.91}$$

$$\sum_{l=1}^{m} \overline{\sigma}_{rl} \overline{\sigma}_{il}(\boldsymbol{H}) = \varepsilon \left\langle \sum_{j,k=1}^{n} \sum_{l,s=1}^{m} \left(2\pi K_{ls} g_{jl} g_{ks} \frac{\partial H_r}{\partial P_j} \frac{\partial H_i}{\partial P_k} \right) \right\rangle_t. \tag{5.92}$$

where P_j have to be replaced by H_r $(r, j = 1, 2, \ldots, n)$ according to the transformation (5.88). Since $Q_j(t)$ are random processes, the time averaging in Eqs. (5.91) and (5.92) is difficult to complete.

Assume that the Hamiltonian system is separable, i.e.,

$$H = \sum_{r=1}^{n} H_r(Q_r, P_r), \quad r = 1, 2, \ldots, n. \tag{5.93}$$

The subsystems with Hamiltonian H_r have periodic solutions with periods T_r. Then the least common multiple of T_r is

$$T = T(H) = \prod_{\mu=1}^{n} T_\mu(H_\mu) = \prod_{\mu=1}^{n} \oint \left(1 \Big/ \frac{\partial H_\mu}{\partial p_\mu}\right) \mathrm{d}q_\mu. \tag{5.94}$$

T can be used as the averaged time in Eqs. (5.91) and (5.92), i.e.,

$$\overline{m}_r(H) = \frac{\varepsilon}{T(H)} \oint \sum_{j,k=1}^{n} \left[-m_{jk} \frac{\partial H}{\partial p_j} \frac{\partial H_r}{\partial p_k} + \sum_{l,s=1}^{m} \left(\pi K_{ls} g_{jl} g_{ks} \frac{\partial^2 H_r}{\partial p_j \partial p_k} \right) \right] \prod_{\mu=1}^{n} \left(1 \Big/ \frac{\partial H_\mu}{\partial p_\mu}\right) \mathrm{d}q_\mu, \tag{5.95}$$

$$\sum_{l=1}^{m} \overline{\sigma}_{rl} \overline{\sigma}_{il}(H) = \frac{\varepsilon}{T(H)} \oint \sum_{j,k=1}^{n} \left(\sum_{l,s=1}^{m} 2\pi K_{ls} g_{jl} g_{ks} \frac{\partial H_r}{\partial p_j} \frac{\partial H_i}{\partial p_k} \right) \times \prod_{\mu=1}^{n} \left(1 \Big/ \frac{\partial H_\mu}{\partial p_\mu}\right) \mathrm{d}q_\mu. \tag{5.96}$$

Finishing averaging with respect to q_i in Eqs. (5.95) and (5.96) and then substituting the results into Eq. (5.90) yield the averaged Itô stochastic differential equations and the associated averaged FPK equation

$$\frac{\partial p}{\partial t} = -\sum_{r=1}^{n} \frac{\partial}{\partial h_r} \left[\overline{a}_r(\boldsymbol{h}) p \right] + \frac{1}{2} \sum_{r,i=1}^{n} \frac{\partial^2}{\partial h_r \partial h_i} \left[\overline{b}_{ri}(\boldsymbol{h}) p \right], \tag{5.97}$$

where

$$\overline{a}_r(\boldsymbol{h}) = \overline{m}_r(\boldsymbol{H})|_{\boldsymbol{H}=\boldsymbol{h}}, \tag{5.98}$$

$$\overline{b}_{ri}(\boldsymbol{h}) = \sum_{l=1}^{m} \overline{\sigma}_{rl} \overline{\sigma}_{il}(\boldsymbol{H})|_{\boldsymbol{H}=\boldsymbol{h}}, \quad r, i = 1, 2, \ldots, n. \tag{5.99}$$

$p = p(\boldsymbol{h}, t|\boldsymbol{h}_0)$ is the transition PDF of $\boldsymbol{H}(t)$ with initial condition

$$p(\boldsymbol{h}, 0|\boldsymbol{h}_0) = \delta(\boldsymbol{h} - \boldsymbol{h}_0). \tag{5.100}$$

p can also be the PDF $p(\boldsymbol{h}, t)$ of $\boldsymbol{H}(t)$ with initial condition

$$p(\boldsymbol{h}, 0) = p(\boldsymbol{h}_0). \tag{5.101}$$

The boundary conditions of Eq. (5.97) depend on the property of corresponding Hamiltonian system and the constraints imposed on system. If $\boldsymbol{h}$ can take any value in the first quadrant of R^n, then the boundary conditions are similar to Eqs. (5.86) and (5.87), i.e.,

$$p = \text{finite}, \quad h_r = 0, \quad r = 1, 2, \ldots, n, \tag{5.102}$$

$$p, \partial p/\partial h_r \to 0, |\ \boldsymbol{h}| \to \infty, \quad r = 1, 2, \ldots, n. \tag{5.103}$$

The diffusion matrices of the averaged Itô stochastic differential equations (5.76) and (5.90) are generally non-degenerate, while the diffusion matrix of the original Itô equation (5.5) is degenerate. This is one of major advantages of the stochastic averaging methods. When the stochastic dynamical programming method is applied to averaged system, the dynamical programming equation (HJB) has classical solution.

One characteristic of the averaged FPK Eqs. (5.91) and (5.97) is that the probability flow contains only potential flow and no circular flow. If Eq. (5.81) or (5.97) has exact stationary solution, it must belong to the solution of stationary potential (Lin and Cai 1995). The relation between $p(\boldsymbol{I})$ or $p(\boldsymbol{h})$ and the approximate stationary PDF $p(\boldsymbol{q}, \boldsymbol{p})$ can be derived using transformations (5.71) and (5.72) or (5.88) as follows

$$\begin{aligned} p(\boldsymbol{q}, \boldsymbol{p}) &= p(\boldsymbol{I}, \boldsymbol{\theta})\left|\frac{\partial(\boldsymbol{I}, \boldsymbol{\theta})}{\partial(\boldsymbol{q}, \boldsymbol{p})}\right| = p(\boldsymbol{\theta}|\boldsymbol{I})p(\boldsymbol{I}) \\ &= \frac{1}{(2\pi)^n} p(\boldsymbol{I})|_{\boldsymbol{I}=\boldsymbol{I}(\boldsymbol{q},\boldsymbol{p})}. \end{aligned} \tag{5.104}$$

or

$$p(\boldsymbol{q}, \boldsymbol{p}) = p(\boldsymbol{q}, \boldsymbol{h})|\partial \boldsymbol{h}/\partial \boldsymbol{p}| = p(\boldsymbol{q}|\boldsymbol{h})p(\boldsymbol{h})|\partial \boldsymbol{h}/\partial \boldsymbol{p}| \tag{5.105}$$

Since the integrable and non-resonant Hamiltonian system with constant I_r ($r = 1, 2, \ldots, n$) is periodic with respective to θ_r, $p(\boldsymbol{\theta}|\boldsymbol{I}) = 1/(2\pi)^n$. Since the transformation from $\boldsymbol{q}, \boldsymbol{p}$ to $\boldsymbol{I}, \boldsymbol{\theta}$ is canonical, the Jacobian determinant $|\partial(\boldsymbol{I}, \boldsymbol{\theta})/\partial(\boldsymbol{q}, \boldsymbol{p})| = 1$. $p(\boldsymbol{q}|\boldsymbol{h})$ is the PDF of $\boldsymbol{q}$ under the condition of constant H_r. $|\partial \boldsymbol{h}/\partial \boldsymbol{p}|$ is the absolute value of Jacobian determinant of transformation from $\boldsymbol{p}$ to $\boldsymbol{h}$. When the Hamiltonian function has the form of Eq. (5.93) and each Hamiltonian subsystem has periodic solution, $p(\boldsymbol{q}|\boldsymbol{h}) = \prod_{r=1}^{n} p(q_r|h_r)$, $p(q_r|h_r) = C_r \Big/ \left|\frac{\partial h_r}{\partial p_r}\right|$, $|\partial \boldsymbol{h}/\partial \boldsymbol{p}| = \prod_{r=1}^{n} |\partial h_r/\partial p_r|$. Similar to Eq. (5.98), it can be proven that

$$p(\boldsymbol{q}, \boldsymbol{p}) = \left.\frac{p(\boldsymbol{h})}{T(\boldsymbol{h})}\right|_{\boldsymbol{h}=\boldsymbol{H}(\boldsymbol{q},\boldsymbol{p})}. \tag{5.106}$$

where $T(\boldsymbol{h})$ is determined by Eq. (5.94). For a SDOF system, the stochastic averaging method described in this section reduces to the stochastic averaging method of energy envelope.

Example 5.3 Consider the van der Pol oscillator and Duffing oscillator coupled by nonlinear damping and excited by Gaussian white noises. The motion equations are

$$\begin{aligned}
&\dot{Q}_1 = P_1,\\
&\dot{P}_1 = -\omega^2 Q_1 - (-\beta_1 + \alpha_1 Q_1^2 + \alpha_2 Q_2^4 + \alpha_3 P_2^2)P_1 + W_{g1}(t) + Q_1 W_{g3}(t),\\
&\dot{Q}_2 = P_2,\\
&\dot{P}_2 = -kQ_2^3 - (\beta_2 + \alpha_4 Q_1^2)P_2 + W_{g2}(t) + Q_2 W_{g4}(t),
\end{aligned} \tag{5.107}$$

where ω, k, α_j, β_i are constants; $W_{gl}(t)$ are independent Gaussian white noises with intensities $2\pi K_l$. For this case, Wong-Zakai correction terms vanish. Similar to Eq. (5.93), the Hamiltonian function associated with Eq. (5.107) is

$$H = H_1 + H_2, \tag{5.108}$$

where

$$H_1 = (P_1^2 + \omega^2 Q_1^2)/2, H_2 = P_2^2/2 + kQ_2^4/4. \tag{5.109}$$

Taking Eq. (5.109) as a transformation from P_1, P_2 to H_1, H_2, the following Itô stochastic differential equations for H_1 and H_2 can be obtained according to Eq. (5.89)

$$\begin{aligned}
\mathrm{d}H_1 =& \{-[-\beta_1 + \alpha_1 Q_1^2 + \alpha_2 Q_2^4 + \alpha_3(2H_2 - kQ_2^4/2)](2H_1 - \omega^2 Q_1^2)\\
&+\pi K_1 + \pi K_3 Q_1^2\}\mathrm{d}t \pm [2\pi K_1(2H_1 - \omega^2 Q_1^2)]^{1/2}\mathrm{d}B_1(t)\\
&\pm [2\pi K_3 Q_1^2(2H_1 - \omega^2 Q_1^2)]^{1/2}\mathrm{d}B_3(t),\\
\mathrm{d}H_2 =& [-(\beta_2 + \alpha_4 Q_1^2)(2H_2 - kQ_2^4/2) + \pi K_2 + \pi K_4 Q_2^2]\mathrm{d}t \pm [2\pi K_2 (2H_2\\
&-kQ_2^4/2)]^{1/2}\mathrm{d}B_2(t) \pm [2\pi K_4 Q_2^2(2H_2 - kQ_2^4/2)]^{1/2}\mathrm{d}B_4(t).
\end{aligned} \tag{5.110}$$

Suppose that β_i $(i = 1, 2)$, α_j, K_j $(j = 1, 2, 3, 4)$ are small of the same order of ε. In new system, Q_1 and Q_2 in Eq. (5.107) are fast-varying processes while H_1 and H_2 in Eq. (5.110) are slowly-varying processes. According to Khasminskii theorem (Khasminskii 1968), $[H_1, H_2]^T$ weakly converges to a 2-dimensional vector Markov diffusion process. The averaged Itô stochastic differential equations are of the form of Eq. (5.90) with $r = 1, 2$. Two Hamiltonian subsystems have period solutions in plane (q_1, p_1) and (q_2, p_2), respectively. Thus, the following first and second derivative moments for the averaged FPK Eq. (5.97) with $r, i = 1, 2$ can be obtained according to Eqs. (5.94)–(5.96) and (5.98)–(5.99)

$$\begin{aligned}
\overline{a}_1(\boldsymbol{h}) =& \beta_1 h_1 - (\alpha_1/2\omega^2)h_1^2 - (4\alpha_2/3k)h_1 h_2\\
&- (4\alpha_3/3)h_1 h_2 + \pi K_1 + (\pi K_3/\omega^2)h_1,
\end{aligned}$$

$$\begin{aligned}
\overline{a}_2(\boldsymbol{h}) &= -(4\beta_2/3)h_2 - \left(4\alpha_4/3\omega^2\right)h_1h_2 + \pi K_2 \\
&\quad + \left[8\Gamma^2(7/4)\pi K_4/9\Gamma^2(5/4)k^{1/2}\right]h_2^{1/2}, \\
\overline{b}_{11}(\boldsymbol{h}) &= 2\pi K_1h_1 + \left(\pi K_3/\omega^2\right)h_1^2, \\
\overline{b}_{22}(\boldsymbol{h}) &= (8\pi K_2/3)h_2 + \left[64\Gamma^2(7/4)\pi K_4/45\Gamma^2(5/4)k^{1/2}\right]h_2^{3/2}, \\
\overline{b}_{12}(\boldsymbol{h}) &= \overline{b}_{21}(\boldsymbol{h}) = 0,
\end{aligned} \tag{5.111}$$

where $\Gamma(\cdot)$ is Gamma function.

Suppose that the FPK Eq. (5.97) with $r, i = 1, 2$ has the following exact stationary solution (Cai and Zhu 2016)

$$p(h_1, h_2) = C\exp[-\lambda(h_1, h_2)]. \tag{5.112}$$

Substituting Eq. (5.111) and (5.112) into the reduced averaged FPK Eq. (5.97) $\left(\partial p/\partial t = 0;\ r, i = 1, 2\right)$ yields

$$\begin{aligned}
\frac{\partial\lambda}{\partial h_1} &= \left[-\beta_1 + \left(\alpha_1/\omega^2\right)h_1 + (4/3)(\alpha_2/k + \alpha_3)h_2\right]\Big/\pi K_1, \\
\frac{\partial\lambda}{\partial h_2} &= \left[4\beta_2/3 + \left(4\alpha_4/3\omega^2\right)h_1\right]\Big/\pi K_2.
\end{aligned} \tag{5.113}$$

If $(\alpha_2/k + \alpha_3)\omega^2K_2 = \alpha_4\pi K_1$, then the compatibility condition $\partial^2\lambda/\partial h_1\partial h_2 = \partial^2\lambda/\partial h_2\partial h_1$ is satisfied. Solving Eq. (5.113) yields

$$\begin{aligned}
\lambda(h_1, h_2) &= \int_0^{h_1}\frac{\partial\lambda}{\partial h_1}\mathrm{d}h_1 + \int_0^{h_2}\frac{\partial\lambda}{\partial h_2}\mathrm{d}h_2 \\
&= -\frac{\beta_1}{\pi K_1}h_1 + \frac{4}{3}\frac{\beta_2}{\pi K_2}h_2 + \frac{\alpha_1}{2\omega^2\pi K_1}h_1^2 + \frac{8\alpha_4}{3\omega^2\pi K_2}h_1h_2.
\end{aligned} \tag{5.114}$$

Substituting Eq. (5.114) into Eq. (5.112) yields $p(h_1, h_2)$. The following stationary PDF of generalized displacement and momenta is then obtained from Eq. (5.112) using Eq. (5.106),

$$p(q_1, q_2, p_1, p_2) = \frac{C}{T(h_1, h_2)}\exp[-\lambda(h_1, h_2)]\big|_{h_i = H_i(q_i, p_i)}, \tag{5.115}$$

where C is a normalization constant, and

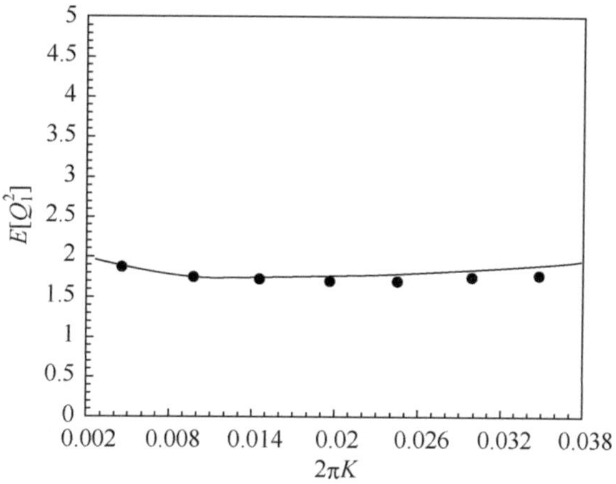

Fig. 5.6 Stationary mean-square displacement of system (5.107). Parameters: $\omega_1 = 1.0, k = 2.0$, $\alpha_1 = \alpha_3 = \alpha_4 = \beta_1 = 0.02$, $\alpha_2 = \beta_2 = 0.01, 2\pi K_i = 2\pi K\ (i = 1, 2, 3, 4)$. ——, result from the stochastic averaging method; ●, result from Monte Carlo simulation (Zhu et al. 1997)

$$
\begin{aligned}
T(h_1, h_2) &= 16\int_0^{A_1}\left(2h_1 - \omega^2 q_1^2\right)^{-1/2}\mathrm{d}q_1\int_0^{A_2}(2h_2 - kq_2^4/2)^{-1/2}\mathrm{d}q_2 \\
&= 4\pi F\left(\frac{\pi}{2}, \frac{1}{\sqrt{2}}\right)\Big/\left[\omega\left(\frac{h_2}{k}\right)^{1/4}\right]
\end{aligned}
\tag{5.116}
$$

where F(·, ·) is the generalized hypergeometric function. The mean-square value of generalized displacement is

$$
\begin{aligned}
E\left[Q_1^2\right] &= \int_{-\infty}^{\infty}\int_{-\infty}^{\infty}\int_{-\infty}^{\infty}\int_{-\infty}^{\infty} q_1^2 p(q_1, q_2, p_1, p_2)\mathrm{d}q_1\mathrm{d}q_2\mathrm{d}p_1\mathrm{d}p_2 \\
&= C\int_0^{\infty}\int_0^{\infty}\exp[-\lambda(h_1, h - h_1)]h_1\mathrm{d}h_1\mathrm{d}h.
\end{aligned}
\tag{5.117}
$$

The numerical results obtained using Eq. (5.117) are in good agreement with those from Monte Carlo simulation, as shown in Fig. 5.6.

5.2.2 *Internal Resonant Case*

Now, assume that the Hamiltonian system associated with Eq. (5.5) is completely integrable with α $(1 \le \alpha \le n - 1)$ weak internal resonance relationships of the form as Eq. (5.75). Introduce combinations of angular variables

$$\Psi_u = \sum_{r=1}^{n} k_r^u \Theta_r, \quad u = 1, 2, \ldots, \alpha. \tag{5.118}$$

The Itô stochastic differential equations for Ψ_u are obtained from the equations of Θ_r in Eq. (5.74) according to Eq. (5.118) as follows

$$\begin{aligned} \mathrm{d}\Psi_u =& \left[O_u(\varepsilon) - \varepsilon \sum_{j,k=1}^{n} \left(m_{jk} \frac{\partial \Psi_u}{\partial P_j} \frac{\partial H}{\partial P_k} - \sum_{l,s=1}^{m} \pi K_{ls} g_{jl} g_{ks} \frac{\partial^2 \Psi_u}{\partial P_j \partial P_k} \right) \right] \mathrm{d}t \\ &+ \varepsilon^{1/2} \sum_{j=1}^{n} \sum_{l=1}^{m} g_{jl} \frac{\partial \Psi_u}{\partial P_j} \mathrm{d}B_l(t), . \\ u =& 1, 2, \ldots, \alpha, \end{aligned} \tag{5.119}$$

where $O_u(\varepsilon)$ are small quantities of ε order. The new system consists of n equations of I_r $(r = 1, 2, \ldots, n)$ in Eq. (5.73), $n - \alpha$ equations of Θ_r $(r = 1, 2, \ldots, n - \alpha)$ in Eq. (5.74) and α equations of Ψ_u $(u = 1, 2, \ldots, \alpha)$ in Eq. (5.119). $I_r(t)$ and $\Psi_u(t)$ are slowly-varying processes while $\Theta_1(t), \ldots, \Theta_{n-\alpha}(t)$ are fast-varying processes. According to the Khasminskii theorem (Khasminskii 1968), when $\varepsilon \to 0$, $[\boldsymbol{I}^T, \boldsymbol{\Psi}^T]^T$ weakly converges to a $(n + \alpha)$–dimensional vector Markov diffusion process. The averaged Itô stochastic differential equations are of the form

$$\begin{aligned} \mathrm{d}I_r &= \overline{\overline{m}}_r(\boldsymbol{I}, \boldsymbol{\Psi})\mathrm{d}t + \sum_{l=1}^{m} \overline{\overline{\sigma}}_{rl}(\boldsymbol{I}, \boldsymbol{\Psi})\mathrm{d}B_l(t), \\ \mathrm{d}\Psi_u &= \overline{\overline{m}}_u(\boldsymbol{I}, \boldsymbol{\Psi})\mathrm{d}t + \sum_{l=1}^{m} \overline{\overline{\sigma}}_{ul}(\boldsymbol{I}, \boldsymbol{\Psi})\mathrm{d}B_l(t), \\ r &= 1, 2, \ldots, n; \quad u = 1, 2, \ldots, \alpha, \end{aligned} \tag{5.120}$$

where $B_l(t)$ $(l = 1, 2, \ldots, m)$ are independent unit Wiener processes. According to Eqs. (4.29) and 4.30), the averaged drift and diffusion coefficients in Eq. (5.120) can be obtained by carrying out time averaging to the drift and diffusion coefficients in Eqs. (5.73) and (5.119). Since the Hamiltonian system associated with this quasi-integrable and resonant Hamiltonian system is ergodic on $(n-\alpha)$-dimensional torus, the time averaging can be replaced by spatial averaging with respective to $\theta_1, \theta_2, \ldots, \theta_{n-\alpha}$. Thus,

$$\begin{aligned} \overline{\overline{m}}_r(\boldsymbol{I}, \boldsymbol{\Psi}) = \frac{\varepsilon}{(2\pi)^{n-\alpha}} \int_0^{2\pi} \cdots \int_0^{2\pi} \sum_{j,k=1}^{n} \Bigg(-m_{jk} \frac{\partial I_r}{\partial p_j} \frac{\partial H}{\partial p_k} \\ + \sum_{l,s=1}^{m} \pi K_{ls} g_{jl} g_{ks} \frac{\partial^2 I_r}{\partial p_j \partial p_k} \Bigg) \mathrm{d}\theta_1 \ldots \mathrm{d}\theta_{n-\alpha}, \end{aligned}$$

$$
\begin{aligned}
\overline{\overline{m}}_u(\boldsymbol{I},\boldsymbol{\Psi}) &= O_u(\varepsilon) + \frac{\varepsilon}{(2\pi)^{n-\alpha}}\int_0^{2\pi}\cdots\int_0^{2\pi}\sum_{j,k=1}^{n}\left(-m_{jk}\frac{\partial \Psi_u}{\partial p_j}\frac{\partial H}{\partial p_k}\right.\\
&\quad\left.+\sum_{l,s=1}^{m}\pi K_{ls}g_{jl}g_{ks}\frac{\partial^2\Psi_u}{\partial p_j\partial p_k}\right)\mathrm{d}\theta_1\ldots\mathrm{d}\theta_{n-\alpha},\\
\sum_{l=1}^{m}\overline{\overline{\sigma}}_{rl}\overline{\overline{\sigma}}_{il}(\boldsymbol{I},\boldsymbol{\Psi}) &= \frac{\varepsilon}{(2\pi)^{n-\alpha}}\\
&\int_0^{2\pi}\cdots\int_0^{2\pi}\sum_{j,k=1}^{n}\sum_{l,s=1}^{m}2\pi K_{ls}g_{jl}g_{ks}\frac{\partial I_r}{\partial p_j}\frac{\partial I_i}{\partial p_k}\mathrm{d}\theta_1\ldots\mathrm{d}\theta_{n-\alpha},\\
\sum_{l=1}^{m}\overline{\overline{\sigma}}_{rl}\overline{\overline{\sigma}}_{ul}(\boldsymbol{I},\boldsymbol{\Psi}) &= \frac{\varepsilon}{(2\pi)^{n-\alpha}}\\
&\int_0^{2\pi}\cdots\int_0^{2\pi}\sum_{j,k=1}^{n}\sum_{l,s=1}^{m}2\pi K_{ls}g_{jl}g_{ks}\frac{\partial I_r}{\partial p_j}\frac{\partial \Psi_u}{\partial p_k}\mathrm{d}\theta_1\ldots\mathrm{d}\theta_{n-\alpha},\\
\sum_{l=1}^{m}\overline{\overline{\sigma}}_{ul}\overline{\overline{\sigma}}_{vl}(\boldsymbol{I},\boldsymbol{\Psi}) &= \frac{\varepsilon}{(2\pi)^{n-\alpha}}\\
&\int_0^{2\pi}\cdots\int_0^{2\pi}\sum_{j,k=1}^{n}\sum_{l,s=1}^{m}2\pi K_{ls}g_{jl}g_{ks}\frac{\partial \Psi_u}{\partial p_j}\frac{\partial \Psi_v}{\partial p_k}\mathrm{d}\theta_1\ldots\mathrm{d}\theta_{n-\alpha},\\
& r,i = 1,2,\ldots,n;\quad u,v=1,2,\ldots,\alpha.
\end{aligned}
\tag{5.121}
$$

The averaged FPK equation associated with averaged Itô Eq. (5.120) is

$$
\begin{aligned}
\frac{\partial p}{\partial t} &= -\sum_{r=1}^{n}\frac{\partial}{\partial I_r}(\overline{\overline{a}}_r p) - \sum_{u=1}^{\alpha}\frac{\partial}{\partial \psi_u}(\overline{\overline{a}}_u p) + \frac{1}{2}\sum_{r,i=1}^{n}\frac{\partial^2}{\partial I_r\partial I_i}(\overline{\overline{b}}_{ri}p)\\
&\quad+\sum_{r=1}^{n}\sum_{u=1}^{\alpha}\frac{\partial^2}{\partial I_r\partial\psi_u}(\overline{\overline{b}}_{ru}p) + \frac{1}{2}\sum_{u,v=1}^{\alpha}\frac{\partial^2}{\partial\psi_u\partial\psi_v}(\overline{\overline{b}}_{uv}p),
\end{aligned}
\tag{5.122}
$$

where

$$
\begin{aligned}
\overline{\overline{a}}_r &= \overline{\overline{a}}_r(\boldsymbol{I},\boldsymbol{\psi}) = \overline{\overline{m}}_r(\boldsymbol{I},\boldsymbol{\Psi})|_{\boldsymbol{\Psi}=\boldsymbol{\psi}},\\
\overline{\overline{a}}_u &= \overline{\overline{a}}_u(\boldsymbol{I},\boldsymbol{\psi}) = \overline{\overline{m}}_u(\boldsymbol{I},\boldsymbol{\Psi})|_{\boldsymbol{\Psi}=\boldsymbol{\psi}},\\
\overline{\overline{b}}_{ri} &= \overline{\overline{b}}_{ri}(\boldsymbol{I},\boldsymbol{\psi}) = \sum_{l=1}^{m}\overline{\overline{\sigma}}_{rl}\overline{\overline{\sigma}}_{il}(\boldsymbol{I},\boldsymbol{\Psi})|_{\boldsymbol{\Psi}=\boldsymbol{\psi}},
\end{aligned}
$$

$$\bar{\bar{b}}_{ru} = \bar{\bar{b}}_{ru}(\boldsymbol{I}, \boldsymbol{\psi}) = \sum_{l=1}^{m} \bar{\bar{\sigma}}_{rl}\bar{\bar{\sigma}}_{ul}(\boldsymbol{I}, \boldsymbol{\Psi})|_{\boldsymbol{\Psi}=\boldsymbol{\psi}},$$
$$\bar{\bar{b}}_{uv} = \bar{\bar{b}}_{uv}(\boldsymbol{I}, \boldsymbol{\psi}) = \sum_{l=1}^{m} \bar{\bar{\sigma}}_{ul}\bar{\bar{\sigma}}_{vl}(\boldsymbol{I}, \boldsymbol{\Psi})|_{\boldsymbol{\Psi}=\boldsymbol{\psi}}. \tag{5.123}$$

In Eq. (5.122), $p = p(\boldsymbol{I}, \boldsymbol{\psi}, t|\boldsymbol{I}_0, \boldsymbol{\psi}_0)$ is the transition PDF of $[\boldsymbol{I}^{\mathrm{T}}, \boldsymbol{\Psi}^{\mathrm{T}}]^{\mathrm{T}}$ with initial conditions

$$p(\boldsymbol{I}, \boldsymbol{\psi}, 0|\boldsymbol{I}_0, \boldsymbol{\psi}_0) = \delta(\boldsymbol{I} - \boldsymbol{I}_0)\delta(\boldsymbol{\psi} - \boldsymbol{\psi}_0), \tag{5.124}$$

or $p = p(\boldsymbol{I}, \boldsymbol{\psi}, t)$ is the joint PDF of $[\boldsymbol{I}^{\mathrm{T}}, \boldsymbol{\Psi}^{\mathrm{T}}]^{\mathrm{T}}$ with initial conditions

$$p(\boldsymbol{I}, \boldsymbol{\psi}, 0) = p(\boldsymbol{I}_0, \boldsymbol{\psi}_0). \tag{5.125}$$

The boundary conditions depend on the system properties and constraints imposed on the system. Generally, the conditions for I_r are similar to Eqs. (5.86) and (5.87). The conditions for ψ_u are periodic, i.e.,

$$p|_{\psi_u+2k\pi} = p|_{\psi_u}, \quad k \text{ is integer}, \quad u = 1, 2, \ldots, \alpha. \tag{5.126}$$

Similar to the non-resonant case, the diffusion matrix of the averaged Itô equation (5.120) is generally non-degenerate. The probability flow of the averaged FPK Eq. (5.122) contains only potential flow and no circular flow. This allows us to find an exact stationary solution belonging to the stationary potential (Cai and Zhu 2016). Once the stationary solution $p(\boldsymbol{I}, \boldsymbol{\psi})$ is obtained, the approximate stationary PDF of the generalized displacements and generalized momenta of system (5.5) can be determined using the following transformation

$$\begin{aligned} p(\boldsymbol{q}, \boldsymbol{p}) &= p(\boldsymbol{I}, \boldsymbol{\psi}, \boldsymbol{\theta}_1)\left|\frac{\partial(\boldsymbol{I}, \boldsymbol{\psi}, \boldsymbol{\theta}_1)}{\partial(\boldsymbol{q}, \boldsymbol{p})}\right| = p(\boldsymbol{\theta}_1|\boldsymbol{I}, \boldsymbol{\psi})p(\boldsymbol{I}, \boldsymbol{\psi})\left|\frac{\partial(\boldsymbol{I}, \boldsymbol{\psi}, \boldsymbol{\theta}_1)}{\partial(\boldsymbol{q}, \boldsymbol{p})}\right| \\ &= \frac{1}{(2\pi)^{n-\alpha}} p(\boldsymbol{I}, \boldsymbol{\psi})\left|\frac{\partial(\boldsymbol{I}, \boldsymbol{\psi}, \boldsymbol{\theta}_1)}{\partial(\boldsymbol{q}, \boldsymbol{p})}\right|. \end{aligned} \tag{5.127}$$

where $|\partial(\boldsymbol{I}, \boldsymbol{\psi}, \boldsymbol{\theta}_1)/\partial(\boldsymbol{q}, \boldsymbol{p})|$ is the Jacobian determinant of the transformation from $\boldsymbol{q}, \boldsymbol{p}$ to $\boldsymbol{I}, \boldsymbol{\psi}, \boldsymbol{\theta}_1$. Due to Eq. (5.118), this determinant is the combination of Jacobian determinant $|\partial(\boldsymbol{I}, \boldsymbol{\theta})/\partial(\boldsymbol{q}, \boldsymbol{p})|$ with integers as coefficients. Since $|\partial(\boldsymbol{I}, \boldsymbol{\theta})/\partial(\boldsymbol{q}, \boldsymbol{p})| = 1$, the transformation from $\boldsymbol{q}, \boldsymbol{p}$ to $\boldsymbol{I}, \boldsymbol{\theta}$ is a canonical transformation.

Finally, it is noted that since $\boldsymbol{I} = \boldsymbol{I}(\boldsymbol{H})$, Eqs. (5.71)–(5.87) and (5.120)–(5.126) still hold after $\boldsymbol{I}$ is replaced by $\boldsymbol{H}$. However, $\boldsymbol{I}, \boldsymbol{\theta}$ are canonical coordinates while $\boldsymbol{H}$, $\boldsymbol{\theta}$ are generally not. The transformation from $\boldsymbol{q}$, $\boldsymbol{p}$ to $\boldsymbol{H}$, $\boldsymbol{\theta}$ is not a canonical transformation, thus $|\partial(\boldsymbol{H}, \boldsymbol{\theta})/\partial(\boldsymbol{q}, \boldsymbol{p})|$ is generally not equal to 1.

Example 5.4 Consider two oscillators coupled by nonlinear dampings and excited by Gaussian white noises. The motion equations are

$$\begin{aligned}
\dot{Q}_1 &= P_1,\\
\dot{P}_1 &= -\omega_1^2 Q_1 - \alpha_{11}P_1 - \alpha_{12}P_2 - \beta_1(Q_1^2 + Q_2^2)P_1 + W_{g1}(t),\\
\dot{Q}_2 &= P_2,\\
\dot{P}_2 &= -\omega_2^2 Q_2 - \alpha_{21}P_1 - \alpha_{22}P_2 - \beta_2(Q_1^2 + Q_2^2)P_2 + W_{g2}(t),
\end{aligned} \tag{5.128}$$

where ω_i, α_{ij}, β_i are constants, $W_{gl}(t)$ are independent Gaussian white noises with intensities $2\pi K_l$. Assuming α_{ij}, β_i, K_l are small quantities of order ε. The Hamiltonian function associated with system (5.128) is

$$H(q,p) = \sum_{i=1}^{2} H_i(q_i,p_i) = \sum_{i=1}^{2} \left(p_i^2 + \omega_i^2 q_i^2\right)/2 = \sum_{i=1}^{2} \omega_i I_i. \tag{5.129}$$

Obviously, system (5.128) is a quasi-integrable Hamiltonian system. Introduce the transformation

$$I_i = \left(P_i^2 + \omega_i^2 Q_i^2\right)/(2\omega_i), \quad \Theta_i = -\arctan\left(P_i/\omega_i Q_i\right), \quad i = 1, 2. \tag{5.130}$$

By applying Itô differential rule and Eq. (5.130), the Itô stochastic differential equations for I_i, Θ_i can be derived from Eq. (5.128) as follows

$$\begin{aligned}
\mathrm{d}I_i &= \left\{-\left[\alpha_{i1}P_1 + \alpha_{i2}P_2 + \beta_i(Q_1^2 + Q_2^2)P_i\right]P_i/\omega_i + \pi K_i/\omega_i\right\}\mathrm{d}t\\
&\quad + \left[(2\pi K_i)^{1/2}P_i/\omega_i\right]\mathrm{d}B_i(t),\\
\mathrm{d}\Theta_i &= \left\{\omega_i + \left[\alpha_{i1}P_1 + \alpha_{i2}P_2 + \beta_i(Q_1^2 + Q_2^2)P_i\right]\omega_i P_i/\left(\omega_i^2 Q_i^2 + P_i^2\right)\right.\\
&\quad \left. + 2\omega_i\pi K_i Q_i P_i/\left(\omega_i^2 Q_i^2 + P_i^2\right)^2\right\}\mathrm{d}t - \left[(2\pi K_i)^{1/2}\omega_i Q_i/\left(\omega_i^2 Q_i^2 + P_i^2\right)\right]\mathrm{d}B_i(t),\\
i &= 1, 2.
\end{aligned} \tag{5.131}$$

The dimension and form of the averaged equations depend on the resonant properties of the Hamiltonian system associated with system (5.128).

(1) **Noninternal resonant case** In this case, the averaged Itô equations are of the form of Eq. (5.76), and the averaged FPK equation is of the form of Eq. (5.81). The first and second derivative moments are obtained using Eqs. (5.79), (5.80), (5.82) and (5.83) as follows

$$\begin{aligned}
a_1 &= -\alpha_{11}I_1 - \beta_1 I_1^2/2\omega_1 - \beta_1 I_1 I_2/\omega_2 + \pi K_1/\omega_1,\\
a_2 &= -\alpha_{22}I_2 - \beta_2 I_2^2/2\omega_2 - \beta_2 I_1 I_2/\omega_1 + \pi K_2/\omega_2,\\
b_{11} &= 2\pi K_1 I_1/\omega_1, \quad b_{22} = 2\pi K_2 I_2/\omega_2, \quad b_{12} = b_{21} = 0.
\end{aligned} \tag{5.132}$$

Assume that the exact stationary solution takes the form

$$p(I_1, I_2) = C\exp[-\lambda(I_1, I_2)]. \tag{5.133}$$

It is shown by substituting Eq. (5.133) into Eq. (5.81) with $\partial p/\partial t = 0$ that probability potential $\lambda(I_1, I_2)$ is a solution of the following two first-order partial differential equations

$$\begin{aligned}
\frac{2\pi K_1 I_1}{\omega_1}\frac{\partial\lambda}{\partial I_1} &= \frac{2\pi K_1}{\omega_1} - 2\left(-\alpha_{11}I_1 - \frac{\beta_1}{2\omega_1}I_1^2 - \frac{\beta_1}{\omega_2}I_1 I_2 + \frac{\pi K_1}{\omega_1}\right)\\
\frac{2\pi K_2 I_2}{\omega_2}\frac{\partial\lambda}{\partial I_2} &= \frac{2\pi K_2}{\omega_2} - 2\left(-\alpha_{22}I_2 - \frac{\beta_2}{2\omega_2}I_2^2 - \frac{\beta_2}{\omega_2}I_1 I_2 + \frac{\pi K_2}{\omega_2}\right)
\end{aligned} \tag{5.134}$$

Under the compatibility condition $(\beta_1/\pi K_1)(\omega_1/\omega_2) = (\beta_2/\pi K_2)(\omega_2/\omega_1) = \gamma$, Eq. (5.134) can be solved to yield

$$\begin{aligned}
\lambda(I_1, I_2) &= \left(\alpha_{11}\omega_1 I_1 + \beta_1 I_1^2/4\right)/\pi K_1\\
&\quad + \left(\alpha_{22}\omega_2 I_2 + \beta_2 I_2^2/4\right)/\pi K_2 + \gamma I_1 I_2
\end{aligned} \tag{5.135}$$

$p(I_1, I_2)$ can be obtained by substituting Eq. (5.135) into (5.133). Then, according to Eq. (5.104), the approximately stationary joint PDF of generalized displacements and generalized momenta for system (5.128) is obtained as follows

$$p(q_1, q_2, p_1, p_2) = \frac{1}{4\pi^2}\, p(I_1, I_2)\big|_{I_i=(p_i^2+\omega_i^2 q_i^2)/2\omega_i} \tag{5.136}$$

(2) **Primary internal resonant case** Suppose that $\omega_1 = \omega_2 = \omega$. Let $\Psi = \Theta_1 - \Theta_2$. The averaged FPK equation is of the form of Eq. (5.122). The averaged first and second derivative moments are obtained according to Eqs. (5.121) and (5.123) as follows

$$\begin{aligned}
\bar{\bar{a}}_1 &= -\alpha_{11}I_1 - \alpha_{12}(I_1 I_2)^{1/2}\cos\psi - (\beta_1/2\omega)I_1^2\\
&\quad - (\beta_1/\omega)I_1 I_2[1 - (1/2)\cos 2\psi] + \pi K_1/\omega,\\
\bar{\bar{a}}_2 &= -\alpha_{22}I_2 - \alpha_{21}(I_1 I_2)^{1/2}\cos\psi - (\beta_2/2\omega)I_2^2\\
&\quad - (\beta_2/\omega)I_1 I_2[1 - (1/2)\cos 2\psi] + \pi K_2/\omega,\\
\bar{\bar{a}}_3 &= \left\{\left[\alpha_{12}(I_2/I_1)^{1/2} + \alpha_{21}(I_1/I_2)^{1/2}\right]/2\right\}\sin\psi\\
&\quad - [(\beta_1 I_2 + \beta_2 I_1)/4\omega]\sin 2\psi,\\
\bar{\bar{b}}_{11} &= 2\pi K_1 I_1/\omega,\quad \bar{\bar{b}}_{22} = 2\pi K_2 I_2/\omega,\quad \bar{\bar{b}}_{33} = (\pi K_1/I_1 + \pi K_2/I_2)/2\omega,\\
\bar{\bar{b}}_{12} &= \bar{\bar{b}}_{21} = \bar{\bar{b}}_{13} = \bar{\bar{b}}_{31} = \bar{\bar{b}}_{23} = \bar{\bar{b}}_{32} = 0.
\end{aligned} \tag{5.137}$$

Assume that the exact stationary solution for the averaged FPK Eq. (5.122) is of the form

$$p(I_1, I_2, \psi) = C\exp[-\lambda(I_1, I_2, \psi)]. \tag{5.138}$$

Substituting Eqs. (5.137) and (5.138) into Eq. (5.122) with $\partial p/\partial t = 0$, Eq. (5.138) is indeed an exact solution of the reduced averaged FPK Eq. (5.122) for system (5.128) if the probability potential $\lambda(I_1, I_2, \psi)$ satisfies the following three first-order partial differential equations

$$\begin{aligned}
\frac{2\pi K_1 I_1}{\omega}\frac{\partial\lambda}{\partial I_1} &= \frac{2\pi K_1}{\omega} - 2\Big[-\alpha_{11}I_1 - \alpha_{12}(I_1I_2)^{1/2}\cos\psi \\
&\quad -\frac{\beta_1}{2\omega}I_1^2 - \frac{\beta_1}{\omega}I_1I_2\left(1-\frac{1}{2}\cos 2\psi\right) + \frac{\pi K_1}{\omega}\Big], \\
\frac{2\pi K_2 I_2}{\omega}\frac{\partial\lambda}{\partial I_2} &= \frac{2\pi K_2}{\omega} - 2\Big[-\alpha_{22}I_2 - \alpha_{21}(I_1I_2)^{1/2}\cos\psi \\
&\quad -\frac{\beta_2}{2\omega}I_2^2 - \frac{\beta_2}{\omega}I_1I_2\left(1-\frac{1}{2}\cos 2\psi\right) + \frac{\pi K_2}{\omega}\Big], \\
\left(\frac{\pi K_1}{2\omega I_1} + \frac{\pi K_2}{2\omega I_2}\right)\frac{\partial\lambda}{\partial\psi} &= -\left[\alpha_{12}\left(\frac{I_2}{I_1}\right)^{1/2} + \alpha_{21}\left(\frac{I_1}{I_2}\right)^{1/2}\right]\sin\psi \\
&\quad + \frac{1}{2\omega}(\beta_1 I_2 + \beta_2 I_1)\sin 2\psi.
\end{aligned} \tag{5.139}$$

Letting

$$\lambda(I_1, I_2, \psi) = \lambda_0(I_1, I_2) + \lambda_1(I_1, I_2)\cos\psi + \lambda_2(I_1, I_2)\cos 2\psi \tag{5.140}$$

and substituting Eq. (5.140) into Eq. (5.139) result in equations for λ_0, λ_1, and λ_2. Under compatible conditions $\beta_1/\pi K_1 = \beta_2/\pi K_2 = \gamma_1$, $\alpha_{12}/\pi K_1 = \alpha_{21}/\pi K_2 = \gamma_2$, the exact stationary solution of probability potential is

$$\begin{aligned}
\lambda(I_1, I_2, \psi) &= (\alpha_{11}\omega/\pi K_1)I_1 + (\alpha_{22}\omega/\pi K_2)I_2 \\
&\quad + (\beta_1/4\pi K_1)I_1^2 + (\beta_2/4\pi K_2)I_2^2 \\
&\quad + \gamma_1 I_1 I_2 - (\gamma_1 I_1 I_2/2)\cos 2\psi + 2\gamma_2\omega(I_1I_2)^{1/2}\cos\psi.
\end{aligned} \tag{5.141}$$

The approximately stationary joint PDF of generalized displacement and generalized momenta for system (5.128) is obtained from Eqs. (5.138) and (5.141) using Eq. (5.127) as follows

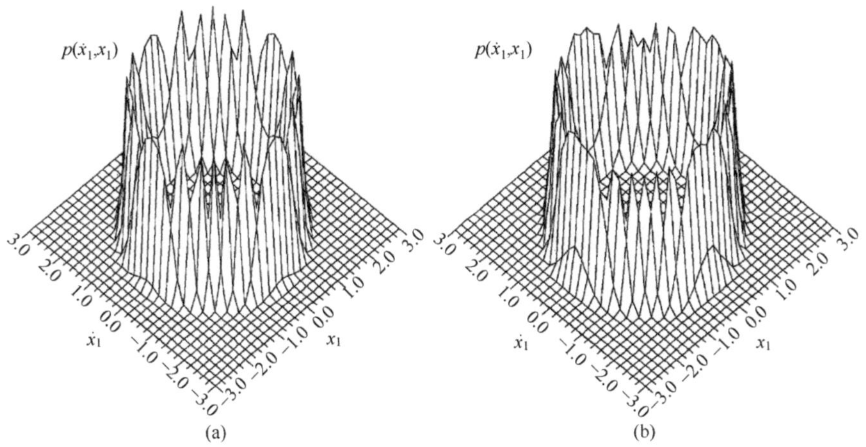

Fig. 5.7 Stationary joint PDF $p(x_1, \dot{x}_1)$ of system (5.128) in the non-resonant case. **a** Result from stochastic averaging method; **b** result from Monte Carlo simulation. System parameters are $\alpha_{11} = -0.05$, $\alpha_{22} = \alpha_{12} = \alpha_{21} = \beta_1 = \beta_2 = 0.05$, $\omega_1 = 1.0$, $\omega_2 = 1.414$, $\pi K_{11} = \pi K_{22} = 0.001$ (Zhu et al. 1997)

$$p(q_1, q_2, p_1, p_2) = \frac{1}{2\pi} p(I_1, I_2, \psi) \Big|_{\substack{I_i = (p_i^2 + \omega_i^2 q_i^2)/2\omega_i \\ \psi = \tan^{-1}(p_2/\omega_2 q_2) - \tan^{-1}(p_1/\omega_1 q_1)}} \quad (5.142)$$

Note that $\left|\partial(I_1, I_2, \psi, \theta_1)/\partial(q_1, q_2, p_1, p_2)\right| = 1$. It is known from comparing Eqs. (5.136) and (5.142) that the two stationary PDFs for non-resonant case and primary resonant case are different.

It is seen from Figs. 5.7–5.9 that the numerical results obtained from Eq. (5.142) are in good agreement with those from Monte Carlo simulation.

5.3 Quasi-Partially Integrable Hamiltonian Systems

Now assume that the Hamiltonian system associated with system (5.5) is partially integrable with $r\,(1 < r < n)$ independent and involutive first integrals, and that the Hamiltonian function consists of an integrable and a non-integrable Hamiltonian sub-functions (Zhu et al. 2002)

$$H = H(\boldsymbol{q}, \boldsymbol{p}) = \sum_{\eta=1}^{r-1} H_\eta(\boldsymbol{q}_1, \boldsymbol{p}_1) + H_r(\boldsymbol{q}_2, \boldsymbol{p}_2), \quad (5.143)$$

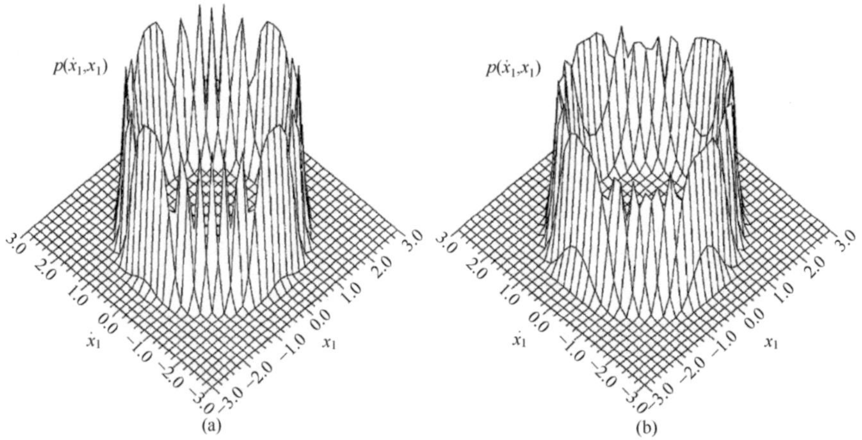

Fig. 5.8 Stationary joint PDF $p(x_1, \dot{x}_1)$ of system (5.128) in the primary resonant case. **a** Result from stochastic averaging method; **b** result from Monte Carlo simulation. System parameters are $\alpha_{11} = \alpha_{22} = -0.05$, $\alpha_{12} = \alpha_{21} = \beta_1 = \beta_2 = 0.05$, $\omega_1 = \omega_2 = 1.0$, $\pi K_{11} = \pi K_{22} = 0.001$ (Zhu et al. 1997)

Fig. 5.9 Stationary PDF $p(\psi)$ of system (5.128) in the primary resonant case. ——, result from the stochastic averaging method; ●, result from Monte Carlo simulation. $\pi K_{11} = \pi K_{22} = 0.01$, other parameters are the same as those in Fig. 5.8 (Zhu et al. 1997)

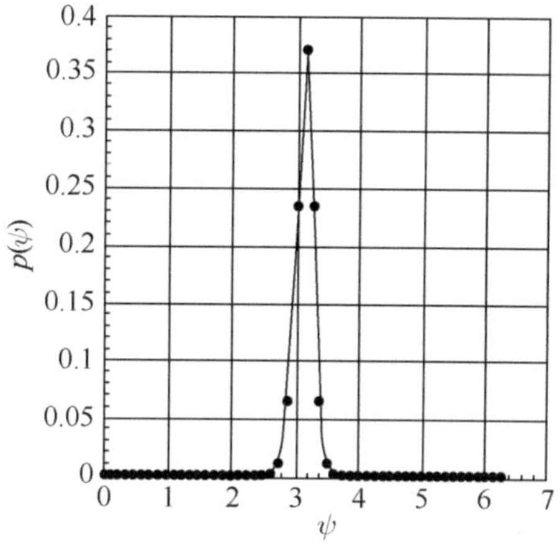

where $\boldsymbol{q}_1 = [q_1, q_2, \ldots, q_{r-1}]^{\mathrm{T}}$, $\boldsymbol{p}_1 = [p_1, p_2, \ldots, p_{r-1}]^{\mathrm{T}}$, $\boldsymbol{q}_2 = [q_r, q_{r+1}, \ldots, q_n]^{\mathrm{T}}$, and $\boldsymbol{p}_2 = [p_r, p_{r+1}, \ldots, p_n]^{\mathrm{T}}$. Introducing action-angle variables I_η and θ_η for the integrable subsystem, Eq. (5.143) can be rewritten as

$$H = H(\boldsymbol{I}', \boldsymbol{q}_2, \boldsymbol{p}_2) = \sum_{\eta=1}^{r-1} H_\eta(\boldsymbol{I}') + H_r(\boldsymbol{q}_2, \boldsymbol{p}_2), \tag{5.144}$$

where $\boldsymbol{I}' = [I_1, I_2, \ldots, I_{r-1}]^{\mathrm{T}}$. Introducing transformation

$$I_\eta = I_\eta(\boldsymbol{Q}_1, \boldsymbol{P}_1),\ \Theta_\eta = \Theta_\eta(\boldsymbol{Q}_1, \boldsymbol{P}_1),\ H_r = H_r(\boldsymbol{Q}_2, \boldsymbol{P}_2), \tag{5.145}$$

and applying Itô differential rule to Eq. (5.5), the following Itô stochastic differential equations governing I_η, Θ_η, H_r are obtained

$$\begin{aligned}
\mathrm{d}I_\eta &= \varepsilon\left(-\sum_{j=1}^{n}\sum_{\eta'=1}^{r-1} m_{\eta' j}\frac{\partial I_\eta}{\partial P_{\eta'}}\frac{\partial H}{\partial P_j} + \sum_{\eta',\eta''=1}^{r-1}\sum_{l,s=1}^{m}\pi K_{ls} g_{\eta' l} g_{\eta'' s}\frac{\partial^2 I_\eta}{\partial P_{\eta'}\partial P_{\eta''}}\right)\mathrm{d}t \\
&\quad + \varepsilon^{1/2}\sum_{\eta'=1}^{r-1}\sum_{l=1}^{m}\frac{\partial I_\eta}{\partial P_{\eta'}} g_{\eta' l}\mathrm{d}B_l'(t), \\
\mathrm{d}\Theta_\eta &= \left[\omega_\eta - \varepsilon\left(\sum_{j=1}^{n}\sum_{\eta'=1}^{r-1} m_{\eta' j}\frac{\partial \Theta_\eta}{\partial P_{\eta'}}\frac{\partial H}{\partial P_j} - \sum_{\eta',\eta''=1}^{r-1}\sum_{l,s=1}^{m}\pi K_{ls} g_{\eta' l} g_{\eta'' s}\frac{\partial^2 \Theta_\eta}{\partial P_{\eta'}\partial P_{\eta''}}\right)\right]\mathrm{d}t \\
&\quad + \varepsilon^{1/2}\sum_{\eta'=1}^{r-1}\sum_{l=1}^{m}\frac{\partial \Theta_\eta}{\partial P_{\eta'}} g_{\eta' l}\mathrm{d}B_l'(t), \\
\mathrm{d}H_r &= \varepsilon\left(-\sum_{j=1}^{n}\sum_{\rho=r}^{n} m_{\rho j}\frac{\partial H_r}{\partial P_\rho}\frac{\partial H}{\partial P_j} + \sum_{\rho,\rho'=r}^{n}\sum_{l,s=1}^{m}\pi K_{ls} g_{\rho l} g_{\rho' s}\frac{\partial^2 H_r}{\partial P_\rho\partial P_{\rho'}}\right)\mathrm{d}t \\
&\quad + \varepsilon^{1/2}\sum_{\rho=r}^{n}\sum_{l=1}^{m}\frac{\partial H_r}{\partial P_\rho} g_{\rho l}\mathrm{d}B_l'(t), \\
\eta &= 1, 2, \ldots, r-1.
\end{aligned} \tag{5.146}$$

The Itô stochastic differential equations of new system consist of Eq. (5.146), $n - r + 1$ equations of Q_i and $n - r$ equations of P_i in Eq. (5.5). The dimension and form of the averaged equations depend on whether there exists internal resonance in the quasi-integrable Hamiltonian subsystem.

5.3.1 Noninternal Resonance Case

In this case, I_η and H_r in Eq. (5.146) are slowly-varying processes while Θ_η in Eq. (5.146) and Q_i, P_i in Eq. (5.5) are fast-varying processes. According to the Khasminskii theorem (Khasminskii 1968), when $\varepsilon \to 0$, $[\boldsymbol{I}'^{\mathrm{T}}, H_r]^{\mathrm{T}}$ weakly converges to a r-dimensional vector Markov diffusion process. The averaged Itô stochastic differential equations are of the form

$$
\begin{aligned}
&\mathrm{d}I_\eta = m_\eta(\boldsymbol{I}', H_r)\mathrm{d}t + \sum_{l=1}^{m}\sigma_{\eta l}(\boldsymbol{I}', H_r)\mathrm{d}B_l(t),\\
&\mathrm{d}H_r = m_r(\boldsymbol{I}', H_r)\mathrm{d}t + \sum_{l=1}^{m}\sigma_{rl}(\boldsymbol{I}', H_r)\mathrm{d}B_l(t),\\
&\eta = 1, 2, \ldots, r-1,
\end{aligned} \tag{5.147}
$$

where $B_l(t)$ $(l = 1, 2, \ldots, m)$ are independent unit Wiener processes. The averaged drift and diffusion coefficients can be obtained from Eq. (5.146) by using Eqs. (4.29) and (4.30) as follows

$$
\begin{aligned}
m_\eta(I', H_r) &= \varepsilon\left\langle -\sum_{j=1}^{n}\sum_{\eta'=1}^{r-1} m_{\eta' j}\frac{\partial I_\eta}{\partial P_{\eta'}}\frac{\partial H}{\partial P_j} + \sum_{\eta',\eta''=1}^{r-1}\sum_{l,s=1}^{m}\pi K_{ls} g_{\eta' l} g_{\eta'' s}\frac{\partial^2 I_\eta}{\partial P_{\eta'}\partial P_{\eta''}}\right\rangle_t,\\
m_r(I', H_r) &= \varepsilon\left\langle -\sum_{j=1}^{n}\sum_{\rho=r}^{n} m_{\rho j}\frac{\partial H_r}{\partial P_\rho}\frac{\partial H}{\partial P_j} + \sum_{\rho,\rho'=r}^{n}\sum_{l,s=1}^{m}\pi K_{ls} g_{\rho l} g_{\rho' s}\frac{\partial^2 H_r}{\partial P_\rho\partial P_{\rho'}}\right\rangle_t,\\
\sum_{l=1}^{m}\sigma_{\eta l}\sigma_{\overline{\eta} l}(I', H_r) &= \varepsilon\left\langle \sum_{\eta',\eta''=1}^{r-1}\sum_{l,s=1}^{m} 2\pi K_{ls} g_{\eta' l} g_{\eta'' s}\frac{\partial I_\eta}{\partial P_{\eta'}}\frac{\partial I_{\overline{\eta}}}{\partial P_{\eta''}}\right\rangle_t,\\
\sum_{l=1}^{m}\sigma_{\eta l}\sigma_{rl}(I', H_r) &= \varepsilon\left\langle \sum_{\eta'=1}^{r-1}\sum_{\rho=r}^{n}\sum_{l,s=1}^{m} 2\pi K_{ls} g_{\eta' l} g_{\rho s}\frac{\partial I_\eta}{\partial P_{\eta'}}\frac{\partial H_r}{\partial P_\rho}\right\rangle_t,\\
\sum_{l=1}^{m}\sigma_{rl}\sigma_{rl}(I', H_r) &= \varepsilon\left\langle \sum_{\rho,\rho'=r}^{n}\sum_{l,s=1}^{m} 2\pi K_{ls} g_{\rho l} g_{\rho' s}\frac{\partial H_r}{\partial P_\rho}\frac{\partial H_r}{\partial P_{\rho'}}\right\rangle_t.
\end{aligned} \tag{5.148}
$$

Note that the integrable Hamiltonian subsystem is ergodic on the $(r-1)$-dimensional torus while the non-integrable Hamiltonian subsystem is ergodic on the $(2(n-r)+1)$-dimensional isoenergetic surface. The time averaging in Eq. (5.148) can be replaced by the following spatial averaging

$$
\begin{aligned}
&\langle[\cdot]\rangle_t = \frac{1}{(2\pi)^{r-1}T(H_r)}\int_{\Omega_1}\int_0^{2\pi}\cdots\int_0^{2\pi}\left([\cdot]\Big/\frac{\partial H_r}{\partial p_r}\right)\mathrm{d}\theta_1\ldots\mathrm{d}\theta_{r-1}\mathrm{d}q_r\ldots\mathrm{d}q_n\mathrm{d}p_{r+1}\ldots\mathrm{d}p_n,\\
&T(H_r) = \int_{\Omega_1}\left(1\Big/\frac{\partial H_r}{\partial p_r}\right)\mathrm{d}q_r\ldots\mathrm{d}q_n\mathrm{d}p_{r+1}\ldots\mathrm{d}p_n,\\
&\Omega_1 = \{(q_r,\ldots,q_n,p_{r+1},\ldots,p_n)\big|H_r(q_r,\ldots,q_n,0,p_{r+1},\ldots,p_n)\le H_r\}.
\end{aligned} \tag{5.149}
$$

The averaged FPK equation associated with averaged Itô Eq. (5.147) is

$$\frac{\partial p}{\partial t} = -\sum_{\eta=1}^{r-1} \frac{\partial}{\partial I_\eta}(a_\eta p) - \frac{\partial}{\partial h_r}(a_r p) + \frac{1}{2}\sum_{\eta,\overline{\eta}=1}^{r-1} \frac{\partial^2}{\partial I_\eta \partial I_{\overline{\eta}}}(b_{\eta\overline{\eta}} p)$$
$$+\sum_{\eta=1}^{r-1} \frac{\partial^2}{\partial I_\eta \partial h_r}(b_{\eta r} p) + \frac{1}{2}\frac{\partial^2}{\partial h_r^2}(b_{rr} p), \tag{5.150}$$

where the first and second derivative moments are

$$a_\eta = a_\eta(I', h_r) = m_\eta(I', H_r)\big|_{H_r=h_r},$$
$$a_r = a_r(I', h_r) = m_r(I', H_r)\big|_{H_r=h_r},$$
$$b_{\eta\overline{\eta}} = b_{\eta\overline{\eta}}(I', h_r) = \sum_{l=1}^{m} \sigma_{\eta l}\sigma_{\overline{\eta} l}(I', H_r)\big|_{H_r=h_r},$$
$$b_{\eta r} = b_{\eta r}(I', h_r) = \sum_{l=1}^{m} \sigma_{\eta l}\sigma_{r l}(I', H_r)\big|_{H_r=h_r},$$
$$b_{rr} = b_{rr}(I', h_r) = \sum_{l=1}^{m} \sigma_{r l}\sigma_{r l}(I', H_r)\big|_{H_r=h_r}. \tag{5.151}$$

In Eq. (5.150), $p = p(\boldsymbol{I}', h_r, t|\boldsymbol{I}'_0, h_{r0})$ is the transition PDF of $[\boldsymbol{I}'^{\mathrm{T}}, H_r]^{\mathrm{T}}$ with initial condition

$$p(\boldsymbol{I}', h_r, 0|\boldsymbol{I}'_0, h_{r0}) = \delta(\boldsymbol{I}' - \boldsymbol{I}'_0)\delta(h_r - h_{r0}), \tag{5.152}$$

or $p = p(\boldsymbol{I}', h_r, t)$ is the probability density of $[\boldsymbol{I}'^{\mathrm{T}}, H_r]^{\mathrm{T}}$ with initial condition

$$p(\boldsymbol{I}', h_r, 0) = p(\boldsymbol{I}'_0, h_{r0}). \tag{5.153}$$

The boundary conditions of Eq. (5.150) depend on the system characteristics and the constraints imposed on the system. When I_η and H_r can vary over $[0, \infty)$, the boundary conditions are

$$p \text{ is finite}, \quad I_\eta = 0 \text{ or } h_r = 0, \tag{5.154}$$

$$p, \ \partial p/\partial I_\eta, \ \ \partial p/\partial h_r \to 0, \quad \left|I'\right| \to \infty \text{ or } h_r \to \infty. \tag{5.155}$$

Similar to the case of quasi-integrable Hamiltonian system, the probability flow in the averaged FPK Eq. (5.150) contains only potential flow and no circular flow. If there exists an exact stationary solution, it belongs to the class of stationary potential (Lin and Cai 1995). Considering the transformation (5.145) and the PDF transformation formulas (5.18) for quasi-non-integrable case and (5.104) for quasi-integrable and non-resonant case, the approximate stationary PDF of displacements and momenta of the original system (5.5) can be obtained from the stationary solution of Eq. (5.150) as follows

$$p(\boldsymbol{q},\boldsymbol{p}) = \left.\frac{p(\boldsymbol{I}', h_r)}{(2\pi)^{r-1}T(h_r)}\right|_{\boldsymbol{I}'=\boldsymbol{I}'(\boldsymbol{q}_1,\boldsymbol{p}_1),h_r=H_r(\boldsymbol{q}_2,\boldsymbol{p}_2)}. \tag{5.156}$$

A common case for Eq. (5.143) is

$$\begin{aligned} H_\eta &= p_\eta^2/2 + U_\eta(q_\eta), \quad \eta = 1, 2, \ldots, r-1, \\ H_r &= \sum_{\rho=r}^{n} p_\rho^2/2 + U_r(\boldsymbol{q}_2, \boldsymbol{p}_2). \end{aligned} \tag{5.157}$$

Introducing action-angle variables

$$I_\eta = f_\eta(H_\eta), \quad \theta_\eta = \omega_\eta(I_\eta)t + \delta_\eta., \tag{5.158}$$

where

$$\omega_\eta(I_\eta) = \frac{\mathrm{d}H_\eta}{\mathrm{d}I_\eta} = \frac{\mathrm{d}f_\eta^{-1}(I_\eta)}{\mathrm{d}I_\eta}. \tag{5.159}$$

Since H_η and I_η are one-to-one correspondence, and independent of θ_η, H_η can be used instead of I_η. The Itô stochastic differential equation for I_η and H_r in Eq. (5.146) can be transformed into those for H_η and H_r as follows

$$\begin{aligned} \mathrm{d}H_\eta &= \varepsilon\left(-\sum_{j=1}^{n} m_{\eta j}P_\eta P_j + \sum_{l,s=1}^{m} \pi K_{ls} g_{\eta l} g_{\eta s}\right)\mathrm{d}t \\ &\quad + \varepsilon^{1/2}\sum_{l=1}^{m} P_\eta g_{\eta l}\mathrm{d}B_l'(t), \\ \mathrm{d}H_r &= \varepsilon\left(-\sum_{\rho=r}^{n}\sum_{j=1}^{n} m_{\rho j}P_\rho P_j + \sum_{\rho=r}^{n}\sum_{l,s=1}^{m} \pi K_{ls} g_{\rho l} g_{\rho s}\right)\mathrm{d}t \\ &\quad + \varepsilon^{1/2}\sum_{\rho=r}^{n}\sum_{l=1}^{m} P_\rho g_{\rho l}\mathrm{d}B_l'(t), \\ \eta &= 1, 2, \ldots, r-1. \end{aligned} \tag{5.160}$$

The averaged Itô equation (5.147) becomes

$$
\begin{aligned}
\mathrm{d}H_\eta &= \overline{m}_\eta(H_1)\mathrm{d}t + \sum_{l=1}^{m} \overline{\sigma}_{\eta l}(H_1)\mathrm{d}B_l(t), \quad \eta = 1, 2, \ldots, r-1, \\
\mathrm{d}H_r &= \overline{m}_r(H_1)\mathrm{d}t + \sum_{l=1}^{m} \overline{\sigma}_{rl}(H_1)\mathrm{d}B_l(t),
\end{aligned} \tag{5.161}
$$

where $\boldsymbol{H}_1 = [H_1, \ldots, H_{r-1}, H_r]^{\mathrm{T}}$. $B_l(t)$ are independent unit Wiener processes. The averaged drift and diffusion coefficients are obtained from Eq. (5.160) according to Eq. (4.29) and (4.30) as follows

$$
\begin{aligned}
\overline{m}_\eta(\boldsymbol{H}_1) &= \varepsilon \left\langle -\sum_{j=1}^{n} m_{\eta j} P_\eta P_j + \sum_{l,s=1}^{m} \pi K_{ls} g_{\eta l} g_{\eta s} \right\rangle_t, \\
\overline{m}_r(\boldsymbol{H}_1) &= \varepsilon \left\langle -\sum_{\rho=r}^{n}\sum_{j=1}^{n} m_{\rho j} P_\rho P_j + \sum_{\rho=r}^{n}\sum_{l,s=1}^{m} \pi K_{ls} g_{\rho l} g_{\rho s} \right\rangle_t, \\
\sum_{l=1}^{m} \overline{\sigma}_{\eta l}\overline{\sigma}_{\overline{\eta} l}(\boldsymbol{H}_1) &= \varepsilon \left\langle \sum_{l,s=1}^{m} 2\pi K_{ls} g_{\eta l} g_{\overline{\eta} s} P_\eta P_{\overline{\eta}} \right\rangle_t, \\
\sum_{l=1}^{m} \overline{\sigma}_{rl}\overline{\sigma}_{rl}(\boldsymbol{H}_1) &= \varepsilon \left\langle \sum_{\rho,\rho'=r}^{n}\sum_{l,s=1}^{m} 2\pi K_{ls} g_{\rho l} g_{\rho' s} P_\rho P_{\rho'} \right\rangle_t, \\
\sum_{l=1}^{m} \overline{\sigma}_{\eta l}\overline{\sigma}_{rl}(\boldsymbol{H}_1) &= \varepsilon \left\langle \sum_{\rho=r}^{n}\sum_{l,s=1}^{m} 2\pi K_{ls} g_{\eta l} g_{\rho s} P_\eta P_\rho \right\rangle_t.
\end{aligned} \tag{5.162}
$$

Assuming that the Hamiltonian subsystem with H_η as Hamiltonian function has a family of periodic solutions on the full plane (q_η, p_η) with period $T(H_\eta) = 2\pi/\omega_\eta = \oint dq_\eta/p_\eta$, the following spatial averaging can be used to replace the time averaging for integrable Hamiltonian subsystem

$$
\frac{1}{T(H_1)\ldots T(H_{r-1})} \oint \cdots \oint [\cdot] \frac{\mathrm{d}q_1 \ldots \mathrm{d}q_{r-1}}{p_1 \ldots p_{r-1}}. \tag{5.163}
$$

Thus, the time averaging in Eq. (5.162) can be replaced by the following spatial averaging

$$
\langle [\cdot] \rangle_t = \frac{1}{T(H_1)\ldots T(H_r)} \int_{\Omega_1} \oint \cdots \oint \left([\cdot] \Big/ \frac{\partial H_r}{\partial p_r} \right) \frac{\mathrm{d}q_1 \ldots \mathrm{d}q_n \mathrm{d}p_{r+1} \ldots \mathrm{d}p_n}{p_1 \ldots p_{r-1}}. \tag{5.164}
$$

The averaged FPK equation associated with averaged Itô Eq. (5.161) is

$$\begin{aligned}\frac{\partial p}{\partial t} &= -\sum_{\eta=1}^{r-1}\frac{\partial}{\partial h_\eta}(\overline{a}_\eta p) - \frac{\partial}{\partial h_r}(\overline{a}_r p)\\ &+ \frac{1}{2}\sum_{\eta,\overline{\eta}=1}^{r-1}\frac{\partial^2}{\partial h_\eta \partial h_{\overline{\eta}}}(\overline{b}_{\eta\overline{\eta}}p)\\ &+ \frac{1}{2}\frac{\partial^2}{\partial h_r^2}(\overline{b}_{rr}p) + \sum_{\eta=1}^{r-1}\frac{\partial^2}{\partial h_\eta \partial h_{\overline{\eta}}}(\overline{b}_{\eta r}p),\end{aligned} \tag{5.165}$$

where the first and second derivative moments are

$$\begin{aligned}\overline{a}_\eta &= \overline{a}_\eta(\boldsymbol{h}_1) = \overline{m}_\eta(\boldsymbol{H}_1)\big|_{\boldsymbol{H}_1=\boldsymbol{h}_1},\\ \overline{a}_r &= \overline{a}_r(\boldsymbol{h}_1) = \overline{m}_r(\boldsymbol{H}_1)|_{\boldsymbol{H}_1=\boldsymbol{h}_1},\\ \overline{b}_{\eta\overline{\eta}} &= \overline{b}_{\eta\overline{\eta}}(\boldsymbol{h}_1) = \sum_{l=1}^{m}\overline{\sigma}_{\eta l}\overline{\sigma}_{\overline{\eta} l}(\boldsymbol{H}_1)\big|_{\boldsymbol{H}_1=\boldsymbol{h}_1},\\ \overline{b}_{rr} &= \overline{b}_{rr}(\boldsymbol{h}_1) = \sum_{l=1}^{m}\overline{\sigma}_{rl}\overline{\sigma}_{rl}(\boldsymbol{H}_1)|_{\boldsymbol{H}_1=\boldsymbol{h}_1},\\ \overline{b}_{\eta r} &= \overline{b}_{\eta r}(\boldsymbol{h}_1) = \sum_{l=1}^{m}\overline{\sigma}_{\eta l}\overline{\sigma}_{rl}(\boldsymbol{H}_1)\big|_{\boldsymbol{H}_1=\boldsymbol{h}_1}.\end{aligned} \tag{5.166}$$

The initial condition and boundary conditions are similar to Eqs. (5.152)–(5.155). Similar to Eq. (5.106), the stationary solution of Eq. (5.165) can be used to obtain the approximate stationary PDF of the original system as follows

$$p(\boldsymbol{q},\boldsymbol{p}) = \left.\frac{p(\boldsymbol{h}_1)}{T(h_1)\ldots T(h_r)}\right|_{h_\eta=H_\eta(q_\eta,p_\eta),h_r=H_r(\boldsymbol{q}_2,\boldsymbol{p}_2)}. \tag{5.167}$$

5.3.2 Internal Resonant Case

Assume that the integrable part of Hamiltonian system has β $(1 \leq \beta \leq r-2)$ weak internal resonant relations between the $r-1$ frequencies ω_η such as Eq. (5.75), i.e.,

$$\sum_{\eta=1}^{r-1} k_\eta^u \omega_\eta = O_u(\varepsilon), \quad u = 1, 2, \ldots, \beta. \tag{5.168}$$

Introduce β combinations of angle variables

$$\Psi_u = \sum_{\eta=1}^{r-1} k_\eta^u \Theta_\eta, \quad \eta = 1, 2, \ldots, \beta. \tag{5.169}$$

The following Itô stochastic differential equations for Ψ_u are obtained by linear combinations of Θ_η equations in Eq. (5.146)

$$\begin{aligned} d\Psi_u = &\left[O_u(\varepsilon) - \varepsilon \left(\sum_{j=1}^{n} \sum_{\eta'=1}^{r-1} m_{\eta' j} \frac{\partial \Psi_u}{\partial P_{\eta'}} \frac{\partial H}{\partial P_j} - \sum_{\eta',\eta''=1}^{r-1} \sum_{l,s=1}^{m} \pi K_{ls} g_{\eta' l} g_{\eta'' s} \frac{\partial^2 \Psi_u}{\partial P_{\eta'} \partial P_{\eta''}} \right) \right] dt \\ &+ \varepsilon^{1/2} \sum_{\eta'=1}^{r-1} \sum_{l=1}^{m} \frac{\partial \Psi_u}{\partial P_{\eta'}} g_{\eta' l} dB_l'(t), \quad u = 1, 2, \ldots, \beta. \end{aligned} \tag{5.170}$$

Now, the Itô stochastic differential equations of the system consist of Eq. (5.170), equations of I_η and H_r, $r - 1 - \beta$ equations of Θ_η in Eq. (5.146), $n - r + 1$ equations of Q_i and $n - r$ equations of P_i in Eq. (5.5). $I_\eta(t), H_r(t), \Psi_u(t)$ are slowly-varying processes while $Q_i(t), P_i(t), \Theta_\eta(t)$ are fast-varying processes. According to the Khasminskii theorem (Khasminskii 1968), when $\varepsilon \to 0$, $[\boldsymbol{I}^T, H_r, \boldsymbol{\Psi}^T]^T$ weakly converges to a $(r + \beta)$-dimensional vector Markov diffusion process. The averaged Itô stochastic differential equations are of the form

$$\begin{aligned} dI_\eta &= \overline{\overline{m}}_\eta(\boldsymbol{I}', \boldsymbol{\Psi}', H_r)dt + \sum_{l=1}^{m} \overline{\overline{\sigma}}_{\eta l}(\boldsymbol{I}', \boldsymbol{\Psi}', H_r)dB_l(t), \\ d\Psi_u &= \overline{\overline{m}}_u(\boldsymbol{I}', \boldsymbol{\Psi}', H_r)dt + \sum_{l=1}^{m} \overline{\overline{\sigma}}_{ul}(\boldsymbol{I}', \boldsymbol{\Psi}', H_r)dB_l(t), \\ dH_r &= \overline{\overline{m}}_r(\boldsymbol{I}', \boldsymbol{\Psi}', H_r)dt + \sum_{l=1}^{m} \overline{\overline{\sigma}}_{rl}(\boldsymbol{I}', \boldsymbol{\Psi}', H_r)dB_l(t), \\ \eta &= 1, 2, \ldots, r - 1; \quad u = 1, 2, \ldots, \beta, \end{aligned} \tag{5.171}$$

where $B_l(t)$ are unit independent Wiener processes, $\boldsymbol{\Psi}' = [\Psi_1, \Psi_2, \ldots, \Psi_\beta]^T$. Using Eqs. (4.29) and (4.30), the averaged drift and diffusion coefficients of Eq. (5.171) are obtained from the time averaging of the coefficients of equations for I_η, H_r in Eq. (5.146) and the coefficients in Eq. (5.170) as follows

$$\begin{aligned} \overline{\overline{m}}_\eta(\boldsymbol{I}', \boldsymbol{\Psi}', H_r) = \varepsilon \Bigg\langle &- \sum_{j=1}^{n} \sum_{\eta'=1}^{r-1} m_{\eta' j} \frac{\partial I_\eta}{\partial P_{\eta'}} \frac{\partial H}{\partial P_j} \\ &+ \sum_{\eta',\eta''=1}^{r-1} \sum_{l,s=1}^{m} \pi K_{ls} g_{\eta' l} g_{\eta'' s} \frac{\partial^2 I_\eta}{\partial P_{\eta'} \partial P_{\eta''}} \Bigg\rangle_t, \end{aligned}$$

$$\overline{\overline{m}}_u(\boldsymbol{I}', \boldsymbol{\Psi}', H_r) = O_u(\varepsilon) + \varepsilon\left\langle -\sum_{j=1}^{n}\sum_{\eta'=1}^{r-1} m_{\eta' j}\frac{\partial \Psi_u}{\partial P_{\eta'}}\frac{\partial H}{\partial P_j} + \sum_{\eta',\eta''=1}^{r-1}\sum_{l,s=1}^{m} \pi K_{ls} g_{\eta' l} g_{\eta'' s}\frac{\partial^2 \Psi_u}{\partial P_{\eta'}\partial P_{\eta''}}\right\rangle_t,$$

$$\overline{\overline{m}}_r(\boldsymbol{I}', \boldsymbol{\Psi}', H_r) = \varepsilon\left\langle -\sum_{j=1}^{n}\sum_{\rho=r}^{n} m_{\rho j}\frac{\partial H_r}{\partial P_\rho}\frac{\partial H}{\partial P_j} + \sum_{\rho,\rho'=r}^{n}\sum_{l,s=1}^{m} \pi K_{ls} g_{\rho l} g_{\rho' s}\frac{\partial^2 H_r}{\partial P_\rho \partial P_{\rho'}}\right\rangle_t,$$

$$\sum_{l=1}^{m}\overline{\overline{\sigma}}_{\eta l}\overline{\overline{\sigma}}_{\overline{\eta} l}(\mathbf{I}', \Psi', H_r) = \varepsilon\left\langle \sum_{\eta',\eta''=1}^{r-1}\sum_{l,s=1}^{m} 2\pi K_{ls} g_{\eta' l} g_{\eta'' s}\frac{\partial I_\eta}{\partial P_{\eta'}}\frac{\partial I_{\overline{\eta}}}{\partial P_{\eta''}}\right\rangle_t,$$

$$\sum_{l=1}^{m}\overline{\overline{\sigma}}_{ul}\overline{\overline{\sigma}}_{vl}(\mathbf{I}', \Psi', H_r) = \varepsilon\left\langle \sum_{\eta',\eta''=1}^{r-1}\sum_{l,s=1}^{m} 2\pi K_{ls} g_{\eta' l} g_{\eta'' s}\frac{\partial \Psi_u}{\partial P_{\eta'}}\frac{\partial \Psi_v}{\partial P_{\eta''}}\right\rangle_t, \tag{5.172}$$

$$\sum_{l=1}^{m}\overline{\overline{\sigma}}_{rl}\overline{\overline{\sigma}}_{rl}(\mathbf{I}', \Psi', H_r) = \varepsilon\left\langle \sum_{\rho,\rho'=r}^{n}\sum_{l,s=1}^{m} 2\pi K_{ls} g_{\rho l} g_{\rho' s}\frac{\partial H_r}{\partial P_\rho}\frac{\partial H_r}{\partial P_{\rho'}}\right\rangle_t,$$

$$\sum_{l=1}^{m}\overline{\overline{\sigma}}_{\eta l}\overline{\sigma}_{rl}(\boldsymbol{I}', \boldsymbol{\Psi}', H_r) = \varepsilon\left\langle \sum_{\rho=r}^{n}\sum_{\eta'=1}^{r-1}\sum_{l,s=1}^{m} 2\pi K_{ls} g_{\eta' l} g_{\rho s}\frac{\partial I_\eta}{\partial P_{\eta'}}\frac{\partial H_r}{\partial P_\rho}\right\rangle_t,$$

$$\sum_{l=1}^{m}\overline{\overline{\sigma}}_{rl}\overline{\overline{\sigma}}_{ul}(\boldsymbol{I}', \boldsymbol{\Psi}', H_r) = \varepsilon\left\langle \sum_{\eta'=1}^{r-1}\sum_{\rho=r}^{n}\sum_{l,s=1}^{m} 2\pi K_{ls} g_{\eta' l} g_{\rho s}\frac{\partial \Psi_u}{\partial P_{\eta'}}\frac{\partial H_r}{\partial P_\rho}\right\rangle_t,$$

$$\sum_{l=1}^{m}\overline{\overline{\sigma}}_{\eta l}\overline{\overline{\sigma}}_{ul}(\boldsymbol{I}', \boldsymbol{\Psi}', H_r) = \varepsilon\left\langle \sum_{\eta',\eta''=1}^{r-1}\sum_{l,s=1}^{m} 2\pi K_{ls} g_{\eta' l} g_{\eta'' s}\frac{\partial I_\eta}{\partial P_{\eta'}}\frac{\partial \Psi_u}{\partial P_{\eta''}}\right\rangle_t,$$

Since the integrable Hamiltonian sub-system is ergodic on the $(r - \beta - 1)$-dimensional torus and non-integrable Hamiltonian sub-system is ergodic on $(2(n - r) + 1)$-dimensional isoenergetic surface, the time averaging in Eq. (5.172) can be replaced by the following spatial averaging

$$\langle[\cdot]\rangle_t = \frac{1}{(2\pi)^{r-\beta-1}T(H_r)}\int_{\Omega_1}\int_0^{2\pi}\cdots\int_0^{2\pi}\left([\cdot]\Big/\frac{\partial H_r}{\partial p_r}\right)\mathrm{d}\theta_1\ldots\mathrm{d}\theta_{r-\beta-1}\mathrm{d}q_r\ldots\mathrm{d}q_n\mathrm{d}p_{r+1}\ldots\mathrm{d}p_n \tag{5.173}$$

where $T(H_r)$ and Ω_1 are determined by Eq. (5.149).

The averaged FPK equation associated with averaged Itô Eq. (5.171) is

$$\begin{aligned}\frac{\partial p}{\partial t} &= -\sum_{\eta=1}^{r-1}\frac{\partial}{\partial I_\eta}(\overline{\overline{a}}_\eta p) - \sum_{u=1}^{\beta}\frac{\partial}{\partial \psi_u}(\overline{\overline{a}}_u p) - \frac{\partial}{\partial h_r}(\overline{\overline{a}}_r p)\\ &+ \frac{1}{2}\sum_{\eta,\overline{\eta}=1}^{r-1}\frac{\partial^2}{\partial I_\eta \partial I_{\overline{\eta}}}(\overline{\overline{b}}_{\eta\overline{\eta}} p) + \frac{1}{2}\sum_{u,v=1}^{\beta}\frac{\partial^2}{\partial \psi_u \partial \psi_v}(\overline{\overline{b}}_{uv} p) + \frac{1}{2}\frac{\partial^2}{\partial h_r^2}(\overline{\overline{b}}_{rr} p)\\ &+ \sum_{\eta=1}^{r-1}\sum_{u=1}^{\beta}\frac{\partial^2}{\partial I_\eta \partial \psi_u}(\overline{\overline{b}}_{\eta u} p) + \sum_{\eta=1}^{r-1}\frac{\partial^2}{\partial I_\eta \partial h_r}(\overline{\overline{b}}_{\eta r} p) + \sum_{u=1}^{\beta}\frac{\partial^2}{\partial h_r \partial \psi_u}(\overline{\overline{b}}_{ru} p), \end{aligned}\quad (5.174)$$

where the averaged first and second derivative moments are

$$\begin{aligned}
\overline{\overline{a}}_\eta &= \overline{\overline{a}}_\eta(\boldsymbol{I}', \boldsymbol{\psi}', h_r) = \overline{\overline{m}}_\eta(\boldsymbol{I}', \boldsymbol{\Psi}', H_r)\big|_{H_r=h_r,\ \boldsymbol{\Psi}'=\boldsymbol{\psi}'},\\
\overline{\overline{a}}_u &= \overline{\overline{a}}_u(\boldsymbol{I}', \boldsymbol{\psi}', h_r) = \overline{\overline{m}}_u(\boldsymbol{I}', \boldsymbol{\Psi}', H_r)\big|_{H_r=h_r,\ \boldsymbol{\Psi}'=\boldsymbol{\psi}'},\\
\overline{\overline{a}}_r &= \overline{\overline{a}}_r(\boldsymbol{I}', \boldsymbol{\psi}', h_r) = \overline{\overline{m}}_r(\boldsymbol{I}', \boldsymbol{\Psi}', H_r)\big|_{H_r=h_r,\ \boldsymbol{\Psi}'=\boldsymbol{\psi}'},\\
\overline{\overline{b}}_{\eta\overline{\eta}} &= \overline{\overline{b}}_{\eta\overline{\eta}}(\boldsymbol{I}', \boldsymbol{\psi}', h_r) = \sum_{l=1}^{m}\overline{\overline{\sigma}}_{\eta l}\overline{\overline{\sigma}}_{\overline{\eta} l}(\boldsymbol{I}', \boldsymbol{\Psi}', H_r)\big|_{H_r=h_r,\ \boldsymbol{\Psi}'=\boldsymbol{\psi}'},\\
\overline{\overline{b}}_{uv} &= \overline{\overline{b}}_{uv}(\boldsymbol{I}', \boldsymbol{\psi}', h_r) = \sum_{l=1}^{m}\overline{\overline{\sigma}}_{ul}\overline{\overline{\sigma}}_{vl}(\boldsymbol{I}', \boldsymbol{\Psi}', H_r)\big|_{H_r=h_r,\ \boldsymbol{\Psi}'=\boldsymbol{\psi}'},\\
\overline{\overline{b}}_{rr} &= \overline{\overline{b}}_{rr}(\boldsymbol{I}', \boldsymbol{\psi}', h_r) = \sum_{l=1}^{m}\overline{\overline{\sigma}}_{rl}\overline{\overline{\sigma}}_{rl}(\boldsymbol{I}', \boldsymbol{\Psi}', H_r)\big|_{H_r=h_r,\ \boldsymbol{\Psi}'=\boldsymbol{\psi}'},\\
\overline{\overline{b}}_{\eta u} &= \overline{\overline{b}}_{\eta u}(\boldsymbol{I}', \boldsymbol{\psi}', h_r) = \sum_{l=1}^{m}\overline{\overline{\sigma}}_{\eta l}\overline{\overline{\sigma}}_{ul}(\boldsymbol{I}', \boldsymbol{\Psi}', H_r)\big|_{H_r=h_r,\ \boldsymbol{\Psi}'=\boldsymbol{\psi}'},\\
\overline{\overline{b}}_{\eta r} &= \overline{\overline{b}}_{\eta r}(\boldsymbol{I}', \boldsymbol{\psi}', h_r) = \sum_{l=1}^{m}\overline{\overline{\sigma}}_{\eta l}\overline{\overline{\sigma}}_{rl}(\boldsymbol{I}', \boldsymbol{\Psi}', H_r)\big|_{H_r=h_r,\ \boldsymbol{\Psi}'=\boldsymbol{\psi}'},\\
\overline{\overline{b}}_{ru} &= \overline{\overline{b}}_{ru}(\boldsymbol{I}', \boldsymbol{\psi}', h_r) = \sum_{l=1}^{m}\overline{\overline{\sigma}}_{rl}\overline{\overline{\sigma}}_{ul}(\boldsymbol{I}', \boldsymbol{\Psi}', H_r)\big|_{H_r=h_r,\ \boldsymbol{\Psi}'=\boldsymbol{\psi}'}.
\end{aligned}\quad (5.175)$$

In (5.174), $p = p(\boldsymbol{I}', h_r, \boldsymbol{\psi}', t|\boldsymbol{I}'_0, h_{r0}, \boldsymbol{\psi}'_0)$ is the transition PDF of $[\boldsymbol{I}'^{\mathrm{T}}, H_r, \boldsymbol{\Psi}'^{\mathrm{T}}]^{\mathrm{T}}$ with the initial condition as

$$p(\boldsymbol{I}', \boldsymbol{\psi}', h_r, 0|\boldsymbol{I}'_0, \boldsymbol{\psi}'_0, h_{r0}) = \delta(\boldsymbol{I}' - \boldsymbol{I}'_0)\delta(\boldsymbol{\psi}' - \boldsymbol{\psi}'_0)\delta(h_r - h_{r0}), \quad (5.176)$$

or $p = p(\boldsymbol{I}', h_r, \boldsymbol{\psi}', t)$ is the joint PDF of $[\boldsymbol{I}'^{\mathrm{T}}, H_r, \boldsymbol{\Psi}'^{\mathrm{T}}]^{\mathrm{T}}$ with the initial condition

$$p(\boldsymbol{I}', \boldsymbol{\psi}', h_r, 0) = p(\boldsymbol{I}'_0, \boldsymbol{\psi}'_0, h_{r0}). \tag{5.177}$$

The boundary conditions about I_η, H_r are the same as those in Eqs. (5.154) and (5.155). The periodic boundary conditions with respect to ψ_u are the same as those in Eq. (5.126).

Considering the transformations (5.145) and (5.169), the relationship between the stationary solution of FPK Eq. (5.174) and the approximate stationary PDF $p(\boldsymbol{q}, \boldsymbol{p})$ of generalized displacement and momenta of original system is

$$p(\boldsymbol{q}, \boldsymbol{p}) = \frac{p(\boldsymbol{I}', \boldsymbol{\psi}', h_r)}{(2\pi)^{r-\beta-1} T(h_r)} \left| \frac{\partial(\boldsymbol{I}', \boldsymbol{\psi}', \boldsymbol{\theta}'')}{\partial(\boldsymbol{q}_1, \boldsymbol{p}_1)} \right|, \tag{5.178}$$

where $\boldsymbol{\theta}'' = [\theta_1, \theta_2, \ldots, \theta_{r-\beta-1}]^{\mathrm{T}}$. Since the determinant of $\partial(\boldsymbol{I}', \boldsymbol{q}')/\partial(\boldsymbol{q}_1, \boldsymbol{p}_1)$ is equal to 1, $\partial(\boldsymbol{I}', \boldsymbol{\psi}', \boldsymbol{\theta}')/\partial(\boldsymbol{q}_1, p_1)$ in Eq. (5.178) is constant and only affects the normalization constant.

If the Hamiltonian sub-functions of a partially integrable Hamiltonian system are of the form of Eq. (5.157), H_η and I_η $(\eta = 1, 2, \ldots, r-1)$ are one-to-one correspondence as those in Eq. (5.158). Then $\partial H / \partial P_j = P_j$, $\sum_{\eta'=1}^{r-1} \partial I_\eta / \partial P_{\eta'} = \sum_{\eta''=1}^{r-1} \partial I_\eta / \partial P_{\eta''} = \partial I_\eta / \partial P_\eta$, and the averaged Itô stochastic differential equations still take the form of Eq. (5.171), but the averaged drift and diffusion coefficients in Eq. (5.172) are simplified as

$$\begin{aligned}
\overline{\overline{m}}_\eta(\boldsymbol{I}', \boldsymbol{\Psi}', H_r) &= \varepsilon \left\langle -\sum_{j=1}^{n} m_{\eta j} \frac{\partial I_\eta}{\partial P_\eta} P_j + \sum_{l,s=1}^{m} \pi K_{ls} g_{\eta l} g_{\eta s} \frac{\partial^2 I_\eta}{\partial P_\eta^2} \right\rangle_t, \\
\overline{\overline{m}}_u(\boldsymbol{I}', \boldsymbol{\Psi}', H_r) &= O_u(\varepsilon) + \varepsilon \left\langle -\sum_{j=1}^{n} \sum_{\eta'=1}^{r-1} m_{\eta' j} \frac{\partial \Psi_u}{\partial P_{\eta'}} P_j \right. \\
&\quad \left. + \sum_{\eta',\eta''=1}^{r-1} \sum_{l,s=1}^{m} \pi K_{ls} g_{\eta' l} g_{\eta'' s} \frac{\partial^2 \Psi_u}{\partial P_{\eta'} \partial P_{\eta''}} \right\rangle_t, \\
\overline{\overline{m}}_r(\boldsymbol{I}', \boldsymbol{\Psi}', H_r) &= \varepsilon \left\langle -\sum_{j=1}^{n} \sum_{\rho=r}^{n} m_{\rho j} P_\rho P_j + \sum_{\rho=r}^{n} \sum_{l,s=1}^{m} \pi K_{ls} g_{\rho l} g_{\rho s} \right\rangle_t, \\
\sum_{l=1}^{m} \overline{\overline{\sigma}}_{\eta l} \overline{\overline{\sigma}}_{\overline{\eta} l}(\boldsymbol{I}', \boldsymbol{\Psi}', H_r) &= \varepsilon \left\langle \sum_{l,s=1}^{m} 2\pi K_{ls} g_{\eta l} g_{\overline{\eta} s} \frac{\partial I_\eta}{\partial P_\eta} \frac{\partial I_{\overline{\eta}}}{\partial P_{\overline{\eta}}} \right\rangle_t, \\
\sum_{l=1}^{m} \overline{\overline{\sigma}}_{ul} \overline{\overline{\sigma}}_{vl}(\boldsymbol{I}', \boldsymbol{\Psi}', H_r) &= \varepsilon \left\langle \sum_{\eta,\eta'=1}^{r-1} \sum_{l,s=1}^{m} 2\pi K_{ls} g_{\eta l} g_{\eta' s} \frac{\partial \Psi_u}{\partial P_\eta} \frac{\partial \Psi_v}{\partial P_{\eta'}} \right\rangle_t,
\end{aligned}$$

$$\sum_{l=1}^{m} \overline{\overline{\overline{\sigma}}}_{rl} \overline{\overline{\overline{\sigma}}}_{rl}(\boldsymbol{I}', \boldsymbol{\Psi}', H_r) = \varepsilon \left\langle \sum_{\rho,\rho'=r}^{n} \sum_{l,s=1}^{m} 2\pi K_{ls} g_{\rho l} g_{\rho' s} P_\rho P_{\rho'} \right\rangle_t, \tag{5.179}$$

$$\sum_{l=1}^{m} \overline{\overline{\overline{\sigma}}}_{\eta l} \overline{\overline{\overline{\sigma}}}_{ul}(\boldsymbol{I}', \boldsymbol{\Psi}', H_r) = \varepsilon \left\langle \sum_{\eta'=1}^{r-1} \sum_{l,s=1}^{m} 2\pi K_{ls} g_{\eta l} g_{\eta' s} \frac{\partial I_\eta}{\partial P_\eta} \frac{\partial \Psi_u}{\partial P_{\eta'}} \right\rangle_t,$$

$$\sum_{l=1}^{m} \overline{\overline{\overline{\sigma}}}_{\eta l} \overline{\overline{\overline{\sigma}}}_{rl}(\boldsymbol{I}', \boldsymbol{\Psi}', H_r) = \varepsilon \left\langle \sum_{\rho=r}^{n} \sum_{l,s=1}^{m} 2\pi K_{ls} g_{\eta l} g_{\rho s} \frac{\partial I_\eta}{\partial P_\eta} P_\rho \right\rangle_t,$$

$$\sum_{l=1}^{m} \overline{\overline{\overline{\sigma}}}_{rl} \overline{\overline{\overline{\sigma}}}_{ul}(\boldsymbol{I}', \boldsymbol{\Psi}', H_r) = \varepsilon \left\langle \sum_{\eta=1}^{r-1} \sum_{\rho=r}^{n} \sum_{l,s=1}^{m} 2\pi K_{ls} g_{\eta l} g_{\rho s} \frac{\partial \Psi_u}{\partial P_\eta} P_\rho \right\rangle_t.$$

The spatial averaging in Eq. (5.173) is simplified as

$$\langle [\cdot] \rangle_t = \frac{1}{(2\pi)^{r-\beta-1} T'(H_r)} \times \int_{\Omega_1} \int_0^{2\pi} \cdots \int_0^{2\pi} ([\cdot]/p_r) \mathrm{d}\theta_1 \ldots \mathrm{d}\theta_{r-\beta-1} \mathrm{d}q_r \ldots \mathrm{d}q_n \mathrm{d}p_{r+1} \ldots \mathrm{d}p_n, \tag{5.180}$$

where

$$T'(H_r) = \int_{\Omega_1} (1/p_r) \mathrm{d}q_r \ldots \mathrm{d}q_n \mathrm{d}p_{r+1} \ldots \mathrm{d}p_n, \tag{5.181}$$

and Ω_1 is defined by Eq. (5.149). The averaged FPK equation is of the form of Eq. (5.174) with the following averaged first and second derivative moments

$$\overline{\overline{\overline{a}}}_\eta = \overline{\overline{\overline{a}}}_\eta(\boldsymbol{I}', \boldsymbol{\psi}', h_r) = \overline{\overline{\overline{m}}}_\eta(\boldsymbol{I}', \boldsymbol{\Psi}', H_r)\Big|_{H_r=h_r,\ \boldsymbol{\Psi}'=\boldsymbol{\psi}'},$$

$$\overline{\overline{a}}_u = \overline{\overline{a}}_u(\boldsymbol{I}', \boldsymbol{\psi}', h_r) = \overline{\overline{m}}_u(\boldsymbol{I}', \boldsymbol{\Psi}', H_r)\Big|_{H_r=h_r,\ \boldsymbol{\Psi}'=\boldsymbol{\psi}'},$$

$$\overline{\overline{\overline{a}}}_r = \overline{\overline{\overline{a}}}_r(\boldsymbol{I}', \boldsymbol{\psi}', h_r) = \overline{\overline{\overline{m}}}_r(\boldsymbol{I}', \boldsymbol{\Psi}', H_r)\Big|_{H_r=h_r,\ \boldsymbol{\Psi}'=\boldsymbol{\psi}'},$$

$$\overline{\overline{\overline{b}}}_{\eta\bar{\eta}} = \overline{\overline{\overline{b}}}_{\eta\bar{\eta}}(\boldsymbol{I}', \boldsymbol{\psi}', h_r) = \sum_{l=1}^{m} \overline{\overline{\overline{\sigma}}}_{\eta l} \overline{\overline{\overline{\sigma}}}_{\bar{\eta} l}(\boldsymbol{I}', \boldsymbol{\Psi}', H_r)\Big|_{H_r=h_r,\ \boldsymbol{\Psi}'=\boldsymbol{\psi}'},$$

$$\overline{\overline{\overline{b}}}_{uv} = \overline{\overline{\overline{b}}}_{uv}(\boldsymbol{I}', \boldsymbol{\psi}', h_r) = \sum_{l=1}^{m} \overline{\overline{\overline{\sigma}}}_{ul} \overline{\overline{\overline{\sigma}}}_{vl}(\boldsymbol{I}', \boldsymbol{\Psi}', H_r)\Big|_{H_r=h_r,\ \boldsymbol{\Psi}'=\boldsymbol{\psi}'},$$

$$\overline{\overline{\overline{b}}}_{rr} = \overline{\overline{\overline{b}}}_{rr}(\boldsymbol{I}', \boldsymbol{\psi}', h_r) = \sum_{l=1}^{m} \overline{\overline{\overline{\sigma}}}_{rl} \overline{\overline{\overline{\sigma}}}_{rl}(\boldsymbol{I}', \boldsymbol{\Psi}', H_r)\Big|_{H_r=h_r,\ \boldsymbol{\Psi}'=\boldsymbol{\psi}'},$$

$$\overline{\overline{\overline{b}}}_{\eta u}=\overline{\overline{\overline{b}}}_{\eta u}(\boldsymbol{I}',\boldsymbol{\psi}',h_r)=\sum_{l=1}^{m}\overline{\overline{\overline{\sigma}}}_{\eta l}\overline{\overline{\overline{\sigma}}}_{ul}(\boldsymbol{I}',\boldsymbol{\Psi}',H_r)\big|_{H_r=h_r,\ \boldsymbol{\Psi}'=\boldsymbol{\psi}'}, \tag{5.182}$$

$$\overline{\overline{\overline{b}}}_{\eta r}=\overline{\overline{\overline{b}}}_{\eta r}(\boldsymbol{I}',\boldsymbol{\psi}',h_r)=\sum_{l=1}^{m}\overline{\overline{\overline{\sigma}}}_{\eta l}\overline{\overline{\overline{\sigma}}}_{rl}(\boldsymbol{I}',\boldsymbol{\Psi}',H_r)\big|_{H_r=h_r,\ \boldsymbol{\Psi}'=\boldsymbol{\psi}'},$$

$$\overline{\overline{\overline{b}}}_{ru}=\overline{\overline{\overline{b}}}_{ru}(\boldsymbol{I}',\boldsymbol{\psi}',h_r)=\sum_{l=1}^{m}\overline{\overline{\overline{\sigma}}}_{rl}\overline{\overline{\overline{\sigma}}}_{ul}(\boldsymbol{I}',\boldsymbol{\Psi}',H_r)\big|_{H_r=h_r,\ \boldsymbol{\Psi}'=\boldsymbol{\psi}'}.$$

Example 5.5 Consider a 4-DOF nonlinear system excited by Gaussian white noises and governed by the following motion equations

$$\begin{aligned}
&\dot{Q}_1=P_1,\\
&\dot{P}_1=-\omega_1^2Q_1-\big[\alpha_{10}+\alpha_{11}P_1^2+\alpha_{12}P_2^2+\alpha_{13}P_3^2+\alpha_{14}P_4^2\\
&\qquad+(\alpha_{13}+\alpha_{14})U(Q_3,Q_4)\big]P_1+W_{g1}(t),\\
&\dot{Q}_2=P_2,\\
&\dot{P}_2=-\omega_2^2Q_2-\big[\alpha_{20}+\alpha_{21}P_1^2+\alpha_{22}P_2^2+\alpha_{23}P_3^2+\alpha_{24}P_4^2\\
&\qquad+(\alpha_{23}+\alpha_{24})U(Q_3,Q_4)\big]P_2+W_{g2}(t),\\
&\dot{Q}_3=P_3,\\
&\dot{P}_3=-\partial U(Q_3,Q_4)/\partial Q_3-\big[\alpha_{30}+\alpha_{31}P_1^2+\alpha_{32}P_2^2+\alpha_{33}P_3^2+\alpha_{34}P_4^2\\
&\qquad+(1/2)(\alpha_{34}+3\alpha_{33})U(Q_3,Q_4)\big]P_3+W_{g3}(t),\\
&\dot{Q}_4=P_4,\\
&\dot{P}_4=-\partial U(Q_3,Q_4)/\partial Q_4-\big[\alpha_{40}+\alpha_{41}P_1^2+\alpha_{42}P_2^2\\
&\qquad+\alpha_{43}P_3^2+\alpha_{44}P_4^2+(1/2)(\alpha_{43}+3\alpha_{44})U(Q_3,Q_4)\big]P_4+W_{g4}(t),
\end{aligned} \tag{5.183}$$

where

$$U(Q_3,Q_4)=\big(\omega_3^2Q_3^2+\omega_4^2Q_4^2\big)/2+b\big(\omega_3^2Q_3^2+\omega_4^2Q_4^2\big)^2\big/4, \tag{5.184}$$

ω_i, α_{ij}, b are positive constants, and $W_{gl}(t)$ are independent Gaussian white noises with intensities $2\pi K_l$. The Hamiltonian function associated with system (5.183) is

$$H=\sum_{\eta=1}^{2}H_\eta+H_3=\sum_{\eta=1}^{2}\omega_\eta I_\eta+H_3, \tag{5.185}$$

where

$$H_\eta = \left(p_\eta^2 + \omega_\eta^2 q_\eta^2\right)/2, \ H_3 = \left(p_3^2 + p_4^2\right)/2 + U(q_3, q_4), \tag{5.186}$$

and $U(q_3, q_4)$ is inseparable. Equation (5.185) is a specific case of Eq. (5.144). Assuming that α_{ij} and K_l are small quantities of ε order, then Eq. (5.183) is a quasi-partially integrable Hamiltonian system. The dimension and form of the averaged equations depend on whether there exists internal resonance between the first two DOF.

(1) **Noninternal resonant case** Transform Eq. (5.184) into the Itô stochastic differential equations for $I_\eta \quad (\eta = 1, 2)$ and H_3 of the form as Eq. (5.146). Performing averaging using Eq. (5.149) yields the following averaged Itô stochastic differential equations

$$\begin{aligned}
&\mathrm{d}I_\eta = m_\eta(I_1, I_2, H_3)\mathrm{d}t + \sum_{l=1}^{4} \sigma_{\eta l}(I_1, I_2, H_3)\mathrm{d}B_l, \\
&\mathrm{d}H_3 = m_3(I_1, I_2, H_3)\mathrm{d}t + \sum_{l=1}^{4} \sigma_{3l}(I_1, I_2, H_3)\mathrm{d}B_l, \\
&\eta = 1, 2,
\end{aligned} \tag{5.187}$$

where the averaged drift and diffusion coefficients are obtained using Eq. (5.148) as follows

$$\begin{aligned}
m_1(I_1, I_2, H_3) &= -\big[\alpha_{10}I_1 + 3\alpha_{11}\omega_1 I_1^2/2 + \alpha_{12}\omega_2 I_1 I_2 \\
&\quad + (\alpha_{13} + \alpha_{14})I_1 H_3\big] + \pi K_1/\omega_1, \\
m_2(I_1, I_2, H_3) &= -\big[\alpha_{20}I_2 + 3\alpha_{22}\omega_2 I_2^2/2 + \alpha_{21}\omega_1 I_1 I_2 \\
&\quad + (\alpha_{23} + \alpha_{24})I_2 H_3\big] + \pi K_2/\omega_2,
\end{aligned}$$

$$\begin{aligned}
&m_3(I_1, I_2, H_3) = -[(\alpha_{30} + \alpha_{40}) + (\alpha_{31} + \alpha_{41})\omega_1 I_1 \\
&+ (\alpha_{32} + \alpha_{42})\omega_2 I_2 + (3\alpha_{33} + 3\alpha_{44} + \alpha_{34} + \alpha_{43})H_3/2] \\
&S(H_3) + \pi K_3 + \pi K_4,
\end{aligned} \tag{5.188}$$

$$\begin{aligned}
&\sigma_{11}^2(I_1, I_2, H_3) = 2\pi K_1 I_1 / \omega_1, \\
&\sigma_{22}^2(I_1, I_2, H_3) = 2\pi K_2 I_2 / \omega_2, \\
&\sigma_{33}^2(I_1, I_2, H_3) = 2\pi (K_3 + K_4) S(H_3), \\
&\sigma_{12} = \sigma_{13} = \sigma_{21} = \sigma_{23} = \sigma_{31} = \sigma_{32} = 0,
\end{aligned}$$

and $S(H_3) = [1 + 8bH_3 - (1 + 4bH_3)^{1/2}] / 12b$.

The averaged FPK equation is of the form of Eq. (5.150). Assume that the exact stationary solution has the form

$$p(I_1, I_2, h_3) = C\exp[-\lambda(I_1, I_2, h_3)] \tag{5.189}$$

It can be shown by substituting Eq. (5.189) into the reduced averaged FPK equation (5.150) with $\frac{\partial p}{\partial t} = 0$ that $\lambda(I_1, I_2, h_3)$ is indeed the solution if the following first-order partial differential equations are satisfied

$$\begin{gathered} \frac{2\pi K_1 I_1}{\omega_1}\frac{\partial\lambda}{\partial I_1} = \frac{2\pi K_1}{\omega_1} - 2a_1, \quad \frac{2\pi K_2 I_2}{\omega_2}\frac{\partial\lambda}{\partial I_2} = \frac{2\pi K_2}{\omega_2} - 2a_2, \\ 2\pi(K_3 + K_4)S(h_3)\frac{\partial\lambda}{\partial h_3} = 2\pi(K_3 + K_4)\frac{dS(h_3)}{dh_3} - 2a_3, \end{gathered} \tag{5.190}$$

where $a_i = m_i$. If the damping coefficients and the random excitation intensity satisfies the compatibility conditions

$$\begin{gathered} \alpha_{12}/K_1 = \alpha_{21}/K_2, (\alpha_{13} + \alpha_{14})/K_1 = (\alpha_{31} + \alpha_{41})/(K_3 + K_4), \\ (\alpha_{23} + \alpha_{24})/K_2 = (\alpha_{32} + \alpha_{42})/(K_3 + K_4), \end{gathered} \tag{5.191}$$

then the exact stationary solution of the reduced averaged FPK equation is

$$\begin{aligned} p(I_1, I_2, h_3) = C\big[(1 + 4bh_3)^{1/2} - 1\big]\exp\{&-[\alpha_{10}\omega_1 I_1/(\pi K_1) + \alpha_{20}\omega_2 I_2/(\pi K_2) \\ &+ (\alpha_{30} + \alpha_{40})h_3/(\pi K_3 + \pi K_4) + 3\alpha_{11}\omega_1^2 I_1^2/(4\pi K_1) + 3\alpha_{22}\omega_2^2 I_2^2/(4\pi K_2) \\ &+ (3\alpha_{33} + 3\alpha_{44} + \alpha_{34} + \alpha_{43})h_3^3/(4\pi K_1 + 4\pi K_4) + \alpha_{12}\omega_1\omega_2 I_1 I_2/(\pi K_1) \\ &+ (\alpha_{13} + \alpha_{14})\omega_1 I_1 h_3/(\pi K_1) + (\alpha_{23} + \alpha_{24})\omega_2 I_2 h_3/(\pi K_2)]\}. \end{aligned} \tag{5.192}$$

Using Eq. (5.156), the approximately stationary PDF of displacements and momenta for system (5.183) is derived as

$$p(\boldsymbol{p}, \boldsymbol{q}) = \left.\frac{p(I_1, I_2, h_3)}{(1 - 4bh_3)^{1/2} - 1}\right|_{I_{1,2} = I_{1,2}(q_{1,2}, p_{1,2}), h_3 = H_3(q_3, q_4, p_3, p_4)}. \tag{5.193}$$

(2) **Principal internal resonance case** Suppose that $\omega_1 = \omega_2$. Let $\Psi = \Theta_1 - \Theta_2$. The averaged Itô equations are of the form of Eq. (5.171), i.e.,

$$\begin{aligned} dI_\eta &= \overline{\overline{m}}_\eta(I_1, I_2, \Psi, H_3)dt + \sum_{l=1}^{4}\overline{\overline{\sigma}}_{\eta l}(I_1, I_2, \Psi, H_3)dB_l(t), \\ d\Psi &= \overline{\overline{m}}_3(I_1, I_2, \Psi, H_3)dt + \sum_{l=1}^{4}\overline{\overline{\sigma}}_{3l}(I_1, I_2, \Psi, H_3)dB_l(t), \end{aligned}$$

$$
\begin{aligned}
&\mathrm{d}H_3 = \overline{\overline{m}}_4(I_1, I_2, \Psi, H_3)\mathrm{d}t + \sum_{l=1}^{4} \overline{\overline{\sigma}}_{4l}(I_1, I_2, \Psi, H_3)\mathrm{d}B_l(t),\\
&\eta = 1, 2.
\end{aligned}
\tag{5.194}
$$

According to Eq. (5.179), the averaged drift and diffusion coefficients of Eq. (5.194) are obtained as follows

$$
\begin{aligned}
\overline{\overline{m}}_1(I_1, I_2, \psi, H_3) = &-\left[\alpha_{10}I_1 + 3\alpha_{11}\omega_1 I_1^2/2 + \alpha_{12}\omega_2 I_1 I_2(1 + (1/2)\cos 2\psi)\right.\\
&\left.+(\alpha_{13} + \alpha_{14})I_1 H_3\right] + \pi K_1/\omega_1,
\end{aligned}
$$

$$
\begin{aligned}
\overline{\overline{m}}_2(I_1, I_2, \psi, H_3) = &-\left[\alpha_{20}I_2 + 3\alpha_{22}\omega_2 I_2^2/2 + \alpha_{21}\omega_1 I_1 I_2\right.\\
&\left.\times(1 + (1/2)\cos 2\psi) + (\alpha_{23} + \alpha_{24})I_2 H_3\right] + \pi K_2/\omega_2,
\end{aligned}
$$

$$
\begin{aligned}
\overline{\overline{m}}_3(I_1, I_2, \psi, H_3) =&[(\alpha_{12}\omega_2 I_2 + \alpha_{21}\omega_1 I_1)/4]\sin 2\psi,\\
\overline{\overline{m}}_4(I_1, I_2, \psi, H_3) = &-[(\alpha_{30} + \alpha_{40}) + (\alpha_{31} + \alpha_{41})\omega_1 I_1 + (\alpha_{32} + \alpha_{42})\omega_2 I_2\\
&+(3\alpha_{33} + 3\alpha_{44} + \alpha_{34} + \alpha_{43})H_3/2]S(H_3) + \pi K_3 + \pi K_4,
\end{aligned}
$$

$$
\begin{aligned}
&\overline{\overline{\sigma}}^2_{11}(I_1, I_2, \psi, H_3) = 2\pi K_1 I_1/\omega_1, \quad \overline{\overline{\sigma}}^2_{22}(I_1, I_2, \psi, H_3) = 2\pi K_2 I_2/\omega_2,\\
&\overline{\overline{\sigma}}^2_{33}(I_1, I_2, \psi, H_3) = \pi K_1/(2\omega_1 I_1) + \pi K_2/(2\omega_2 I_2),\\
&\overline{\overline{\sigma}}^2_{44}(I_1, I_2, \psi, H_3) = 2\pi(K_3 + K_4)S(H_3),
\end{aligned}
\tag{5.195}
$$

$$
\begin{aligned}
&\overline{\overline{\sigma}}_{12} = \overline{\overline{\sigma}}_{13} = \overline{\overline{\sigma}}_{14} = \overline{\overline{\sigma}}_{21} = \overline{\overline{\sigma}}_{23} = \overline{\overline{\sigma}}_{24},\\
&= \overline{\overline{\sigma}}_{31} = \overline{\overline{\sigma}}_{32} = \overline{\overline{\sigma}}_{34} = \overline{\overline{\sigma}}_{41} = \overline{\overline{\sigma}}_{42} = \overline{\overline{\sigma}}_{43} = 0.
\end{aligned}
$$

The averaged FPK equation associated with Itô Eq. (5.194) is of the form of Eq. (5.174) and the averaged first and second derivative moments are derived from Eq. (5.195) using the relationships in Eq. (5.182). When the compatibility conditions (5.191) are satisfied, the exact stationary solution of FPK equation can be obtained as follows

$$
\begin{aligned}
p(I_1, I_2, \psi, h_3) = &C\left[(1 + 4bh_3)^{1/2} - 1\right]\exp\{-[\alpha_{10}\omega_1 I_1/(\pi K_1)\\
&+ \alpha_{20}\omega_2 I_2/(\pi K_2) + (\alpha_{30} + \alpha_{40})h_3/(\pi K_3 + \pi K_4) + 3\alpha_{11}\omega_1^2 I_1^2/(4\pi K_1)\\
&+ 3\alpha_{22}\omega_2^2 I_2^2/(4\pi K_2) + (3\alpha_{33} + 3\alpha_{44} + \alpha_{34} + \alpha_{43})h_3^2/(4\pi K_3 + 4\pi K_4)\\
&+ (\alpha_{12}\omega_1\omega_2 I_1 I_2/(\pi K_1))(1 + (1/2)\cos 2\psi) + (\alpha_{13} + \alpha_{14})\omega_1 I_1 h_3/(\pi K_1)\\
&+ (\alpha_{23} + \alpha_{24})\omega_2 I_2 h_3/(\pi K_2)]\}.
\end{aligned}
\tag{5.196}
$$

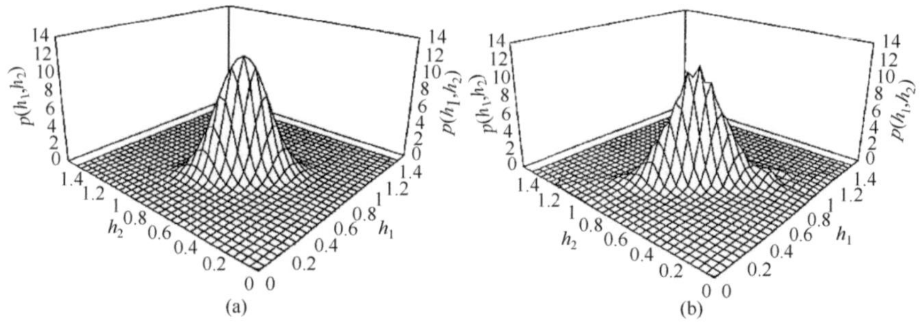

Fig. 5.10 Stationary PDF $p(h_1, h_2)$ of first integrals H_1, H_2 of system (5.183) in resonant case. **a** Result from stochastic averaging method; **b** result from Monte Carlo simulation. System parameters are $\omega_1 = \omega_2 = 1.0$, $\omega_3 = 1.11$, $\omega_4 = 1.732$, $\alpha_{10} = \alpha_{20} = \alpha_{30} = \alpha_{40} = -0.08$, $\alpha_{11} = \alpha_{22} = \alpha_{33} = \alpha_{44} = \alpha_{34} = \alpha_{43} = 0.04$, $\alpha_{12} = \alpha_{21} = 0.08$, $\alpha_{31} = \alpha_{41} = \alpha_{23} = \alpha_{42} = 0.02$, $\alpha_{13} = \alpha_{14} = \alpha_{23} = \alpha_{24} = 0.01$, $\pi K_1 = \pi K_2 = \pi K_3 = \pi K_4 = 0.05$ (Zhu et al. 2002)

Applying Eq. (5.178) and considering $\left|\partial(I_1, I_2, \psi, \theta_1)/\partial(q_1, q_2, p_1, p_2)\right| = 1$, the approximate stationary PDF for the displacements and momenta of the original system (5.183) is

$$p(\boldsymbol{q}, \boldsymbol{p}) = \left.\frac{2\pi p(I_1, I_2, \psi, h_3)}{(1 - 4bh_3)^{1/2} - 1}\right|_{I_\eta = I_\eta(q_\eta, p_\eta), h_3 = H_3(q_3, q_4, p_3, p_4), \psi = \psi(q_1, q_2, p_1, p_2)} . \tag{5.197}$$

For both the non-resonant case and resonant case, the numerical results obtained using the stochastic averaging method are in good agreement with those from Monte Carlo simulation, see Fig. 5.10.

5.4 Stationary Response of 2-DOF Vibration-Impact System

As indicated in Sects. 5.1 and 5.2 that the stochastic averaging method of quasi-non-integrable Hamiltonian systems and that of quasi-integrable Hamiltonian systems are different. The non-integrability of a quasi-Hamiltonian system may change from weak to strong as the system parameters change. The problem is how to determine which stochastic averaging method should be applied. In this section, this problem is illustrated through an example (Liu et al. 2002; Huang et al. 2004). Consider a 2-DOF vibration-impact system excited by Gaussian white noises shown in Fig. 5.11., where m_1, m_2 are the masses of two objects; k_1, k_2 are the linear spring stiffnesses; c_1, c_2 are the linear damping coefficients. Mass m_2 has limiting walls at its both sides with distances δ_l, δ_r from the static equilibrium position. The walls are assumed to be elastic bodies. The interaction between walls and m_2 abided by the Hertz contact

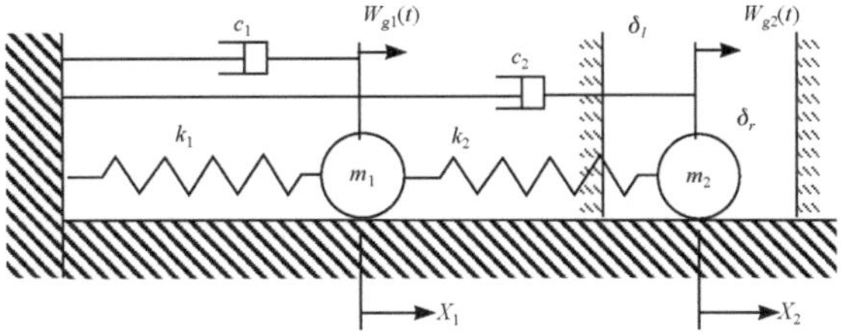

Fig. 5.11 Schematic diagram of 2-DOF vibration-impact system

law. $W_{g1}(t)$ and $W_{g2}(t)$ are independent Gaussian white noises with intensities $2\pi K_1$ and $2\pi K_2$, respectively. c_1, c_2, K_1, K_2 are assumed to be small parameters of ε order.

The motion equations of the system are

$$\begin{aligned} m_1\ddot{X}_1 + c_1\dot{X}_1 + k_1X_1 + k_2(X_1 - X_2) &= W_{g1}(t), \\ m_2\ddot{X}_2 + c_2\dot{X}_2 + k_2(X_2 - X_1) + F(X_2) &= W_{g2}(t), \end{aligned} \tag{5.198}$$

where $F(X_2)$ is the impact force acting on mass m_2. It is determined by the Hertz contact law as follows

$$F(X_2) = \begin{cases} B_r(X_2 - \delta_r)^{3/2} & X_2 \geq \delta_r \\ 0 & -\delta_l < X_2 < \delta_r \\ -B_l(-X_2 - \delta_l)^{3/2} & X_2 \leq -\delta_l \end{cases} \tag{5.199}$$

where B_r, B_l are constants which are related to mass m_2, material and size of the walls. Let

$$Q_i = X_i, \quad P_i = m_i\dot{X}_i, \quad i = 1, 2. \tag{5.200}$$

System (5.198) can be rebuilt to a quasi-Hamiltonian system

$$\dot{Q}_i = \frac{\partial H}{\partial P_i}, \quad \dot{P}_i = -\frac{\partial H}{\partial Q_i} - \frac{c_i}{m_i}\frac{\partial H}{\partial P_i} + W_{gi}(t), \quad i = 1, 2, \tag{5.201}$$

where the Hamiltonian function is

$$\begin{aligned} H &= H(Q_1, Q_2, P_1, P_2) \\ &= \frac{P_1^2}{2m_1} + \frac{P_2^2}{2m_2} + \frac{1}{2}k_1Q_1^2 + \frac{1}{2}k_2(Q_2 - Q_1)^2 + G(Q_2) \\ &= \frac{P_1^2}{2m_1} + \frac{P_2^2}{2m_2} + U(Q_1, Q_2), \end{aligned} \tag{5.202}$$

in which

$$G(Q_2) = \begin{cases} \frac{2}{5}B_r(Q_2 - \delta_r)^{5/2} & (Q_2 > \delta_r) \\ 0 & (-\delta_l \le Q_2 \le \delta_r) \\ \frac{2}{5}B_l(-Q_2 - \delta_l)^{5/2} & (Q_2 < -\delta_l) \end{cases}. \tag{5.203}$$

Equation (5.201) can be transformed into the following Itô stochastic differential equations

$$\begin{aligned} dQ_1 &= \frac{\partial H}{\partial P_1}dt \\ dP_1 &= -\left(\frac{\partial H}{\partial Q_1} + \frac{c_1}{m_1}\frac{\partial H}{\partial P_1}\right)dt + \sqrt{2\pi K_1}dB_1(t) \\ dQ_2 &= \frac{\partial H}{\partial P_2}dt \\ dP_2 &= -\left(\frac{\partial H}{\partial Q_2} + \frac{c_2}{m_2}\frac{\partial H}{\partial P_2}\right)dt + \sqrt{2\pi K_2}dB_2(t) \end{aligned} \tag{5.204}$$

where $B_1(t)$ and $B_2(t)$ are independent unit Wiener processes. Under certain condition, exact stationary solution exists for quasi-Hamiltonian system (5.204). Otherwise, approximate stationary solutions can be obtained by using the stochastic averaging methods.

5.4.1 Exact Stationary Solution

When the system parameters satisfy condition $\frac{c_1}{\pi K_1} = \frac{c_2}{\pi K_2} = \beta = const,$, the FPK equation associated with Eq. (5.204) has exact stationary solution. The stationary joint PDF of system state q_1, q_2, p_1, p_2, the marginal stationary PDFs of q_2, p_2 and q_2 are

$$p(q_1, q_2, p_1, p_2) = C\exp[-\beta h]|_{h=H(q_1,q_2,p_1,p_2)} \tag{5.205}$$

$$\begin{aligned} p(q_2, p_2) &= \int_{-\infty}^{\infty}\int_{-\infty}^{\infty} p(q_1, q_2, p_1, p_2)dq_1dp_1 \\ &= C_1\exp\left\{-\beta\left[\frac{1}{2m_2}p_2^2 + \frac{k_1k_2}{2(k_1+k_2)}q_2^2 + G\right]\right\}, \end{aligned} \tag{5.206}$$

$$p(q_2) = \int_{-\infty}^{\infty} p(q_2, p_2)dp_2 = C_2\exp\left\{-\beta\left[\frac{k_1k_2}{2(k_1+k_2)}q_2^2 + G\right]\right\} \tag{5.207}$$

where C, C_1, C_2 are normalization constants; $H(q_1, q_2, p_1, p_2)$ is determined by Eq. (5.202); $G = G(q_2)$ is determined by Eq. (5.203).

Numerical calculations were carried for two cases of double-wall and right-wall only under different combinations of system parameters B, δ, β. Figures 5.12–5.17 depict the stationary PDF of displacement Q_2 for system with parameters $m_1 = m_2 = k_1 = k_2 = 1$. It is seen from these figures that the influence of impact between mass m_2 and elastic walls on the stationary response increases, and the PDF deviated more from Gaussian distribution when (i) the stiffness B_r and B_l (or B) of elastic walls increases, (ii) the distance δ_r and δ_l (or δ) between the elastic walls and mass m_2 decreases, and (iii) when the ratio of excitation intensity to damping increases. On the contrary, as the influence of impact on system stationary response decreases, the PDF is closer to Gaussian distribution as shown in Fig. 5.13 for smaller δ values.

5.4.2 Application of Stochastic Averaging Method of Quasi-Non-Integrable Hamiltonian Systems

When the condition $\frac{c_1}{\pi K_1} = \frac{c_2}{\pi K_2} = \beta$ is not satisfied, the exact stationary solution can not be obtained. Since $U(Q_1, Q_2)$ in Eq. (5.202) is inseparable, the stochastic averaging method of quasi-non-integrable Hamiltonian systems may be applied to study the stationary response of system (5.204) (Fig. 5.14).

By applying the stochastic averaging method of quasi-non-integrable Hamiltonian systems described in Sect. 5.1, the following averaged Itô stochastic differential equation for H can be obtained

$$dH = m(H)dt + \sigma(H)dB(t). \tag{5.208}$$

According to Eqs. (5.10) to (5.13) and completing the integral over p_2, the drift and diffusion coefficients of averaged Itô equation (5.208) are obtained as follows

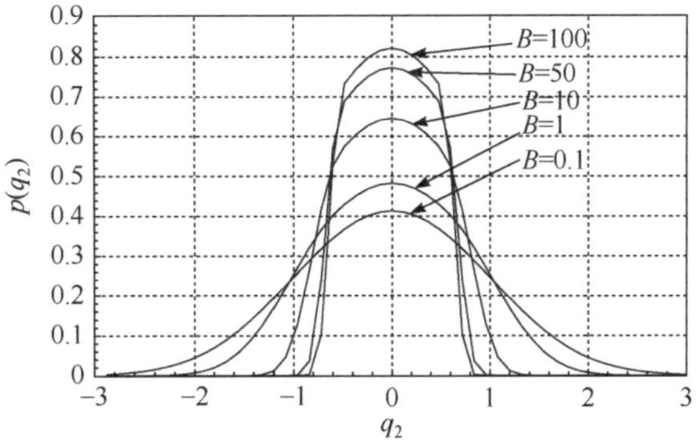

Fig. 5.12 Stationary PDF of displacement Q_2 of double-wall system. Parameters are $\beta = 2.0$, $\delta = \delta_l = \delta_r = 0.5$, $B = B_l = B_r = 0.1,\ 1,\ 10,\ 50,\ 100$ (Huang et al. 2004)

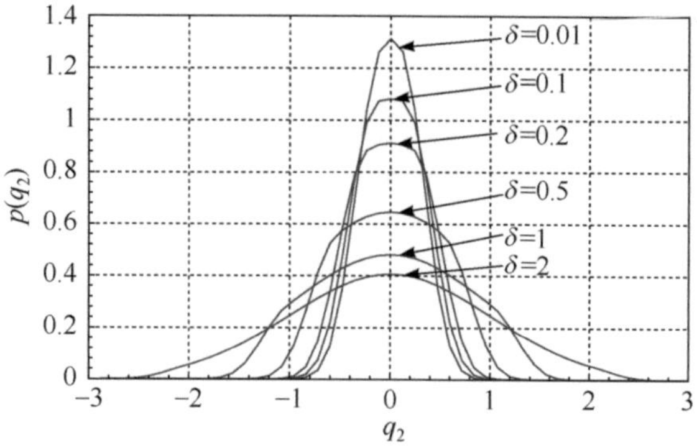

Fig. 5.13 Stationary PDF of displacement Q_2 of double-wall system. Parameters are $\beta = 2.0$, $B = B_l = B_r = 10$, $\delta = \delta_l = \delta_r = 0.01,\ 0.1,\ 0.2,\ 0.5,\ 1, 2$ (Huang et al. 2004)

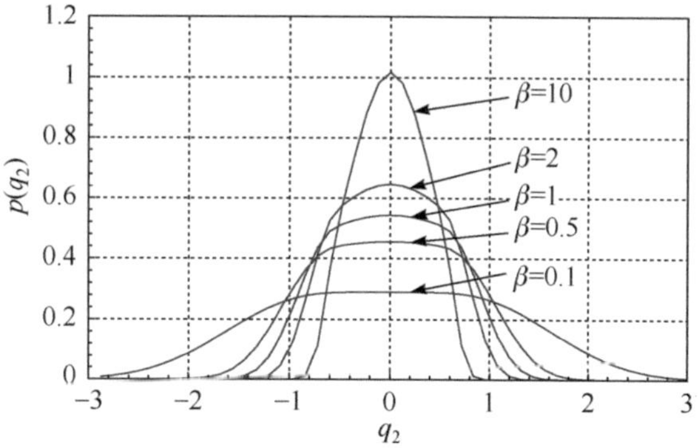

Fig. 5.14 Stationary PDF of displacement Q_2 of double-wall system. Parameters are $B = B_l = B_r = 10$, $\delta = \delta_l = \delta_r = 0.5$, $\beta = 0.1,\ 0.5,\ 1,\ 2,\ 10$ (Huang et al. 2004)

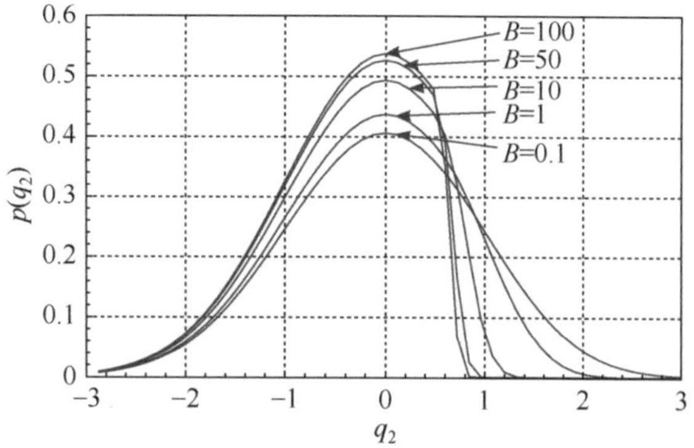

Fig. 5.15 Stationary PDF of displacement Q_2 of right-wall system. Parameters are $\beta = 2.0$, $\delta = \delta_r = 0.5, B = B_r = 0.1,\ 1,\ 10,\ 50,\ 100$ (Huang et al. 2004)

$$m(H) = \frac{1}{T(H)} \int\limits_{\Omega'} \left[-\left(\frac{c_1}{m_1} + \frac{c_2}{m_2} \right)(H - U) + \pi K_1 + \pi K_2 \right] \mathrm{d}q_1 \mathrm{d}q_2,$$

$$\sigma^2(H) = \frac{2(\pi K_1 + \pi K_2)}{T(H)} \int\limits_{\Omega'} (H - U) \mathrm{d}q_1 \mathrm{d}q_2,$$

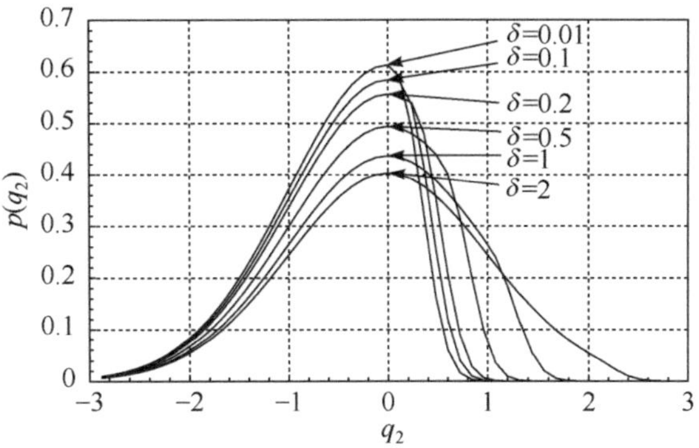

Fig. 5.16 Stationary PDF of displacement Q_2 of right-wall system. Parameters are $\beta = 2.0$, $B = B_r = 10$, $\delta = \delta_r = 0.01,\ 0.1,\ 0.2,\ 0.5,\ 1,\ 2$ (Huang et al. 2004)

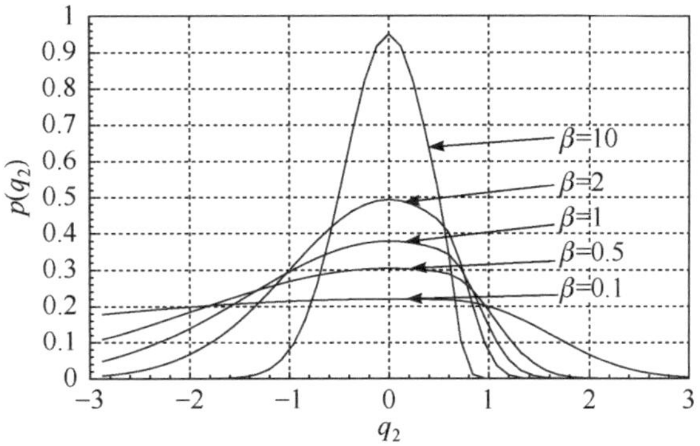

Fig. 5.17 Stationary PDF of displacement Q_2 of right-wall system. Parameters are $B = B_r = 10$, $\delta = \delta_r = 0.5$, $h = 0.1,\ 0.5,\ 1,\ 2,\ 10$ (Huang et al. 2004)

$$T(H) = \int\limits_{\Omega'} \mathrm{d}q_1 \mathrm{d}q_2, \quad \Omega' = \{(q_1, q_2) | H(q_1, q_2, 0, 0) \le H\}, \tag{5.209}$$

where the integral domain Ω' is shown in Fig. 5.18, in which Q_2^+, Q_2^- are the two real roots of equation $\frac{k_1 k_2}{2(k_1+k_2)} q_2^2 + G(q_2) - H = 0$, and

$$Q_1^{\pm} = (k_2 q_2 \pm Z)/(k_1 + k_2). \tag{5.210}$$

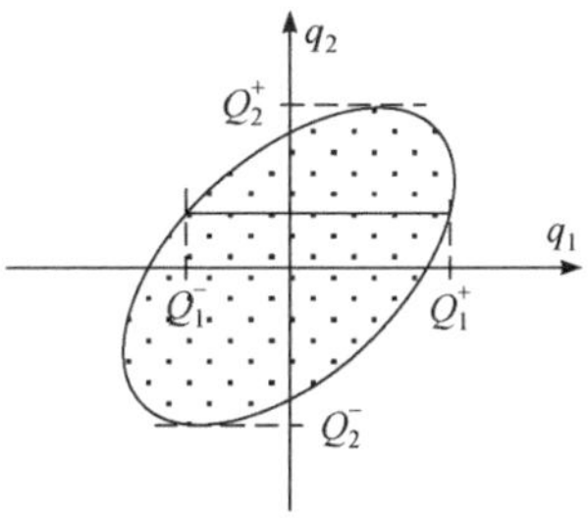

Fig. 5.18 Integral domain Ω' in Eq. (5.209)

Equation (5.209) can be further simplified to

$$
\begin{aligned}
m(H) &= -\frac{\frac{c_1}{m_1} + \frac{c_2}{m_2}}{3(k_1 + k_2)} \frac{I_2(H)}{I_1(H)} + \pi K_1 + \pi K_2, \\
\sigma^2(H) &= \frac{2\pi(K_1 + K_2)}{3(k_1 + k_2)} \frac{I_2(H)}{I_1(H)}, \\
I_1(H) &= \int_{Q_2^-}^{Q_2^+} Z(q_2, H)\mathrm{d}q_2, \quad I_2(H) = \int_{Q_2^-}^{Q_2^+} Z^3(q_2, H)\mathrm{d}q_2, \\
Z &= Z(q_2, H) = \sqrt{k_2^2 q_2^2 - 2(k_1 + k_2)\left[\frac{1}{2}k_2 q_2^2 + G(q_2) - H\right]}.
\end{aligned}
\tag{5.211}
$$

The stationary solution of reduced averaged FPK equation associated with averaged Itô equation (5.208) is

$$
p(h) = \frac{C}{\sigma^2(h)} \exp\left[\int \frac{2m(h)}{\sigma^2(h)} \mathrm{d}h\right]. \tag{5.212}
$$

where C is a normalization constant. According to Eq. (5.18), the following approximately stationary PDF functions can be obtained from Eq. (5.212)

$$
p(q_1, q_2, p_1, p_2) = \frac{(k_1 + k_2)p(h)}{2\pi I_1(h)}\Big|_{h=H(q_1,q_2,p_1,p_2)} \tag{5.213}
$$

$$
p(q_2, p_2) = \int_{-\infty}^{+\infty}\int_{-\infty}^{+\infty} p(q_1, q_2, p_1, p_2)\mathrm{d}q_1\mathrm{d}p_1 \tag{5.214}
$$

$$
p(q_2) = \int_{-\infty}^{+\infty} p(q_2, p_2)\mathrm{d}p_2 \tag{5.215}
$$

Figures 5.18–5.26 show the stationary PDFs of system total energy H and displacement Q_2 of mass m_2 for the cases of double-wall and right-wall. System parameters are $m_1 = m_2 = k_1 = k_2 = 1$, $c_1 = 0.08$, $c_2 = 0.04$. Other parameters are shown in figures. Numerical results were calculated using both the stochastic averaging method and Monte Carlos simulations for comparison. In the cases of large stiffness of elastic wall, strong excitation intensity, and/or short distance between mass m_2 and elastic wall, the effect of impact become strong, and the stochastic averaging method of quasi-non-integrable Hamiltonian systems can yield good results. On the contrary, relatively large errors of the stochastic averaging method appear when the effect of impact become weak (Fig. 5.15).

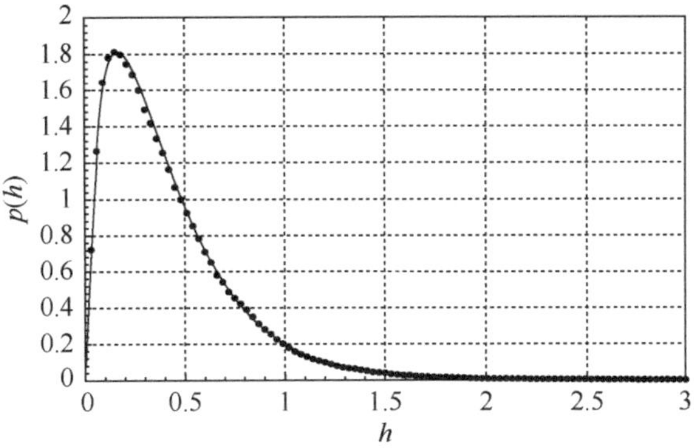

Fig. 5.19 Stationary PDF of Hamiltonian of double-wall system. ——, result from Eq. (5.212); ●, result from Monte Carlo simulation. System parameters are $\pi K_1 = 0.01$, $\pi K_2 = 0.02$, $\delta_l = \delta_r = 0.5$, $B_l = B_r = 100$ (Huang et al. 2004)

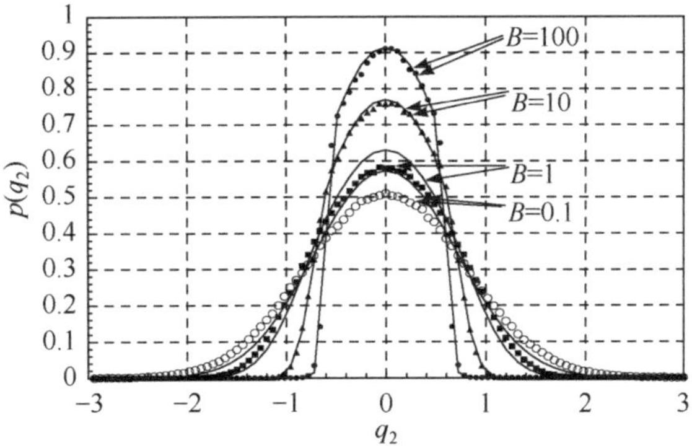

Fig. 5.20 Stationary PDF of displacement Q_2 of double-wall system. ——, results from Eq. (5.215); ●▲○■, results from Monte Carlo simulation. System parameters are $\pi K_1 = 0.01$, $\pi K_2 = 0.02$, $\delta_l = \delta_r = 0.5$, $B_l = B_r = 0.1,\ 1,\ 10,\ 100$ (Huang et al. 2004)

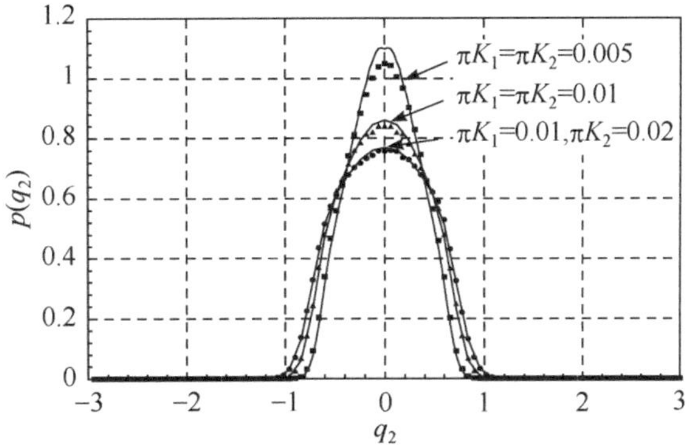

Fig. 5.21 Stationary PDF of displacement Q_2 of double-wall system. ——, results from Eq. (5.215); ●▲■, results from Monte Carlo simulation. System parameters are $B_l = B_r = 10$, $\delta_l = \delta_r = 0.5$, $\pi K_1 = 0.01$, $\pi K_2 = 0.02$, $\pi K_1 = \pi K_2 = 0.01$, $\pi K_1 = \pi K_2 = 0.005$ (Huang et al. 2004)

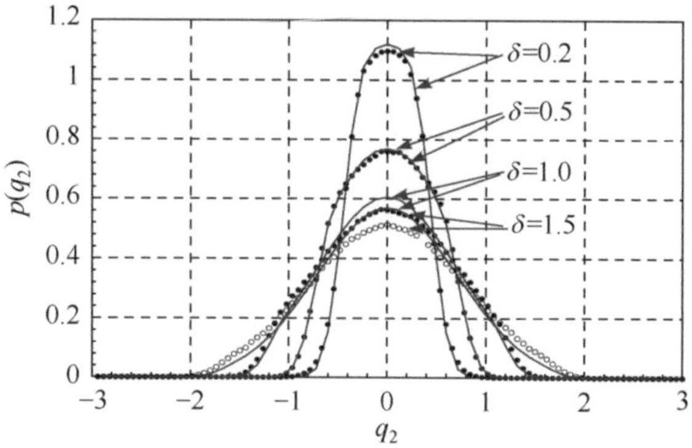

Fig. 5.22 Stationary PDF of displacement Q_2 of double-wall system. ——, results from Eq. (5.215); ●▲○■, results from Monte Carlo simulation. System parameters are $\pi K_1 = 0.01$, $\pi K_2 = 0.02$, $B_l = B_r = 10$, $\delta_l = \delta_r = 0.2,\ 0.5,\ 1.0,\ 1.5$ (Huang et al. 2004)

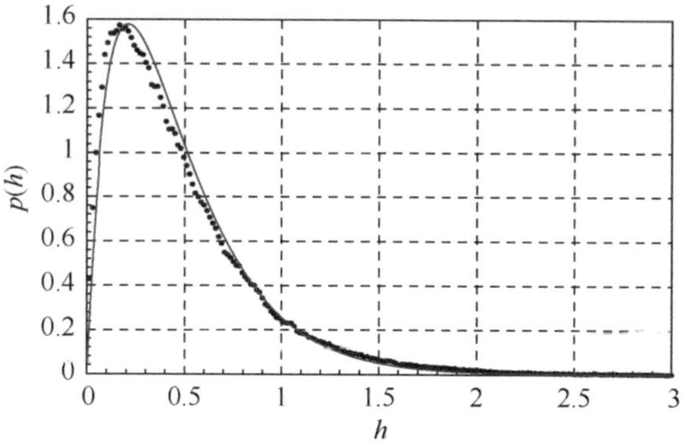

Fig. 5.23 Stationary PDF of Hamiltonian of right-wall system. ——, result from Eq. (5.212); ●, result from Monte Carlo simulation. System parameters are $\pi K_1 = 0.01$, $\pi K_2 = 0.02$, $\delta_r = 0.5$, $B_r = 100$ (Liu et al. 2002)

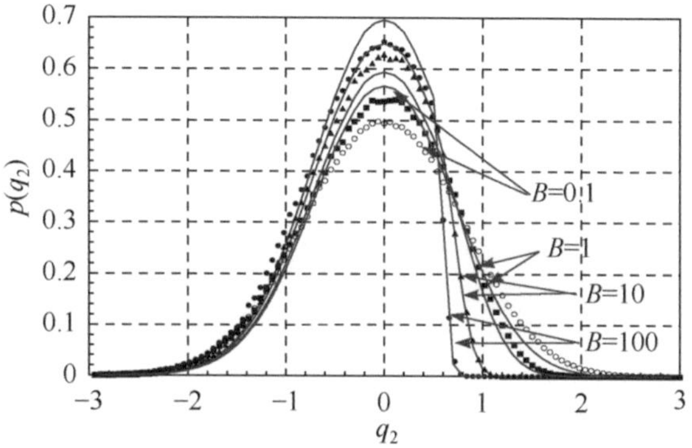

Fig. 5.24 Stationary PDF of displacement Q_2 of right-wall system. ——, results from Eq. (5.215); ●▲○■, results from Monte Carlo simulation. System parameters are $\pi K_1 = 0.01$, $\pi K_2 = 0.02$, $\delta_r = 0.5$, $B_r = 0.1,\ 1,\ 10,\ 100$ (Liu et al. 2002)

5.4.3 *Application of Stochastic Averaging Method of Quasi-Integrable Hamiltonian Systems*

Assume that the effect of impact is small and can be neglected. In this case, system (5.198) can be simplified as linear one governed by the following equations

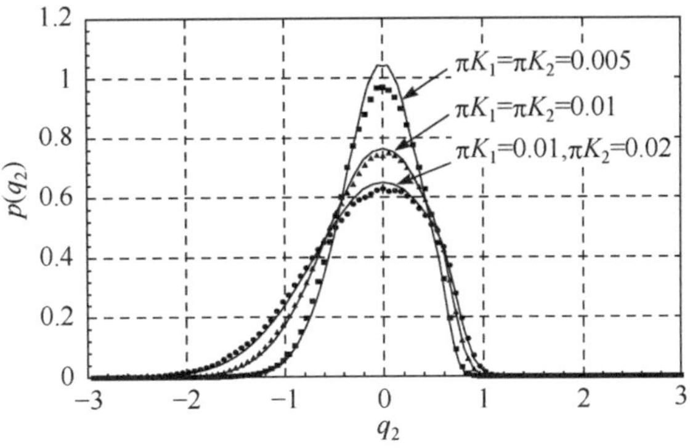

Fig. 5.25 Stationary PDF of displacement Q_2 of right-wall system. ——, results from Eq. (5.215); ●▲■, results from Monte Carlo simulation. System parameters are $B_r = 10$, $\delta_r = 0.5$, $\pi K_1 = 0.01$, $\pi K_2 = 0.02$, $\pi K_1 = \pi K_2 = 0.01$, $\pi K_1 = \pi K_2 = 0.005$ (Liu et al. 2002)

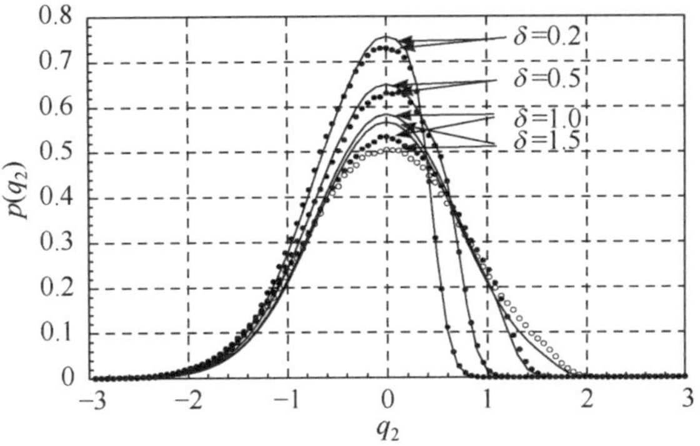

Fig. 5.26 Stationary PDF of displacement Q_2 of right-wall system. ——, results from Eq. (5.215); ●▲○■, results from Monte Carlo simulation. System parameters are $\pi K_1 = 0.01$, $\pi K_2 = 0.02$, $B_r = 10$, $\delta_r = 0.2,\ 0.5,\ 1.0,\ 1.5$ (Liu et al. 2002)

$$\begin{aligned} m_1\ddot{X}_1 + c_1\dot{X}_1 + k_1X_1 + k_2(X_1 - X_2) &= W_{g1}(t), \\ m_2\ddot{X}_2 + c_2\dot{X}_2 + k_2(X_2 - X_1) &= W_{g2}(t). \end{aligned} \tag{5.216}$$

Equation (5.216) can be rewritten as the following vector equation

$$\boldsymbol{M}\ddot{\boldsymbol{X}} + \boldsymbol{C}\dot{\boldsymbol{X}} + \boldsymbol{K}\boldsymbol{X} = \boldsymbol{W}_g(t), \tag{5.217}$$

where

$$\begin{aligned} \boldsymbol{M} &= \begin{bmatrix} m_1 & 0 \\ 0 & m_2 \end{bmatrix}, \boldsymbol{K} = \begin{bmatrix} k_1 + k_2, & -k_2 \\ -k_2, & k_2 \end{bmatrix}, \\ \boldsymbol{C} &= \begin{bmatrix} C_1, & 0 \\ 0 & C_2 \end{bmatrix}, \boldsymbol{W}_g = \begin{bmatrix} W_{g1}(t) \\ W_{g2}(t) \end{bmatrix}, \boldsymbol{X} = \begin{bmatrix} X_1 \\ X_2 \end{bmatrix}. \end{aligned} \tag{5.218}$$

Introduce a transformation

$$\boldsymbol{X} = \boldsymbol{T}\boldsymbol{Q}, \tag{5.219}$$

where

$$
\begin{aligned}
&\boldsymbol{T}=\begin{bmatrix} T_{11}, T_{12} \\ T_{21}, T_{22} \end{bmatrix}, \\
&T_{11}=-\frac{-k_2m_1+k_1m_2+k_2m_2-\sqrt{-4k_1k_2m_1m_2+(-k_2m_1-k_1m_2-k_2m_2)^2}}{2k_2m_1}, \\
&T_{12}=-\frac{-k_2m_1+k_1m_2+k_2m_2+\sqrt{-4k_1k_2m_1m_2+(-k_2m_1-k_1m_2-k_2m_2)^2}}{2k_2m_1}, \\
&T_{21}=T_{22}=1.
\end{aligned}
\tag{5.220}
$$

Substituting transformation (5.219) into Eq. (5.217) and multiplying Eq. (5.217) with $\boldsymbol{T}^{\mathrm{T}}$ from left yield

$$
\boldsymbol{T}^{\mathrm{T}}\boldsymbol{M}\boldsymbol{T}\dot{\boldsymbol{Q}}+\boldsymbol{T}^{\mathrm{T}}\boldsymbol{K}\boldsymbol{T}\boldsymbol{Q}=-\boldsymbol{T}^{\mathrm{T}}\boldsymbol{C}\boldsymbol{T}\dot{\boldsymbol{Q}}+\boldsymbol{T}^{\mathrm{T}}\boldsymbol{W}_g(t), \tag{5.221}
$$

where

$$
\begin{aligned}
&\boldsymbol{T}^T\boldsymbol{M}\boldsymbol{T}=\boldsymbol{M}^*=\begin{bmatrix} m_1^*, 0 \\ 0, m_2^* \end{bmatrix}, \boldsymbol{T}^T\boldsymbol{K}\boldsymbol{T}=\boldsymbol{K}^*=\begin{bmatrix} k_1^*, 0 \\ 0, k_2^* \end{bmatrix}, \\
&\boldsymbol{T}^T\boldsymbol{C}\boldsymbol{T}=\boldsymbol{C}^*=\begin{bmatrix} c_{11}^*, c_{12}^* \\ c_{21}^*, c_{22}^* \end{bmatrix}.
\end{aligned}
\tag{5.222}
$$

The scalar equations of Eq. (5.221) are

$$
\begin{aligned}
m_1^*\ddot{Q}_1+k_1^*Q_1&=-[c_{11}^*\dot{Q}_1+c_{12}^*\dot{Q}_2]+T_{11}W_{g1}(t)+T_{21}W_{g2}(t), \\
m_2^*\ddot{Q}_2+k_2^*Q_2&=-[c_{21}^*\dot{Q}_1+c_{22}^*\dot{Q}_2]+T_{12}W_{g1}(t)+T_{22}W_{g2}(t).
\end{aligned}
\tag{5.223}
$$

Let

$$
P_i=m_i^*\dot{Q}_i, \quad i=1,2, \tag{5.224}
$$

System (5.223) can be rewritten as the following quasi-Hamiltonian system

$$
\begin{aligned}
\dot{Q}_i=\frac{\partial H}{\partial P_i}, \quad \dot{P}_i&=-\frac{\partial H}{\partial Q_i}-\sum_{j=1}^{2}c_{ij}^*\frac{\partial H}{\partial P_j} \\
&+\sum_{k=1}^{2}T_{ki}W_{gk}(t), \quad i=1,2,
\end{aligned}
\tag{5.225}
$$

where

$$H = H_1 + H_2,$$
$$H_1 = \frac{P_1^2}{2m_1^*} + \frac{1}{2}k_1^* Q_1^2, \quad H_2 = \frac{P_2^2}{2m_2^*} + \frac{1}{2}k_2^* Q_2^2. \tag{5.226}$$

Transform system (5.225) into Itô stochastic differential equations. Applying the Itô differential rule to Eq. (5.226), the following Itô stochastic differential equations for H_1, H_2 can be obtained from Itô equations of Q_1, Q_2, P_1, P_2

$$\begin{aligned}
\mathrm{d}H_1 &= \left(-c_{11}^* \frac{P_1 P_1}{m_1^* m_1^*} - c_{12}^* \frac{P_1 P_2}{m_1^* m_2^*} + \frac{\pi K_1 T_{11}^2}{m_1^*} + \frac{\pi K_2 T_{21}^2}{m_1^*}\right)\mathrm{d}t \\
&\quad + \frac{T_{11} P_1}{m_1^*}\mathrm{d}B_1(t) + \frac{T_{21} P_1}{m_1^*}\mathrm{d}B_2(t), \\
\mathrm{d}H_2 &= \left(-c_{21}^* \frac{P_1 P_2}{m_1^* m_2^*} - c_{22}^* \frac{P_2 P_2}{m_2^* m_2^*} + \pi K_1 T_{12}^2 \frac{1}{m_2^*} + \frac{\pi K_2 T_{22}^2}{m_2^*}\right)\mathrm{d}t \\
&\quad + \frac{T_{12} P_2}{m_2^*}\mathrm{d}B_1(t) + \frac{T_{22} P_2}{m_2^*}\mathrm{d}B_2(t).
\end{aligned} \tag{5.227}$$

Assume that the damping coefficients c_{jk} $(j, k = 1, 2)$ and excitation intensities $2\pi K_l$ $(l = 1, 2)$ are small quantities of the same order. Applying the stochastic averaging method of quasi-integrable and non-resonant Hamiltonian systems described in Sect. 5.2.1 yields the following averaged Itô stochastic differential equations

$$\begin{aligned}
\mathrm{d}H_1 &= a_1 \mathrm{d}t + \sigma_{11}\mathrm{d}B_1(t) + \sigma_{12}\mathrm{d}B_2(t), \\
\mathrm{d}H_2 &= a_2 \mathrm{d}t + \sigma_{21}\mathrm{d}B_1(t) + \sigma_{22}\mathrm{d}B_2(t),
\end{aligned} \tag{5.228}$$

where

$$\begin{aligned}
a_1 &= -\frac{c_{11}^*}{m_1^*} H_1 + (\pi K_1 T_{11}^2 + \pi K_2 T_{21}^2)\frac{1}{m_1^*}, \\
a_2 &= -\frac{c_{22}^*}{m_2^*} H_2 + (\pi K_1 T_{12}^2 + \pi K_2 T_{22}^2)\frac{1}{m_2^*},
\end{aligned}$$

$$\begin{aligned}
b_{11} &= 2\left[\sigma\sigma^T\right]_{11} = 2(\pi K_1 T_{11}^2 + \pi K_2 T_{21}^2)\frac{1}{m_1^*} H_1, \\
b_{12} &= 2\left[\sigma\sigma^T\right]_{12} = 0, \\
b_{21} &= 2\left[\sigma\sigma^T\right]_{21} = 0, \\
b_{22} &= 2\left[\sigma\sigma^T\right]_{22} = 2(\pi K_1 T_{12}^2 + \pi K_2 T_{22}^2)\frac{1}{m_2^*} H_2.
\end{aligned} \tag{5.229}$$

The reduced averaged FPK equation associated with averaged Itô Eq. (5.228) is

$$-\sum_{r=1}^{2}\frac{\partial}{\partial h_r}(a_r p)+\frac{1}{2}\sum_{r,s=1}^{2}\frac{\partial^2}{\partial h_r\partial h_s}(b_{rs}p)=0, \tag{5.230}$$

where a_r, b_{rs} are obtained from Eq. (5.229) by replacing H_i with h_i. Assume the stationary solution of reduced FPK equation (5.230) is of the form

$$p(h_1,h_2)=C\exp[-\lambda(h_1,h_2)]. \tag{5.231}$$

Substituting Eq. (5.231) into Eq. (5.232) yields the following stationary probability density

$$p(h_1,h_2)=C\exp\left[-\left(\frac{c_{11}^*}{\pi K_{11}T_{11}^2+\pi K_{22}T_{21}^2}h_1+\frac{c_{22}^*}{\pi K_{11}T_{12}^2+\pi K_{22}T_{22}^2}h_2\right)\right], \tag{5.232}$$

where

$$C=\left(\frac{c_{11}^*}{\pi K_{11}T_{11}^2+\pi K_{22}T_{21}^2}\right)\left(\frac{c_{22}^*}{\pi K_{11}T_{12}^2+\pi K_{22}T_{22}^2}\right). \tag{5.233}$$

Using Eq. (5.226) and making the transformation

$$Q=T^{-1}X,\quad P=M^*T^{-1}\dot{X}, \tag{5.234}$$

the following joint stationary PDF of displacements and velocities of the system (5.216) can be obtained from Eq. (5.232) using relationship (5.242)

$$p(x_1,x_2,\dot{x}_1,\dot{x}_2)=Z\exp[-(Z_1h_1+Z_2h_2)], \tag{5.235}$$

where

$$\begin{aligned}
h_1&=\frac{1}{2}k_1^*(T_{11}^{-1}x_1+T_{12}^{-1}x_2)^2+\frac{1}{2}m_1^*(T_{11}^{-1}\dot{x}_1+T_{12}^{-1}\dot{x}_2)^2,\\
h_2&=\frac{1}{2}k_2^*(T_{21}^{-1}x_1+T_{22}^{-1}x_2)^2+\frac{1}{2}m_2^*(T_{21}^{-1}\dot{x}_1+T_{22}^{-1}\dot{x}_2)^2,\\
Z_1&=\frac{c_{11}^*}{\pi K_1T_{11}^2+\pi K_2T_{21}^2},\\
Z_2&=\frac{c_{22}^*}{\pi K_1T_{12}^2+\pi K_2T_{22}^2},
\end{aligned} \tag{5.236}$$

$$\begin{aligned}
Z&=\frac{1}{4\pi^2}Z_1Z_2\sqrt{k_1^*k_2^*m_1^*m_2^*}(T_{12}^{-1}T_{21}^{-1}-T_{11}^{-1}T_{22}^{-1})^2\\
&=\frac{1}{4\pi^2}Z_1Z_2\sqrt{k_1^*k_2^*m_1^*m_2^*}\left|T^{-1}\right|^2.
\end{aligned} \tag{5.237}$$

Finally, the marginal stationary PDF of displacement and velocity of mass m_2 and the marginal stationary PDF of displacement of mass m_2 are obtained as

$$p(x_2, \dot{x}_2) = \int_{-\infty}^{\infty}\int_{-\infty}^{\infty} p(x_1, x_2, \dot{x}_1, \dot{x}_2)\mathrm{d}x_1\mathrm{d}\dot{x}_1, \tag{5.238}$$

$$p(x_2) = \int_{-\infty}^{\infty} p(x_2, \dot{x}_2)\mathrm{d}\dot{x}_2. \tag{5.239}$$

In order to compare the results in Eqs. (5.235), (5.238) and (5.239) with those obtained from quasi-non-integrable Hamiltonian systems, make transformation $q_2 = X_2, p_2 = m_2\dot{X}_2$ and obtain $p(q_2, p_2), p(q_2)$ from Eqs. (5.238) and (5.239) (Fig. 5.16).

For the two cases of double-wall and right-wall, the stationary PDF of displacement Q_2 is obtained using both the stochastic averaging methods of quasi-integrable Hamiltonian systems and that for quasi-non-integrable Hamiltonian systems and shown in Figs. 5.27–5.32, along with simulation results. It is seen that the stochastic averaging method of quasi-integrable Hamiltonian system is more accurate than that for quasi-non-integrable Hamiltonian system in the cases of small elastic wall stiffness, weak excitation intensity, and/or large distance between mass and wall. In these cases, the impact effect is weak, the response of 2-DOF vibration-impact system can be more accurately predicted by applying the stochastic averaging method of quasi-integrable Hamilton systems.

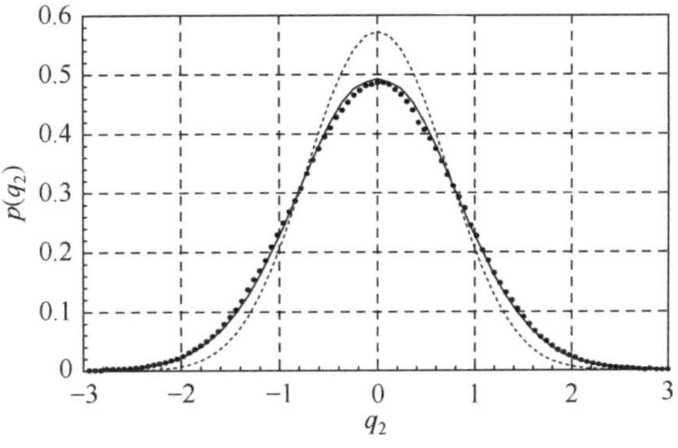

Fig. 5.27 Stationary PDF of displacement Q_2 of double-wall system. ——, result from Eq. (5.239); -----, result from Eq. (5.215); ●, result from Monte Carlo simulation. System parameters are $\pi K_1 = 0.01$, $\pi K_2 = 0.02$, $\delta_l = \delta_r = 0.5$, $B_l = B_r = 0.1$ (Huang et al. 2004)

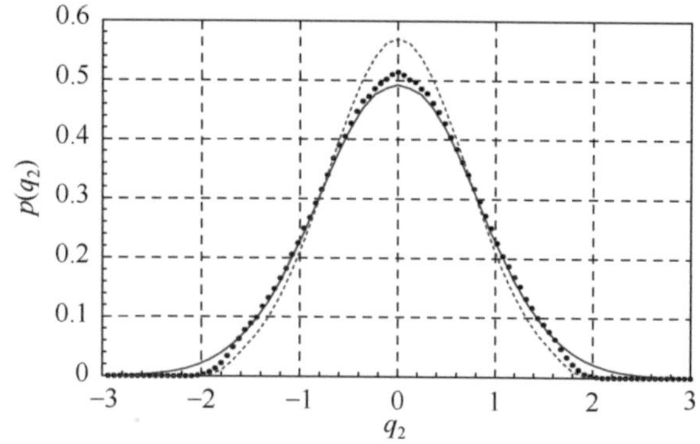

Fig. 5.28 Stationary PDF of displacement Q_2 of double-wall system. ——, result from Eq. (5.239); -----, result from Eq. (5.215); ●, result from Monte Carlo simulation. System parameters are $\pi K_1 = 0.01$, $\pi K_2 = 0.02$, $B_l = B_r = 10$, $\delta_l = \delta_r = 1.5$ (Huang et al. 2004)

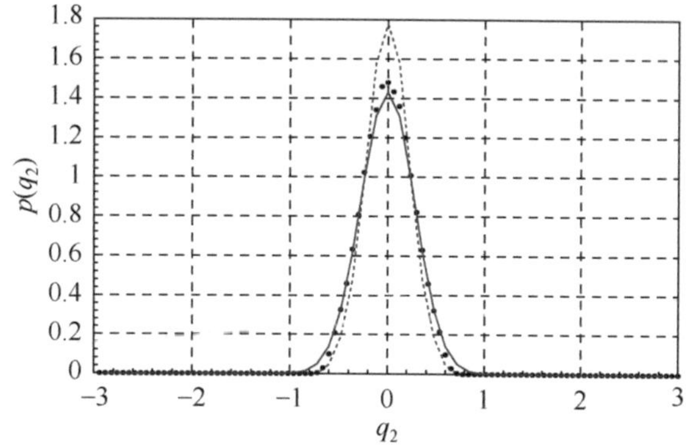

Fig. 5.29 Stationary PDF of displacement Q_2 of double-wall system. ——, result from Eq. (5.239); -----, result from Eq. (5.215); ●, result from Monte Carlo simulation. System parameters are $\delta_l = \delta_r = 0.5$, $B_l = B_r = 10$, $\pi K_1 = \pi K_2 = 0.002$ (Liu et al. 2002)

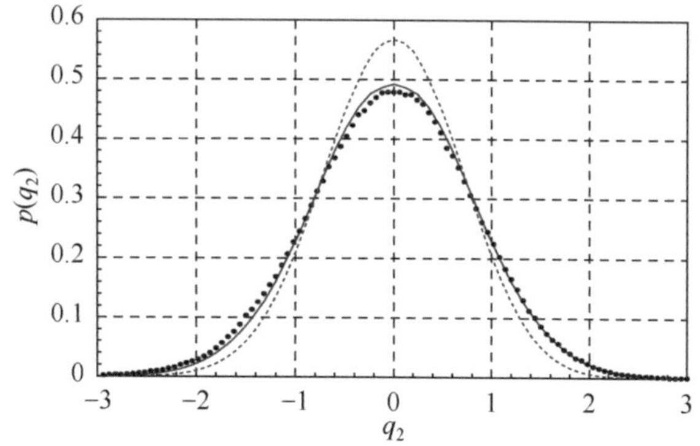

Fig. 5.30 Stationary PDF of displacement Q_2 of right-wall system. ——, result from Eq. (5.239); -----, result from Eq. (5.215); ●, result from Monte Carlo simulation. System parameters are $\pi K_1 = 0.01$, $\pi K_2 = 0.02$, $\delta_r = 0.5$, $B_r = 0.1$ (Liu et al. 2002)

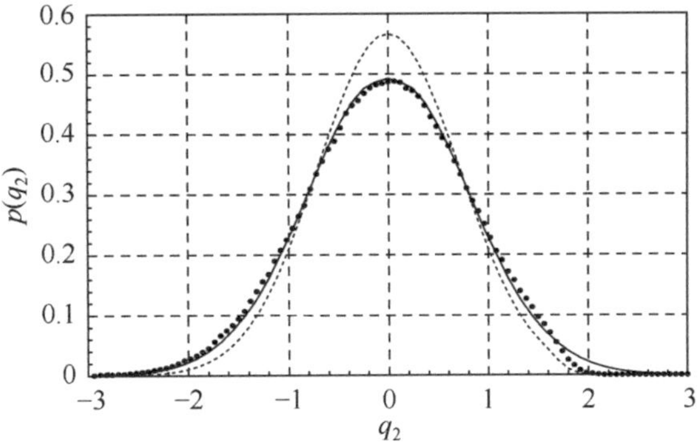

Fig. 5.31 Stationary PDF of displacement Q_2 of right-wall system. ——, result from Eq. (5.239); -----, result from Eq. (5.215); ●, result from Monte Carlo simulation. System parameters are $\pi K_1 = 0.01$, $\pi K_2 = 0.02$, $B_r = 10$, $\delta_r = 1.5$ (Liu et al. 2002)

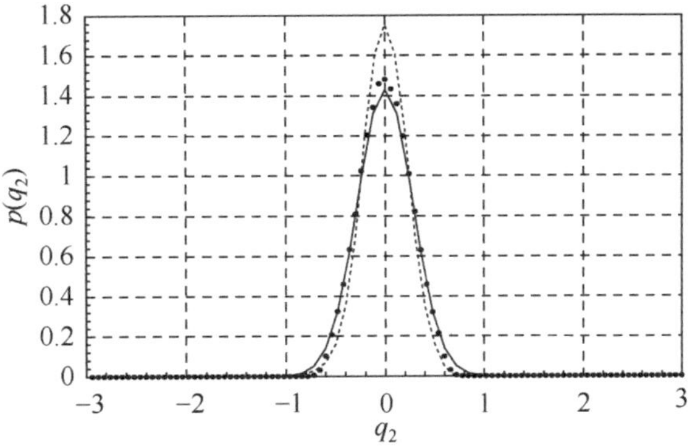

Fig. 5.32 Stationary PDF of displacement Q_2 of right-wall system. ——, result from Eq. (5.239); -----, result from Eq. (5.215); ●, result from Monte Carlo simulation. System parameters are $\delta_r = 0.5$, $B_r = 10$, $\pi K_1 = \pi K_2 = 0.002$ (Liu et al. 2002)

5.4.4 Combined Application of Both Stochastic Averaging Methods of Quasi-Non-Integrable and Quasi-Integrable Hamiltonian Systems

It is known from Sects. 5.4.2 and 5.4.3 that for system (5.204) the stochastic averaging method of quasi-non-integrable Hamiltonian systems yields more accurate when the impact effect is strong, i.e., high excitation intensity, large stiffness of elastic wall, and/or short distance between mass and wall. On the contrary, the stochastic averaging method of quasi-integrable Hamiltonian system yields more accurate results when the impact is weak. When the impact effect is middle, either method may generate larger errors. In the following, a procedure of combined application of the two stochastic averaging methods for middle impact effect is proposed. For the vibration-impact system shown in Fig. 5.11, the total energy H of the system can be used to measure the intensity of the impact effect. Introduce a critical energy H_c so that the system may be regarded as a quasi-integrable Hamiltonian system when $H < H_c$ and as a quasi-non-integrable Hamiltonian system when $H > H_c$. First, the stochastic averaging method of quasi-integrable Hamiltonian systems is applied to obtain a non-normalized stationary PDF $C_{in}p_{in}(\boldsymbol{q}, \boldsymbol{p})$, for the domain of

$H(\boldsymbol{q},\boldsymbol{p}) \le H_c$. Then the stochastic averaging method of quasi-non-integrable Hamiltonian system is applied to obtain a non-normalized stationary PDF $C_{out}p_{out}(\boldsymbol{q},\boldsymbol{p})$ for the domain of $H(\boldsymbol{q},\boldsymbol{p}) > H_c$. At the interface $H(\boldsymbol{q},\boldsymbol{p}) = H_c$, the two stationary PDFs are usually different. In order to minimize the difference, select s typical points $(\boldsymbol{q}_i,\boldsymbol{p}_i)$ $(i = 1, 2, \ldots, s)$ on the interface $H(\boldsymbol{q},\boldsymbol{p}) = H_c$ and apply the least square method. Define a mean square function of the differences of the two stationary PDFs at s points

$$
\begin{aligned}
E &= \sum_{i=1}^{s} k_i[C_{in}p_{in}(\boldsymbol{q}_i,\boldsymbol{p}_i) - C_{out}p_{out}(\boldsymbol{q}_i,\boldsymbol{p}_i)]^2 \\
&= C_{in}\sum_{i=1}^{s} k_i[p_{in}(\boldsymbol{q}_i,\boldsymbol{p}_i) - \varepsilon p_{out}(\boldsymbol{q}_i,\boldsymbol{p}_i)]^2,
\end{aligned}
\tag{5.240}
$$

where k_i is the weight coefficient representing the importance of i th point. To minimize E, let $\partial E/\partial\varepsilon = 0$. This results in

$$
\varepsilon = \frac{\sum\limits_{i=1}^{s} k_i[p_{out}(\boldsymbol{q}_i,\boldsymbol{p}_i)p_{in}(\boldsymbol{q}_i,\boldsymbol{p}_i)]}{\sum\limits_{i=1}^{s}[k_i p_{out}^2(\boldsymbol{q}_i,\boldsymbol{p}_i)]}. \tag{5.241}
$$

Introduce a combined stationary probability density

$$
p(\boldsymbol{q},\boldsymbol{p}) = \begin{cases} Cp_{in}(\boldsymbol{q},\boldsymbol{p}), & H(\boldsymbol{q},\boldsymbol{p}) \le H_c, \\ C\varepsilon p_{out}(\boldsymbol{q},\boldsymbol{p}), & H(\boldsymbol{q},\boldsymbol{p}) > H_c, \end{cases} \tag{5.242}
$$

for system (5.204) in middle impact effect domain. In Eq. (5.242), C is the normalization constant. The marginal stationary PDFs $p(q_2,p_2)$ and $p(q_2)$ are then

$$
p(q_2,p_2) = \int_{-\infty}^{\infty}\int_{-\infty}^{\infty} p(q_1,p_1,q_2,p_2)\mathrm{d}q_1\mathrm{d}p_1, \tag{5.243}
$$

$$
p(q_2) = \int_{-\infty}^{\infty} p(q_2,p_2)\mathrm{d}p_2. \tag{5.244}
$$

For the present vibration-impact system in Fig. 5.11, we are mainly interested in the statistics of displacement Q_2. Select $H_c(q_1,p_1,q_2,p_2) = H_c(0,0,\delta,0)$ and $H_c(0,0,-\delta,0)$, where $\delta = \min(|\delta_l|,|\delta_r|)$ and choosing $q_2 = \pm\delta$ as the primary points and other points on interface as the secondary points for calculations.

The combined PDFs in Eqs. (5.242)–(5.244) are applied to system (5.204) in the two cases of double-wall and right-wall. The stationary PDFs of displacement Q_2 for different parameter values are calculated and shown in Figs. 5.33–5.38 together with results from Monte Carlo simulations. It is seen from Figs. 5.33–5.38 that, in the case of middle impact effect, the combined averaging method yields more accurate results.

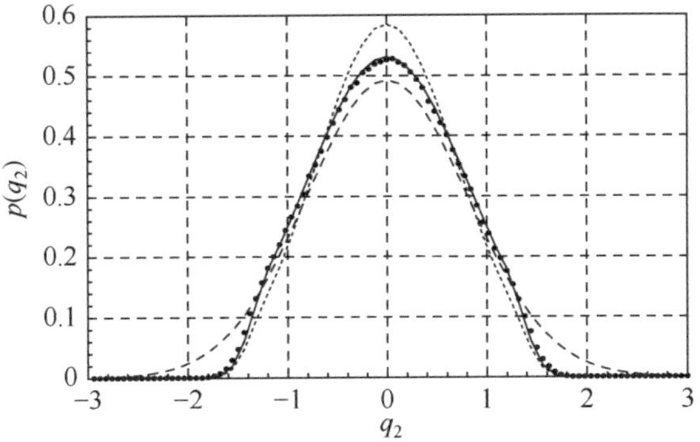

Fig. 5.33 Stationary PDF of displacement Q_2 of double-wall system. ——, result from Eq. (5.244); -----, result from Eq. (5.239);......, result from Eq. (5.215); ●, result from Monte Carlo simulation. System parameters are $\pi K_1 = 0.01$, $\pi K_2 = 0.02$, $B_l = B_r = 10$, $\delta_l = \delta_r = 1.2$ (Liu et al. 2002)

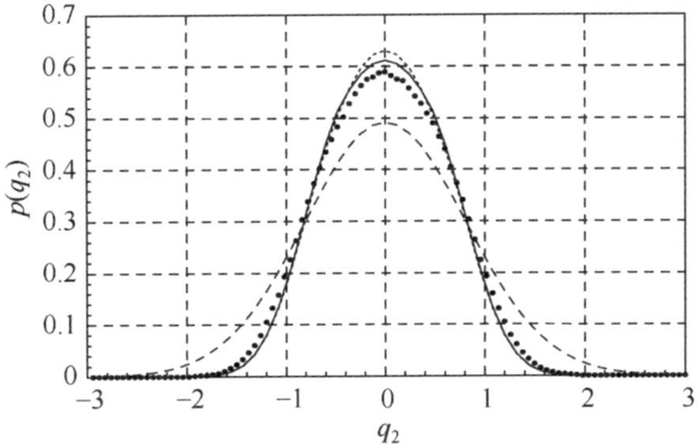

Fig. 5.34 Stationary PDF of displacement Q_2 of double-wall system. ——, result from Eq. (5.244); -----, result from Eq. (5.239);......, result from Eq. (5.215); ●, result from Monte Carlo simulation. System parameters are $\pi K_1 = 0.01$, $\pi K_2 = 0.02$, $\delta_l = \delta_r = 0.5$, $B_l = B_r = 1.0$ (Liu et al. 2002)

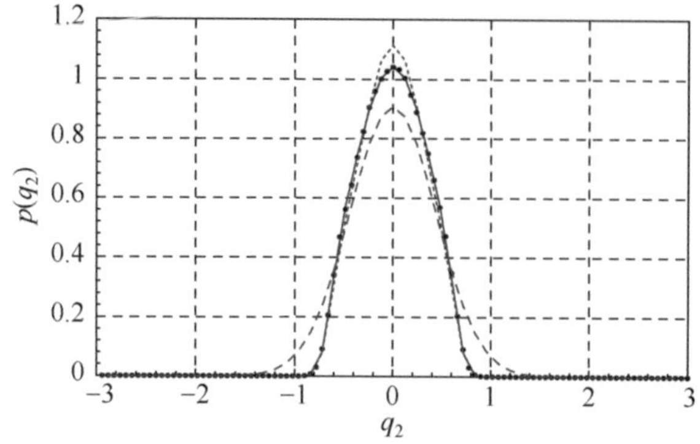

Fig. 5.35 Stationary PDF of displacement Q_2 of double-wall system. ——, result from Eq. (5.244); -----, result from Eq. (5.239); ……, result from Eq. (5.215); ●, result from Monte Carlo simulation. System parameters are $\delta_l = \delta_r = 0.5, B_l = B_r = 10, \pi K_1 = 0.005, \pi K_2 = 0.005$ (Liu et al. 2002)

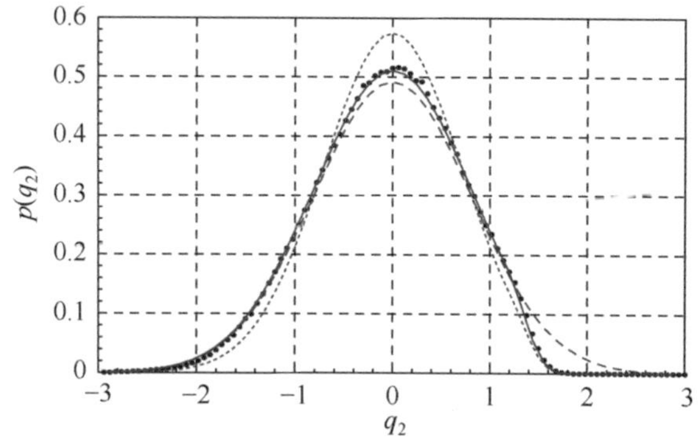

Fig. 5.36 Stationary PDF of displacement Q_2 of right-wall system. ——, result from Eq. (5.244); -----, result from Eq. (5.239); ……, result from Eq. (5.215); ●, result from Monte Carlo simulation. System parameters are $\pi K_1 = 0.01, \pi K_2 = 0.02, B_r = 10, \delta_r = 1.2$ (Liu et al. 2002)

5.5 Quasi-Non-Integrable Hamiltonian Systems with Markov Jump Parameters

Let $s(t)$ be a Markov jump parameter, which is a time-continuous Markov jump process taking discrete values at finite set $S = \{1, 2, 3, \ldots, l\}$, where l is the number of all the possible states. The transition probability of the Markov jump process $s(t)$ is given by

$$\begin{aligned}\text{Prob}[s(t+\Delta t) = j|s(t) = i] &= P(j, t+\Delta t|i, t)\\ &= \begin{cases} \lambda_{ij}\Delta t + \text{O}(\Delta t), & i \neq j, \\ 1 + \lambda_{ii}\Delta t + \text{O}(\Delta t), & i = j. \end{cases}\end{aligned} \tag{5.245}$$

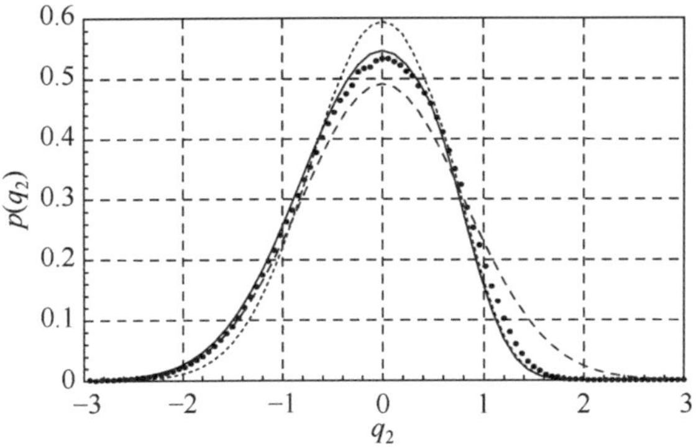

Fig. 5.37 Stationary PDF of displacement Q_2 of right-wall system. ——, result from Eq. (5.244); -----, result from Eq. (5.239); ......, result from Eq. (5.215); ●, result from Monte Carlo simulation. System parameters are $\pi K_1 = 0.01$, $\pi K_2 = 0.02$, $\delta_r = 0.5$, $B_r = 1.0$ (Liu et al. 2002)

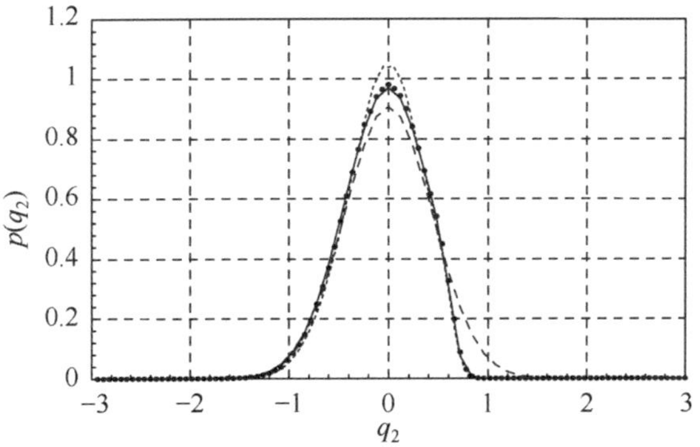

Fig. 5.38 Stationary PDF of displacement Q_2 of right-wall system. ——, result from Eq. (5.244); -----, result from Eq. (5.239); ......, result from Eq. (5.215); ●, result from Monte Carlo simulation. System parameters are $B_r = 10$, $\delta_r = 0.5$, $\pi K_1 = 0.005$, $\pi K_2 = 0.005$ (Liu et al. 2002)

where $P(j, t + \Delta t | i, t)$ is the transition probability from state i at time t to state j at time $t + \Delta t$. $\lambda_{ij} \geq 0$ $(i \neq j)$ are called jump rate from state i to j. Obviously,

$$\lambda_{ii} = - \sum_{j=1, j \neq i}^{l} \lambda_{ij}. \tag{5.246}$$

which makes the transition occurs only within the state set S. Let $\Lambda = [\lambda_{ij}]_{l \times l}$ be the jump rate matrix, and all its elements λ_{ij} are known. The Markov jump process $s(t)$ considered in this section has the following properties: (i) $s(t)$ is independent of system motion state, i.e., an independent jump process; (ii) The jump is very slow, i.e., $\lambda_{ij} \Delta t \ll 1$.

5.5.1 Single-DOF Systems

Consider a single-DOF nonlinear system with Markov jump parameter and excited by Gaussian white noises (Pan et al. 2017). The motion equation is given by

$$\begin{aligned} &\ddot{X} + \varepsilon c(X, \dot{X}, s(t)) + g(X) = \varepsilon^{1/2} \sum_{k=1}^{m} f_k(X, s(t)) W_{gk}(t), \\ &s \in \{1, 2, \ldots, l\}, \end{aligned} \tag{5.247}$$

where $g(X)$ is nonlinear restoring force; ε is a small parameter; $\varepsilon c(X, \dot{X}, s(t))$ represents weak damping with jump parameter; $\varepsilon^{1/2} f_k(X, s(t))$ represents the excitation amplitude with jump parameters; parameter $s(t)$ is a Markov jump process; $W_{gk}(t)$ are independent Gaussian white noises with zero mean-value and intensities $2D_k$.

Firstly, consider system (5.247) with a fixed jump parameter value. Let $Q = X$, $P = \dot{X}$. System (5.247) can be transformed into the following Itô stochastic differential equations

$$\begin{aligned} &\mathrm{d}Q = P\mathrm{d}t, \\ &\mathrm{d}P = \left[-g(Q) - \varepsilon c(Q, P, s)\right]\mathrm{d}t + \varepsilon^{1/2} \sum_{k=1}^{m} \sigma_k(Q, s)\mathrm{d}B_k(t), \end{aligned} \tag{5.248}$$

where $\sigma_k = \sqrt{2D_k} f_k$, $B_k(t)$ are independent unit Wiener processes. The Hamiltonian function or total energy of system (5.248) reads

$$H(Q, P) = \frac{1}{2}P^2 + U(Q), \quad U(Q) = \int_0^Q g(u)\mathrm{d}u. \tag{5.249}$$

Applying Itô differential rule and Eq. (5.249), the Itô stochastic differential equation for $H(t)$ can be derived from Eq. (5.248). By replacing the second equation in Eq. (5.248) with that for H, a new system of Itô stochastic differential equations are obtained

$$\begin{aligned} &\mathrm{d}Q = P\mathrm{d}t, \\ &\mathrm{d}H = \varepsilon\left[-c(Q, P, s)P + \frac{1}{2}\sum_{k=1}^{m} \sigma_k(Q, s)\sigma_k(Q, s)\right]\mathrm{d}t \\ &\quad + \varepsilon^{1/2} P \sum_{k=1}^{m} \sigma_k(Q, s)\mathrm{d}B_k(t). \end{aligned} \tag{5.250}$$

where P should be replaced by Q, H according to Eq. (5.249). In system (5.250), the generalized displacement $Q(t)$ is fast-varying process while $H(t)$ is a slowly-varying process under the conditions of weak damping and weak excitations. Applying the stochastic averaging method of quasi non-integrable Hamiltonian systems described in Sect. 5.1, the averaged Itô stochastic differential equation governing the slowly-varying process $H(t)$ is obtained as follows

$$\mathrm{d}H = m(H, s)\mathrm{d}t + \overline{\sigma}(H, s)\mathrm{d}B(t), \quad s \in \{1, 2, \ldots, l\}. \tag{5.251}$$

where the drift and diffusion coefficients are

$$\begin{aligned} m(H, s) &= \frac{\varepsilon}{T(H)} \int_{\Omega} \left\{ \left[-c(q, p, s)p + \frac{1}{2} \sum_{k=1}^{m} \sigma_k(q, s)\sigma_k(q, s) \right] / p \right\} \mathrm{d}q, \\ \overline{\sigma}^2(H, s) &= \frac{\varepsilon}{T(H)} \int_{\Omega} \left\{ \left[\sum_{k=1}^{m} \sigma_k(q, s)\sigma_k(q, s) \right] / p \right\} \mathrm{d}q, \\ T(H) &= \int_{\Omega} 1/p\mathrm{d}q, \quad \Omega = \{q | H(q, 0) \le H\}. \end{aligned} \tag{5.252}$$

Since the jump parameter $s(t)$ in system (5.247) has l states, l averaged equations of the form as Eq. (5.251) can be obtained.

Let $p(h, s, t + \tau | h', s, t)$ be the conditional transition PDF of $H(t)$ without parameter jump. It satisfies the following FPK equation associated with Itô Eq. (5.251)

$$\begin{aligned} \frac{\partial p(h, s, t + \tau | h', s, t)}{\partial \tau} &= -\frac{\partial}{\partial h}\left[m(h, s)p(h, s, t + \tau | h', s, t)\right] \\ &+ \frac{1}{2}\frac{\partial^2}{\partial h^2}\left[\overline{\sigma}^2(h, s)p(h, s, t + \tau | h', s, t)\right]. \end{aligned} \tag{5.253}$$

Rewrite the left term of Eq. (5.253) as

$$\frac{\partial p(h, s, t + \tau | h', s, t)}{\partial \tau} = \lim_{\Delta t \to 0} \frac{1}{\Delta t}\left[p(h, s, t + \Delta t | h', s, t) - p(h, s, t | h', s, t)\right]. \tag{5.254}$$

Substituting Eq. (5.254) into Eq. (5.253) yields

$$\begin{aligned} p(h, s, t + \Delta t | h', s, t) &= p(h, s, t | h', s, t) \\ &- \Delta t \frac{\partial}{\partial h}[m(h, s)p(h, s, t + \Delta t | h', s, t)] \\ &+ \frac{1}{2}\Delta t \frac{\partial^2}{\partial h^2}[\overline{\sigma}^2(h, s)p(h, s, t + \Delta t | h', s, t)]. \end{aligned} \tag{5.255}$$

The Taylor expansion of $p\big(h, s, t + \Delta t\big|h', s, t\big)$ at t is

$$p(h, s, t + \Delta t\big|h', s, t) = p(h, s, t\big|h', s, t) + \Delta t\frac{\partial p(h, s, t\big|h', s, t)}{\partial t} + \ldots. \tag{5.256}$$

Substituting Eq. (5.256) into the right side of Eq. (5.253) and neglecting the terms of order higher than Δt yield

$$\begin{aligned}p\big(h, s, t + \Delta t\big|h', s, t\big) &= p\big(h, s, t\big|h', s, t\big)\\ &- \Delta t\frac{\partial}{\partial h}\big[m(h, s)p(h, s, t\big|h', s, t)\big]\\ &+ \frac{1}{2}\Delta t\frac{\partial^2}{\partial h^2}\big[\overline{\sigma}^2(h, s)p(h, s, t\big|h', s, t)\big].\end{aligned} \tag{5.257}$$

The initial value of the transition PDF $p(h, s, t + \tau\big|h', s, t)$ is

$$p(h, s, t\big|h', s, t) = \delta(h - h'). \tag{5.258}$$

Substituting Eq. (5.258) into Eq. (5.257) yields

$$\begin{aligned}&p(h, s, t + \Delta t\big|h', s, t) = \delta(h - h')\\ &- \Delta t\frac{\partial}{\partial h}\big[m(h, s)\delta(h - h')\big] + \frac{1}{2}\Delta t\frac{\partial^2}{\partial h^2}\big[\overline{\sigma}^2(h, s)\delta(h - h')\big].\end{aligned} \tag{5.259}$$

Equation (5.259) is the expression of the conditional transition PDF $p(h, s, t + \Delta t\big|h', s, t)$ of $H(t)$ in small time interval Δt, in which system parameter does not jump.

Note that the state of the Markov jump process and Hamiltonian function $H(t)$ at time $t + \Delta t$ is evolved from the previous state at t due to two complete events. One event is that Hamiltonian function evolves from $H'(t)$ at time t into $H(t)$ at $t + \Delta t$ while jump state keeps at s. The transition PDF of this event is $p(h, s, t+\Delta t\big|h', s, t)$. Another event is that the Markov jump process jumps from state r $(r \neq s)$ to state s while the Hamiltonian function changes from $H'(t)$ at time t to $H(t)$ at time $t + \Delta t$. The transition PDF of this event is $p(h, s, t+\Delta t\big|h', r, t)$. Thus, the PDF $p(h, s, t+\Delta t)$ of system state at time $t + \Delta t$ can be expressed as

$$\begin{aligned}&p(h, s, t + \Delta t) = P(s, t + \Delta t|s, t)\\ &\int_0^\infty p(h', s, t)p(h, s, t + \Delta t\big|h', s, t)\mathrm{d}h'\\ &+ \sum_{r=1, r\neq s}^{l} P(s, t + \Delta t|r, t)\int_0^\infty p(h', r, t)p(h, s, t + \Delta t\big|h', r, t)\mathrm{d}h'.\end{aligned} \tag{5.260}$$

Substituting Eqs. (5.245) and (5.259) into Eq. (5.260) leads to

$$
\begin{aligned}
p(h, s, t + \Delta t) &= p(h, s, t) - \Delta t \frac{\partial}{\partial h}[m(h, s)p(h, s, t)] \\
&+ \frac{1}{2}\Delta t \frac{\partial^2}{\partial h^2}[\overline{\sigma}^2(h, s)p(h, s, t)] \\
&- \Delta t \sum_{r=1, r\neq s}^{l} \left[\lambda_{sr} p(h, s, t) - \lambda_{rs} \int_0^{\infty} p(h', r, t)p(h, s, t + \Delta t \big| h', r, t) \right] \mathrm{d}h'. \quad (5.261)
\end{aligned}
$$

Taylor expanding $p(h, s, t + \Delta t \big| h', r, t)$ in Eq. (5.261) and neglecting the higher order terms yield

$$
\begin{aligned}
p(h, s, t + \Delta t \big| h', r, t) &= p(h, s, t \big| h', r, t) \\
&+ \Delta t \frac{\partial p(h, s, t \big| h', r, t)}{\partial t} + \ldots . \quad (5.262)
\end{aligned}
$$

The initial value of transition PDF $p(h, s, t \big| h', r, t)$ is

$$
p\big(h, s, t \big| h', r, t\big) = \delta(h - h'). \quad (5.263)
$$

Substituting Eqs. (5.262) and (5.263) into Eq. (5.261), neglecting terms of order higher than Δt, and taking $\Delta t \to 0$, the following averaged FPK equation governing the response PDF of the stochastic dynamical system (5.248) with Markov jump parameter is obtained

$$
\begin{aligned}
\frac{\partial p(h, s, t)}{\partial t} &= -\frac{\partial}{\partial h}[m(h, s)p(h, s, t)] + \frac{1}{2}\frac{\partial^2}{\partial h^2}[\overline{\sigma}^2(h, s)p(h, s, t)] \\
&- \sum_{r=1, r\neq s}^{l} [\lambda_{sr} p(h, s, t) - \lambda_{rs} p(h, r, t)], \quad s \in \{1, 2, \ldots, l\}. \quad (5.264)
\end{aligned}
$$

Equation (5.264) consists of l coupled averaged FPK equations, each of which corresponds to one jump state. Compared with the averaged FPK equation for quasi-non-integrable Hamiltonian systems without jump, Eq. (5.264) has two additional terms, i.e., the probability $\sum_{r=1, r\neq s}^{l} \lambda_{sr} p(h, s, t)$ (also known as absorption term) of jumping from state s to state r, and the probability $\sum_{r=1, r\neq s}^{l} \lambda_{rs} p(h, r, t)$ (also known as recovery term) of jumping from state r to state s.

The initial condition of FPK Eq. (5.264) is

$$
p(h, s, 0) = p(h_0, s), \quad s \in \{1, 2, \ldots, l\}. \quad (5.265)
$$

The boundary conditions are

$$p(0, s, t) = \text{finite},$$
$$p(h, s, t)|_{h\to\infty} \to 0, \quad \left.\frac{\partial p(h, s, t)}{\partial h}\right|_{h\to\infty} \to 0. \tag{5.266}$$

Generally, it is difficult to obtain the exact analytical solution for the FPK Eq. (5.264). In practical applications, only the following reduced averaged FPK is of concern,

$$0 = -\frac{\partial}{\partial h}[m(h, s)p(h, s, t)] + \frac{1}{2}\frac{\partial^2}{\partial h^2}[\overline{\sigma}^2(h, s)p(h, s, t)]$$
$$-\sum_{r=1,r\neq s}^{l}[\lambda_{sr}p(h, s, t) - \lambda_{rs}p(h, r, t)], \quad s \in \{1, 2, \ldots, l\}. \tag{5.267}$$

The family of partial differential equations of (5.267) can be solved using numerical methods such as the finite difference method or the Runge–Kutta method to obtain the stationary PDF$p(h, s)$..

Subsequently, the stationary PDF $p(h)$ of Hamiltonian or total energy, as well as the stationary joint PDF $p(q, p)$ of generalized displacement and momentum, can be derived as

$$p(h) = C\sum_{s=1}^{l}p(h, s), \quad p(q, p) = \left.\frac{p(h)}{T(h)}\right|_{h=H(q,p)}. \tag{5.268}$$

where C is a normalization constant. The marginal PDF $p(q)$ of generalized displacement, and the mean-square value $E[Q^2]$, can then be obtained as follows

$$p(q) = \int_{-\infty}^{\infty} p(q, p)\mathrm{d}p, \quad E[Q^2] = \int_{-\infty}^{\infty} q^2 p(q)\mathrm{d}q. \tag{5.269}$$

Example 5.6 Consider a Duffing oscillator with Markov jumping parameter subjected to Gaussian white noise excitation. The governing differential equation is

$$\ddot{X} + \beta(s(t))\dot{X} + \omega^2 X + \alpha X^3 = f(s(t))W_g(t),$$
$$s \in \{1, 2, \ldots, l\}, \tag{5.270}$$

where $\beta(s(t))$ and $f(s(t))$ are the jumping damping coefficient and excitation amplitude, respectively; $s(t)$ is the Markov jump process with transition probability defined in Eq. (5.245); $W_g(t)$ is a Gaussian white noises with intensity $2D$.

Let $Q = X$ and $P = \dot{X}$. The Hamiltonian function or total energy of system (5.270) is

$$H(Q, P) = \frac{1}{2}P^2 + \frac{1}{2}\omega^2 Q^2 + \frac{1}{4}\alpha Q^4. \tag{5.271}$$

System (5.270) can be transformed into the following Itô stochastic differential equations

$$\begin{aligned} \mathrm{d}Q &= P\mathrm{d}t, \\ \mathrm{d}P &= \left[-\omega^2 Q - \alpha Q^3 - \beta(s)P\right]\mathrm{d}t + \sqrt{2D}f(s)\mathrm{d}B(t). \end{aligned} \tag{5.272}$$

Deriving the Itô equation for $H(t)$ from Eq. (5.272) according to Eq. (5.271) using Itô differential rule and replacing the equation for P by that for H lead to

$$\begin{aligned} \mathrm{d}Q &= P\mathrm{d}t, \\ \mathrm{d}H &= [-\beta(s)P^2 + Df^2(s)]\mathrm{d}t + \sqrt{2D}Pf(s)\mathrm{d}B(t). \end{aligned} \tag{5.273}$$

Applying the stochastic averaging method of quasi-non-integrable Hamiltonian systems to Eq. (5.273) yields the averaged Itô stochastic differential equation (5.251). The drift and diffusion coefficients are obtained by using Eq. (5.252) as follows

$$\begin{aligned} m(H, s) &= f^2(s)D - \beta(s)G(H), \quad \overline{\sigma}^2(H, s) = 2f^2(s)DG(H), \\ G(H) &= H + \frac{3H^2\alpha}{8\omega^4} - \frac{13H^3\alpha^2}{16\omega^8} + \frac{2211H^4\alpha^3}{1024\omega^{12}} + \ldots. \end{aligned} \tag{5.274}$$

After setting the jumping rules of system parameters, the corresponding reduced FPK equation family, the stationary PDF and statistics can be obtained according to Eqs. (5.267) to (5.269). In the following, numerical calculations have been performed for two different jumping rules of system parameter.

(i) **Two-State Jumping Parameters**

Let the system parameters are $\omega = 1$, $\alpha = 1$, $D = 0.1$. The two-state damping coefficients and excitation amplitudes are $\beta(s = 1) = 0.02$, $\beta(s = 2) = 0.04$ and $f(s = 1) = 2$, $f(s = 2) = 1$, respectively. It can be seen that the system in state $s = 2$ has larger damping force and weak excitation, implying that the system in state $s = 2$ has littler energy.

Suppose that there are 3 jumping rules

$$\Lambda_1 = \begin{bmatrix} -2 & 2 \\ 2 & -2 \end{bmatrix}, \quad \Lambda_2 = \begin{bmatrix} -1 & 1 \\ 2 & -2 \end{bmatrix}, \quad \Lambda_3 = \begin{bmatrix} -2 & 2 \\ 1 & -1 \end{bmatrix}. \tag{5.275}$$

In Λ_1, $\lambda_{21} = \lambda_{12}$, which means symmetry jump. In Λ_2, $\lambda_{21} < \lambda_{12}$, the jump rate from state 1 to state 2 is less than the jump rate from state 2 to state 1, that is, the

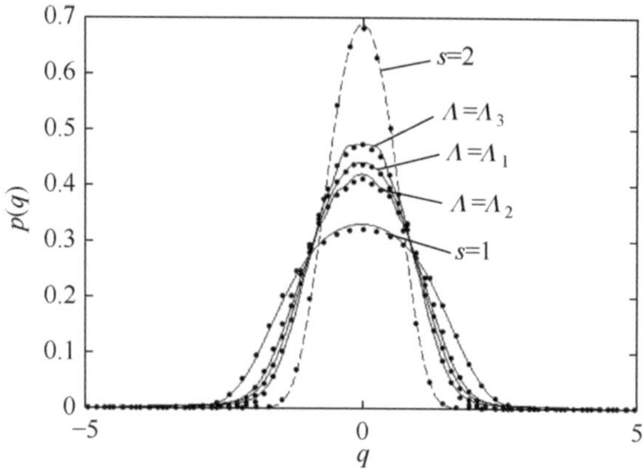

Fig. 5.39 Stationary PDF of generalized displacement Q of two-state system (5.270) under different jump laws. -----, results from solving Eq. (5.269); ●, results from Monte Carlo simulation (Pan et al. 2017)

probability of system staying in state 1 is larger. The jump in Λ_3 is just opposite of that in Λ_2.

The stationary PDF of displacement of two-state system (5.270) for jumping rules $\Lambda = \Lambda_1$, $\Lambda = \Lambda_2$, $\Lambda = \Lambda_3$ and non-jump case is shown in Fig. 5.39. It is seen from Fig. 5.39 that, when $s = 2$, the stationary PDF of system displacement has highest peak value. According to the order $s = 2$, $\Lambda = \Lambda_3$, $\Lambda = \Lambda_1$, $\Lambda = \Lambda_2$, $s = 1$, the peak of PDF curve reduces one by one. The main reason for this is, the probability of system staying in state $s = 1$ is getting higher and higher. It's known from the above analysis, the higher the probability of system staying in state $s = 1$, the larger the system energy, making the PDF curve more flat. In Fig. 5.39, the lines represent theoretical results obtained from solving FPK Eqs. (5.267), the dots are results from Monte Carlo simulation of system (5.270). Obviously, theoretical results agree with the simulation results very well. Figure 5.40 shows a sample of jump parameter $s(t)$ with rule $\Lambda = \Lambda_2$. Figure 5.41a shows the theoretical result of stationary joint PDF $p(q, p)$ of system (5.270) with jumping rule $\Lambda = \Lambda_2$ Fig. 5.41b shows the Monte Carlo simulation result of $p(q, p)$ with jump rule $\Lambda = \Lambda_2$. It is seen that the two results agree very well.

(ii) Three-State Jumping Parameters

Suppose that system (5.270) has three-state damping coefficients and excitation amplitudes $\beta(s = 1) = 0.004$, $\beta(s = 2) = 0.008$, $\beta(s = 3) = 0.01$ and $f(s = 1) = 2, f(s = 2) = 1, f(s = 3) = 0.5$, respectively. It is seen that system in state $s = 3$ has maximum damping coefficient and minimum excitation amplitude, which imply system in state 3 has minimum energy.

Suppose that there are 3 jump rules

$$\Lambda_1 = \begin{bmatrix} -1.4 & 0.7 & 0.7 \\ 1.5 & -3.0 & 1.5 \\ 1.5 & 1.5 & -3.0 \end{bmatrix}, \quad \Lambda_2 = \begin{bmatrix} -3.0 & 1.5 & 1.5 \\ 0.7 & -1.4 & 0.7 \\ 1.5 & 1.5 & -3.0 \end{bmatrix},$$

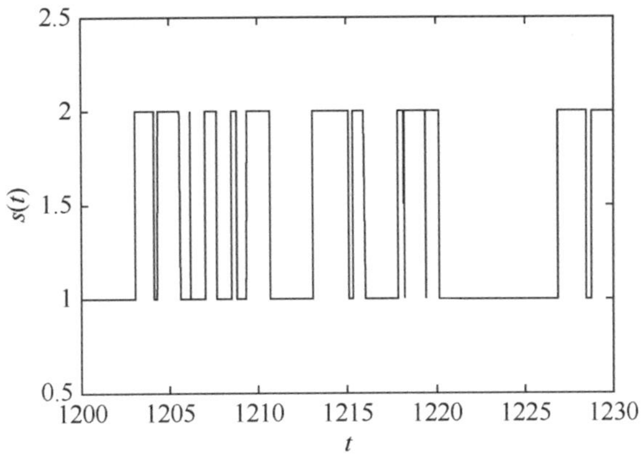

Fig. 5.40 A sample of jump process $s(t)$ in two-state system (5.270) with jump rule $\Lambda = \Lambda_2$ (Pan et al., 2017)

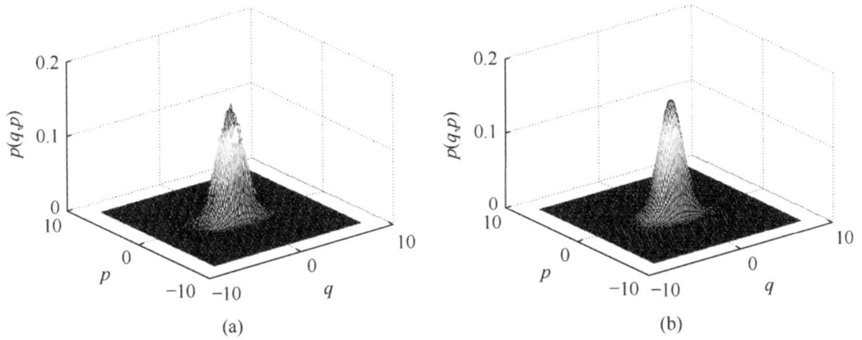

Fig. 5.41 Stationary joint PDF of two-state system (5.270) with jump rule $\Lambda = \Lambda_2$. **a** Theoretical result, **b** result from Monte Carlo simulation (Pan et al. 2017)

$$\Lambda_3 = \begin{bmatrix} -3.0 & 1.5 & 1.5 \\ 1.5 & -3.0 & 1.5 \\ 0.7 & 0.7 & -1.4 \end{bmatrix}, \tag{5.276}$$

It is seen that, according to the rule Λ_1, the probability of system staying in state $s = 1$ is the largest, while according to the rules Λ_2 and Λ_3, the probability of system staying in states $s = 2$ and $s = 3$ are the largest, respectively.

Figure 5.42 depicts the stationary PDF of displacement of system under jumping rules $\Lambda = \Lambda_1$, $\Lambda = \Lambda_2$, $\Lambda = \Lambda_3$, and staying at states $s = 1$, $s = 2$ and $s = 3$. When $s = 3$, the stationary PDF of displacement has highest peak value. According the order of $\Lambda = \Lambda_3$, $\Lambda = \Lambda_2$, $\Lambda = \Lambda_1$, the peak of PDF curve reduces one by one The results from Monte Carlo simulation agree with the theoretical results very well. In Fig. 5.43, a sample of jump parameter $s(t)$ following the rule $\Lambda = \Lambda_2$ is shown. In Fig. 5.44a and b, the theoretical results and Monte Carlo simulation results of the stationary PDF $p(q, p)$ of system with jump rule $\Lambda = \Lambda_2$ are shown. The two results also agree very well.

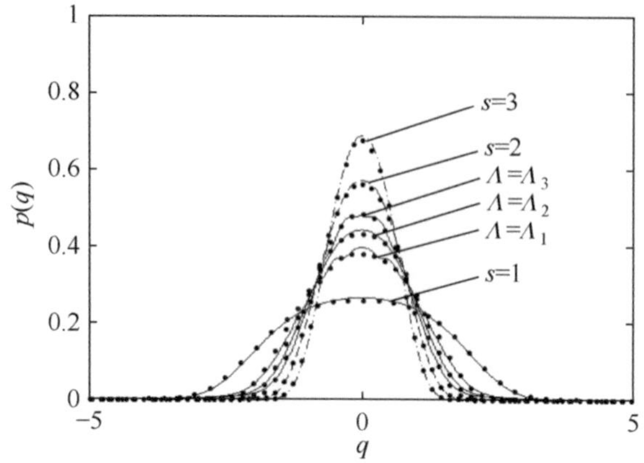

Fig. 5.42 Stationary PDF of generalized displacement Q of system (5.270) with different jump rules. -----, theoretical results; ●, results from Monte Carlo simulation (Pan et al., 2017)

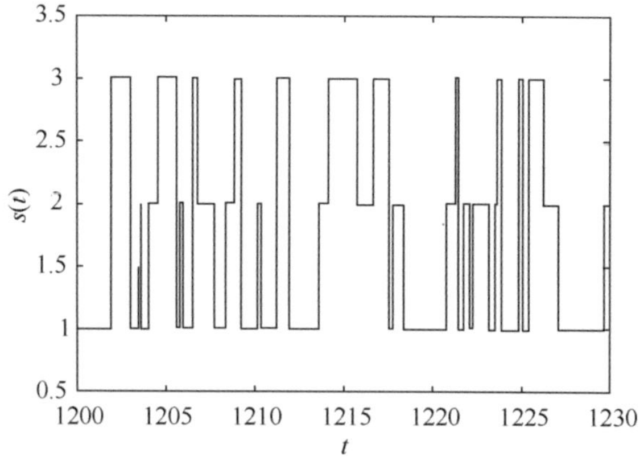

Fig. 5.43 A sample of jump process $s(t)$ in system (5.270) with jump rule $\Lambda = \Lambda_2$ (Pan et al. 2017)

5.5.2 Multi-DOF Systems

Consider the following n-DOF stochastically excited and dissipated Hamiltonian system with Markov jump parameter (Huan et al. 2015)

$$\dot{Q}_i = \frac{\partial H}{\partial P_i},$$
$$\dot{P}_i = -\frac{\partial H}{\partial Q_i} - \varepsilon \sum_{j=1}^{n} c_{ij}(\boldsymbol{Q}, \boldsymbol{P}, s(t)) \frac{\partial H}{\partial P_j}$$

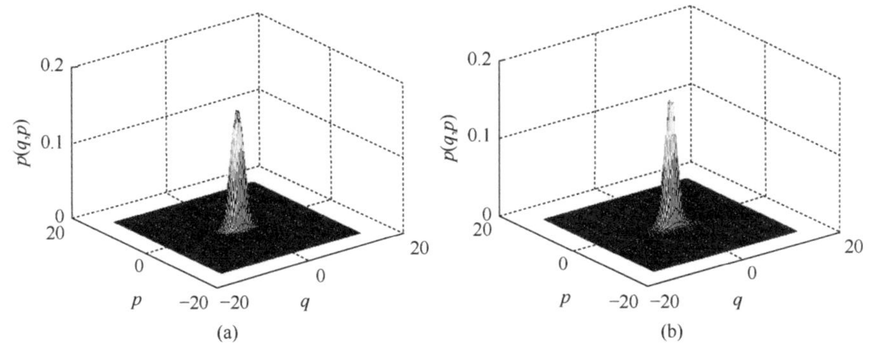

Fig. 5.44 Stationary joint PDF of system (5.270) with jump rule $\Lambda = \Lambda_2$. **a** Theoretical result, **b** result from Monte Carlo simulation (Pan et al. 2017)

$$
+\varepsilon^{1/2}\sum_{k=1}^{m} f_{ik}(\boldsymbol{Q}, s(t))W_{gk}(t),
$$

$$
i = 1, 2, \ldots, n; \quad s \in \{1, 2, \ldots, l\}, \tag{5.277}
$$

where $\boldsymbol{Q} = [Q_1, Q_2, \ldots, Q_n]$ is a vector of generalized displacements; $\boldsymbol{P} = [P_1, P_2, \ldots, P_n]$ is a vector of generalized momenta; $H = H(\boldsymbol{Q}, \boldsymbol{P})$ is Hamiltonian function; ε is a small parameter; $\varepsilon c_{ij}(\boldsymbol{Q}, \boldsymbol{P}, s(t))$ is the small damping coefficients with jump parameter $s(t)$; $\varepsilon^{1/2} f_{ik}(\boldsymbol{Q}, s(t))$ are the weak amplitudes of excitations with jump parameter $s(t)$; $W_{gk}(t)$ $(k = 1, 2, \ldots, m)$ are Gaussian white noises with zero mean and correlation functions $E[W_{gk}(t)W_{gl}(t+\tau)] = 2D_{kl}\delta(\tau)$; $s(t)$ is a time-continuous and state-discrete Markov process with transition probability in Eq. (5.245) and with state in set $S = \{1, 2, 3, \ldots, l\}$; $s(t)$ is assumed to be independent of $W_{gk}(t)$ and system state, and is a slowly-varying process.

System (5.277) firstly is transformed into Stratonovich stochastic differential equations and then transformed into Itô stochastic differential equations

$$
\begin{aligned}
\mathrm{d}Q_i &= \frac{\partial H}{\partial P_i}\mathrm{d}t,\\
\mathrm{d}P_i &= -\left[\frac{\partial H}{\partial Q_i} + \varepsilon\sum_{j=1}^{n} c_{ij}(Q, P, s(t))\frac{\partial H}{\partial P_j}\right]\mathrm{d}t\\
&\quad + \varepsilon^{1/2}\sum_{k=1}^{m}\sigma_{ik}(Q, s(t))\mathrm{d}B_k(t),\\
i &= 1, 2, \ldots, n; \quad s \in \{1, 2, \ldots, l\},
\end{aligned} \tag{5.278}
$$

where $B_k(t)$ are independent unit Wiener processes, $\sigma_{ik} = (\boldsymbol{fL})_{ik}$, and $\boldsymbol{LL}^T = 2\boldsymbol{D}$. Assume the functions $f_{ik}(\boldsymbol{Q}, s(t))$ do not contain $\boldsymbol{P}$, so that there is no Wong-Zakai correction term.

Assume that the Hamiltonian system associated with system (5.277) is non-integrable. First, set $s(t) = s$. The following Itô stochastic differential equation for Hamiltonian process $H(t)$ is obtained from Eq. (5.278) by using relation $H = H(\boldsymbol{Q}, \boldsymbol{P})$ and applying Itô differential rule

$$\begin{aligned}\mathrm{d}H =&\varepsilon\left[-\sum_{i,j=1}^{n} c_{ij}(\boldsymbol{Q}, \boldsymbol{P}, s)\frac{\partial H}{\partial P_i}\frac{\partial H}{\partial P_j}\right.\\&\left.+\frac{1}{2}\sum_{i,j=1}^{n}\sum_{k=1}^{m}\sigma_{ik}(\boldsymbol{Q}, s)\sigma_{jk}(\boldsymbol{Q}, s)\frac{\partial^2 H}{\partial P_i \partial P_j}\right]\mathrm{d}t\\&+\varepsilon^{1/2}\sum_{i=1}^{n}\sum_{k=1}^{m}\frac{\partial H}{\partial P_i}\sigma_{ik}(\boldsymbol{Q}, s)\mathrm{d}B_k(t),\\&s \in \{1, 2, \ldots, l\}.\end{aligned} \tag{5.279}$$

Replacing the equation of $\mathrm{d}P_1$ in Eq. (5.278) with Eq. (5.279) to constitute a new system. In this new system, $\boldsymbol{Q}(t), P_2(t), P_3(t), \ldots, P_n(t)$ are fast-varying processes, while $H(t)$ is slowly-varying process. After stochastic averaging, the averaged Itô stochastic differential equation governing the diffusion process $H(t)$ is established as

$$\mathrm{d}H = m(H, s)\mathrm{d}t + \overline{\sigma}(H, s)\mathrm{d}B(t). \tag{5.280}$$

As described in Sect. 5.1, the non-integrable Hamiltonian systems are ergodic on the isoenergetic surface. Thus, time averaging in the averaged drift and diffusion coefficients of Eq. (5.280) can be replaced by spatial averaging (see Eqs. (5.10)–(5.13)). So, the drift and diffusion coefficients of Eq. (5.280) are

$$\begin{aligned}m(H, s) =& \frac{1}{T(H)}\int_{\Omega}\left[\left(-\sum_{i,j=1}^{n} c_{ij}\frac{\partial H}{\partial p_i}\frac{\partial H}{\partial p_j} + \frac{1}{2}\sum_{i,j=1}^{n}\sum_{k=1}^{m}\sigma_{ik}\sigma_{jk}\frac{\partial^2 H}{\partial p_i \partial p_j}\right)\Big/\frac{\partial H}{\partial p_1}\right]\\&\mathrm{d}\boldsymbol{q}\mathrm{d}p_2\ldots\mathrm{d}p_n,\\ \overline{\sigma}^2(H, s) =& \frac{1}{T(H)}\int_{\Omega}\left[\sum_{i,j=1}^{n}\sum_{k=1}^{m}\sigma_{ik}\sigma_{jk}\frac{\partial H}{\partial p_i}\frac{\partial H}{\partial p_j}\Big/\frac{\partial H}{\partial p_1}\right]\mathrm{d}\boldsymbol{q}\mathrm{d}p_2\ldots\mathrm{d}p_n,\\ T(H) =& \int_{\Omega}\left[1\Big/\frac{\partial H}{\partial p_1}\right]\mathrm{d}\boldsymbol{q}\mathrm{d}p_2\ldots\mathrm{d}p_n,\\ \Omega =& \{(\boldsymbol{q}, p_2, \ldots, p_n)|H(\boldsymbol{q}, 0, p_2, \ldots, p_n) \le H\}.\end{aligned} \tag{5.281}$$

Now consider $s(t)$ is a Markov jump process among l states. Similar to the derivations from Eqs. (5.253) to (5.269), the transient averaged FPK equation corresponding to Eq. (5.280) is similar to Eq. (5.264), its initial condition is similar to Eq. (5.265) and boundary conditions similar to Eq. (5.266). The reduced averaged FPK equations are of the same form of those in Eq. (5.267), and the expressions for other PDFs and statistics are the same as those in Eq. (5.268) and (5.269).

Example 5.7 Consider a 2-DOF quasi-non-integrable Hamiltonian system with Markov jump parameter. The motion equations are given by

$$\begin{aligned}
&\ddot{X}_1 + c_1(s(t))\dot{X}_1 + \beta\omega_1^2 X_1(\omega_1^2 X_1^2 + \omega_2^2 X_2^2)^2 = f_1(s(t))W_{g1}(t),\\
&\ddot{X}_2 + c_2(s(t))\dot{X}_2 + \beta\omega_2^2 X_2(\omega_1^2 X_1^2 + \omega_2^2 X_2^2)^2 = f_2(s(t))W_{g2}(t),\\
&s \in \{1, 2, \ldots, l\},
\end{aligned} \tag{5.282}$$

where $c_1(s(t))$, $c_2(s(t))$ and $f_1(s(t)), f_2(s(t))$ are damping coefficients and excitation amplitudes with jump parameter $s(t)$, respectively; $s(t)$ is a time-continuous and state-discrete Markov jumping process with transition probability defined in Eq. (5.245); $W_{g1}(t)$, $W_{g2}(t)$ are independent Gaussian white noises with intensities $2D_1, 2D_2$; $s(t)$ is independent of $W_{g1}(t)$, $W_{g2}(t)$.

Letting $Q_1 = X_1$, $Q_2 = X_2$, $P_1 = \dot{X}_1$, $P_2 = \dot{X}_2$, the Hamiltonian function or total energy of system (5.282) read

$$\begin{aligned}
H(\boldsymbol{Q}, \boldsymbol{P}) &= \frac{1}{2}(P_1^2 + P_2^2) + U(\boldsymbol{Q}),\\
U(\boldsymbol{Q}) &= \frac{1}{6}\beta(\omega_1^2 Q_1^2 + \omega_2^2 Q_2^2)^3.
\end{aligned} \tag{5.283}$$

System (5.282) can be transformed into the following Itô stochastic differential equations

$$\begin{aligned}
\mathrm{d}Q_1 &= P_1 \mathrm{d}t,\\
\mathrm{d}Q_2 &= P_2 \mathrm{d}t,\\
\mathrm{d}P_1 &= -\left(\frac{\partial U}{\partial Q_1} + c_1(s)P_1\right)\mathrm{d}t + \sqrt{2D_1} f_1(s)\mathrm{d}B_1(t),\\
\mathrm{d}P_2 &= -\left(\frac{\partial U}{\partial Q_2} + c_2(s)P_2\right)\mathrm{d}t + \sqrt{2D_2} f_2(s)\mathrm{d}B_2(t),\\
s &\in \{1, 2, \ldots, l\}.
\end{aligned} \tag{5.284}$$

By using the Itô stochastic differential rule, the Itô stochastic differential equation governing $H(t)$ can be obtained from Eqs. (5.283) and (5.284) as

$$\begin{aligned}
\mathrm{d}H = &(-c_1 P_1^2 - c_2 P_2^2 + f_1^2 D_1 + f_2^2 D_2)\mathrm{d}t\\
&+ \sqrt{2D_1} P_1 \mathrm{d}B_1(t) + \sqrt{2D_2} P_2 \mathrm{d}B_2(t).
\end{aligned} \tag{5.285}$$

Note that $U(Q)$ in Eq. (5.283) is non-separable and system (5.284) is a quasi-non-integrable Hamiltonian system if c_1, f_i, D_i are small. By performing the stochastic averaging of quasi-non-integrable Hamiltonian systems, the averaged Itô equation governing $H(t)$ of the form of Eq. (5.280) is obtained. According to Eq. (5.281), the drift coefficient $m(H, s)$ and the diffusion coefficient $\overline{\sigma}^2(H, s)$ are obtained as

$$\begin{aligned}
m(H, s) &= \frac{1}{T(H)} \int\limits_{\Omega} (-c_1 p_1^2 - c_2 p_2^2 + f_1^2 D_1 + f_2^2 D_2) \frac{1}{P_1} \mathrm{d}q_1 \mathrm{d}q_2 \mathrm{d}p_2, \\
\overline{\sigma}^2(H, s) &= \frac{1}{T(H)} \int\limits_{\Omega} (2f_1^2 D_1 p_1^2 + 2f_2^2 D_2 p_2^2) \frac{1}{p_1} \mathrm{d}q_1 \mathrm{d}q_2 \mathrm{d}p_2, \\
T(H) &= \int\limits_{\Omega} \frac{1}{p_1} \mathrm{d}q_1 \mathrm{d}q_2 \mathrm{d}p_2, \\
\Omega &= \{(q_1, q_2, p_2) | p_2^2/2 + \beta(\omega_1^2 q_1^2 + \omega_2^2 q_2^2)^3/6 \le H\}.
\end{aligned} \tag{5.286}$$

Making transformation $q_1 = r\cos\theta / \omega_1$, $q_2 = r\cos\theta / \omega_2$ and completing the integrals in Eq. (5.286) lead to

$$\begin{aligned}
m(h, s) &= f_1^2(s) D_1 + f_2^2(s) D_2 - \frac{3}{4}(c_1(s) + c_2(s))h, \\
\overline{\sigma}^2(h, s) &= \frac{3}{2}(f_1^2(s) D_1 + f_2^2(s) D_2)h, \\
T(h) &= \frac{2\pi^2}{\omega_1 \omega_2} \left(\frac{6h}{\beta}\right)^{1/3}.
\end{aligned} \tag{5.287}$$

Solving the reduced averaged FPK Eq. (5.267) with derivative moments in Eq. (5.287) numerically yields the stationary PDF $p(h, s)$, $(s \in \{1, 2, \ldots, l\})$. The stationary PDF $p(h)$ and the joint stationary PDF $p(q_1, q_2, p_1, p_2)$ are obtained from $p(h, s)$ following Eq. (5.268) as follows

$$p(q_1, q_2, p_1, p_2) = \frac{\omega_1 \omega_2}{2\pi^2} \left(\frac{\beta}{6h}\right)^{1/3} p(h) \bigg|_{h = \frac{1}{2}(p_1^2 + p_2^2) + \frac{1}{6}\beta(\omega_1^2 q_1^2 + \omega_2^2 q_2^2)^3}. \tag{5.288}$$

The marginal stationary PDFs and statistics can then be calculated from Eq. (5.288) using Eq. (5.269). In the following, two kinds of jump parameters are used for numerical calculation.

(i) Two-State Jumping Parameter

Let $\omega_1 = 1$, $\omega_2 = 2.2$, $\beta = 1$, $D_1 = 0.04$, $D_2 = 0.04$, $c_1(s = 1) = 0.01$, $c_2(s = 1) = 0.03$, $c_1(s = 2) = 0.03$, $c_2(s = 2) = 0.05$, $f_1(s = 1) = 2$, $f_2(s = 1) = 2$, $f_1(s = 2) = 1$, $f_2(s = 2) = 1$. It is seen that system in state $s = 2$ has larger damping coefficient and smaller excitation amplitude, implying the system in state $s = 2$ has littler energy. The jump rate matrices of $s(t)$ are

$$\Lambda_1 = \begin{bmatrix} -2 & 2 \\ 2 & -2 \end{bmatrix}, \quad \Lambda_2 = \begin{bmatrix} -1 & 1 \\ 2 & -2 \end{bmatrix}, \quad \Lambda_3 = \begin{bmatrix} -2 & 2 \\ 1 & -1 \end{bmatrix}. \tag{5.289}$$

The numerical results of the stationary PDF of total energy of system (5.282) with $\Lambda = \Lambda_1$, $\Lambda = \Lambda_2$, $\Lambda = \Lambda_3$ and $s = 1$, $s = 2$ respectively are shown in Fig. 5.45. It is seen that the peak of PDF $p(h)$ of system in $s = 2$ is highest and it gradually decreases in the order of Λ_3, Λ_1, Λ_2, $s = 1$. The solid line denotes theoretical results while the black dot denotes Monte Carlo simulation results. It is seen that the two results agree very well. The numerical results of stationary PDFs $p(q_1)$ and $p(q_2)$ of the system with different jumping rule are shown in Fig. 5.46. It is seen that the curves of $p(q_1), p(q_2)$ become flatten gradually in the order of $s = 2$, Λ_3, Λ_1, Λ_2, $s = 1$, indicating an increasing probability of the system moving away from the equilibrium point. Also, the theoretical results agree well with the simulation results. A sample of parameter $s(t)$ with $\Lambda = \Lambda_2$ is shown in Fig. 5.47.

(ii) **Three-State Jumping Parameter**

Now suppose that more state jumping parameters, i.e., $c_1(s = 3) = 0.05$, $c_2(s = 3) = 0.07$, $f_1(s = 3) = 1$ and $f_2(s = 3) = 1$ are added to the Case (i). Consider the following 4 jumping rules

$$\Lambda_1 = \begin{bmatrix} -3 & 1.5 & 1.5 \\ 1.5 & -3 & 1.5 \\ 1.5 & 1.5 & -3 \end{bmatrix}, \quad \Lambda_2 = \begin{bmatrix} -2 & 1 & 1 \\ 1.5 & -3 & 1.5 \\ 1.5 & 1.5 & -3 \end{bmatrix},$$
$$\Lambda_3 = \begin{bmatrix} -3 & 1.5 & 1.5 \\ 1 & -2 & 1 \\ 1.5 & 1.5 & -3 \end{bmatrix}, \quad \Lambda_4 = \begin{bmatrix} -3 & 1.5 & 1.5 \\ 1.5 & -3 & 1.5 \\ 1 & 1 & -2 \end{bmatrix}. \tag{5.290}$$

The stationary PDFs $p(h), p(q_1), p(q_2)$ of system (5.280) obtained from the stochastic averaging method and Monte Carlo simulation for different jumping rules are shown in Figs. 5.48 and 5.49. It is seen that the two results agree well. The system in $s = 3$ has largest damping coefficient and least excitation amplitude. So it has least stationary energy. Thus, all the PDFs in $s = 3$ have highest peak. On the contrary, the system in $s = 1$ has least damping coefficient and largest excitation amplitude. So it has largest stationary energy. Thus, all the PDFs in $s = 1$ have lowest peak. The system in $s = 2$ and all 4 jumping rules has middle stationary energy and so all the peaks of PDFs locate between the highest and lowest ones. The specific location of peak depends on the jumping rule. A sample of jumping parameter $s(t)$ is shown in Fig. 5.50.

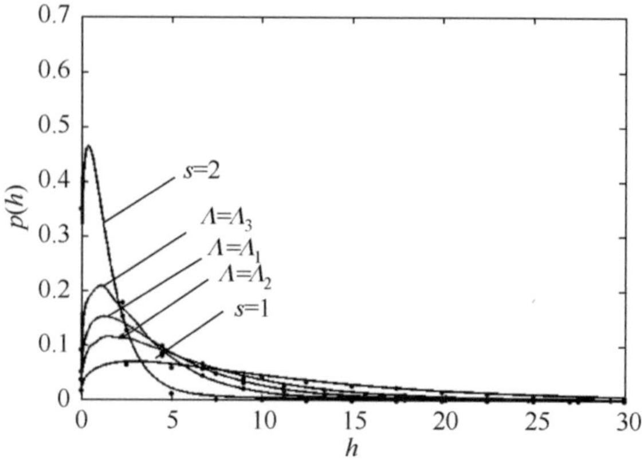

Fig. 5.45 Stationary PDF of total energy of two-state system (5.282) under different jump rules. ——, theoretical results; ●, results from Monte Carlo simulation (Hao 2015)

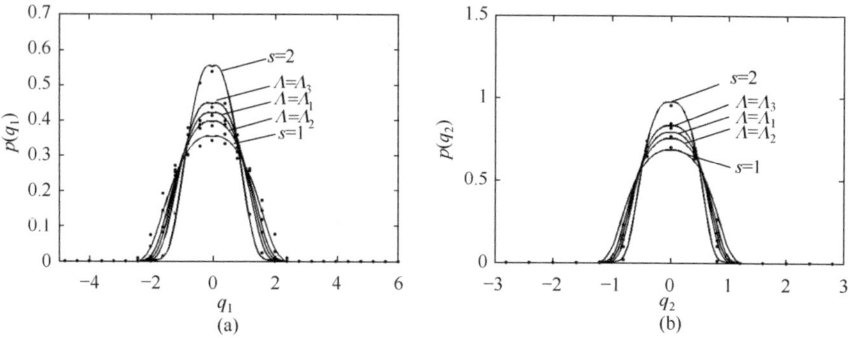

Fig. 5.46 Stationary PDF of generalized displacement of two-state system (5.282) under different jump rules. ——, theoretical results; ●, results from Monte Carlo simulation. **a** $p(q_1)$, **b** $p(q_2)$ (Hao 2015)

Fig. 5.47 A sample of jump process $s(t)$ in two-state system (5.282) with jump rule $\Lambda = \Lambda_2$ (Hao 2015)

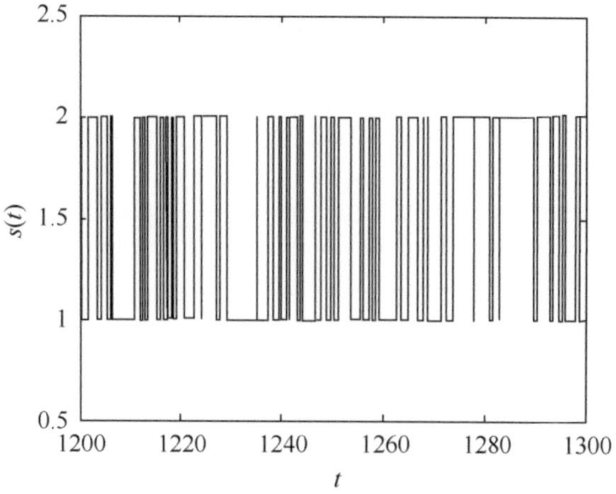

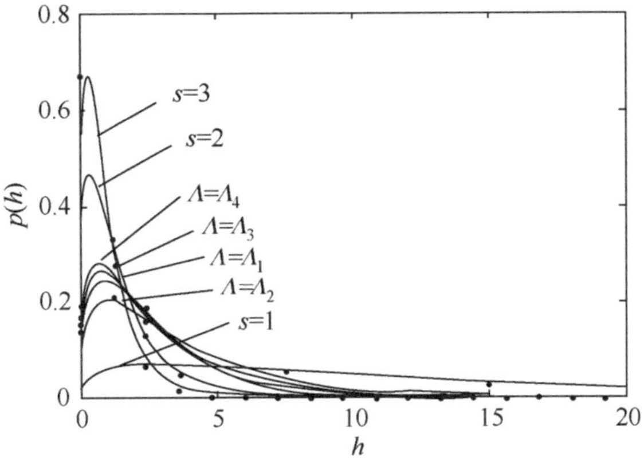

Fig. 5.48 Stationary PDF of total energy of three-state system (5.282) under different jump rules. ——, theoretical results; ●, results from Monte Carlo simulation (Hao 2015)

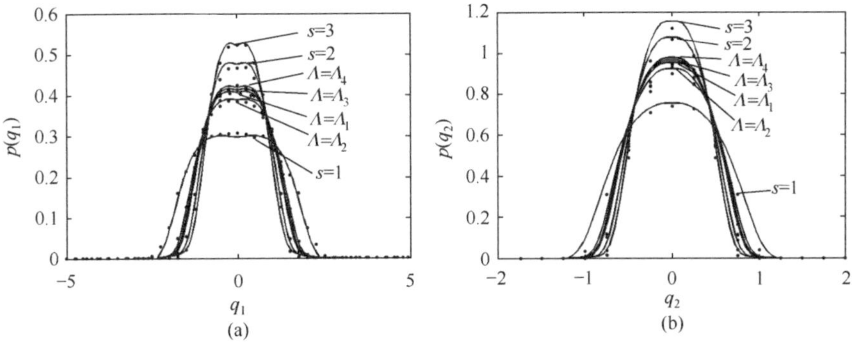

Fig. 5.49 Stationary PDFs of generalized displacements of three-state system (5.282) under different jump laws. ——, theoretical results; ●, results from Monte Carlo simulation (Hao 2015)

Fig. 5.50 A sample of jump process $s(t)$ in three-state system (Hao 2015)

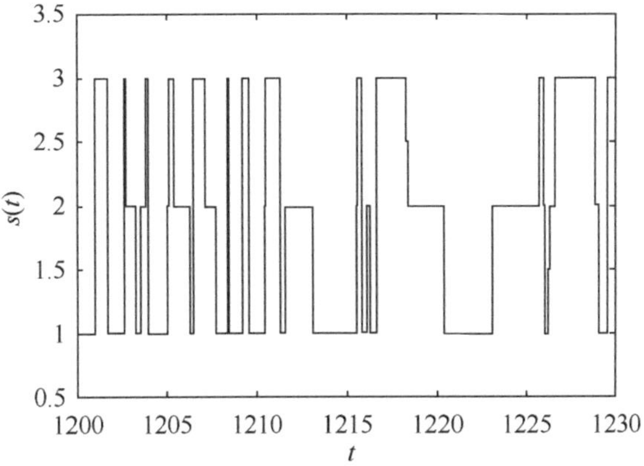

References

Blumenson L E. 1960. A derivation of n-dimensional spherical coordinates. American Mathematical Monthly, 67(1): 63–66.

Cai G Q, Zhu W Q. 2016. Elements of Stochastic Dynamics. Singapore: World Scientific Publisher.

Gan C B, Zhu W Q. 2001. First-passage failure of quasi-non-integrable-Hamiltonian systems. International Journal of Non-Linear Mechanics, 36: 209–220.

Hao Q. 2015. Stationary Response of Stochastically Excited Nonlinear Systems Undergoing Markovian Jump. M.S. Thesis. Hangzhou: Zhejiang University. (in Chinese)

Huan R H, Zhu W Q, Ma F, Ying Z G. 2015. Stationary response of a class of nonlinear stochastic systems undergoing Markovian jumps. ASME Journal of Applied Mechanics, 82: 051008.

Huang Z L, Liu Z H, Zhu W Q. 2004. Stationary response of multi-degree-of-freedom vibro-impact systems under white noise excitations. Journal of Sound and Vibration, 275: 223–240.

Khasminskii R Z. 1968. On the averaging principle for Itô stochastic differential equations. Kibernetika, 3(4): 260–279. (in Russian)

Liu Z H, Huang Z L, Zhu W Q. 2002. Stochastic response of a 2-DOF collision-vibration system. Journal of Vibration Engineering, 15(3): 257–261. (in Chinese)

Pan S S, Zhu W Q, Hu R C, Huan R H. 2017. Stationary response of stochastically excited nonlinear systems with continuous-time Markov jump. Journal of Zhejiang University-Science A, 18(2): 83–91.

Sun J J, Huan R H, Deng M L, Zhu W Q. 2021. A novel method for evaluating the averaged drift and diffusion coefficients of high DOF quasi-non-integrable Hamiltonian systems. Nonlinear Dynamics, 106: 2975–2989.

Zhu W Q, Deng M L. 2004. Optimal bounded control for minimizing the response of quasi non-integrable Hamiltonian systems. Nonlinear Dynamics, 35: 81–100.

Zhu W Q, Huang Z L. 2004. Stochastic stabilization of quasi non-integrable Hamiltonian systems. International Journal of Non-Linear Mechanics, 39: 879–895.

Zhu W Q, Yang Y Q. 1997. Stochastic averaging of quasi-nonintegrable -Hamiltonian systems. ASME Journal of Applied Mechanics, 64: 157–164.

Zhu W Q, Huang Z L, Yang Y Q. 1997. Stochastic averaging of quasi integrable-Hamiltonian systems. ASME Journal of Applied Mechanics, 64(4): 975–984.

Zhu W Q, Huang Z L, Suzuki Y. 2002. Stochastic averaging and Lyapunov exponent of quasi partially integrable Hamiltonian systems. International Journal of Non-Linear Mechanics, 37(3): 419–437.

Chapter 6
Stochastic Averaging Methods of Quasi-Hamiltonian Systems Excited by Gaussian and Poisson White Noises

In the study of stochastic dynamics, it is assumed that excitations are continuous random processes in general, and random jump processes in rare cases. However, in some practical problems in science, engineering and society, random excitations may be composed of both continuous random processes and random jump processes, such as uneven terrains and wind turbulences. So far, there have been limited researches on the stochastic dynamics of nonlinear systems excited by both continuous and jump random processes. Since the stochastic averaging methods are powerful approximate analytical techniques for studying the dynamics of MDOF nonlinear systems, in the present chapter the stochastic averaging methods of quasi-Hamiltonian systems excited by combined Gaussian white noises and Poisson white noises are presented. It is shown that the effects of the two kinds of white noises can be separated during the averaging procedure. Hence, the stochastic averaging methods described in this chapter can also be applied to quasi-Hamiltonian systems excited solely by Poisson white noises, just deleting the terms caused by Gaussian white noises.

6.1 Quasi-Hamiltonian Systems Excited by Gaussian and Poisson White Noises

Consider an n-DOF quasi-Hamiltonian system excited by both Gaussian and Poisson white noises. The motion equations are

$$\dot{Q}_i = \frac{\partial H'}{\partial P_i},$$

W.-Q. Zhu et al., *Stochastic Averaging*, https://doi.org/10.1007/978-981-96-3829-1_6

$$
\begin{aligned}
\dot{P}_i = {} & \frac{\partial H'}{\partial Q_i} - \varepsilon^2 \sum_{j=1}^{n} c_{ij}(\boldsymbol{Q}, \boldsymbol{P}) \frac{\partial H'}{\partial P_j} + \varepsilon \sum_{k=1}^{n_g} g_{ik}(\boldsymbol{Q}, \boldsymbol{P}) W_{gk}(t) \\
& + \sum_{l=1}^{n_p} f_{il}(\boldsymbol{Q}, \boldsymbol{P}) W_{pl}(t), \\
& i = 1, 2, \ldots, n,
\end{aligned} \tag{6.1}
$$

where $\boldsymbol{Q} = [Q_1, Q_2, \ldots, Q_n]^{\mathrm{T}}$ and $\boldsymbol{P} = [P_1, P_2, \ldots, P_n]^{\mathrm{T}}$ are vectors of generalized displacements and momenta, respectively, $H' = H'(\boldsymbol{Q}, \boldsymbol{P})$ is an infinitely differentiable Hamiltonian, $c_{ij}(\boldsymbol{Q}, \boldsymbol{P})$ denote coefficients of quasi-linear dampings which are differentiable functions of $\boldsymbol{Q}$ and $\boldsymbol{P}$, $g_{ik}(\boldsymbol{Q}, \boldsymbol{P})$ are twice-differentiable functions representing the amplitudes of Gaussian white noise excitations, $f_{il}(\boldsymbol{Q}, \boldsymbol{P})$ are infinitely differentiable functions representing the amplitudes of Poisson white noise excitations, ε is a small positive parameter, $W_{gk}(t),\ k = 1, 2, \ldots, n_g$ are Gaussian white noises with correlation functions defined in Eq. (2.128), and $W_{pl}(t),\ l = 1, 2, \ldots, n_p$ are mutually independent Poisson white noises with average arrival rates λ_l and Gaussian distributed pulse amplitudes.

As shown in Chap. 2, Gaussian white noises $W_{gk}(t)$ and Poisson white noises $W_{pl}(t)$ can be treated as derivative processes of the Wiener processes (Brownian motion processes) $B_k(t)$ and the compound Poisson processes $C_l(t)$, respectively, Eq. (6.1) can be rewritten as the following Stratonovich stochastic differential equations (SDEs)

$$
\begin{aligned}
dQ_i = {} & \frac{\partial H'}{\partial P_i} dt, \\
dP_i = {} & \left[\frac{\partial H'}{\partial Q_i} - \varepsilon^2 \sum_{j=1}^{n} c_{ij}(\boldsymbol{Q}, \boldsymbol{P}) \frac{\partial H'}{\partial P_j} \right] dt + \varepsilon \sum_{k=1}^{n_g} g_{ik}(\boldsymbol{Q}, \boldsymbol{P}) \mathrm{d}^{\circ} B_k(t) \\
& + \sum_{l=1}^{n_p} f_{il}(\boldsymbol{Q}, \boldsymbol{P}) \mathrm{d}^{\circ} C_l(t), \\
& i = 1, 2, \ldots, n,
\end{aligned} \tag{6.2}
$$

and further transformed into the following Itô SDEs (Di Paola and Falsone 1993a, b)

$$
\begin{aligned}
\mathrm{d}Q_i &= \frac{\partial H'}{\partial P_i}\mathrm{d}t,\\
\mathrm{d}P_i &= \left[-\frac{\partial H'}{\partial Q_i} - \varepsilon^2\sum_{j=1}^{n} c_{ij}(\mathbf{Q},\mathbf{P})\frac{\partial H'}{\partial P_j} + \varepsilon^2\sum_{k,k'=1}^{n_g}\sum_{j=1}^{n}\pi K_{kk'}g_{jk'}(\mathbf{Q},\mathbf{P})\frac{\partial g_{ik}}{\partial P_j}\right]\mathrm{d}t\\
&\quad +\varepsilon\sum_{k=1}^{n_g}\sigma_{ik}(\mathbf{Q},\mathbf{P})\mathrm{d}B_k(t) + \sum_{l=1}^{n_p}\left(\sum_{r=1}^{\infty}\frac{\varepsilon^r}{r!}f_{il}^{(r)}(\mathbf{Q},\mathbf{P})(\mathrm{d}C_l(t))^r\right),\\
i &= 1,2,\ldots,n,
\end{aligned}
\tag{6.3}
$$

where $B_k(t)$ are independent unit Wiener processes, $\sigma_{ik} = (\boldsymbol{gL})_{ik}$, $\boldsymbol{g} = \left[g_{ik}\right]_{n\times n_g}$, $\boldsymbol{LL}^T = 2\pi\boldsymbol{K}$ is the intensity matrix of Gaussian white noises, $\varepsilon^2\sum_{k,k'=1}^{n_g}\sum_{j=1}^{n}\pi K_{kk'}g_{jk'}\frac{\partial g_{ik}}{\partial P_j}$ are Wong-Zakai correction terms due to Gaussian white noise excitations, $\sum_{l=1}^{n_p}\sum_{r=2}^{\infty}\frac{\varepsilon^r}{r!}f_{il}^{(r)}(\boldsymbol{Q},\boldsymbol{P})(dC_l(t))^r$ are correction terms due to Poisson white noise excitations, and

$$
f_{il}^{(r)}(\boldsymbol{Q},\boldsymbol{P}) = \sum_{j=1}^{n}\frac{\partial f_{il}^{(r-1)}(\boldsymbol{Q},\boldsymbol{P})}{\partial P_j}f_{jl};\ f_{il}^{(1)}(\boldsymbol{Q},\boldsymbol{P}) = f_{il}(\boldsymbol{Q},\boldsymbol{P})
$$

The Wong-Zakai correction terms are functions of generalized displacements Q_i and generalized momenta P_i, and can be divided into conservative and dissipative parts. The conservative part can be merged with $-\partial H'/\partial Q_i$ to form $-\partial H\big/\partial Q_i$ with $-\partial H/\partial Q_i = -\partial H'/\partial Q_i$, where $H = H(\boldsymbol{Q},\boldsymbol{P})$ is the modified Hamiltonian. The dissipative parts can be merged with $-\varepsilon^2 c_{ij}\partial H'/\partial P_j$ to form the modified damping $-\varepsilon^2 m_{ij}(\boldsymbol{Q},\boldsymbol{P})\partial H/\partial P_j$. . Thus, system (6.3) can be rewritten as

$$
\begin{aligned}
\mathrm{d}Q_i &= \frac{\partial H}{\partial P_i}\mathrm{d}t,\\
\mathrm{d}P_i &= -\left[\frac{\partial H}{\partial Q_i} + \varepsilon^2\sum_{j=1}^{n} m_{ij}(\boldsymbol{Q},\boldsymbol{P})\frac{\partial H}{\partial P_j}\right]\mathrm{d}t\\
&\quad +\varepsilon\sum_{k=1}^{n_g}\sigma_{ik}(\boldsymbol{Q},\boldsymbol{P})\mathrm{d}B_k(t)\\
&\quad +\sum_{l=1}^{n_p}\left(\sum_{r=1}^{\infty}\frac{\varepsilon^r}{r!}f_{il}^{(r)}(\boldsymbol{Q},\boldsymbol{P})(\mathrm{d}C_l(t))^r\right),\\
i &= 1,2,\ldots,n.
\end{aligned}
\tag{6.4}
$$

Based on the integral representation of the compound Poisson process in Eq. (2.159), Eq. (6.4) can also be rewritten as the following equivalent stochastic integral differential equations (SIDEs)

$$
\begin{aligned}
\mathrm{d}Q_i &= \frac{\partial H}{\partial P_i}\mathrm{d}t, \\
\mathrm{d}P_i &= -\left[\frac{\partial H}{\partial Q_i} + \varepsilon^2 \sum_{j=1}^{n} m_{ij}(\boldsymbol{Q}, \boldsymbol{P})\frac{\partial H}{\partial P_j}\right]\mathrm{d}t \\
&\quad + \varepsilon \sum_{k=1}^{n_g} \sigma_{ik}(\boldsymbol{Q}, \boldsymbol{P})\mathrm{d}B_k(t) \\
&\quad + \sum_{l=1}^{n_p} \int_{\mathcal{Y}_l} \gamma_{il}(\boldsymbol{Q}, \boldsymbol{P}, Y_l)\mathcal{P}_l(\mathrm{d}t, \mathrm{d}Y_l), \\
i &= 1, 2, \ldots, n,
\end{aligned}
\tag{6.5}
$$

where $\mathcal{Y}_i$ is the Poisson symbol space, $\mathcal{P}_l(dt, dY_l)$ is the Poisson stochastic measure, and

$$
\begin{aligned}
\gamma_{il}(\boldsymbol{Q}, \boldsymbol{P}, Y_l) &= \sum_{r=1}^{\infty} \frac{\varepsilon^r}{r!} f_{il}^{(r)}(\boldsymbol{Q}, \boldsymbol{P})Y_l^r, \\
i &= 1, 2, \ldots, n; \quad l = 1, 2, \ldots, n_p.
\end{aligned}
\tag{6.6}
$$

Based on the integrability and resonance of Hamiltonian systems associated with Eq. (6.5), quasi-Hamiltonian system (6.4) or (6.5) can be categorized into five classes: quasi-nonintegrable Hamiltonian system, quasi-integrable and non-resonant Hamiltonian system, quasi-integrable and resonant Hamiltonian system, quasi-partially integrable and non-resonant Hamiltonian system, and quasi-partially integrable and resonant Hamiltonian system. In the following sections, the averaged SIDEs and FPK equations for these five classes of quasi-Hamiltonian systems are derived, and the averaged FPK equations are solved by using perturbation methods to obtain the approximate stationary PDFs (PDFs) and statistics.

6.2 Quasi-Non-Integrable Hamiltonian Systems

6.2.1 Combined Gaussian and Poisson White Noise Excitations

Assume that the Hamiltonian system associated with system (6.5) is non-integrable, i.e., the Hamiltonian $H(t)$ is the unique first integral of the Hamiltonian system associated with Eq. (6.5). Introduce the transformation

$$
H(t) = H(\boldsymbol{Q}(t), \boldsymbol{P}(t)) \tag{6.7}
$$

According to the Itô differential rule and the stochastic jump-diffusion chain rule (Hanson 2007), the Hamiltonian $H(t)$ satisfies the following SIDE

$$
\begin{aligned}
\mathrm{d}H &= \varepsilon^2\left(-\sum_{i,j=1}^{n} m_{ij}\frac{\partial H}{\partial P_i}\frac{\partial H}{\partial P_j}+\frac{1}{2}\sum_{i,j=1}^{n}\sum_{k=1}^{n_g}\sigma_{ik}\sigma_{jk}\frac{\partial^2 H}{\partial P_i\partial P_j}\right)\mathrm{d}t\\
&+\varepsilon\sum_{j=1}^{n}\sum_{k=1}^{n_g}\sigma_{jk}\frac{\partial H}{\partial P_j}\mathrm{d}B_k(t)\\
&+\sum_{l=1}^{n_p}\int_{\mathcal{Y}_l}\left[H\left(\boldsymbol{Q},\boldsymbol{P}+\hat{\boldsymbol{\gamma}}_l\right)-H(\boldsymbol{Q},\boldsymbol{P})\right]\mathcal{P}_l(\mathrm{d}t,\mathrm{d}Y_l),
\end{aligned}
\tag{6.8}
$$

where $\hat{\boldsymbol{\gamma}}_l$ is the l-th column of the matrix $\left[\gamma_{il}\right]_{n\times n_p}$.

Performing Taylor expansion for $H(\boldsymbol{Q},\boldsymbol{P}+\hat{\gamma}_l)-H(\boldsymbol{Q},\boldsymbol{P}),\ l=1,\ldots,n_p$ yields

$$
\begin{aligned}
&H\left(\boldsymbol{Q},\boldsymbol{P}+\hat{\boldsymbol{\gamma}}_l\right)-H(\boldsymbol{Q},\boldsymbol{P})\\
&=\sum_{r=1}^{\infty}\frac{1}{r!}\sum_{i_1=1}^{n}\cdots\sum_{i_r=1}^{n}\frac{\partial^r H}{\partial P_{i_1}\cdots\partial P_{i_r}}\sum_{j_1=1}^{\infty}\frac{\varepsilon^{j_1}}{j_1!}f_{i_1 l}^{(j_1)}Y_l^{j_1}\cdots\sum_{j_r=1}^{\infty}\frac{\varepsilon^{j_r}}{j_r!}f_{i_r l}^{(j_r)}Y_l^{j_r}\\
&=\sum_{r=1}^{\infty}\frac{1}{r!}\sum_{i_1=1}^{n}\cdots\sum_{i_r=1}^{n}\frac{\partial^r H}{\partial P_{i_1}\cdots\partial P_{i_r}}\sum_{j_1=1}^{\infty}\cdots\sum_{j_r=1}^{\infty}\frac{\varepsilon^{j_1+\cdots+j_r}}{j_1!\cdots j_r!}f_{i_1 l}^{(j_1)}\cdots f_{i_r l}^{(j_r)}Y_l^{j_1+\cdots+j_r}\\
&=\sum_{k=1}^{\infty}\varepsilon^k Y_l^k K_{1,k,l}(\boldsymbol{Q},\boldsymbol{P}),
\end{aligned}
\tag{6.9}
$$

Thus,

$$
\begin{aligned}
\left[H\left(\boldsymbol{Q},\boldsymbol{P}+\hat{\boldsymbol{\gamma}}_l\right)-H(\boldsymbol{Q},\boldsymbol{P})\right]^j &=\left[\sum_{k=1}^{\infty}\varepsilon^k K_{1,k,l}Y_l^k\right]^j\\
&=\sum_{k_1=1}^{\infty}\cdots\sum_{k_j=1}^{\infty}\varepsilon^{k_1+\cdots+k_j}Y_l^{k_1+\cdots+k_j}\left(\prod_{i=1}^{j}K_{1,k_i,l}\right)\\
&=\sum_{k=j}^{\infty}\varepsilon^k Y_l^k\left(\sum_{\substack{k_1+\cdots+k_j=k\\ k_s\geq 1,s=1,\cdots,j}}\left(\prod_{i=1}^{j}K_{1,k_i,l}\right)\right)\\
&=\sum_{k=j}^{\infty}\varepsilon^k Y_l^k K_{j,k,l}(\boldsymbol{Q},\boldsymbol{P}),
\end{aligned}
\tag{6.10}
$$

where $K_{j,k,l}=K_{j,k,l}(\boldsymbol{Q},\boldsymbol{P})$. Therefore, Eq. (6.8) can be rewritten as

$$
\begin{aligned}
\mathrm{d}H = {} & \varepsilon^2\left(-\sum_{i,j=1}^{n} m_{ij}\frac{\partial H}{\partial P_i}\frac{\partial H}{\partial P_j} + \frac{1}{2}\sum_{i,j=1}^{n}\sum_{k=1}^{n_g}\sigma_{ik}\sigma_{jk}\frac{\partial^2 H}{\partial P_i \partial P_j}\right)\mathrm{d}t \\
& + \varepsilon\sum_{j=1}^{n}\sum_{k=1}^{n_g}\sigma_{jk}\frac{\partial H}{\partial P_j}\mathrm{d}B_k(t) \\
& + \sum_{l=1}^{n_p}\int_{\mathcal{Y}_l}\left[\sum_{k=1}^{\infty}\varepsilon^k K_{1,k,l}(\boldsymbol{Q},\boldsymbol{P})Y_l^k\right]\mathcal{P}_l(\mathrm{d}t,\mathrm{d}Y_l).
\end{aligned}
\tag{6.11}
$$

Replace the equation of P_1 in Eq. (6.5) by Eq. (6.11) and express P_1 with $\boldsymbol{Q}, H, P_2, \cdots, P_n$ according to Eq. (6.7). The new system equations are then composed of Eq. (6.11) and the equations in Eq. (6.5) except for equation of P_1. In the new set of equations, $\boldsymbol{Q}(t)$, $P_2(t), \cdots, P_n(t)$ are fast-varying processes while $H(t)$ is slowly-varying process. According to the stochastic averaging principle (Khasminskii 1968; Xu et al. 2011), $H(t)$ weakly converges to a one-dimensional Markov process as $\varepsilon \to 0$. The averaged SIDE for $H(t)$ can be obtained by performing time averaging over Eq. (6.11). Replace $\mathrm{d}t$ with $(\partial H / \partial P_1)^{-1}\mathrm{d}Q_1$ using the first equation of Eq. (6.5). Note that since the non-integrable Hamiltonian system is ergodic on the $(2n-1)$-dimensional isoenergetic surface (Binney et al. 1992), the time averaging can be replaced by spatial averaging with respect to $Q_2, \cdots, Q_n, P_2, \cdots, P_n$.

It is seen from Eq. (6.11) that the averaged SIDE for the Hamiltonian contains infinite terms. To obtain a closed-form averaged SIDE, it is necessary to rearrange Eq. (6.11) in terms of power of ε and then truncate it. Neglecting terms of orders equal to or higher than ε^{u+1}, the following truncated averaged SIDE can be obtained

$$
\begin{aligned}
\mathrm{d}H = {} & \left(\varepsilon^2\overline{m}(H) + \sum_{i=0}^{u-2}\varepsilon^{i+2}U_i(H)\right)\mathrm{d}t + \varepsilon\overline{\sigma}(H)\mathrm{d}B(t) \\
& + \sum_{l=1}^{n_p}\sum_{k=1}^{u-1}\int_{\mathcal{Y}_l}\left(\sum_{s=1}^{k}\varepsilon^s V_{k,s,l}(H)Y_{k,l}\right)\mathcal{P}_{k,l}\big(\mathrm{d}t,\mathrm{d}Y_{k,l}\big),
\end{aligned}
\tag{6.12}
$$

where $B(t)$ is a unit Wiener process, and

$$
\begin{aligned}
\overline{m}(H) = {} & \frac{1}{T(H)}\int_{\Omega}\left\{\left[-\sum_{i,j=1}^{n} m_{ij}\frac{\partial H}{\partial p_i}\frac{\partial H}{\partial p_j}\right.\right. \\
& \left.\left. + \frac{1}{2}\sum_{i,j=1}^{n}\sum_{k=1}^{n_g}\sigma_{ik}\sigma_{jk}\frac{\partial^2 H}{\partial p_i \partial p_j}\right] \Big/ \frac{\partial H}{\partial p_1}\right\}\mathrm{d}q_1\cdots\mathrm{d}q_n\mathrm{d}p_2\cdots\mathrm{d}p_n,
\end{aligned}
\tag{6.13}
$$

$$
U_0(H) = \sum_{l=1}^{n_p}\frac{\lambda_l E\left[Y_l^2\right]}{T(H)}\int_{\Omega}\left(K_{1,2,l}\Big/\frac{\partial H}{\partial p_1}\right)\mathrm{d}q_1\cdots\mathrm{d}q_n\mathrm{d}p_2\cdots\mathrm{d}p_n
\tag{6.14}
$$

$$U_k(H) = \sum_{l=1}^{n_p} \frac{\lambda_l E\left[Y_l^{k+2}\right]}{T(H)} \int_\Omega \left(K_{1,k+2,l} \bigg/ \frac{\partial H}{\partial p_1} \right) \mathrm{d}q_1 \cdots \mathrm{d}q_n \mathrm{d}p_2 \cdots \mathrm{d}p_n, \quad k = 1, \cdots, u-2, \tag{6.15}$$

$$\overline{\sigma}^2(H) = \frac{1}{T(H)} \int_\Omega \left[\sum_{i,j=1}^{n} \sum_{k=1}^{n_g} \sigma_{jk}\sigma_{ik} \frac{\partial H}{\partial p_i} \frac{\partial H}{\partial p_j} \bigg/ \frac{\partial H}{\partial p_1} \right] \mathrm{d}q_1 \cdots \mathrm{d}q_n \mathrm{d}p_2 \cdots \mathrm{d}p_n \tag{6.16}$$

The expression for $V_{k,s,l}(h),\ k = 1, \cdots, u-1;\ s = 1, \cdots, k;\ l = 1, \cdots, n_p$ needs to be determined from the following equation for the terms of each power of ε,

$$\sum_{k=1}^{u-1} \left(\sum_{s=1}^{k} \varepsilon^s V_{k,s,l}(H) \right)^j \lambda_{kl} E\left[Y_{kl}^j\right] = \sum_{m=j}^{u} \frac{\varepsilon^m \lambda_l E\left[Y_l^m\right]}{T(H)} \int_\Omega \left(K_{j,m,l} \bigg/ \frac{\partial H}{\partial p_1} \right) \mathrm{d}q_1 \cdots \mathrm{d}q_n \mathrm{d}p_2 \cdots \mathrm{d}p_n, \quad l = 1, \cdots, n_p, \tag{6.17}$$

where

$$T(H) = \int_\Omega \left(1 \bigg/ \frac{\partial H}{\partial p_1} \right) \mathrm{d}q_1 \cdots \mathrm{d}q_n \mathrm{d}p_2 \cdots \mathrm{d}p_n,$$
$$\Omega = \{ (q_1, \cdots, q_n, p_2, \cdots, p_n) | H(q_1, \cdots, q_n, 0, p_2, \cdots, p_n) \le H \}. \tag{6.18}$$

Furthermore, $\mathcal{P}_{k,l}(dt, dY_{k,l})$ represents independent Poisson random measure with the following properties

$$\lambda_{k,l} E\left[Y_{k,l}^j\right] = \begin{cases} 0 & r = 1, \\ \overline{M}_{k,r,l} & r = 2, \cdots, u-k+1, \\ \overline{m}_{k,r,l} & r = u-k+2, u-k+3, \cdots, \end{cases} \tag{6.19}$$

where $\overline{M}_{k,r,l} \gg \varepsilon$ and $\overline{m}_{k,l,r} < \varepsilon^u$.

The truncated averaged FPK equation associated with the truncated averaged SIDE (6.12) is

$$\frac{\partial}{\partial t} p = \sum_{k=1}^{u} (-1)^k \frac{1}{k!} \frac{\partial^k}{\partial h^k} (\overline{a}_k(h) p) + O\left(\varepsilon^{u+1}\right) \tag{6.20}$$

where

$$
\begin{aligned}
\overline{a}_1(h) &= \varepsilon^2\overline{m}(h) + \sum_{i=0}^{u-2}\varepsilon^{i+2}U_i(h) \\
&= \frac{1}{T(h)}\int_{\Omega}\left\{\left[-\sum_{i=1}^{n}\sum_{j=1}^{n}m_{ij}\frac{\partial h}{\partial p_i}\frac{\partial h}{\partial p_j}\right.\right. \\
&\left.\left.+\frac{1}{2}\sum_{i,j=1}^{n}\sum_{k=1}^{n_g}\sigma_{ik}\sigma_{jk}\frac{\partial^2 h}{\partial p_i \partial p_j}\right]\Big/\frac{\partial h}{\partial p_1}\right\}\mathrm{d}q_1\cdots\mathrm{d}q_n\mathrm{d}p_2\cdots\mathrm{d}p_n \\
&+\sum_{k=2}^{u}\varepsilon^k\left\{\sum_{l=1}^{n_p}\frac{\lambda_l E\left[Y_l^k\right]}{T(h)}\int_{\Omega}\left(K_{1,k,l}\Big/\frac{\partial h}{\partial p_1}\right)\mathrm{d}q_1\cdots\mathrm{d}q_n\cdots\mathrm{d}p_2\cdots\mathrm{d}p_n\right\},
\end{aligned}
\tag{6.21}
$$

$$
\begin{aligned}
\overline{a}_2(h) &= \varepsilon^2\overline{\sigma}^2(h) + \frac{1}{\mathrm{d}t}E\left[\left\{\sum_{l=1}^{n_p}\sum_{k=1}^{u-1}\sum_{s=1}^{k}\varepsilon^s\int_{\mathcal{Y}_l}V_{k,s,l}(H)Y_{kl}\mathcal{P}_{kl}(\mathrm{d}t,\mathrm{d}Y_{kl})\right\}^2\middle| H=h\right] \\
&= \frac{\varepsilon^2}{T(h)}\int_{\Omega}\left[\sum_{i,j=1}^{n}\sum_{k=1}^{n_g}\sigma_{jk}\sigma_{ik}\frac{\partial h}{\partial p_i}\frac{\partial h}{\partial p_j}\Big/\frac{\partial h}{\partial p_1}\right]\mathrm{d}q_1\cdots\mathrm{d}q_n\mathrm{d}p_2\cdots\mathrm{d}p_n \\
&+\sum_{m=2}^{u}\varepsilon^m\left\{\sum_{l=1}^{n_p}\frac{\lambda_l E\left[Y_l^m\right]}{T(h)}\int_{\Omega}\left(K_{2,m,l}\Big/\frac{\partial h}{\partial p_1}\right)\mathrm{d}q_1\cdots\mathrm{d}q_n\mathrm{d}p_2\cdots\mathrm{d}p_n\right\},
\end{aligned}
\tag{6.22}
$$

$$
\begin{aligned}
\overline{a}_j(h) &= \frac{1}{\mathrm{d}t}E\left[\left\{\sum_{l=1}^{n_p}\sum_{k=1}^{u-1}\sum_{s=1}^{k}\varepsilon^s\int_{\mathcal{Y}_l}V_{k,s,l}(H)Y_{kl}\mathcal{P}_{kl}(\mathrm{d}t,\mathrm{d}Y_{kl})\right\}^j\middle| H=h\right] \\
&= \sum_{m=j}^{u}\varepsilon^m\left\{\sum_{l=1}^{n_p}\frac{\lambda_l E\left[Y_l^m\right]}{T(h)}\int_{\Omega}\left(K_{j,m,l}\Big/\frac{\partial h}{\partial p_1}\right)\mathrm{d}q_1\cdots\mathrm{d}q_n\mathrm{d}p_2\cdots\mathrm{d}p_n\right\}, \\
j &= 3,\cdots,u.
\end{aligned}
\tag{6.23}
$$

In Eq. (6.20), $p = p(h, t|h_0)$ represents the transition PDF of the Hamiltonian with initial condition given by

$$
p(h, 0|h_0) = \delta(h - h_0) \tag{6.24}
$$

or $p = p(h, t)$ represents the PDF of the Hamiltonian with initial condition given by

$$
p(h, 0) = p(h_0) \tag{6.25}
$$

Since Hamiltonian $H(t)$ varies within interval $[0, \infty)$, Eq. (6.20) must satisfy the following boundary conditions

$$\begin{aligned} & p(h)|_{h=0} = \text{finite}, \quad \left.\frac{\partial^n p(h)}{\partial h^n}\right|_{h=0} = \text{finite}, \\ & \lim_{h\to\infty} p(h) = 0\,, \quad \lim_{h\to\infty} \frac{\partial^n p}{\partial h^n} = 0\,, n = 1, 2, ... \end{aligned} \tag{6.26}$$

Additionally, p must satisfy the normalization condition

$$\int_0^\infty p(h)\mathrm{d}h = 1 \tag{6.27}$$

The solution of the truncated FPK Eq. (6.20) with $\partial p/\partial t=0$ can be obtained by using the perturbation methods. Assume that Eq. (6.20) has a stationary solution in the form of a power series

$$p(h) = p_0(h) + \varepsilon p_1(h) + \varepsilon^2 p_2(h) + \cdots \tag{6.28}$$

Substituting Eq. (6.28) into Eq. (6.20) with $\partial p/\partial t = 0$ and letting the sum of all terms with same power of ε to be zero, a series of ordinary differential equations for $p_0(h), p_1(h), p_2(h), \cdots$ can be established. Solving these equations successively and substituting the obtained solutions into Eq. (6.28) yield the stationary solution of truncated FPK Eq. (6.20). Based on Eq. (5.18), the approximate joint stationary PDF of generalized displacements and generalized momenta can be obtained as follows

$$p(\boldsymbol{p}, \boldsymbol{q}) = p(h)/T(h)|_{h=h(\boldsymbol{q},\boldsymbol{p})} \tag{6.29}$$

In practical applications, since the coefficients in Eq. (6.12) rapidly decrease with increasing k, it is sufficient to take $u = 4$, resulting in the following closed-form averaged SIDE

$$\begin{aligned} \mathrm{d}H = & \left(\varepsilon^2\overline{m}(H) + \sum_{i=0}^{2} \varepsilon^{i+2} U_i(H)\right)\mathrm{d}t + \varepsilon^2\overline{\sigma}^2(H)\mathrm{d}B(t) \\ & + \sum_{l=1}^{n_p}\sum_{k=1}^{3} \int_{\mathcal{Y}_l} \left(\sum_{s=1}^{k} \varepsilon^s V_{k,s,l}(H) Y_{k,l}\right) \mathcal{P}_{k,l}\big(\mathrm{d}t, \mathrm{d}Y_{k,l}\big), \end{aligned} \tag{6.30}$$

where

$$\overline{m}(H) = \frac{1}{T(H)} \int_\Omega \left\{\left[-\sum_{i,j=1}^{n} m_{ij} \frac{\partial H}{\partial p_i}\frac{\partial H}{\partial p_j}\right.\right.$$

$$+\frac{1}{2}\sum_{i,j=1}^{n}\sum_{k=1}^{n_g}\sigma_{ik}\sigma_{jk}\frac{\partial^2 H}{\partial p_i \partial p_j}\Bigg]\Bigg/\frac{\partial H}{\partial p_1}\Bigg\}\mathrm{d}q_1\cdots\mathrm{d}q_n\mathrm{d}p_2\cdots\mathrm{d}p_n\,, \tag{6.31}$$

$$U_0(H)=\sum_{l=1}^{n_p}\frac{\lambda_l E\left[Y_l^2\right]}{T(H)}\int_\Omega\left(K_{1,2,l}\Bigg/\frac{\partial H}{\partial p_1}\right)\mathrm{d}q_1\cdots\mathrm{d}q_n\mathrm{d}p_2\cdots\mathrm{d}p_n\,, \tag{6.32}$$

$$U_1(H)=\sum_{l=1}^{n_p}\frac{\lambda_l E\left[Y_l^3\right]}{T(H)}\int_\Omega\left(K_{1,3,l}\Bigg/\frac{\partial H}{\partial p_1}\right)\mathrm{d}q_1\cdots\mathrm{d}q_n\mathrm{d}p_2\cdots\mathrm{d}p_n\,, \tag{6.33}$$

$$U_2(H)=\sum_{l=1}^{n_p}\frac{\lambda_l E\left[Y_l^4\right]}{T(H)}\int_\Omega\left(K_{1,4,l}\Bigg/\frac{\partial H}{\partial p_1}\right)\mathrm{d}q_1\cdots\mathrm{d}q_n\mathrm{d}p_2\cdots\mathrm{d}p_n, \tag{6.34}$$

$$\overline{\sigma}^2(H)=\frac{1}{T(H)}\int_\Omega\left[\sum_{i,j=1}^{n}\sum_{k=1}^{n_g}\sigma_{jk}\sigma_{ik}\frac{\partial H}{\partial p_i}\frac{\partial H}{\partial p_j}\Bigg/\frac{\partial H}{\partial p_1}\right]\mathrm{d}q_1\cdots\mathrm{d}q_n\mathrm{d}p_2\cdots\mathrm{d}p_n. \tag{6.35}$$

$$l=1,\cdots,n_p$$

$$\lambda_{1,l}E\left[Y_{1,l}^4\right]V_{1,1,l}^4=\frac{\lambda_l E\left[Y_l^4\right]}{T(H)}\int_\Omega\left(K_{4,4,l}\Bigg/\frac{\partial H}{\partial p_1}\right)\mathrm{d}q_1\cdots\mathrm{d}q_n\mathrm{d}p_2\cdots\mathrm{d}p_n \tag{6.36}$$

$$\lambda_{1,l}E\left[Y_{1,l}^3\right]V_{1,1,l}^3+\lambda_{2,l}E\left[Y_{2,l}^3\right]V_{2,1,l}^3=\frac{\lambda_l E\left[Y_l^3\right]}{T(H)}\int_\Omega\left(K_{3,3,l}\Bigg/\frac{\partial H}{\partial p_1}\right)\mathrm{d}q_1\cdots\mathrm{d}q_n\mathrm{d}p_2\cdots\mathrm{d}p_n \tag{6.37}$$

$$3\lambda_{2,l}E\left[Y_{2,l}^3\right]V_{2,1,l}^2V_{2,2,l}=\frac{\lambda_l E\left[Y_l^4\right]}{T(H)}\int_\Omega\left(K_{3,4,l}\Bigg/\frac{\partial H}{\partial p_1}\right)\mathrm{d}q_1\cdots\mathrm{d}q_n\mathrm{d}p_2\cdots\mathrm{d}p_n \tag{6.38}$$

$$\begin{aligned}&\lambda_{1,l}E\left[Y_{1,l}^2\right]V_{1,1,l}^2+\lambda_{2,l}E\left[Y_{2,l}^2\right]V_{2,1,l}^2+\lambda_{3,l}E\left[Y_{3,l}^2\right]V_{3,1,l}^2\\&=\frac{\lambda_l E\left[Y_l^2\right]}{T(H)}\int_\Omega\left(K_{2,2,l}\Bigg/\frac{\partial H}{\partial p_1}\right)\mathrm{d}q_1\cdots\mathrm{d}q_n\mathrm{d}p_2\cdots\mathrm{d}p_n,\end{aligned} \tag{6.39}$$

$$\begin{aligned}&2\lambda_{2,l}E\left[Y_{2,l}^2\right]V_{2,1,l}V_{2,2,l}+2\lambda_{3,l}E\left[Y_{3,l}^2\right]V_{3,1,l}V_{3,2,l}\\&=\frac{\lambda_l E\left[Y_l^3\right]}{T(H)}\int_\Omega\left(K_{2,3,l}\Bigg/\frac{\partial H}{\partial p_1}\right)\mathrm{d}q_1\cdots\mathrm{d}q_n\mathrm{d}p_2\cdots\mathrm{d}p_n,\end{aligned} \tag{6.40}$$

$$\begin{aligned}&\lambda_{2,l}E\left[Y_{2,l}^2\right]V_{2,2,l}^2+\lambda_{3,l}E\left[Y_{3,l}^2\right]\left(V_{3,2,l}^2+2V_{3,1,l}V_{3,3,l}\right)\\&=\frac{\lambda_l E\left[Y_l^4\right]}{T(H)}\int\limits_{\Omega}\left(K_{2,4,l}/\frac{\partial H}{\partial p_1}\right)\mathrm{d}q_1\cdots\mathrm{d}q_n\mathrm{d}p_2\cdots\mathrm{d}p_n,\end{aligned}\tag{6.41}$$

where Eqs. (6.36) to (6.41) are used to calculate the coefficients $V_{k,s,l}$ in Eq. (6.30).

The truncated averaged FPK Eq. (6.10) can be rewritten as

$$\frac{\partial p}{\partial t}=\sum_{j=1}^{4}\frac{(-1)^j}{j!}\frac{\partial^j}{\partial h^j}\left(\overline{a}_j(h)p\right)+O\left(\varepsilon^5\right)\tag{6.42}$$

where

$$\overline{a}_1=\varepsilon^2\overline{m}(h)+\varepsilon^2U_0(h)+\varepsilon^3U_1(h)+\varepsilon^4U_2(h)\tag{6.43}$$

$$\begin{aligned}\overline{a}_2=\ &\varepsilon^2\overline{\sigma}^2(h)+\sum_{l=1}^{n_p}\left[\varepsilon^2\left(\lambda_{1,l}E\left[Y_{1,l}^2\right]V_{1,1,l}^2+\lambda_{2,l}E\left[Y_{2,l}^2\right]V_{2,1,l}^2+\lambda_{3,l}E\left[Y_{3,l}^2\right]V_{3,1,l}^2\right)\right.\\&+2\varepsilon^3\left(\lambda_{2,l}E\left[Y_{2,l}^2\right]V_{2,1,l}V_{2,2,l}+\lambda_{3,l}E\left[Y_{3,l}^2\right]V_{3,1,l}V_{3,2,l}\right)+\varepsilon^4\left(\lambda_{2,l}E\left[Y_{2,l}^2\right]V_{2,2,l}^2\right.\\&\left.\left.+2\lambda_{3,l}E\left[Y_{3,l}^2\right]\left(V_{3,2,l}^2+2V_{3,1,l}V_{3,3,l}\right)\right)\right],\end{aligned}\tag{6.44}$$

$$\overline{a}_3=\sum_{l=1}^{n_p}\left[\varepsilon^3\left(\lambda_{1,l}E\left[Y_{1,l}^3\right]V_{1,1,l}^3+\lambda_{2,l}E\left[Y_{2,l}^3\right]V_{2,1,l}^3\right)+3\varepsilon^4\lambda_{2,l}E\left[Y_{2,l}^3\right]V_{2,1,l}^3V_{2,2,l}\right]\tag{6.45}$$

$$\overline{a}_4=\varepsilon^4\sum_{l=1}^{n_p}\lambda_{1,l}E\left[Y_{1,l}^4\right]V_{1,1,l}^4\tag{6.46}$$

Example 6.1 Consider linearly and nonlinearly coupled two van der Pol oscillators excited by both Gaussian and Poisson white noises (Jia et al. 2013). The motion equations are.

$$\begin{aligned}&\ddot{X}_1-\beta_1\dot{X}_1+\alpha_1X_1^2\dot{X}_1+\omega_1^2X_1+aX_2+b(X_1-X_2)^3\\&\quad=f_1X_1\left(W_{g1}(t)+W_{p1}(t)\right),\\&\ddot{X}_2-(\beta_1-\beta_2)\dot{X}_2+\alpha_2X_2^2\dot{X}_2+\omega_2^2X_2+aX_1+b(X_2-X_1)^3\\&\quad=f_2X_2\left(W_{g2}(t)+W_{p2}(t)\right),\end{aligned}\tag{6.47}$$

where a, b, ω_1, ω_2 are constants, α_i, β_i $(i=1,2)$ are constants of ε^2 order, f_i $(i=1,2)$ are constants of ε order, $W_{gi}(t)$ $(i=1,2)$ are two independent Gaussian white noises with power spectral densities K_i, and $W_{pi}(t)$ $(i=1,2)$ are two independent Poisson white noises with zero mean and Gaussian distributed pulse amplitudes.

Introducing the transformation $Q_1 = X_1, Q_2 = X_2, P_1 = \dot{X}_1, P_2 = \dot{X}_2$, Eq. (6.47) can be rewritten in the form of quasi-Hamiltonian system

$$\begin{aligned}
\dot{Q}_1 &= P_1, \\
\dot{P}_1 &= \beta_1 P_1 - \alpha_1 Q_1^2 P_1 - \omega_1^2 Q_1 - aQ_2 - b(Q_1 - Q_2)^3 \\
&\quad + f_1 Q_1 W_{g1}(t) + f_1 Q_1 W_{p1}(t), \\
\dot{Q}_2 &= P_2, \\
\dot{P}_2 &= (\beta_1 - \beta_2) P_2 - \alpha_2 Q_2^2 P_2 - \omega_2^2 Q_2 - aQ_1 - b(Q_2 - Q_1)^3 \\
&\quad + f_2 Q_2 W_{g2}(t) + f_2 Q_2 W_{p2}(t).
\end{aligned} \tag{6.48}$$

Equation (6.48) represents a quasi-non-integrable Hamiltonian system jointly excited by Gaussian and Poisson white noises. The Hamiltonian system associated with Eq. (6.48) has only one first integral, namely the Hamiltonian,

$$H(\boldsymbol{Q}, \boldsymbol{P}) = P_1^2/2 + P_2^2/2 + U(Q_1, Q_2) \tag{6.49}$$

where

$$U(Q_1, Q_2) = \omega_1^2 Q_1^2/2 + \omega_2^2 Q_2^2/2 + aQ_1 Q_2 + b(Q_1 - Q_2)^4/4 \tag{6.50}$$

Equation (6.48) can be further rewritten in the following SIDEs

$$\begin{aligned}
dQ_1 &= \frac{\partial H}{\partial P_1} dt, \\
dP_1 &= -\left(\frac{\partial H}{\partial Q_1} + m_{11}\frac{\partial H}{\partial P_1}\right) dt + \sigma_{11} dB_1(t) + \int_{\mathcal{Y}_1} f_1 Q_1 Y_1 \mathcal{P}_1(dt, dY_1), \\
dQ_2 &= \frac{\partial H}{\partial P_2} dt, \\
dP_2 &= -\left(\frac{\partial H}{\partial Q_2} + m_{22}\frac{\partial H}{\partial P_2}\right) dt + \sigma_{22} dB_2(t) + \int_{\mathcal{Y}_2} f_2 Q_2 Y_2 \mathcal{P}_2(dt, dY_2),
\end{aligned} \tag{6.51}$$

where

$$\begin{aligned}
m_{11} &= -\beta_1 + \alpha_1 Q_1^2, \quad m_{22} = -(\beta_1 - \beta_2) + \alpha_2 Q_2^2, \\
\sigma_{11}^2 &= 2\pi K_1 f_1^2 Q_1^2, \quad \sigma_{22}^2 = 2\pi K_2 f_2^2 Q_2^2.
\end{aligned} \tag{6.52}$$

Following the stochastic averaging method of quasi-non-integrable Hamiltonian systems described in Sect. 6.1 and neglecting the terms of the 4th and higher order of ε, the following truncated averaged FPK equation can be obtained

$$\frac{\partial}{\partial t}p = -\frac{\partial}{\partial h}(\overline{a}_1(h)p) + \frac{1}{2}\frac{\partial^2}{\partial h^2}(\overline{a}_2(h)p) - \frac{1}{3!}\frac{\partial^3}{\partial h^3}(\overline{a}_3(h)p) + \frac{1}{4!}\frac{\partial^4}{\partial h^4}(\overline{a}_4(h)p) \tag{6.53}$$

where

$$\begin{aligned}\overline{a}_1(h) = &\frac{1}{T(h)}\int_\Omega \{[(\beta_1 - \alpha_1 q_1^2)p_1^2 + (\beta_1 - \beta_2 - \alpha_2 q_2^2)p_2^2 \\ &+\pi K_1 f_1^2 q_1^2 + \pi K_2 f_2^2 q_2^2]/p_1\}\mathrm{d}q_1\mathrm{d}q_2\mathrm{d}p_2 \\ &+ \frac{\lambda_1 E[Y_1^2]}{2T(h)}\int_\Omega (f_1^2 q_1^2/p_1)\mathrm{d}q_1\mathrm{d}q_2\mathrm{d}p_2 \\ &+ \frac{\lambda_2 E[Y_2^2]}{2T(h)}\int_\Omega (f_2^2 q_2^2/p)\mathrm{d}q_1\mathrm{d}q_2\mathrm{d}p_2\,,\end{aligned} \tag{6.54}$$

$$\begin{aligned}\overline{a}_2(h) = &\frac{2\pi}{T(h)}\int_\Omega [(K_1 f_1^2 q_1^2 + K_2 f_2^2 q_2^2/p_1)]\mathrm{d}q_1\mathrm{d}q_2\mathrm{d}p_2 \\ &+ \sum_{l=1}^{2}\left[\frac{\lambda_l E[Y_l^2]}{T(h)}\int_\Omega (f_l^2 p_l^2 q_l^2/p_1)\mathrm{d}q_1\mathrm{d}q_2\mathrm{d}p_2 \right. \\ &\left. + \frac{\lambda_l E[Y_l^4]}{4T(h)}\int_\Omega (f_l^4 q_l^4/p_1)\mathrm{d}q_1\mathrm{d}q_2\mathrm{d}p_2\right],\end{aligned} \tag{6.55}$$

$$\begin{aligned}\overline{a}_3(h) = &\frac{3\lambda_1 E[Y_1^4]}{2T(h)}\int_\Omega (f_1^4 p_1^4 q_1^4/p_1)\mathrm{d}q_1\mathrm{d}q_2\mathrm{d}p_2 \\ &+ \frac{3\lambda_2 E[Y_2^4]}{2T(h)}\int_\Omega (f_2^4 p_2^4 q_2^4/p_1)\mathrm{d}q_1\mathrm{d}q_2\mathrm{d}p_2,\end{aligned} \tag{6.56}$$

$$\begin{aligned}\overline{a}_4(h) = &\frac{\lambda_1 E[Y_1^4]}{T(h)}\int_\Omega (f_1^4 p_1^4 q_1^4/p_1)\mathrm{d}q_1\mathrm{d}q_2\mathrm{d}p_2 \\ &+ \frac{\lambda_2 E[Y_2^4]}{T(h)}\int_\Omega (f_2^4 p_2^4 q_2^4/p_1)\mathrm{d}q_1\mathrm{d}q_2\mathrm{d}p_2,\end{aligned} \tag{6.57}$$

$$T(h) = \int_\Omega (1/p_1)\mathrm{d}q_1\mathrm{d}q_2\mathrm{d}p_2 \tag{6.58}$$

$$\Omega = \{(q_1, q_2, p_2)|h(q_1, q_2, 0, p_2) \le h\} \tag{6.59}$$

The calculation procedures for the integrals in Eqs. (6.54) to (6.58) are provided in the Appendix B of the Reference (Jia et al. 2013). Solving the averaged FPK Eq. (6.53) with $\partial p/\partial t = 0$ by using the perturbation method yields the stationary solution $p(h)$, and then applying Eq. (6.29) to obtain the joint stationary PDF of the generalized displacements and momenta

$$p(q_1, q_2, p_1, p_2) = p(h)/T(h)\big|_{h=H(q_1,q_2,p_1,p_2)} \tag{6.60}$$

The marginal PDFs and statistics can then be calculated from Eq. (6.60). For example, the marginal PDF $p(q_1)$ and the mean square value $E[Q_1^2]$ of the generalized displacement q_1 can be calculated as follows

$$p(q_1) = \int_{-\infty}^{\infty}\int_{-\infty}^{\infty}\int_{-\infty}^{\infty} p(q_1, q_2, p_1, p_2)\mathrm{d}p_1\mathrm{d}p_2\mathrm{d}q_2 \tag{6.61}$$

$$E\left[Q_1^2\right] = \int_{-\infty}^{\infty} q_1^2 p(q_1)\mathrm{d}q_1 \tag{6.62}$$

The approximate stationary PDFs $p(h), p(q_1)$ of the Hamiltonian H and generalized displacement Q_1 for system (6.47) with different values of nonlinear damping coefficients $\alpha_1 = \alpha_2 = \alpha$ are shown in Figs. 6.1 and 6.2. The values of other system parameters in the calculation are $\alpha = 0.009, 0.0075, 0.006, 0.0045$,$\varepsilon = 0.1$,$\lambda_1 = \lambda_2 = 1.0$, $E[Y_1^2] = E[Y_2^2] = 81.0$, $K_1 = K_2 = 3.0$, $\omega_1 = 1.0$, $\omega_2 = 2.0$, $a = 0.01$, $b = 1.0$, $f_1 = f_2 = 0.1$, and $\beta_1 = \beta_2 = 0.008$. The solid lines represent the second-order perturbation solution, and symbols of ▲◆● the results from Monte Carlo simulation. It can be seen from Figs. 6.1 and 6.2 that for different values of $\alpha(= \alpha_1 = \alpha_2)$, the second-order perturbation solution for the stationary PDF closely matches the results from Monte Carlo simulation, and is more accurate than the Gaussian truncation solution.

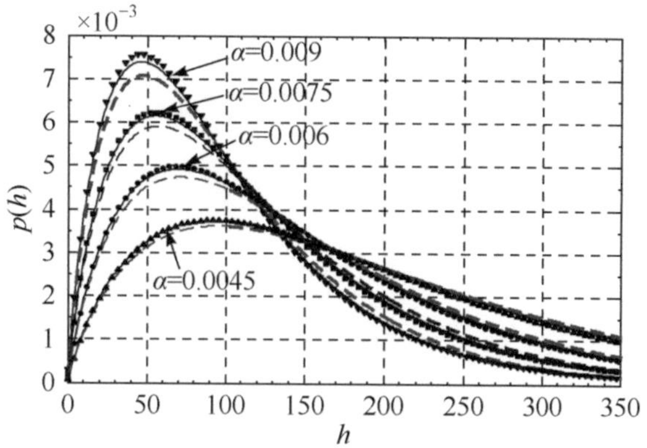

Fig. 6.1 $p(h)$ of system (6.47) with different $\alpha_1 = \alpha_2 = \alpha$. ——, second-order perturbation solution; ------, Gaussian truncation solution; ▲■●▼, results from Monte Carlo simulation (Jia et al. 2013)

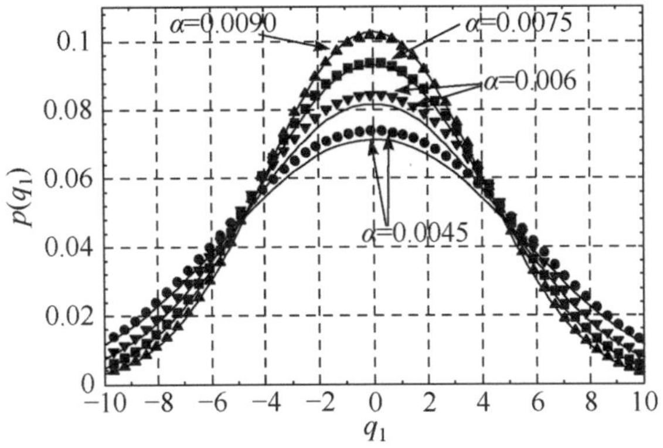

Fig. 6.2 $p(q_1)$ of system (6.47) with different $\alpha_1 = \alpha_2 = \alpha$. ——, second-order perturbation solution; ▲■●▼, results from Monte Carlo simulation (Jia et al. 2013)

The stationary PDFs of the Hamiltonian for system (6.47) under the excitations of the same total excitation intensity, but with different proportions of Gaussian and Poisson white noises are shown in Fig. 6.3. The values of system parameters in calculations are $\varepsilon = 0.1$, $\omega_1 = 1.0$, $\omega_2 = 2.0$, $a = 0.01$, $b = 1.0$, $f_1 = f_2 = 0.1$, $\beta_1 = \beta_2 = 0.008$, $\alpha_1 = \alpha_2 = 0.015$. For the case of only Poisson white noise, $E[Y_1^2] = E[Y_2^2] = 81.0$ and $\lambda_1 = \lambda_2 = 1.0$. For the case of only Gaussian white noise, $2\pi K_1 = 2\pi K_2 = 81.0$. For the case of combined excitation of both noises, $E[Y_1^2] = E[Y_2^2] = 60.0$, $\lambda_1 = \lambda_2 = 1.0$, $2\pi K_1 = 2\pi K_2 = 21.0$. It can be observed from Fig. 6.3 that the stationary PDF of the Hamiltonian has the highest peak for the case of Poisson white noise alone, and the lowest peak for the case of Gaussian white noise alone. For the case of combined noise excitations, the peak of the stationary PDF of the Hamiltonian falls between those of the former two cases. It is evident that under the same noise intensity, the effect of Gaussian white noise is more significant than that of Poisson white noise.

Figure 6.4 illustrates the impact of different average arrival rates of Poisson white noise on the stationary PDF of the Hamiltonian for system (6.47) with the same total excitation intensity. The solid lines represent the stationary PDF of the Hamiltonian

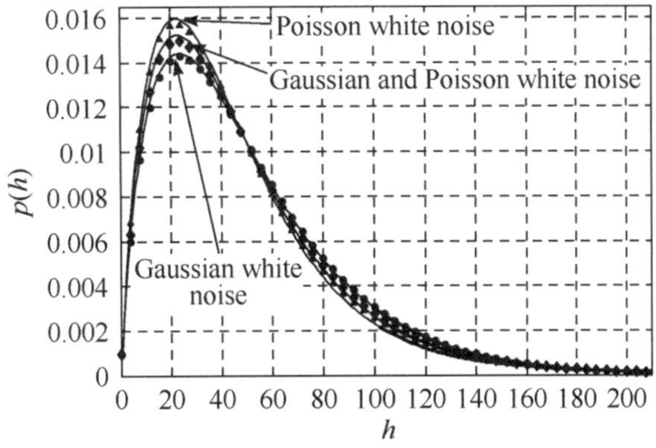

Fig. 6.3 $p(h)$ of system (6.47) with different noise intensity ratios under same total noise intensity. ——, second-order perturbation solution; ▲■●▼, results from Monte Carlo simulation (Jia et al. 2013)

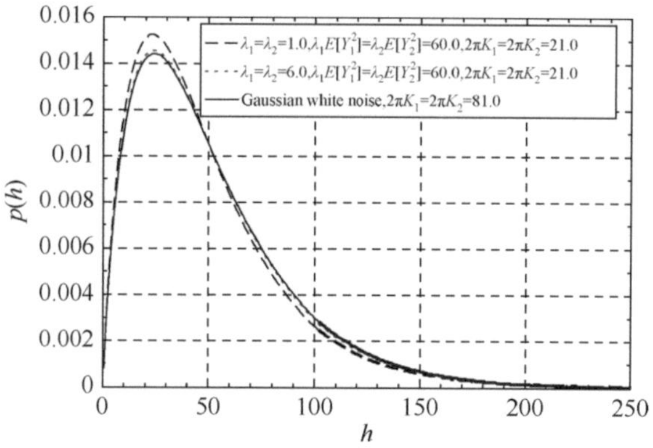

Fig. 6.4 Impact of the average arrival rate of Poisson white noise on $p(h)$ of system (6.47). Other parameters are the same as those in Fig. 6.3 (Jia et al. 2013)

in the case of only Gaussian white noise. The dashed lines represent the stationary PDF of the Hamiltonian for the case of both Gaussian and Poisson white noises, with the same Poisson white noise intensity but different average arrival rates. It can be seen from the figure that, as the average arrival rate λ_i increases from 1.0 to 6.0, the stationary PDF of the Hamiltonian for the case of combined noise excitations approaches to that of the case of only Gaussian white noise. This verifies that, under the condition of constant $\lambda E[Y^2]$, Poisson white noise approaches to Gaussian white noise as the average arrival rate λ approaches to infinite.

6.2.2 Poisson White Noise Excitation

It is seen from the original system (6.5) and the averaged Eqs. (6.12) and (6.20) that the effects of Gaussian white noise and Poisson white noise can be separated. The averaged equation for quasi-non-integrable Hamiltonian equation under only Poisson white noise can be obtained by just removing the terms in Eq. (6.12) or Eq. (6.20) generated by Gaussian white noise. As shown in Eqs. (6.4) and (6.5), the quasi-non-integrable Hamiltonian system under Poisson white noise excitation can have two equivalent forms. One is SDEs (6.4) and the other is SIDEs (6.5). In Sect. 6.2.1, the averaged SIDE (6.12) was derived by performing averaging over Eq. (6.5). In this section, the averaged equation for the quasi-non-integrable Hamiltonian system under Poisson white noise from the SDEs (6.4) will be derived (Zeng and Zhu 2011).

Consider a n-DOF quasi-non-integrable Hamiltonian system excited by Poisson white noises. The motion equations are

$$\dot{Q}_i = \frac{\partial H}{\partial P_i},$$

$$\dot{P}_i = -\frac{\partial H}{\partial Q_i} - \varepsilon^2 \sum_{j=1}^{n} c_{ij} \frac{\partial H}{\partial P_j} + \varepsilon \sum_{k=1}^{m} f_{ik} W_{pk}(t),$$
$$i = 1, 2, \cdots, n. \tag{6.63}$$

Following the derivation from Eq. (6.1) to Eq. (6.4), the following SDEs for system (6.63) can be obtained

$$\mathrm{d}Q_i = \frac{\partial H}{\partial P_i} \mathrm{d}t$$
$$\mathrm{d}P_i = -\left(\frac{\partial H}{\partial Q_i} + \varepsilon^2 \sum_{j=1}^{n} c_{ij} \frac{\partial H}{\partial P_j} \right) \mathrm{d}t + \sum_{s=1}^{\infty} \frac{\varepsilon^s}{s!} G_{is}$$
$$i = 1, 2, \ldots, n, \tag{6.64}$$

where

$$G_{is} = \sum_{k=1}^{m} f_{ik}^{(s)} (\mathrm{d}C_k)^s, \tag{6.65}$$

$$f_{ik}^{(s)} = \sum_{l=1}^{n} \frac{\partial f_{ik}^{(s-1)}}{\partial P_l} f_{lk}, \ f_{ik}^{(1)} = f_{ik}. \tag{6.66}$$

Based on the stochastic differential rule of Poisson white noise given by Di Paola and Falsone (Di Paola and Falsone 1993a, b), the SDE for the Hamiltonian $H(t)$ can be derived from Eq. (6.64) as follows

$$\mathrm{d}H = -\varepsilon^2 \sum_{j=1}^{n} \sum_{k=1}^{n} c_{jk} \frac{\partial H}{\partial P_j} \frac{\partial H}{\partial P_k} \mathrm{d}t$$
$$+ \sum_{r=1}^{\infty} \frac{1}{r!} \sum_{j_1=1}^{n} \sum_{j_2=1}^{n} \cdots \sum_{j_r=1}^{n} \frac{\partial^r H}{\partial P_{j_1} \partial P_{j_2} \cdots \partial P_{j_r}}$$
$$\sum_{s_1=1}^{\infty} \frac{\varepsilon^{s_1}}{s_1!} G_{j_1 s_1} \sum_{s_2=1}^{\infty} \frac{\varepsilon^{s_2}}{s_2!} G_{j_2 s_2} \cdots \sum_{s_r=1}^{\infty} \frac{\varepsilon^{s_r}}{s_r!} G_{j_r s_r}. \tag{6.67}$$

Replace the equation for $\mathrm{d}P_1$ in Eq. (6.64) with Eq. (6.67) and replace P_1 in all equations with $\boldsymbol{Q}, H, P_2, \ldots, P_n$ according to Eq. (6.7). In the new system of Eqs. (6.67) and (6.64) excluding the equation for $\mathrm{d}P_1$, Hamiltonian $H(t)$ is a slowly-varying process while $\boldsymbol{Q}(t), P_2(t), \cdots, P_n(t)$ are fast-varying processes. According to the stochastic averaging principle (Xu et al. 2011), Hamiltonian $H(t)$ weakly converges to a one-dimensional Markov process as $\varepsilon \to 0$. The SDE governing $H(t)$ is obtained by performing time averaging over Eq. (6.67). Based on the same reasoning as that in the derivation of Eq. (6.12), the time averaging can be replaced by the spatial averaging with respect to the generalized displacements $Q_1, Q_2, \cdots, Q_n$ and momenta $P_2, P_3, \cdots, P_n$. Since the averaged SDE of Hamiltonian $H(t)$ contains infinite terms, to obtain a closed-form averaged SDE, the following simplification is made.

First, construct a series in powers of ε as follows

$$\frac{1}{\mathrm{d}t}\left\langle \sum_{r=1}^{\infty}\frac{1}{r!}\sum_{j_1=1}^{n}\sum_{j_2=1}^{n}\cdots\sum_{j_r=1}^{n}\frac{\partial^r H}{\partial P_{j_1}\partial P_{j_2}\cdots\partial P_{j_r}} \sum_{s_1=1}^{\infty}\frac{\varepsilon^{s_1}}{s_1!}G_{j_1 s_1}\sum_{s_2=1}^{\infty}\frac{\varepsilon^{s_2}}{s_2!}G_{j_2 s_2}\cdots\sum_{s_r=1}^{\infty}\frac{\varepsilon^{s_r}}{s_r!}G_{j_r s_r}\right\rangle$$
$$= \sum_{k=2}^{u}\varepsilon^k A_{1,k}(q_1, q_2, \ldots, q_n, H, p_2, \ldots, p_n) + O\left(\varepsilon^{u+1}\right) \tag{6.68}$$

$$\frac{1}{\mathrm{d}t}\left\langle\left\{ \sum_{r=1}^{\infty}\frac{1}{r!}\sum_{j_1=1}^{n}\sum_{j_2=1}^{n}\cdots\sum_{j_r=1}^{n}\frac{\partial^r H}{\partial P_{j_1}\partial P_{j_2}\cdots\partial P_{j_r}} \sum_{s_1=1}^{\infty}\frac{\varepsilon^{s_1}}{s_1!}G_{j_1 s_1}\sum_{s_2=1}^{\infty}\frac{\varepsilon^{s_2}}{s_2!}G_{j_2 s_2}\cdots\sum_{s_r=1}^{\infty}\frac{\varepsilon^{s_r}}{s_r!}G_{j_r s_r}\right\}^j\right\rangle$$
$$= \sum_{k=j}^{u}\varepsilon^k A_{j,k}(q_1, q_2, \ldots, q_n, H, p_2, \ldots, p_n) + O\left(\varepsilon^{u+1}\right), \quad j = 2, \cdots, \infty, \tag{6.69}$$

where $\langle[\bullet]\rangle$ is equivalent to the expectation operator $E[\bullet]$. Then, based on Eqs. (6.68) and (6.69), the coefficients in Eq. (6.67) are rearranged in terms of powers of ε, the terms higher than order ε^u are neglected, and then the averaging is performed. Thus result in the following truncated averaged SDE

$$\mathrm{d}H = \sum_{i=0}^{u-2}\varepsilon^{i+2}U_i(H)\mathrm{d}t + \sum_{k=1}^{u-1}\sum_{l=1}^{k}\varepsilon^l V_{kl}(H)\mathrm{d}\overline{C}_k \tag{6.70}$$

where

$$U_0 = \frac{1}{T(H)} \int_{\Omega} \left[-\sum_{j=1}^{n}\sum_{k=1}^{n} c_{jk}(q_1, q_2, \cdots, q_n, H, p_2, \cdots, p_n) \frac{\partial H}{\partial p_j}\frac{\partial H}{\partial p_k} \Big/ \frac{\partial H}{\partial p_1} \right]$$
$$\mathrm{d}q_1 \cdots \mathrm{d}q_n \mathrm{d}p_2 \cdots \mathrm{d}p_n$$
$$+ \frac{1}{T(H)} \int_{\Omega} \left[A_{1,2}(q_1, q_2, \cdots, q_n, H, p_2, \cdots, p_n) \Big/ \frac{\partial H}{\partial p_1} \right] \mathrm{d}q_1 \cdots \mathrm{d}q_n \mathrm{d}p_2 \cdots \mathrm{d}p_n, \tag{6.71}$$

$$U_i = \frac{1}{T(H)} \int_{\Omega} \left[A_{1,i+2}(q_1, q_2, \cdots, q_n, H, p_2, \cdots, p_n) \Big/ \frac{\partial H}{\partial p_1} \right]$$
$$\mathrm{d}q_1 \cdots \mathrm{d}q_n \mathrm{d}p_2 \cdots \mathrm{d}p_n,$$
$$i = 1, \cdots, u-2. \tag{6.72}$$

The expressions for V_{kl} are obtained by equating the coefficients for each power of ε as follows

$$\frac{1}{\mathrm{d}t} \left\langle \left\{ \sum_{k=1}^{u-1} \sum_{l=1}^{k} \varepsilon^l V_{kl}(H) d\overline{C}_k \right\}^j \right\rangle$$
$$= \sum_{m=j}^{u} \frac{\varepsilon^m}{T(H)} \int_{\Omega} \left[A_{j,m}(q_1, \cdots, q_n, H, p_2, \cdots, p_n) \Big/ \frac{\partial H}{\partial p_1} \right]$$
$$\mathrm{d}q_1 \cdots \mathrm{d}q_n \mathrm{d}p_2 \cdots \mathrm{d}p_n + O\left(\varepsilon^{u+1}\right),$$
$$j = 2, \cdots, u, \tag{6.73}$$

$$T(H) = \int_{\Omega} \left(1 \Big/ \frac{\partial H}{\partial p_1} \right) \mathrm{d}q_1 \cdots \mathrm{d}q_n \mathrm{d}p_2 \cdots \mathrm{d}p_n, \tag{6.74}$$

$$\Omega = \{ (q_1, \cdots, q_n, p_2, \cdots, p_n) | H(q_1, \cdots, q_n, 0, p_2, \cdots, p_n) \le H \}. \tag{6.75}$$

Furthermore, the homogeneous compound Poisson processes $\overline{C}_k(t)$ in Eq. (6.70) have the following property

$$E\left[\left(\mathrm{d}\overline{C}_k \right)^r \right] = \lambda_k E\left[\overline{Y}_k^r \right] = \begin{cases} 0 & r = 1, \\ M_{k,r} & r = 2, 3, \cdots, u-k+1, \\ m_{k,r} & r = u-k+2, u-k+3, \cdots, \end{cases} \tag{6.76}$$

where $M_{k,r} \gg \varepsilon$ and $m_{k,r} < \varepsilon^u$.

The truncated averaged FPK equation associated with the truncated averaged SDE (6.70) takes the form

$$\frac{\partial p}{\partial t} = \sum_{j=1}^{u} \frac{(-1)^j}{j!} \frac{\partial^j}{\partial h^j}(\overline{a}_j p) + O(\varepsilon^{u+1}) \tag{6.77}$$

where

$$\begin{aligned}
\overline{a}_1 &= \sum_{i=0}^{u-2} \varepsilon^{i+2} U_i(h) \\
&= \frac{1}{T(h)} \int_{\Omega} \left[-\varepsilon^2 \sum_{j=1}^{n} \sum_{k=1}^{n} c_{jk}(q_1, q_2, \cdots, q_n, h, p_2, \cdots, p_n) \frac{\partial h}{\partial p_j} \frac{\partial h}{\partial p_k} \Big/ \frac{\partial h}{\partial p_1} \right] \\
&\quad \mathrm{d}q_1 \cdots \mathrm{d}q_n \mathrm{d}p_2 \cdots \mathrm{d}p_n \\
&\quad + \sum_{k=2}^{\infty} \varepsilon^k \frac{1}{T(h)} \int_{\Omega} \left[A_{1,k}(q_1, q_2, \ldots, q_n, h, p_2, \ldots, p_n) \Big/ \frac{\partial h}{\partial p_1} \right] \\
&\quad \mathrm{d}q_1 \cdots \mathrm{d}q_n \mathrm{d}p_2 \cdots \mathrm{d}p_n
\end{aligned} \tag{6.78}$$

$$\begin{aligned}
\overline{a}_j &= \frac{1}{\mathrm{d}t} \left\langle \left\{ \sum_{k=1}^{u-1} \sum_{l=1}^{k} \varepsilon^l V_{kl}(H) d\overline{C}_k \right\}^j \middle| H = h \right\rangle \\
&= \sum_{k=j}^{u} \frac{\varepsilon^k}{T(h)} \int_{\Omega} \left[A_{j,k}(q_1, q_2, \ldots, q_n, h, p_2, \ldots, p_n) \Big/ \frac{\partial h}{\partial p_1} \right] \\
&\quad \mathrm{d}q_1 \cdots \mathrm{d}q_n \mathrm{d}p_2 \cdots \mathrm{d}p_n \\
&\quad \mathrm{d}q_1 \cdots \mathrm{d}q_n \mathrm{d}p_2 \cdots \mathrm{d}p_n \\
&j = 2, \ldots, u.
\end{aligned} \tag{6.79}$$

In Eq. (6.77), $p = p(h, t|h_0)$ is the transition PDF of Hamiltonian with the initial condition

$$p(h, 0|h_0) = \delta(h - h_0) \tag{6.80}$$

or, $p = p(h, t)$ is the PDF of Hamiltonian with the initial condition

$$p(h, 0) = p(h_0) \tag{6.81}$$

In addition, Eq. (6.77) must satisfy the following boundary conditions

$$\begin{aligned} &p(0) = \text{finite}, \quad \left.\frac{\partial^s p}{\partial h^s}\right|_{h=0} = \text{finite}, \quad s = 1, 2, \cdots, \\ &\lim_{h\to\infty} p(h) = 0, \quad \lim_{h\to\infty} \frac{\partial^s p}{\partial h^s} = 0, \quad s = 1, 2, \cdots, \end{aligned} \tag{6.82}$$

and the normalization condition.

To find the stationary solution of the truncated averaged FPK Eq. (6.77) with $\partial p/\partial t = 0$, assume that the stationary solution takes the form of power series of ε as follows

$$p(h) = p_0(h) + \varepsilon p_1(h) + \varepsilon^2 p_2(h) + \cdots \tag{6.83}$$

Substituting Eq. (6.83) into Eq. (6.77) and letting the coefficients with same ε order be zero lead to a series of ordinary differential equations for $p_0(h),\ p_1(h),\ p_2(h), \cdots$. Solving these equations successively and then substituting the results into Eq. (6.83) yield the stationary solution. According to Eq. (5.18), the approximate stationary PDF of the generalized displacements and generalized momenta is obtained as follows

$$p(\boldsymbol{q}, \boldsymbol{p}) = \left.\frac{p(h)}{T(h)}\right|_{h=h(\boldsymbol{q},\boldsymbol{p})} \tag{6.84}$$

Example 6.2 Consider two linearly and nonlinearly coupled van der Pol oscillators excited by Poisson white noises (Zeng and Zhu 2011). The motion equations are

$$\begin{aligned} &\ddot{X}_1 - \beta_1 \dot{X}_1 + \alpha_1 X_1^2 \dot{X}_1 + \omega_1^2 X_1 \\ &\quad + aX_2 + b(X_1 - X_2)^3 = c_1 X_1 W_{p1}(t), \\ &\ddot{X}_2 - (\beta_1 - \beta_2)\dot{X}_2 + \alpha_2 X_2^2 \dot{X}_2 + \omega_2^2 X_2 \\ &\quad + aX_1 + b(X_2 - X_1)^3 = c_2 X_2 W_{p2}(t), \end{aligned} \tag{6.85}$$

where $\beta_1, \beta_2, \alpha_1, \alpha_2$ are of ε^2 order; c_1, c_2 are of ε order; $W_{p1}(t)$ and $W_{p2}(t)$ are two independent Poisson white noises with zero-mean, identically Gaussian distributed impulse magnitudes and impulse average arrival rates λ_1 and λ_2, respectively. The Hamiltonian system associated with Eq. (6.85) has only one first integral, namely the Hamiltonian. Let $Q_1 = X_1, Q_2 = X_2, P_1 = \dot{X}_1, P_2 = \dot{X}_2$, the Hamiltonian is

$$H = P_1^2/2 + P_2^2/2 + U(Q_1, Q_2) \tag{6.86}$$

where

$$U(Q_1, Q_2) = \omega_1^2 Q_1^2/2 + \omega_2^2 Q_2^2/2 + aQ_1Q_2 + b(Q_1 - Q_2)^4/4 \tag{6.87}$$

Equation (6.85) can be transformed into the following SDEs

$$
\begin{aligned}
\mathrm{d}Q_1 &= \frac{\partial H}{\partial P_1}\mathrm{d}t, \\
\mathrm{d}P_1 &= -\left(\frac{\partial H}{\partial Q_1} - (\beta_1 - \alpha_1 Q_1^2)\frac{\partial H}{\partial P_1}\right)\mathrm{d}t + c_1 Q_1 \mathrm{d}C_1(t), \\
\mathrm{d}Q_2 &= \frac{\partial H}{\partial P_2}\mathrm{d}t, \\
\mathrm{d}P_2 &= -\left(\frac{\partial H}{\partial Q_2} - [(\beta_1 - \beta_2) - \alpha_2 Q_2^2]\frac{\partial H}{\partial P_2}\right)\mathrm{d}t + c_2 Q_2 \mathrm{d}C_2(t).
\end{aligned}
\tag{6.88}
$$

The SDE of Hamiltonian can be derived from Eq. (6.88) by using the differential rule of Di Paola and Falsone as follows

$$
\begin{aligned}
\mathrm{d}H = &\left\{(\beta_1 - \alpha_1 Q_1^2)\left(\frac{\partial H}{\partial P_1}\right)^2 + (\beta_1 - \beta_2 - \alpha_2 Q_2^2)\left(\frac{\partial H}{\partial P_2}\right)^2\right\}\mathrm{d}t \\
&+ \frac{\partial H}{\partial P_1} c_1 Q_1 \mathrm{d}C_1(t) + \frac{\partial H}{\partial P_2} c_2 Q_2 \mathrm{d}C_2(t) \\
&+ \frac{1}{2}\left[c_1 Q_1 \mathrm{d}C_1(t)\right]^2 + \frac{1}{2}\left[c_2 Q_2 \mathrm{d}C_2(t)\right]^2 + \cdots
\end{aligned}
\tag{6.89}
$$

Completing averaging over Eq. (6.89) and taking $u = 4$ yield the truncated averaged SDE and averaged FPK equation in the forms of Eqs. (6.70) and (6.77), respectively. The coefficients are obtained according to Eqs. (6.71) to (6.76) and Eqs. (6.78), (6.79) as follows

$$
\begin{aligned}
\varepsilon^2 U_0 = \frac{1}{T(H)} \int\limits_{\Omega} &\{[(\beta_1 - \alpha_1 q_1^2) p_1^2 + (\beta_1 - \beta_2 - \alpha_2 q_2^2) p_2^2 \\
&+ \frac{1}{2}(c_1^2 q_1^2 \lambda_1 E[Y_1^2] + c_2^2 q_2^2 \lambda_2 E[Y_2^2])\Big]\Big/ p_1\Big\}\mathrm{d}q_1 \mathrm{d}q_2 \mathrm{d}p_2,
\end{aligned}
\tag{6.90}
$$

$$
U_1 = U_2 = 0 \tag{6.91}
$$

$$
\begin{aligned}
&\varepsilon^4 \lambda_1 E\left[\overline{Y_1^4}\right] V_{11}^4 = \frac{1}{T(H)} \\
&\int\limits_{\Omega} \{(c_1^4 p_1^4 q_1^4 \lambda_1 E[Y_1^4] + c_2^4 p_2^4 q_2^4 \lambda_2 E[Y_2^4]) / p_1\}\mathrm{d}q_1 \mathrm{d}q_2 \mathrm{d}p_2,
\end{aligned}
\tag{6.92}
$$

$$
\varepsilon^3\left(\lambda_1 E\left[\overline{Y_1^3}\right] V_{11}^3 + \lambda_2 E\left[\overline{Y_2^3}\right] V_{21}^3\right) = 0 \tag{6.93}
$$

$$
3\varepsilon^4 \lambda_2 E\left[\overline{Y_2^3}\right] V_{21}^2 V_{22} = \frac{1}{T(H)}
$$

$$\int_{\Omega} \left\{ \left[\frac{3}{2} (c_1^4 p_1^2 q_1^4 \lambda_1 E[Y_1^4] + c_2^4 p_2^2 q_2^4 \lambda_2 E[Y_2^4]) \right] \Big/ p_1 \right\} \mathrm{d}q_1 \mathrm{d}q_2 \mathrm{d}p_2, \tag{6.94}$$

$$\begin{aligned} &\varepsilon^2 \left(\lambda_1 E\left[\overline{Y}_1^2\right] V_{11}^2 + \lambda_2 E\left[\overline{Y}_2^2\right] V_{21}^2 + \lambda_3 E\left[\overline{Y}_3^2\right] V_{31}^2 \right) \\ &= \frac{1}{T(H)} \int_{\Omega} \{ (c_1^2 p_1^2 q_1^2 \lambda_1 E[Y_1^2] + c_2^2 p_2^2 q_2^2 \lambda_2 E[Y_2^2]) / p_1 \} \mathrm{d}q_1 \mathrm{d}q_2 \mathrm{d}p_2, \end{aligned} \tag{6.95}$$

$$\varepsilon^3 \left(\lambda_2 E\left[\overline{Y}_2^2\right] V_{21} V_{22} + \lambda_3 E\left[\overline{Y}_3^2\right] V_{31} V_{32} \right) = 0, \tag{6.96}$$

$$\begin{aligned} &\varepsilon^4 \left\{ \lambda_2 E\left[\overline{Y}_2^2\right] V_{22}^2 + \lambda_3 E\left[\overline{Y}_3^2\right] (V_{32}^2 + 2 V_{31} V_{33}) \right\} \\ &= \frac{1}{T(H)} \int_{\Omega} \left\{ \left(\frac{1}{4} (c_1^4 q_1^4 \lambda_1 E[Y_1^4] + c_2^4 q_2^4 \lambda_2 E[Y_2^4]) \right) \Big/ p_1 \right\} \mathrm{d}q_1 \mathrm{d}q_2 \mathrm{d}p_2, \end{aligned} \tag{6.97}$$

$$\begin{aligned} \overline{a}_1 = \frac{1}{T(h)} \int_{\Omega} &\{ [(\beta_1 - \alpha_1 q_1^2) p_1^2 + (\beta_1 - \beta_2 - \alpha_2 q_2^2) p_2^2 \\ &+ \frac{1}{2} (c_1^2 q_1^2 \lambda_1 E[Y_1^2] + c_2^2 q_2^2 \lambda_2 E[Y_2^2]) \Big] \Big/ p_1 \Big\} \mathrm{d}q_1 \mathrm{d}q_2 \mathrm{d}p_2, \end{aligned} \tag{6.98}$$

$$\begin{aligned} \overline{a}_2 = \frac{1}{T(h)} \int_{\Omega} &\{ [c_1^2 p_1^2 q_1^2 \lambda_1 E[Y_1^2] + c_2^2 p_2^2 q_2^2 \lambda_2 E[Y_2^2] \\ &+ \frac{1}{4} (c_1^4 q_1^4 \lambda_1 E[Y_1^4] + c_2^4 q_2^4 \lambda_2 E[Y_2^4]) \Big] \Big/ p_1 \Big\} \mathrm{d}q_1 \mathrm{d}q_2 \mathrm{d}p_2, \end{aligned} \tag{6.99}$$

$$\begin{aligned} \overline{a}_3 = \frac{1}{T(h)} \int_{\Omega} &\left\{ \left[\frac{3}{2} (c_1^4 p_1^2 q_1^4 \lambda_1 E[Y_1^4] + c_2^4 p_2^2 q_2^4 \lambda_2 E[Y_2^4]) \right] \Big/ p_1 \right\} \\ &\mathrm{d}q_1 \mathrm{d}q_2 \mathrm{d}p_2, \end{aligned} \tag{6.100}$$

$$\overline{a}_4 = \frac{1}{T(h)} \int_{\Omega} \{ (c_1^4 p_1^4 q_1^4 \lambda_1 E[Y_1^4] + c_2^4 p_2^4 q_2^4 \lambda_2 E[Y_2^4]) / p_1 \} \mathrm{d}q_1 \mathrm{d}q_2 \mathrm{d}p_2, \tag{6.101}$$

where

$$T(h) = \int_{\Omega} (1/p_1) \mathrm{d}q_1 \mathrm{d}q_2 \mathrm{d}p_2, \tag{6.102}$$

$$\Omega = \{ (q_1, q_2, p_2) | H(q_1, q_2, 0, p_2) \le h \} \tag{6.103}$$

Note that in Eqs. (6.90) to (6.102), p_1 should be replaced by q_1, q_2, H, p_2 according to Eq. (6.86), $T(H) = T(h)|_{h=H}$. The integration operations in Eqs. (6.90) to (6.102) can be found in (Zeng and Zhu 2011). Substituting Eqs. (6.98)–(6.103) into Eq. (6.77) and solving it using the perturbation method yield the approximate stationary solution for the PDF of Hamiltonian H. Finally, the approximate joint stationary PDF of the generalized displacements and generalized momenta is obtained according to Eq. (6.84) as follows

$$p(q_1, q_2, p_1, p_2) = \left.\frac{p(h)}{T(h)}\right|_{h=H(q_1,q_2,p_1,p_2)} \tag{6.104}$$

In numerical calculation, the parameters of system (6.85) are $\varepsilon = 0.1, \lambda_1 = \lambda_2 = \lambda = 1.0, E[Y_1^2] = E[Y_2^2] = E[Y^2] = 81.0, \omega_1 = 1.0, \omega_2 = 2.0, a = 0.01, b = 1.0, c_1 = c_2 = \varepsilon = 0.1, \beta_1 = \beta_2 = 0.8\varepsilon^2 = 0.008$, and $\alpha_1 = \alpha_2 = \varepsilon^2\alpha = 0.01\alpha$. The stationary PDF of Hamiltonian H and the stationary PDF of the generalized displacement Q_1 are shown in Figs. 6.5 and 6.6, respectively. For comparison, the method of Gaussian approximation, i.e., replacing the Poisson white noises in Eq. (6.85) with Gaussian white noises of the same variance and assuming that the Hamiltonian H is a diffusion process is used. In Figs. 6.5 and 6.6, the solid lines represent the approximate stationary solutions obtained by using the perturbation method, the dashed lines represent the results from Gaussian approximation, and symbols ▲■●▼ represent results from Monte Carlo simulation. It can be seen from these figures that the approximate stationary solutions obtained by using the stochastic averaging method described in this section are in good agreement with the results from Monte Carlo simulation and more accurate than the Gaussian approximation solution. Besides, the Gaussian approximation solution slightly overestimates the response.

Example 6.3 Consider a 2-DOF vibration-impact system excited by Poisson white noises (Zeng and Zhu 2011). The motion equations are

$$\begin{aligned} &\ddot{X}_1 + c_1\dot{X}_1 + k_1X_1 + k_2(X_1 - X_2) = f_1W_{p1}(t), \\ &\ddot{X}_2 + c_2\dot{X}_2 + k_2(X_2 - X_1) + g(X_2) = f_2W_{p2}(t), \end{aligned} \tag{6.105}$$

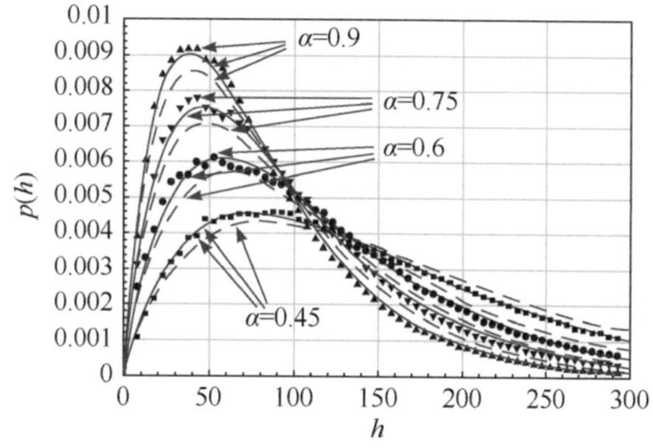

Fig. 6.5 $p(h)$ of system (6.85). ——, second-order perturbation solution; ------, Gaussian approximation solution; ▲■●▼, results from Monte Carlo simulation (Zeng and Zhu 2011)

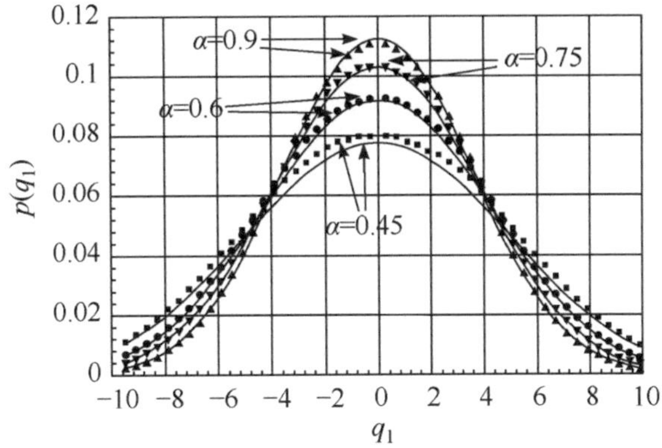

Fig. 6.6 $p(q_1)$ of system (6.85). ——, second-order perturbation solution; -----, Gaussian approximation solution; ▲■●▼, results from Monte Carlo simulation (Zeng and Zhu, 2011)

where

$$g(X_2) = \begin{cases} -B_l(-X_2 - \delta_l)^{3/2}, & X_2 < -\delta_l, \\ 0, & -\delta_l \le X_2 \le \delta_r, \\ B_r(X_2 - \delta_r)^{3/2}, & X_2 > \delta_r, \end{cases} \tag{6.106}$$

and $B_l, B_r, \delta_l, \delta_r > 0, c_1, c_2$ are small parameters of ε^2 order; f_1, f_2 are small parameters of ε order; $W_{p1}(t)$ and $W_{p2}(t)$ are Poisson white noises with zero mean, independent and Gaussian distributed pulse amplitudes, and pulse average arrival rates of λ_1 and λ_2, respectively. The Hamiltonian system associated with system (6.105) has only one independent first integral, namely the Hamiltonian. Letting $Q_1 = X_1, Q_2 = X_2$, $P_1 = \dot{X}_1, P_2 = \dot{X}_2$, the Hamiltonian is

$$H(Q_1, Q_2, P_1, P_2) = \frac{P_1^2}{2} + \frac{P_2^2}{2} + \frac{k_1 Q_1^2}{2} + \frac{k_2(Q_1 - Q_2)^2}{2} + U(Q_2) \tag{6.107}$$

where

$$U(Q_2) = \int_0^{Q_2} g(x)\mathrm{d}x = \begin{cases} (2/5)B_l(-Q_2 - \delta_l)^{5/2}, & Q_2 < -\delta_l, \\ 0, & -\delta_l \le Q_2 \le \delta_r, \\ (2/5)B_r(Q_2 - \delta_r)^{5/2}, & Q_2 > \delta_r. \end{cases} \tag{6.108}$$

System (6.105) can be transformed into the following SDEs

$$\begin{aligned} \mathrm{d}Q_1 &= \frac{\partial H}{\partial P_1}\mathrm{d}t, \quad \mathrm{d}P_1 = -\left(\frac{\partial H}{\partial Q_1} + c_1\frac{\partial H}{\partial P_1}\right)\mathrm{d}t + f_1\mathrm{d}C_1, \\ \mathrm{d}Q_2 &= \frac{\partial H}{\partial P_2}\mathrm{d}t, \quad \mathrm{d}P_2 = -\left(\frac{\partial H}{\partial Q_2} + c_2\frac{\partial H}{\partial P_2}\right)\mathrm{d}t + f_2\mathrm{d}C_2. \end{aligned} \tag{6.109}$$

Applying Eq. (6.107) and the Di Paola and Falsone rule, the SDE for the Hamiltonian is derived from Eq. (6.109) as follows

$$\begin{aligned} \mathrm{d}H = & -\left[c_1\left(\frac{\partial H}{\partial P_1}\right)^2 + c_2\left(\frac{\partial H}{\partial P_2}\right)^2\right]\mathrm{d}t \\ & + \frac{\partial H}{\partial P_1} f_1 \mathrm{d}C_1 + \frac{\partial H}{\partial P_2} f_2 \mathrm{d}C_2 + \frac{1}{2} f_1^2 (\mathrm{d}C_1)^2 + \frac{1}{2} f_2^2 (\mathrm{d}C_2)^2. \end{aligned} \tag{6.110}$$

Completing the averaging in Eq. (6.110) and setting $u = 4$ result in the truncated averaged SDE and averaged FPK equation of the form of Eqs. (6.70) and (6.77), respectively, with the following coefficients

$$\begin{aligned} \varepsilon^2 U_0 = & \frac{1}{T(H)} \\ & \int_{\Omega} \left\{\left[-c_1 p_1^2 - c_2 p_2^2 + \frac{1}{2}\left(f_1^2 \lambda_1 E\left[Y_1^2\right] + f_2^2 \lambda_2 E\left[Y_2^2\right]\right)\right] \Big/ p_1 \right\} \mathrm{d}q_1 \mathrm{d}q_2 \mathrm{d}p_2, \end{aligned} \tag{6.111}$$

$$U_1 = U_2 = 0, \tag{6.112}$$

$$\begin{aligned} \varepsilon^4 \lambda_1 E\left[\overline{Y}_1^4\right] V_{11}^4 = & \frac{1}{T(H)} \\ & \int_{\Omega} \left\{\left(f_1^4 p_1^4 \lambda_1 E\left[Y_1^4\right] + f_2^4 p_2^4 \lambda_2 E\left[Y_2^4\right]\right) / p_1 \right\} \mathrm{d}q_1 \mathrm{d}q_2 \mathrm{d}p_2, \end{aligned} \tag{6.113}$$

$$\varepsilon^3 \left(\lambda_1 E\left[\overline{Y}_1^3\right] V_{11}^3 + \lambda_2 E\left[\overline{Y}_2^3\right] V_{21}^3\right) = 0, \tag{6.114}$$

$$\begin{aligned} 3\varepsilon^4 \lambda_2 E\left[\overline{Y}_2^3\right] V_{21}^2 V_{22} = & \frac{1}{T(H)} \\ & \int_{\Omega} \left\{\left[\frac{3}{2}\left(f_1^4 p_1^2 \lambda_1 E\left[Y_1^4\right] + f_2^4 p_2^2 \lambda_2 E\left[Y_2^4\right]\right)\right] \Big/ p_1 \right\} \mathrm{d}q_1 \mathrm{d}q_2 \mathrm{d}p_2, \end{aligned} \tag{6.115}$$

$$\begin{aligned} \varepsilon^2 \left(\lambda_1 E\left[\overline{Y}_1^2\right] V_{11}^2 + \lambda_2 E\left[\overline{Y}_2^2\right] V_{21}^2 + \lambda_3 E\left[\overline{Y}_3^2\right] V_{31}^2\right) = & \frac{1}{T(H)} \\ & \int_{\Omega} \left\{\left[f_1^2 p_1^2 \lambda_1 E\left[Y_1^2\right] + f_2^2 p_2^2 \lambda_2 E\left[Y_2^2\right]\right] / p_1 \right\} \mathrm{d}q_1 \mathrm{d}q_2 \mathrm{d}p_2, \end{aligned} \tag{6.116}$$

$$\varepsilon^3 \left(\lambda_2 E\left[\overline{Y}_2^2\right] V_{21} V_{22} + \lambda_3 E\left[\overline{Y}_3^2\right] V_{31} V_{32}\right) = 0, \tag{6.117}$$

$$\varepsilon^4 \left\{\lambda_2 E\left[\overline{Y}_2^2\right] V_{22}^2 + \lambda_3 E\left[\overline{Y}_3^2\right] \left(V_{32}^2 + 2 V_{31} V_{33}\right)\right\} = \frac{1}{T(H)}$$

$$\int_{\Omega} \left\{ \left[\frac{1}{4}(f_1^4 \lambda_1 E[Y_1^4] + f_2^4 \lambda_2 E[Y_2^4]) \right] \Big/ p_1 \right\} dq_1 dq_2 dp_2, \tag{6.118}$$

$$\bar{a}_1 = \frac{1}{T(h)} \int_{\Omega} \left\{ \left[-c_1 p_1^2 - c_2 p_2^2 + \frac{1}{2}(f_1^2 \lambda_1 E[Y_1^2] + f_2^2 \lambda_2 E[Y_2^2]) \right] \Big/ p_1 \right\}$$
$$dq_1 dq_2 dp_2, \tag{6.119}$$

$$\bar{a}_2 = \frac{1}{T(h)} \int_{\Omega} \{ [p_1^2 f_1^2 \lambda_1 E[Y_1^2] + p_2^2 f_2^2 \lambda_2 E[Y_2^2]$$
$$+ \frac{1}{4}(f_1^4 \lambda_1 E[Y_1^4] + f_2^4 \lambda_2 E[Y_2^4]) \Big] \Big/ p_1 \Big\} dq_1 dq_2 dp_2, \tag{6.120}$$

$$\bar{a}_3 = \frac{1}{T(h)} \int_{\Omega} \left\{ \left[\frac{3}{2}(p_1^2 f_1^4 \lambda_1 E[Y_1^4] + p_2^2 f_2^4 \lambda_2 E[Y_2^4]) \right] \Big/ p_1 \right\} dq_1 dq_2 dp_2, \tag{6.121}$$

$$\bar{a}_4 = \frac{1}{T(h)} \int_{\Omega} \{ (p_1^4 f_1^4 \lambda_1 E[Y_1^4] + p_2^4 f_2^4 \lambda_2 E[Y_2^4]) / p_1 \} dq_1 dq_2 dp_2, \tag{6.122}$$

where

$$T(h) = \int_{\Omega} (1/p_1) dq_1 dq_2 dp_2, \tag{6.123}$$

$$\Omega = \{(q_1, q_2, p_2) | H(q_1, q_2, 0, p_2) \le h\}. \tag{6.124}$$

Note that in Eqs. (6.111) to (6.123), p_1 should be replaced with q_1, q_2, H, p_2 according to Eq. (6.107). The integration procedures for Eq. (6.111) to (6.123) are referred to reference (Zeng and Zhu 2011). Substituting the obtained coefficients of the averaged FPK equation into Eq. (6.77) and solving (6.77) using the perturbation method, an approximate stationary PDF $p(h)$ of the Hamiltonian is obtained. The joint stationary PDF $p(q_1, q_2, p_1, p_2)$ of the generalized displacements and generalized momenta can then be obtained using Eq. (6.84), i.e.,

$$p(q_1, q_2, p_1, p_2) = \left. \frac{p(h)}{T(h)} \right|_{h=H(q_1, q_2, p_1, p_2)} \tag{6.125}$$

In numerical calculations, the parameters of system (6.105) are $k_1 = 1.0, k_2 = 2.0$, $B_l = B_r = 10.0$, $\delta_l = \delta_r = 0.5$, $c_1 = c_2 = \varepsilon^2, f_1 = f_2 = \varepsilon$, $\lambda E[Y^2] = 4.0$ $\lambda_1 = \lambda_2 = \lambda$, and $E[Y_1^2] = E[Y_2^2] = E[Y^2]$. The stationary PDF $p(h)$ and the stationary PDF $p(q_2)$ are shown in Figs. 6.7 and 6.8, respectively. In the figures, the solid line represents the approximate stationary solution obtained using the perturbation method, and the dashed line represents the Gaussian approximation solution, and

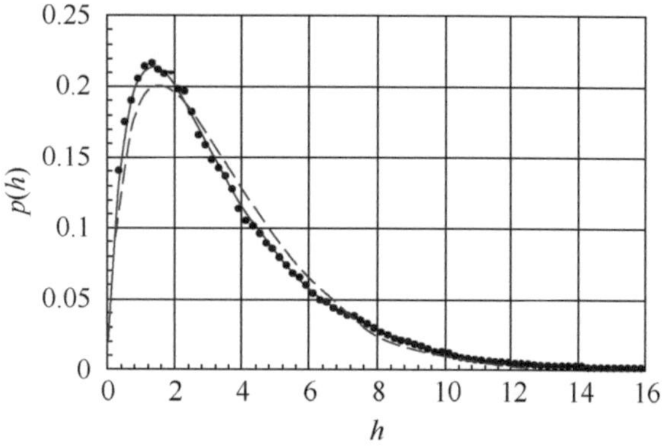

Fig. 6.7 $p(h)$ of 2-DOF vibration-impact system (6.105), $\varepsilon = 0.1, \lambda = 0.1$. ——, second-order perturbation solution; -----, Gaussian approximation solution; ●, result from Monte Carlo simulation (Zeng and Zhu 2011)

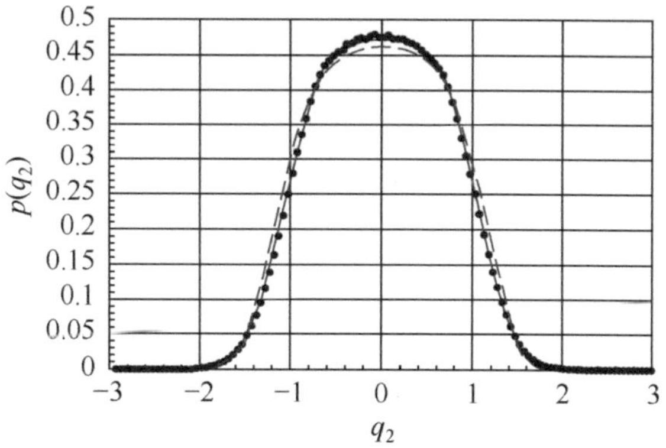

Fig. 6.8 $p(q_2)$ of 2-DOF vibration-impact system (6.105), $\varepsilon = 0.1, \lambda = 0.1$. ——, second-order perturbation solution; -----, Gaussian approximation solution; ●, result from Monte Carlo simulation (Zeng and Zhu 2011)

symbol ● denotes the results from Monte Carlo simulation. It is seen from Figs. 6.7 and 6.8 that the approximate stationary solution obtained by using the stochastic averaging method are in good agreement with those from Monte Carlo simulation, and more accurate than those obtained by using the Gaussian approximation solution.

6.3 Quasi-Integrable Hamiltonian Systems

Assume that the Hamiltonian system associated with Eq. (6.5) is completely integrable, and the action-angle variables are available. Introducing the transformation

$$I_r = I_r(\boldsymbol{Q}, \boldsymbol{P}), \quad \Theta_r = \Theta_r(\boldsymbol{Q}, \boldsymbol{P}), \quad r = 1, 2, \cdots, n, \tag{6.126}$$

and using the stochastic jump-diffusion chain rule (Hanson 2007), the SIDEs for I_r, Θ_r can be derived from Eq. (6.5) as follows

$$
\begin{aligned}
\mathrm{d}I_r &= \varepsilon^2\left(-\sum_{i,j=1}^{n} m_{ij}\frac{\partial H}{\partial P_j}\frac{\partial I_r}{\partial P_i}+\frac{1}{2}\sum_{i,j=1}^{n}\sum_{k=1}^{n_g}\sigma_{ik}\sigma_{jk}\frac{\partial^2 I_r}{\partial P_i \partial P_j}\right)\mathrm{d}t \\
&\quad+\varepsilon\sum_{i=1}^{n}\sum_{k=1}^{n_g}\frac{\partial I_r}{\partial P_i}\sigma_{ik}\mathrm{d}B_k(t) \\
&\quad+\sum_{l=1}^{n_p}\int_{\mathcal{Y}_l}\left[I_r(\boldsymbol{Q},\boldsymbol{P}+\hat{\boldsymbol{\gamma}}_l)-I_r(\boldsymbol{Q},\boldsymbol{P})\right]\mathcal{P}_l(\mathrm{d}t,\mathrm{d}Y_l), \\
\mathrm{d}\Theta_r &= \left(\omega_r-\varepsilon^2\sum_{i,j}^{n} m_{ij}\frac{\partial H}{\partial P_j}\frac{\partial \Theta_r}{\partial P_i}+\frac{\varepsilon^2}{2}\sum_{i,j=1}^{n}\sum_{k=1}^{n_g}\sigma_{ik}\sigma_{jk}\frac{\partial^2 \Theta_r}{\partial P_i \partial P_j}\right)\mathrm{d}t \\
&\quad+\varepsilon\sum_{i=1}^{n}\sum_{k=1}^{n_g}\frac{\partial \Theta_r}{\partial P_i}\sigma_{ik}\mathrm{d}B_k(t) \\
&\quad+\sum_{l=1}^{n_p}\int_{\mathcal{Y}_l}\left[\Theta_r(\boldsymbol{Q},\boldsymbol{P}+\hat{\boldsymbol{\gamma}}_l)-\Theta_r(\boldsymbol{Q},\boldsymbol{P})\right]\mathcal{P}_l(\mathrm{d}t,\mathrm{d}Y_l), \\
r &= 1,2,\cdots,n,
\end{aligned}
\tag{6.127}
$$

where $\hat{\gamma}_l$ represents the l-th column of the matrix $[\gamma_{il}]_{n\times n_p}$ and the expression for γ_{il} is given by Eq. (6.6). To obtain truncated averaged SIDEs and the averaged FPK equation, first expand $I_r(\boldsymbol{Q},\boldsymbol{P}+\hat{\boldsymbol{\gamma}}_l)-I_r(\boldsymbol{Q},\boldsymbol{P})$ and $\Theta_r(\boldsymbol{Q},\boldsymbol{P}+\hat{\boldsymbol{\gamma}}_l)-\Theta_r(\boldsymbol{Q},\boldsymbol{P})$, $(r=1,\cdots,n;\, l=1,\cdots,n_p)$ into Taylor series. Thus, Eq. (6.127) can be rewritten as the following equations.

$$
\begin{aligned}
\mathrm{d}I_r &= \varepsilon^2\left(-\sum_{i,j=1}^{n} m_{ij}\frac{\partial H}{\partial P_j}\frac{\partial I_r}{\partial P_i}+\frac{1}{2}\sum_{i,j=1}^{n}\sum_{k=1}^{n_g}\sigma_{ik}\sigma_{jk}\frac{\partial^2 I_r}{\partial P_i \partial P_j}\right)\mathrm{d}t \\
&\quad+\varepsilon\sum_{i=1}^{n}\sum_{k=1}^{n_g}\frac{\partial I_r}{\partial P_i}\sigma_{ik}\mathrm{d}B_k(t)+\sum_{l=1}^{n_p}\int_{\mathcal{Y}_l}\left[\sum_{s=1}^{\infty}\varepsilon^s Y_l^s A_{r;s;l}(\boldsymbol{Q},\boldsymbol{P})\right]\mathcal{P}_l(\mathrm{d}t,\mathrm{d}Y_l), \\
\mathrm{d}\Theta_r &= \left(\omega_r-\varepsilon^2\sum_{i,j=1}^{n} m_{ij}\frac{\partial H}{\partial P_j}\frac{\partial \Theta_r}{\partial P_i}+\frac{\varepsilon^2}{2}\sum_{i,j=1}^{n}\sum_{k=1}^{n_g}\sigma_{ik}\sigma_{jk}\frac{\partial^2 \Theta_r}{\partial P_i \partial P_j}\right)\mathrm{d}t \\
&\quad+\varepsilon\sum_{i=1}^{n}\sum_{k=1}^{n_g}\frac{\partial \Theta_r}{\partial P_i}\sigma_{ik}\mathrm{d}B_k(t)+\sum_{l=1}^{n_p}\int_{\mathcal{Y}_l}\left[\sum_{s=1}^{\infty}\varepsilon^s Y_l^s D_{r;s;l}(\boldsymbol{Q},\boldsymbol{P})\right]\mathcal{P}_l(\mathrm{d}t,\mathrm{d}Y_l), \\
r &= 1,2,\cdots,n.
\end{aligned}
\tag{6.128}
$$

For detailed derivations, refer to Appendix 2 in (Jia and Zhu 2014a).

The dimension and form of the averaged SIDEs depend on the resonance characteristics of the Hamiltonian system associated with Eq. (6.5).

6.3.1 Non-Internal Resonant Case

Suppose that the n frequencies of the Hamiltonian system do not satisfy the following weak internal resonance relation

$$\sum_{r=1}^{n} k_r \omega_r = O(\varepsilon^2) \tag{6.129}$$

where k_r are integers and $O(\varepsilon^2)$ represents a small quantity of ε^2 order. In Eq. (6.128), I_r are slowly-varying processes while Θ_r are fast-varying process. According to the stochastic averaging theorem (Khasminskii 1968; Xu et al. 2011), when $\varepsilon \to 0$, the action vector $\boldsymbol{I}(t) = [I_1(t), I_2(t), \cdots, I_n(t)]^{\mathrm{T}}$ weakly converges to a n-dimensional vector Markov process. The governing SIDEs for this n-dimensional vector Markov process can be obtained by taking time averaging over the equations for I_r in Eq. (6.128). Since the integrable and non-resonant Hamiltonian system is ergodic on the n-dimensional torus with constant I_r, time averaging can be replaced by spatial averaging with respect to Θ_r. Neglecting the terms of order ε^{u+1} and higher order, the truncated averaged SIDEs for this limiting vector process can be obtained as follows (Jia and Zhu 2014a)

$$\begin{aligned} \mathrm{d}I_r = & \left(\varepsilon^2 \overline{m}_r(\boldsymbol{I}) + \sum_{i=0}^{u-2} \varepsilon^{i+2} U_{r,i}(\boldsymbol{I}) \right) \mathrm{d}t \\ & + \varepsilon \sum_{k=1}^{n_g} \overline{\sigma}_{rk}(\boldsymbol{I}) \mathrm{d}B_k(t) + \sum_{l=1}^{u-1} G_{r,l}, \\ r = & 1, 2, \cdots, n, \end{aligned} \tag{6.130}$$

where

$$\begin{aligned} G_{r,l} = & \sum_{s_1=1}^{n} \sum_{s_2=1}^{s_1} \cdots \sum_{s_{u-l+1}=1}^{s_{u-l}} \left[\int_{\mathcal{Y}_{s_1,s_2,\ldots,s_{u-l+1};l}} \left(\sum_{k=1}^{l} \varepsilon^k V_{r;s_1,s_2,\ldots,s_{u-l+1};l,k}(\boldsymbol{I}) \right) \right. \\ & \left. \times Y_{s_1,s_2,\ldots,s_{u-l+1};l} \mathcal{P}_{s_1,s_2,\ldots,s_{u-l+1};l}\left(\mathrm{d}t, \mathrm{d}Y_{s_1,s_2,\ldots,s_{u-l+1};l}\right) \right], \end{aligned} \tag{6.131}$$

$$\overline{m}_r(\boldsymbol{I}) = \frac{1}{(2\pi)^n} \int_0^{2\pi} \left(-\sum_{i,j=1}^{n} m_{ij} \frac{\partial H}{\partial p_j} \frac{\partial I_r}{\partial p_i} \right.$$

$$+\frac{1}{2}\sum_{i,j=1}^{n}\sum_{k=1}^{n_g}\sigma_{ik}\sigma_{jk}\frac{\partial^2 I_r}{\partial p_i \partial p_j}\Bigg)\mathrm{d}\boldsymbol{\theta}, \tag{6.132}$$

$$U_{r,0}(\boldsymbol{I})=\sum_{l=1}^{n_p}\frac{\lambda_l E\left[Y_l^2\right]}{(2\pi)^n}\int_0^{2\pi}A_{r;2;l}\mathrm{d}\boldsymbol{\theta}, \tag{6.133}$$

$$U_{r,k}(\boldsymbol{I})=\sum_{l=1}^{n_p}\frac{\lambda_l E\left[Y_l^{k+2}\right]}{(2\pi)^n}\int_0^{2\pi}A_{r;k+2;l}\mathrm{d}\boldsymbol{\theta},$$
$$k=1,\cdots,u-2 \tag{6.134}$$

$$\sum_{k=1}^{n_g}\overline{\sigma}_{r_1k}(\boldsymbol{I})\overline{\sigma}_{r_2k}(\boldsymbol{I})=\frac{1}{(2\pi)^n}\int_0^{2\pi}\left(\sum_{i,j=1}^{n}\sum_{k=1}^{n_g}\frac{\partial I_{r_1}}{\partial p_i}\frac{\partial I_{r_2}}{\partial p_j}\sigma_{ik}\sigma_{jk}\right)\mathrm{d}\boldsymbol{\theta}, \tag{6.135}$$

$$\begin{aligned}&\sum_{k=1}^{u-1}\left\{\sum_{s_1=1}^{n}\sum_{s_2=1}^{s_1}\cdots\sum_{s_{u-k+1}=1}^{s_{u-k}}\left[\left(\sum_{k_1=1}^{k}\varepsilon^{k_1}V_{r_1;s_1,\ldots,s_{u-k+1};k;k_1}(\boldsymbol{I})\right)\right.\right.\\&\times\left(\sum_{k_2=1}^{k}\varepsilon^{k_2}V_{r_2;s_1,\ldots,s_{u-k+1};k;k_2}(\boldsymbol{I})\right)\cdots\\&\left.\left.\times\left(\sum_{k_j=1}^{k}\varepsilon^{k_j}V_{r_j;s_1,\ldots,s_{u-k+1};k;k_j}(\boldsymbol{I})\right)\lambda_{s_1,s_2,\ldots,s_{u-k+1};k}E\left[Y^j_{s_1,s_2,\ldots,s_{u-k+1};k}\right]\right]\right\}\\&=\sum_{k=j}^{u}\varepsilon^k\sum_{l=1}^{n_p}\frac{\lambda_l E\left[Y_l^k\right]}{(2\pi)^n}\int_0^{2\pi}\left(\sum_{k_1+k_2+\cdots+k_j=k}A_{r_1;k_1;l}A_{r_2;k_2;l}\cdots A_{r_j;k_j;l}\right)\mathrm{d}\boldsymbol{\theta},\end{aligned} \tag{6.136}$$

in which $\boldsymbol{\theta}=[\theta_1,\theta_2,\cdots,\theta_n]^{\mathrm{T}}$, $A_{r;k;l}=A_{r;k;l}(\boldsymbol{Q},\boldsymbol{P})$, $\boldsymbol{Q},\boldsymbol{P}$ in Eqs. (6.130)–(6.136) should be replaced by $\boldsymbol{I}$, $\boldsymbol{\Theta}$ according to Eq. (6.126). The form of function $V_{r_1;s_1,\cdots,s_{u-k+1};k;k_i}(\boldsymbol{I})$ can be determined by equating the same-order terms of ε on both sides of Eq. (6.136). The $\mathcal{P}_{s_1,s_2,\cdots,s_{u-l+1};l}(\mathrm{d}t,\mathrm{d}Y_{s_1,s_2,\cdots,s_{u-l+1};l})$ in Eq. (6.131) are mutually independent Poisson random measures with the following properties

$$\lambda_{r_1,r_2,\cdots,r_{u-s+1};s}E\left[Y^r_{r_1,r_2,\cdots,r_{u-s+1};s}\right]=\begin{cases}0 & r=1,\\ \overline{M}_{r_1,r_2,\cdots,r_{u-s+1};s,r} & r=2,3,\cdots,u-s+1,\\ \overline{m}_{r_1,r_2,\cdots,r_{u-s+1};s,r} & r=u-s+2,u-s+3,\cdots,\end{cases} \tag{6.137}$$

$$\overline{M}_{r_1,r_2,\cdots,r_{u-s+1};s,r}\gg\varepsilon,\ \overline{m}_{r_1,r_2,\cdots,r_{u-s+1};s,r}<\varepsilon^u.$$

The truncated averaged FPK equation associated with Eq. (6.130) is

$$\begin{aligned}\frac{\partial p}{\partial t} &= -\sum_{r_1=1}^{n}\frac{\partial}{\partial I_{r_1}}\left(\overline{a}_{r_1}(\boldsymbol{I})p\right)+\frac{1}{2!}\sum_{r_1,r_2=1}^{n}\frac{\partial^2}{\partial I_{r_1}\partial I_{r_2}}\left(\overline{a}_{r_1,r_2}(\boldsymbol{I})p\right)\\ &-\frac{1}{3!}\sum_{r_1,r_2,r_3=1}^{n}\frac{\partial^3}{\partial I_{r_1}\partial I_{r_2}\partial I_{r_3}}\left(\overline{a}_{r_1,r_2,r_3}(\boldsymbol{I})p\right)\\ &+\cdots+(-1)^u\frac{1}{u!}\sum_{r_1,r_2,\cdots,r_u}^{n}\frac{\partial^u}{\partial I_{r_1}\partial I_{r_2}\cdots\partial I_{r_u}}\left(\overline{a}_{r_1,r_2,\cdots,r_u}(\boldsymbol{I})p\right)+\mathrm{O}\left(\varepsilon^{u+1}\right),\end{aligned} \tag{6.138}$$

where

$$\begin{aligned}\overline{a}_{r_1}(\boldsymbol{I}) &= \varepsilon^2\overline{m}_{r_1}(\boldsymbol{I})+\sum_{i=0}^{u-2}\varepsilon^{i+2}U_{r_1,i}(\boldsymbol{I})\\ &=\frac{\varepsilon^2}{(2\pi)^n}\int_0^{2\pi}\left(-\sum_{i,j=1}^{n}\frac{\partial H}{\partial p_j}\frac{\partial I_{r_1}}{\partial p_i}+\frac{1}{2}\sum_{i,j=1}^{n}\sum_{k=1}^{n_g}\sigma_{ik}\sigma_{jk}\frac{\partial^2 I_{r_1}}{\partial p_i\partial p_j}\right)\mathrm{d}\boldsymbol{\theta}\\ &+\sum_{k=2}^{u}\varepsilon^k\sum_{l=1}^{n_p}\frac{\lambda_l E[Y_l^k]}{(2\pi)^n}\int_0^{2\pi}A_{r;k;l}\mathrm{d}\boldsymbol{\theta},\end{aligned} \tag{6.139}$$

$$\begin{aligned}\overline{a}_{r_1,r_2}(\boldsymbol{I}) &= \frac{\varepsilon^2}{(2\pi)^n}\int_0^{2\pi}\left(\sum_{i,j=1}^{n}\sum_{k=1}^{n_g}\frac{\partial I_{r_1}}{\partial p_i}\frac{\partial I_{r_2}}{\partial p_j}\sigma_{ik}\sigma_{jk}\right)\mathrm{d}\boldsymbol{\theta}\\ &+\sum_{k=2}^{u}\varepsilon^k\sum_{l=1}^{n_p}\frac{\lambda_l E[Y_l^k]}{(2\pi)^n}\int_0^{2\pi}\left(\sum_{k_1+k_2=k}A_{r_1;k_1;l}A_{r_2;k_2;l}\right)\mathrm{d}\boldsymbol{\theta},\end{aligned} \tag{6.140}$$

$$\begin{aligned}\overline{a}_{r_1,r_2,\cdots,r_j}(\boldsymbol{I}) &= \sum_{k=j}^{u}\varepsilon^k\sum_{l=1}^{n_p}\frac{\lambda_l E[Y_l^k]}{(2\pi)^n}\\ &\int_0^{2\pi}\left(\sum_{k_1+k_2+\cdots+k_j=k}A_{r_1;k_1;l}A_{r_2;k_2;l}\cdots A_{r_j;k_j;l}\right)\mathrm{d}\boldsymbol{\theta}\\ &j=1,2,\cdots,u,\end{aligned} \tag{6.141}$$

$\int_0^{2\pi}[\bullet]\mathrm{d}\boldsymbol{\theta}=\int_0^{2\pi}\int_0^{2\pi}\cdots\int_0^{2\pi}[\bullet]\mathrm{d}\theta_1\mathrm{d}\theta_2\cdots\mathrm{d}\theta_n$represents n-fold integral.

In Eq. (6.138), $p=p(\boldsymbol{I},t|\boldsymbol{I}_0)$ represents the transition PDF of the action vector $\boldsymbol{I}(t)$ with initial condition

$$p(\boldsymbol{I}, 0|\boldsymbol{I}_0) = \delta(\boldsymbol{I} - \boldsymbol{I}_0), \tag{6.142}$$

or $p = p(\boldsymbol{I}, t)$ represents the PDF of $\boldsymbol{I}(t)$ with initial condition

$$p(\boldsymbol{I}, 0) = p(\boldsymbol{I}_0), \tag{6.143}$$

The boundary conditions for the truncated averaged FPK Eq. (6.138) depend on the nature of the corresponding Hamiltonian system and the constraints applied to the system. If $\boldsymbol{I}$ can vary arbitrarily within the first quadrant of R^n, then the boundary conditions are

$$\begin{aligned} &p(\boldsymbol{I})|_{I_r=0} = \text{finite}, \qquad \left.\frac{\partial^k p(\boldsymbol{I})}{\partial I_r^k}\right|_{I_r=0} = \text{finite}, \\ &\lim_{I_r\to\infty} p(\boldsymbol{I}) = 0, \qquad \lim_{I_r\to\infty} \frac{\partial^k}{\partial I_r^k} p(\boldsymbol{I}) = 0, \\ &r = 1, 2, \cdots, n; \quad k = 1, 2, \cdots, \end{aligned} \tag{6.144}$$

$p(\boldsymbol{I}, t)$ and $p(\boldsymbol{I})$ must also satisfy normalization condition.

The stationary solution of the averaged FPK Eq. (6.19) can be obtained by using the perturbation method. The approximate joint stationary PDF of the generalized displacements and momenta can be obtained by Eq. (5.104) as follows

$$p(\boldsymbol{q}, \boldsymbol{p}) = \left.\frac{1}{(2\pi)^n} p(\boldsymbol{I})\right|_{\boldsymbol{I}=\boldsymbol{I}(\boldsymbol{q},\boldsymbol{p})} \tag{6.145}$$

Assume that the Hamiltonian system associated with Eq. (6.5) is integrable and non-resonant, the action-angle variables are not available but the n independent, mutually involutory first integrals $H_r(\boldsymbol{q}, \boldsymbol{p})$ $(r = 1, 2, \cdots, n)$ are known. Introduce the transformation

$$H_r = H_r(\boldsymbol{Q}, \boldsymbol{P}), \quad r = 1, 2, \cdots, n. \tag{6.146}$$

The SIDEs for H_r can be derived from Eq. (6.5) by using the stochastic jump-diffusion chain rule (Hanson 2007) as follows

$$\begin{aligned} \mathrm{d}H_r = {} & \varepsilon^2\left(-\sum_{i,j=1}^{n} m_{ij}\frac{\partial H}{\partial P_j}\frac{\partial H_r}{\partial P_i} + \frac{1}{2}\sum_{i,j=1}^{n}\sum_{k=1}^{n_g}\sigma_{ik}\sigma_{jk}\frac{\partial^2 H_r}{\partial P_i \partial P_j}\right)\mathrm{d}t \\ & + \varepsilon\sum_{i=1}^{n}\sum_{k=1}^{n_g}\frac{\partial H_r}{\partial P_i}\sigma_{ik}\mathrm{d}B_k(t) \end{aligned}$$

$$
\begin{aligned}
&+\sum_{l=1}^{n_p}\int_{\mathcal{Y}_l}\big[H_r(\boldsymbol{Q},\boldsymbol{P}+\hat{\boldsymbol{\gamma}}_l)-H_r(\boldsymbol{Q},\boldsymbol{P})\big]\mathcal{P}_l(\mathrm{d}t,\mathrm{d}Y_l),\\
&r=1,2,\cdots,n.
\end{aligned}
\tag{6.147}
$$

Replacing the equations of P_i $(i = 1, 2, \cdots, n)$ in Eq. (6.5) with Eq. (6.147), and replacing P_i $(i = 1, 2, \cdots, n)$ with $Q_1, \cdots, Q_n, H_1, \cdots, H_n$ according to Eq. (6.146) result in a new system of SIDEs for $Q_1, \cdots, Q_n, H_1, \cdots, H_n$. Expanding $H_r(\boldsymbol{Q}, \boldsymbol{P}+\hat{\boldsymbol{\gamma}}_l)-H_r(\boldsymbol{Q}, \boldsymbol{P})$ into Taylor series and following the derivations in Appendix 1 of Reference (Jia and Zhu 2014a), Eq. (6.147) can be rewritten as

$$
\begin{aligned}
\mathrm{d}H_r=&\,\varepsilon^2\left(-\sum_{i,j=1}^{n}m_{ij}\frac{\partial H}{\partial P_j}\frac{\partial H_r}{\partial P_i}+\frac{1}{2}\sum_{i,j=1}^{n}\sum_{k,l=1}^{n_g}\sigma_{ik}\sigma_{jl}\frac{\partial^2 H_r}{\partial P_i\partial P_j}\right)\mathrm{d}t\\
&+\varepsilon\sum_{i=1}^{n}\sum_{k=1}^{n_g}\frac{\partial H_r}{\partial P_i}\sigma_{ik}\mathrm{d}B_k(t)\\
&+\sum_{l=1}^{n_p}\int_{\mathcal{Y}_l}\left[\sum_{s=1}^{\infty}\varepsilon^s Y_l^s B_{r;s;l}(\boldsymbol{Q},\boldsymbol{P})\right]\mathcal{P}_l(\mathrm{d}t,\mathrm{d}Y_l),\\
r=&\,1,2,\cdots,n.
\end{aligned}
\tag{6.148}
$$

In the new system formed by Eq. (6.148) and equations for Q_i in Eq. (6.5), H_r are slowly-varying processes, and Q_i are fast-varying processes. According to the stochastic averaging principle (Khasminskii 1968; Xu et al. 2011), when $\varepsilon \to 0$, $\boldsymbol{H}(t) = [H_1(t), H_2(t), \cdots, H_n(t)]^{\mathrm{T}}$ weakly converges to a n-dimensional vector Markov process.

The following truncated SIDEs governing the Markov process can be obtained by completing time averaging on Eq. (6.148) and neglecting terms of order ε^{u+1} and higher.

$$
\begin{aligned}
\mathrm{d}H_r=&\left(\varepsilon^2\overline{m}_r(\boldsymbol{H})+\sum_{i=0}^{u-2}\varepsilon^{i+2}U_{r,i}(\boldsymbol{H})\right)\mathrm{d}t\\
&+\varepsilon\sum_{k=1}^{n_g}\overline{\sigma}_{rk}(\boldsymbol{H})\mathrm{d}B_k(t)+\sum_{l=1}^{u-1}G_{r,l},\\
r=&\,1,2,\cdots,n,
\end{aligned}
\tag{6.149}
$$

where

$$
\overline{m}_r(\mathbf{H})=\left\langle-\sum_{i,j=1}^{n}m_{ij}\frac{\partial H}{\partial P_j}\frac{\partial H_r}{\partial P_i}+\frac{1}{2}\sum_{i,j=1}^{n}\sum_{k=1}^{n_g}\sigma_{ik}\sigma_{jk}\frac{\partial^2 H_r}{\partial P_i\partial P_j}\right\rangle_t,
\tag{6.150}
$$

$$U_{r,0}(\boldsymbol{H}) = \sum_{l=1}^{n_p} \left\langle C_{r;2;l}\lambda_l E[Y_l^2] \right\rangle_t, \tag{6.151}$$

$$U_{r,k}(\boldsymbol{H}) = \sum_{l=1}^{n_p} \left\langle C_{r;k+2;l}\lambda_l E[Y_l^{k+2}] \right\rangle_t,$$
$$k = 1, \cdots, u-2, \tag{6.152}$$

$$\sum_{k=1}^{n_g} \overline{\sigma}_{rk}(\boldsymbol{H})\overline{\sigma}_{sk}(\boldsymbol{H}) = \left\langle \sum_{i,j=1}^{n} \sum_{k=1}^{n_g} \sigma_{ik}\sigma_{jk} \frac{\partial H_r}{\partial P_i}\frac{\partial H_s}{\partial P_j} \right\rangle_t,$$
$$r, s = 1, 2, \cdots, n, \tag{6.153}$$

$$G_{r,l} = \sum_{s_1=1}^{n}\sum_{s_2=1}^{s_1} \cdots \sum_{s_{u-l+1}=1}^{s_{u-l}} \int_{\mathcal{Y}_{s_1,s_2,\cdots,s_{u-l+1};l}} \left(\sum_{k=1}^{l} \varepsilon^k V_{r;s_1,s_2,\cdots,s_{u-l+1};l,k}(\boldsymbol{H}) \right)$$
$$\times\ Y_{s_1,s_2,\cdots,s_{u-l+1};l} \mathcal{P}_{s_1,s_2,\cdots,s_{u-l+1};l}\left(\mathrm{d}t, \mathrm{d}Y_{s_1,s_2,\cdots,s_{u-l+1};l}\right), \tag{6.154}$$

where $\mathcal{P}_{s_1,s_2,\cdots,s_{u-l+1};l}(\mathrm{d}t, \mathrm{d}Y_{s_1,s_2,\cdots,s_{u-l+1};l})$ are independent Poisson random measures given by Eq. (6.137). The functions $V_{r_1;s_1,\cdots,s_{u-k+1};k;k_i}(\boldsymbol{H})$ in Eq. (6.154) can be determined from the following equation by vanishing the terms with same order of ε.

$$\sum_{k=1}^{u-1} \left\{ \sum_{s_1=1}^{n}\sum_{s_2=1}^{s_1} \cdots \sum_{s_{u-k+1}=1}^{s_{u-k}} \left[\left(\sum_{k_1=1}^{k} \varepsilon^{k_1} V_{r_1;s_1,\ldots,s_{u-k+1};k;k_1}(\boldsymbol{H}) \right) \right.\right.$$
$$\times \left(\sum_{k_2=1}^{k} \varepsilon^{k_2} V_{r_2;s_1,\ldots,s_{u-k+1};k;k_2}(\boldsymbol{H}) \right) \cdots$$
$$\left.\left. \times \left(\sum_{k_j=1}^{k} \varepsilon^{k_j} V_{r_j;s_1,\ldots,s_{u-k+1};k;k_j}(\boldsymbol{H}) \right) \lambda_{s_1,s_2,\ldots,s_{u-k+1};k} E\left[Y^j_{s_1,s_2,\ldots,s_{u-k+1};k} \right] \right] \right\}$$
$$= \sum_{k=j}^{u} \varepsilon^k \sum_{l=1}^{n_p} \lambda_l E[Y_l^k] \left\langle \sum_{k_1+k_2+\cdots+k_j=k} B_{r_1;k_1;l} B_{r_2;k_2;l} \cdots B_{r_j;k_j;l} \right\rangle_t, \tag{6.155}$$

The truncated averaged FPK equation associated with SIDEs Eq. (6.149) is

$$\frac{\partial p}{\partial t} = -\sum_{r_1=1}^{n} \frac{\partial}{\partial h_{r_1}}(\overline{a}_{r_1}(\boldsymbol{h})p) + \frac{1}{2!}\sum_{r_1,r_2=1}^{n} \frac{\partial^2}{\partial h_{r_1}\partial h_{r_2}}(\overline{a}_{r_1,r_2}(\boldsymbol{h})p)$$

$$
-\frac{1}{3!}\sum_{r_1,r_2,r_3=1}^{n}\frac{\partial^3}{\partial h_{r_1}\partial h_{r_2}\partial h_{r_3}}\left(\overline{a}_{r_1,r_2,r_3}(\boldsymbol{h})p\right)+\cdots
$$

$$
+(-1)^u\frac{1}{u!}\sum_{r_1,r_2,\cdots,r_u=1}^{n}\frac{\partial^u}{\partial h_{r_1}\partial h_{r_2}\cdots\partial h_{r_u}}\left(\overline{a}_{r_1,r_2,r_3,\cdots,r_u}(\boldsymbol{h})p\right)+\mathrm{O}\left(\varepsilon^{u+1}\right), \quad (6.156)
$$

where $\boldsymbol{h}=[h_1,h_2,\cdots,h_n]^{\mathrm{T}}$ and

$$
\overline{a}_{r_1}(\boldsymbol{h})=\varepsilon^2\left\langle-\sum_{i,j=1}^{n}\frac{\partial h}{\partial p_j}\frac{\partial h_{r_1}}{\partial p_i}+\frac{1}{2}\sum_{i,j=1}^{n}\sum_{k=1}^{n_g}\sigma_{ik}\sigma_{jk}\frac{\partial^2 h_{r_1}}{\partial p_i\partial p_j}\right\rangle_t
+\sum_{k=2}^{u}\varepsilon^k\left\langle\sum_{l=1}^{n_p}\lambda_l E[Y_l^k]B_{r;k;l}\right\rangle_t, \quad (6.157)
$$

$$
\overline{a}_{r_1,r_2}(\boldsymbol{h})=\varepsilon^2\left\langle\sum_{i,j=1}^{n}\sum_{k=1}^{n_g}\sigma_{ik}\sigma_{jk}\frac{\partial h_{r_1}}{\partial p_i}\frac{\partial h_{r_2}}{\partial p_j}\right\rangle_t
+\sum_{k=2}^{u}\varepsilon^k\left\langle\sum_{l=1}^{n_p}\lambda_l E[Y_l^k]\sum_{k_1+k_2=k}B_{r_1;k_1;l}B_{r_2;k_2;l}\right\rangle_t, \quad (6.158)
$$

$$
\overline{a}_{r_1,r_2,\cdots,r_j}(\boldsymbol{h})=\sum_{k=j}^{u}\varepsilon^k\left\langle\sum_{l=1}^{n_p}\lambda_l E\left[Y_l^k\right]\sum_{k_1+k_2+\cdots+k_j=k}B_{r_1;k_1;l}B_{r_2;k_2;l}\cdots B_{r_j;k_j;l}\right\rangle_t, \quad (6.159)
$$

in which

$$
\langle[\bullet]\rangle_t=\lim_{T\to\infty}\int_{t_0}^{t_0+T}[\bullet]\mathrm{d}t \quad (6.160)
$$

represents time averaging.

If the Hamiltonian is separable, i.e.,

$$
H=\sum_{r=1}^{n}H_r(Q_r,P_r) \quad (6.161)
$$

and each Hamiltonian subsystem with Hamiltonian H_r has a periodic solution with period T_r. Then the time averaging in terms of t can be replaced by spatial averaging with respect to q_r on the constant H_r surfaces. The derivative moments in averaged FPK Eq. (6.156) are then

$$\overline{a}_{r_1}(\boldsymbol{h}) = \frac{\varepsilon^2}{T(\boldsymbol{h})}\oint\left(-\sum_{i,j=1}^{n}\frac{\partial h}{\partial p_j}\frac{\partial h_{r_1}}{\partial p_i}+\frac{1}{2}\sum_{i,j=1}^{n}\sum_{k=1}^{n_g}\sigma_{ik}\sigma_{jk}\frac{\partial^2 h_{r_1}}{\partial p_i \partial p_j}\right)$$

$$\times\prod_{\mu=1}^{n}\left(1\Big/\frac{\partial h_\mu}{\partial p_\mu}\right)\mathrm{d}q_\mu+\sum_{k=2}^{u}\frac{\varepsilon^k}{T(\boldsymbol{h})}\sum_{l=1}^{n_p}\oint\left(\lambda_l E[Y_l^k]B_{r;k;l}\right)\times\prod_{\mu=1}^{n}\left(1\Big/\frac{\partial h_\mu}{\partial p_\mu}\right)\mathrm{d}q_\mu, \tag{6.162}$$

$$\overline{a}_{r_1,r_2}(\boldsymbol{h}) = \frac{\varepsilon^2}{T(\boldsymbol{h})}\oint\left(\sum_{i,j=1}^{n}\sum_{k=1}^{n_g}\sigma_{ik}\sigma_{jk}\frac{\partial h_{r_1}}{\partial p_i}\frac{\partial h_{r_2}}{\partial p_j}\right)\times\prod_{\mu=1}^{n}\left(1\Big/\frac{\partial h_\mu}{\partial p_\mu}\right)\mathrm{d}q_\mu$$

$$+\sum_{k=2}^{u}\frac{\varepsilon^k}{T(\boldsymbol{h})}\oint\left(\sum_{l=1}^{n_p}\lambda_l E[Y_l^k]\sum_{k_1+k_2=k}B_{r_1;k_1;l}B_{r_2;k_2;l}\right)\times\prod_{\mu=1}^{n}\left(1\Big/\frac{\partial h_\mu}{\partial p_\mu}\right)\mathrm{d}q_\mu, \tag{6.163}$$

$$\overline{a}_{r_1,r_2,\cdots,r_j}(\boldsymbol{h}) = \sum_{k=j}^{u}\frac{\varepsilon^k}{T(\boldsymbol{h})}\oint\left(\sum_{l=1}^{n_p}\lambda_l E[Y_l^k]\sum_{k_1+k_2+\cdots+k_j=k}B_{r_1;k_1;l}B_{r_2;k_2;l}\cdots B_{r_j;k_j;l}\right)$$

$$\times\prod_{\mu=1}^{n}\left(1\Big/\frac{\partial h_\mu}{\partial p_\mu}\right)\mathrm{d}q_\mu, \tag{6.164}$$

where $\oint[\bullet]\prod_{\mu=1}^{n}\mathrm{d}q_\mu$ represents n-fold integral, and the least common multiple of T_r is

$$T(\boldsymbol{h}) = \prod_{\mu=1}^{n}T_\mu = \oint\prod_{\mu=1}^{n}\left(1\Big/\frac{\partial h_\mu}{\partial p_\mu}\right)\mathrm{d}q_\mu. \tag{6.165}$$

In the averaged FPK Eq. (6.156), $p = p(\boldsymbol{h}, t|\boldsymbol{h}_0)$ represents the transition PDF of $H(t)$ with the initial condition

$$p(\boldsymbol{h}, 0|\boldsymbol{h}_0) = \delta(\boldsymbol{h}-\boldsymbol{h}_0) \tag{6.166}$$

or $p = p(\boldsymbol{h}, t)$ represent PDF of $H(t)$ with the initial condition

$$p(\boldsymbol{h}, 0|\boldsymbol{h}_0) = p(\boldsymbol{h}_0). \tag{6.167}$$

If $\boldsymbol{h}$ can vary arbitrarily within the first quadrant of R^n, then boundary conditions for the averaged FPK Eq. (6.156) are

$$
\begin{aligned}
&p(\boldsymbol{h})|_{h_r=0} = \text{finite}, \quad \left.\frac{\partial^k p(\boldsymbol{h})}{\partial h_r^k}\right|_{h_r=0} = \text{finite}, \\
&\lim_{|\boldsymbol{h}|\to\infty} p(\boldsymbol{h}) = 0, \quad \lim_{|\boldsymbol{h}|\to\infty} \frac{\partial^k}{\partial h_r^k} p(\boldsymbol{h}) = 0, \\
&r = 1, 2, \ldots, n; \quad k = 1, 2, \ldots,
\end{aligned} \tag{6.168}
$$

Equation (6.156) also must satisfies normalization condition.

The stationary solution $p(\boldsymbol{h})$ of the truncated averaged FPK of Eq. (6.156) can be obtained using the perturbation method, and the approximate joint stationary PDF $p(\boldsymbol{q}, \boldsymbol{p})$ can be obtained following Eq. (5.106) as follows

$$
p(\boldsymbol{q}, \boldsymbol{p}) = \left.\frac{p(\boldsymbol{h})}{T(\boldsymbol{h})}\right|_{\boldsymbol{h}=H(\boldsymbol{q},\boldsymbol{p})} \tag{6.169}
$$

6.3.2 *Internal Resonant Case*

Now suppose that the Hamiltonian system associated with Eq. (6.5) is completely integrable and there exist the following α $(1 \le \alpha \le n-1)$ weak internal resonance relations

$$
\sum_{r=1}^{n} k_r^v \omega_r = O_v(\varepsilon^2), \quad v = 1, 2, \cdots, \alpha \tag{6.170}
$$

where k_r^v are integers and $O_v(\varepsilon^2)$ are small quantities of ε^2 order. Introduce the following combinations of angle variables

$$
\Psi_v = \sum_{r=1}^{n} k_r^v \Theta_r, \quad v = 1, 2, \cdots, \alpha \tag{6.171}
$$

The SIDEs for Ψ_v can be obtained according to Eq. (6.171) by combining the SIDEs for Θ_r in Eq. (6.127) as follows (Jia et al. 2014)

$$
\begin{aligned}
\mathrm{d}\Psi_v = & \left[O_v(\varepsilon^2) - \varepsilon^2 \sum_{i,j=1}^{n} m_{ij} \frac{\partial H}{\partial P_j} \frac{\partial \Psi_v}{\partial P_i} \right. \\
& \left. + \frac{\varepsilon^2}{2} \sum_{i,j=1}^{n} \sum_{k=1}^{n_g} \sigma_{ik} \sigma_{jk} \frac{\partial^2 \Psi_v}{\partial P_i \partial P_j} \right] \mathrm{d}t \\
& + \varepsilon \sum_{i=1}^{n} \sum_{k=1}^{n_g} \frac{\partial \Psi_v}{\partial P_i} \sigma_{ik} \mathrm{d}B_k(t) \\
& + \sum_{l=1}^{n_p} \int_{\mathcal{Y}_l} \left[\Psi_v(\boldsymbol{Q}, \boldsymbol{P} + \hat{\boldsymbol{\gamma}}_l) - \Psi_v(\boldsymbol{Q}, \boldsymbol{P}) \right] \mathcal{P}_l(\mathrm{d}t, \mathrm{d}Y_l).
\end{aligned} \tag{6.172}
$$

Expanding $\Psi_v(\boldsymbol{Q}, \boldsymbol{P} + \hat{\boldsymbol{\gamma}}_l) - \Psi_v(\boldsymbol{Q}, \boldsymbol{P})$ into Taylor series, Eq. (6.172) can be rewritten as follows

$$
\begin{aligned}
\mathrm{d}\Psi_v = & \left[O_v(\varepsilon^2) - \varepsilon^2 \sum_{i,j=1}^{n} m_{ij} \frac{\partial H}{\partial P_j} \frac{\partial \Psi_v}{\partial P_i} + \frac{\varepsilon^2}{2} \sum_{i,j=1}^{n} \sum_{k=1}^{n_g} \sigma_{ik} \sigma_{jk} \frac{\partial^2 \Psi_v}{\partial P_i \partial P_j} \right] \mathrm{d}t \\
& + \varepsilon \sum_{i=1}^{n} \sum_{k=1}^{n_g} \frac{\partial \Psi_v}{\partial P_i} \sigma_{ik} \mathrm{d}B_k(t) \\
& + \sum_{l=1}^{n_p} \int_{\mathcal{Y}_l} \left[\varepsilon^k Y_l^k C_{v;k;l}(\boldsymbol{Q}, \boldsymbol{P}) \right] \mathcal{P}_l(\mathrm{d}t, \mathrm{d}Y_l), \\
v = & \ 1, 2, \cdots, \alpha.
\end{aligned} \tag{6.173}
$$

The new system consists of n equations for I_r and $n - \alpha$ equations for $\Theta_1, \cdots, \Theta_{n-\alpha}$ in Eq. (6.128) and α equations for Ψ_v in Eq. (6.173). In these equations, Q_i, P_i should be replaced by I_r, Θ_r, Ψ_v according to the transformations (6.126) and (6.171). $I_1, \cdots, I_n, \Psi_1, \cdots, \Psi_\alpha$ are slowly-varying processes while $\Theta_1, \cdots, \Theta_{n-\alpha}$ are fast-varying processes. Based on the stochastic averaging theorem (Khasminskii 1968; Xu et al. 2011), $\left[\boldsymbol{I}^{\mathrm{T}}, \boldsymbol{\Psi}^{\mathrm{T}}\right]^{\mathrm{T}}$ converges weakly to a $(n+\alpha)$-dimensional vector Markov process as $\varepsilon \to 0$.

The SIDEs governing $n+\alpha$-dimensional vector Markov process $\left[\boldsymbol{I}^{\mathrm{T}}, \boldsymbol{\Psi}^{\mathrm{T}}\right]^{\mathrm{T}}$ can be obtained by time averaging over the first n equations for I_r in Eq. (6.128) and α equations for Ψ_v in Eq. (6.173) under the condition of constant I_r and Ψ_v. Due to the ergodicity of integrable and resonant Hamiltonian system on $(n-\alpha)$-dimensional sub-torus, time averaging can be replaced by spatial averaging with respective to $\Theta_1, \Theta_2, \cdots, \Theta_{n-\alpha}$. Since there are infinite terms in the SIDEs for $I_r(t)$, $\Psi_v(t)$, truncation is necessary when deriving the finite-term averaged SIDEs and the averaged FPK equation. Neglecting ε^{u+1} and higher-order terms leads to the following truncated averaged SIDEs

$$
\begin{aligned}
\mathrm{d}I_r = &\left(\varepsilon^2 \overline{m}_r(\boldsymbol{I}, \mathbf{Y}) + \sum_{i=1}^{u-2} \varepsilon^{i+2} U_{r,i}(\boldsymbol{I}, \mathbf{Y})\right)\mathrm{d}t \\
&+ \varepsilon \sum_{k=1}^{n_g} \overline{\sigma}_{rk}(\boldsymbol{I}, \mathbf{Y})\mathrm{d}B_k(t) + \sum_{l=1}^{u-1} G_{r,l}, \\
&r = 1, 2, \cdots, n,
\end{aligned} \tag{6.174}
$$

$$
\begin{aligned}
\mathrm{d}\Psi_v = &\left(\varepsilon^2 \overline{m}_v(\boldsymbol{I}, \mathbf{Y}) + \sum_{i=1}^{u-2} \varepsilon^{i+2} U_{v,i}(\boldsymbol{I}, \mathbf{Y})\right)\mathrm{d}t \\
&+ \varepsilon \sum_{k=1}^{n_g} \overline{\sigma}_{vk}(\boldsymbol{I}, \mathbf{Y})\mathrm{d}B_k(t) + \sum_{l=1}^{u-1} G_{v,l}, \\
&v = 1, 2, \cdots, \alpha,
\end{aligned} \tag{6.175}
$$

where

$$
\begin{aligned}
\overline{m}_r(\boldsymbol{I}, \boldsymbol{\Psi}) = &\frac{1}{(2\pi)^{n-\alpha}} \\
&\int_0^{2\pi} \left(-\sum_{i,j=1}^{n} m_{ij} \frac{\partial H}{\partial p_j}\frac{\partial I_r}{\partial p_i} + \frac{1}{2}\sum_{i,j=1}^{n}\sum_{k=1}^{n_g} \sigma_{ik}\sigma_{jk} \frac{\partial^2 I_r}{\partial p_i \partial p_j}\right)\mathrm{d}\boldsymbol{\theta}_1,
\end{aligned} \tag{6.176}
$$

$$
U_{r,0}(\boldsymbol{I}, \boldsymbol{\Psi}) = \sum_{l=1}^{n_p} \frac{\lambda_l E[Y_l^2]}{(2\pi)^{n-\alpha}} \int_0^{2\pi} A_{r;2;l}\mathrm{d}\boldsymbol{\theta}_1, \tag{6.177}
$$

$$
U_{r,k}(\boldsymbol{I}, \boldsymbol{\Psi}) = \sum_{l=1}^{n_p} \frac{\lambda_l E[Y_l^{k+2}]}{(2\pi)^{n-\alpha}} \int_0^{2\pi} A_{r;k+2;l}\mathrm{d}\boldsymbol{\theta}_1, \tag{6.178}
$$

$$
\begin{aligned}
\overline{m}_v(\boldsymbol{I}, \boldsymbol{\Psi}) = &\frac{1}{(2\pi)^{n-\alpha}} \\
&\int_0^{2\pi} \left(\frac{\mathrm{O}(\varepsilon^2)}{\varepsilon^2} - \sum_{i,j=1}^{n} m_{ij} \frac{\partial H}{\partial p_j}\frac{\partial \Psi_v}{\partial p_i} + \frac{1}{2}\sum_{i,j=1}^{n}\sum_{k=1}^{n_g} \sigma_{ik}\sigma_{jk} \frac{\partial^2 \Psi_u}{\partial p_i \partial p_j}\right)\mathrm{d}\boldsymbol{\theta}_1,
\end{aligned} \tag{6.179}
$$

$$
U_{v,0}(\boldsymbol{I}, \boldsymbol{\Psi}) = \sum_{l=1}^{n_p} \frac{\lambda_l E[Y_l^2]}{(2\pi)^{n-\alpha}} \int_0^{2\pi} C_{v;2;l}\mathrm{d}\boldsymbol{\theta}_1, \tag{6.180}
$$

$$
U_{v,k}(\boldsymbol{I}, \boldsymbol{\Psi}) = \sum_{l=1}^{n_p} \frac{\lambda_l E[Y_l^{k+2}]}{(2\pi)^{n-\alpha}} \int_0^{2\pi} C_{v;k+2;l}\mathrm{d}\boldsymbol{\theta}_1, \tag{6.181}
$$

$$\overline{\sigma}_{r_1,k}(\boldsymbol{I},\boldsymbol{\Psi})\overline{\sigma}_{r_2,l}(\boldsymbol{I},\boldsymbol{\Psi}) = \frac{1}{(2\pi)^{n-\alpha}} \int_0^{2\pi} \left(\sum_{i,j=1}^{n} \sum_{k=1}^{n_g} \sigma_{ik}\sigma_{jk} \frac{\partial I_{r_1}}{\partial p_i} \frac{\partial I_{r_2}}{\partial p_j} \right) \mathrm{d}\boldsymbol{\theta}_1, \tag{6.182}$$

$$\overline{\sigma}_{r_1,k}(\boldsymbol{I},\boldsymbol{\Psi})\overline{\sigma}_{v_1,l}(\boldsymbol{I},\boldsymbol{\Psi}) = \frac{1}{(2\pi)^{n-\alpha}} \int_0^{2\pi} \left(\sum_{i,j=1}^{n} \sum_{k=1}^{n_g} \sigma_{ik}\sigma_{jk} \frac{\partial I_{r_1}}{\partial p_i} \frac{\partial \Psi_{v_1}}{\partial p_j} \right) \mathrm{d}\boldsymbol{\theta}_1, \tag{6.183}$$

$$\sum_{k=1}^{n_g} \overline{\sigma}_{v_1,k}(\boldsymbol{I},\boldsymbol{\Psi})\overline{\sigma}_{v_1,l}(\boldsymbol{I},\boldsymbol{\Psi}) = \frac{1}{(2\pi)^{n-\alpha}} \int_0^{2\pi} \left(\sum_{i,j=1}^{n} \sum_{k=1}^{n_g} \sigma_{ik}\sigma_{jk} \frac{\partial \Psi_{v_1}}{\partial p_i} \frac{\partial \Psi_{v_2}}{\partial p_j} \right) \mathrm{d}\theta_1, \tag{6.184}$$

$$G_{r,l} = \sum_{s_1=1}^{n+\alpha} \sum_{s_2=1}^{s_1} \cdots \sum_{s_{u-l+1}=1}^{s_{u-l}} \left[\int_{\mathcal{Y}_{s_1,\cdots,s_{u-l+1};l}} \left(\sum_{k=1}^{l} \varepsilon^k V_{r;s_1,\cdots,s_{u-l+1};l,k}(\boldsymbol{I},\boldsymbol{\Psi}) \right) Y_{s_1,\cdots,s_{u-l+1};l} \mathcal{P}_{s_1,\cdots,s_{u-l+1};l}\big(dt, dY_{s_1,\cdots,s_{u-l+1};l}\big) \right], \tag{6.185}$$

$$G_{v,l} = \sum_{s_1=1}^{n+\alpha} \sum_{s_2=1}^{s_1} \cdots \sum_{s_{u-l+1}=1}^{s_{u-l}} \left[\int_{\mathcal{Y}_{s_1,\cdots,s_{u-l+1};l}} \left(\sum_{k=1}^{l} \varepsilon^k V_{v;s_1,\cdots,s_{u-l+1};l,k}(\boldsymbol{I},\boldsymbol{\Psi}) \right) Y_{s_1,\cdots,s_{u-l+1};l} \mathcal{P}_{s_1,\cdots,s_{u-l+1};l}(dt, dY_{s_1,\cdots,s_{u-l+1};l}) \right], \tag{6.186}$$

where $\boldsymbol{I} = [I_1, I_2, \cdots, I_n]^{\mathrm{T}}$, $\boldsymbol{\Psi} = [\Psi_1, \Psi_2, \cdots, \Psi_\alpha]^{\mathrm{T}}$. Function $V_{r;s_1,s_2,,s_{u-k+1};l;k}(\boldsymbol{I},\boldsymbol{\Psi})$ can be determined by balancing the following equations with same order of ε,

$$\sum_{k=1}^{u-1} \left\{ \sum_{s_1=1}^{n} \sum_{s_2=1}^{s_1} \cdots \sum_{s_{u-k+1}=1}^{s_{u-k}} \left[\left(\sum_{k_1=1}^{k} \varepsilon^{k_1} V_{r_1;s_1,\cdots,s_{u-k+1};k;k_1}(\boldsymbol{I},\boldsymbol{\Psi}) \right) \times \left(\sum_{k_2=1}^{k} \varepsilon^{k_2} V_{r_2;s_1,\cdots,s_{u-k+1};k;k_2}(\boldsymbol{I},\boldsymbol{\Psi}) \right) \cdots \times \left(\sum_{k_j=1}^{k} \varepsilon^{k_j} V_{r_j;s_1,\cdots,s_{u-k+1};k;k_j}(\boldsymbol{I},\boldsymbol{\Psi}) \right) \lambda_{s_1,s_2,\cdots,s_{u-k+1};k} E\Big[Y^j_{s_1,s_2,\cdots,s_{u-k+1;k}} \Big] \right] \right\}$$

$$= \sum_{k=j}^{u} \varepsilon^k \sum_{l=1}^{n_p} \frac{\lambda_l E\left[Y_l^k\right]}{(2\pi)^{n-\alpha}} \int_0^{2\pi} \left(\sum_{k_1+k_2+\cdots+k_j=k} A_{r_1;k_1;l} A_{r_2;k_2;l} \cdots A_{r_j;k_j;l} \right) \mathrm{d}\boldsymbol{\theta}_1 \quad (6.187)$$

$$\sum_{k=1}^{u-1} \left\{ \sum_{s_1=1}^{n} \sum_{s_2=1}^{s_1} \cdots \sum_{s_{u-k+1}=1}^{s_{u-k}} \left[\left(\sum_{k_1=1}^{k} \varepsilon^{k_1} V_{r_1;s_1,\cdots,s_{u-k+1};k;k_1}(\boldsymbol{I}, \Psi) \right) \right.\right.$$
$$\times \left(\sum_{k_2=1}^{k} \varepsilon^{k_2} V_{r_2;s_1,\cdots,s_{u-k+1};k;k_2}(\boldsymbol{I}, \Psi) \right) \cdots \times \left(\sum_{k_s=1}^{k} \varepsilon^{k_s} V_{r_s;s_1,\cdots,s_{u-k+1};k;k_s}(\boldsymbol{I}, \Psi) \right)$$
$$\times \left(\sum_{k_{s+1}=1}^{k} \varepsilon^{k_{s+1}} V_{n+v_1;s_1,\cdots,s_{u-k+1};k;k_{s+1}}(\boldsymbol{I}, \Psi) \right) \cdots$$
$$\left.\left. \times \left(\sum_{k_j=1}^{k} \varepsilon^{k_j} V_{n+v_{j-s};s_1,\cdots,s_{u-k+1};k;k_j}(\boldsymbol{I}, \Psi) \right) \lambda_{s_1,s_2,\cdots,s_{u-k+1};k} E[Y^j_{s_1,s_2,\cdots,s_{u-k+1};k}] \right] \right\}$$
$$= \sum_{k=j}^{u} \varepsilon^k \sum_{l=1}^{n_p} \frac{\lambda_l E[Y_l^k]}{(2\pi)^{n-\alpha}}$$
$$\int_0^{2\pi} \left(\sum_{k_1+k_2+\cdots+k_j=k} A_{r_1;k_1;l} A_{r_2;k_2;l} \cdots A_{r_t;k_t;l} C_{v_1;k_{t+1};l} \cdots C_{v_{j-t};k_j;l} \right) \mathrm{d}\boldsymbol{\theta}_1, \quad (6.188)$$

where $t = 1, 2, \cdots, j-1$,

$$\sum_{k=1}^{u-1} \left\{ \sum_{s_1=1}^{n} \sum_{s_2=1}^{s_1} \cdots \sum_{s_{u-k+1}=1}^{s_{u-k}} \left[\left(\sum_{k_1=1}^{k} \varepsilon^{k_1} V_{n+v_1;s_1,\cdots,s_{u-k+1};k;k_1}(\boldsymbol{I}, \boldsymbol{\Psi}) \right) \right.\right.$$
$$\times \left(\sum_{k_2=1}^{k} \varepsilon^{k_2} V_{n+v_2;s_1,\cdots,s_{u-k+1};k;k_2}(\boldsymbol{I}, \boldsymbol{\Psi}) \right)$$
$$\left.\left. \cdots \times \left(\sum_{k_j=1}^{k} \varepsilon^{k_j} V_{n+v_j;s_1,\cdots,s_{u-k+1};k;k_j}(\boldsymbol{I}, \boldsymbol{\Psi}) \right) \lambda_{s_1,s_2,\cdots,s_{u-k+1};k} E[Y^j_{s_1,s_2,\cdots,s_{u-k+1};k}] \right] \right\}$$
$$= \sum_{k=j}^{u} \varepsilon^k \sum_{l=1}^{n_p} \frac{\lambda_l E[Y_l^k]}{(2\pi)^{n-\alpha}} \int_0^{2\pi} \left(\sum_{k_1+k_2+\cdots+k_j=k} C_{v_1;k_1;l} C_{v_2;k_2;l} \cdots C_{v_j;k_j;l} \right) \mathrm{d}\boldsymbol{\theta}_1, \quad (6.189)$$

$$r, r_1, \cdots; \quad r_j = 1, 2, \cdots, n; \quad v, v_1, \cdots; \quad v_j = 1, 2, \cdots \alpha; \quad j = 2, \cdots, u.$$

where $\boldsymbol{\theta}_1 = \left[\theta_1, \theta_2, \cdots, \theta_{n-\alpha}\right]^{\mathrm{T}}$, $A_{r;k;l} = A_{r;k;l}(\boldsymbol{I}, \boldsymbol{\Psi}, \boldsymbol{\theta}_1)$, $C_{r;k;l} = C_{r;k;l}(\boldsymbol{I}, \boldsymbol{\Psi}, \boldsymbol{\theta}_1)$, $\mathcal{P}_{s_1,s_2,\cdots,s_{u-l+1};l}(\mathrm{d}t, \mathrm{d}Y_{s_1,s_2,\cdots,s_{u-l+1};l})$ are independent Poisson random measure with the following property

$$E\left[\left(\mathrm{d}C_{r_1,r_2,,r_{u-s+1};s}\right)^r\right]=\lambda_{r_1,r_2,\cdots,r_{u-s+1};s}E\left[Y^r_{r_1,r_2,\cdots,r_{u-s+1};s}\right]$$
$$=\begin{cases}0 & r=1,\\ \overline{M}_{r_1,r_2,\cdots,r_{u-s+1};s,r} & r=2,3,\cdots,u-s+1,\\ \overline{m}_{r_1,r_2,\cdots,r_{u-s+1};s,r} & r=u-s+2,u-s+3,\cdots,\end{cases}\tag{6.190}$$
$$\overline{M}_{r_1,r_2,\cdots,r_{u-s+1};s,r}\gg\varepsilon,\ \overline{m}_{r_1,r_2,\cdots,r_{u-s+1};s,r}<\varepsilon^u$$

The averaged FPK equation associated with the truncated averaged SIDEs. (6.174) and (6.175) is

$$\begin{aligned}\frac{\partial}{\partial t}p=&-\sum_{r_1=1}^{n}\frac{\partial}{\partial I_{r_1}}(\overline{a}_{r_1}p)-\sum_{v_1=1}^{\alpha}\frac{\partial}{\partial\psi_{v_1}}(\overline{a}_{v_1}p)+\frac{1}{2}\sum_{r_1,r_2=1}^{n}\frac{\partial^2}{\partial I_{r_1}\partial I_{r_2}}(\overline{a}_{r_1,r_2}p)\\&+\frac{C_2^1}{2}\sum_{r_1=1}^{n}\sum_{v_1=1}^{\alpha}\frac{\partial^2}{\partial I_{r_1}\partial\psi_{v_1}}(\overline{a}_{r_1,v_1}p)+\frac{1}{2}\sum_{v_1,v_2=1}^{\alpha}\frac{\partial^2}{\partial\psi_{v_1}\partial\psi_{v_2}}(\overline{a}_{v_1,v_2}p)\\&-\frac{1}{3!}\sum_{r_1,r_2,r_3=1}^{n}\frac{\partial^3}{\partial I_{r_1}\partial I_{r_2}\partial I_{r_3}}(\overline{a}_{r_1,r_2,r_3}p)\\&-\frac{C_3^1}{3!}\sum_{r_1,r_2=1}^{n}\sum_{v_1=1}^{\alpha}\frac{\partial^3}{\partial I_{r_1}\partial I_{r_2}\partial\psi_{v_1}}(\overline{a}_{r_1,r_2,v_1}p)\\&-\frac{C_3^2}{3!}\sum_{r_1=1}^{n}\sum_{v_1,v_2=1}^{\alpha}\frac{\partial^3}{\partial I_{r_1}\partial\psi_{v_1}\partial\psi_{v_2}}(\overline{a}_{r_1,v_1,v_2}p)\\&-\frac{1}{3!}\sum_{v_1,v_2,v_3=1}^{\alpha}\frac{\partial^3}{\partial\psi_{v_1}\partial\psi_{v_2}\partial\psi_{v_3}}(\overline{a}_{v_1,v_2,v_3}p)+\cdots+(-1)^j\\&\sum_{s=0}^{j}\sum_{r_1,\cdots,r_s=1}^{n}\sum_{v_1,\cdots,v_{j-s}=1}^{\alpha}\frac{C_j^s}{j!}\frac{\partial^j}{\partial I_{r_1}\cdots\partial I_{r_s}\partial\psi_{v_1}\cdots\partial\psi_{v_{j-s}}}(\overline{a}_{r_1,\cdots,r_s,v_1,\cdots,v_{j-s}}p)+\cdots\\&+(-1)^u\sum_{s=0}^{u}\sum_{r_1,\cdots,r_s=1}^{n}\sum_{v_1,\cdots,v_{u-s}=1}^{\alpha}\frac{C_u^s}{u!}\frac{\partial^u}{\partial I_{r_1}\cdots\partial I_{r_s}\partial\psi_{v_1}\cdots\partial\psi_{v_{u-s}}}(\overline{a}_{r_1,\cdots,r_s,v_1,\cdots,v_{u-s}}p),\end{aligned}\tag{6.191}$$

where $C_j^s=\frac{j!}{s!(m-s)!}$ and

$$\begin{aligned}\overline{a}_{r_1}&=\overline{a}_{r_1}(\boldsymbol{I},\boldsymbol{\Psi})\big|_{\boldsymbol{\Psi}=\boldsymbol{\psi}}\\&=\frac{\varepsilon^2}{(2\pi)^{n-\alpha}}\int_0^{2\pi}\left(-\sum_{i,j=1}^{n}m_{ij}\frac{\partial H}{\partial p_j}\frac{\partial I_{r_1}}{\partial p_i}+\frac{1}{2}\sum_{i,j=1}^{n}\sum_{k=1}^{n_g}\sigma_{ik}\sigma_{jk}\frac{\partial^2 I_{r_1}}{\partial p_i\partial p_j}\right)\mathrm{d}\boldsymbol{\theta}_1\end{aligned}$$

$$+\sum_{k=1}^{u}\varepsilon^{k}\sum_{l=1}^{n_p}\frac{\lambda_l E[Y_l^k]}{(2\pi)^{n-\alpha}}\int_0^{2\pi}A_{r_1;k;l}\mathrm{d}\boldsymbol{\theta}_1, \tag{6.192}$$

$$\begin{aligned}\overline{a}_{v_1} &= \overline{a}_{v_1}(\boldsymbol{I},\boldsymbol{\Psi})\big|_{\boldsymbol{\Psi}=\boldsymbol{\psi}}\\ &= \frac{1}{(2\pi)^{n-\alpha}}\int_0^{2\pi}\left[\mathrm{O}(\varepsilon^2)+\varepsilon^2\left(-\sum_{i,j=1}^{n}m_{ij}\frac{\partial H}{\partial p_j}\frac{\partial \psi_{v_1}}{\partial p_i}\right.\right.\\ &\quad\left.\left.+\frac{1}{2}\sum_{i,j=1}^{n}\sum_{k=1}^{n_g}\sigma_{ik}\sigma_{jk}\frac{\partial^2\psi_{v_1}}{\partial p_i\partial p_j}\right)\right]\mathrm{d}\boldsymbol{\theta}_1\\ &\quad+\sum_{k=1}^{u}\varepsilon^{k}\sum_{l=1}^{n_p}\frac{\lambda_l E[Y_l^k]}{(2\pi)^{n-\alpha}}\int_0^{2\pi}C_{v_1;k;l}\mathrm{d}\boldsymbol{\theta}_1,\end{aligned} \tag{6.193}$$

$$\begin{aligned}\overline{a}_{r_1,r_2} = \overline{a}_{r_1,r_2}(\boldsymbol{I},\boldsymbol{\Psi})\big|_{\boldsymbol{\Psi}=\boldsymbol{\psi}} &= \frac{\varepsilon^2}{(2\pi)^{n-\alpha}}\int_0^{2\pi}\left(\sum_{i,j=1}^{n}\sum_{l=1}^{n_g}\sigma_{ik}\sigma_{jk}\frac{\partial I_{r_1}}{\partial p_i}\frac{\partial I_{r_2}}{\partial p_j}\right)\mathrm{d}\boldsymbol{\theta}_1\\ &\quad+\sum_{k=2}^{u}\varepsilon^{k}\frac{\lambda_l E[Y_l^k]}{(2\pi)^{n-\alpha}}\int_0^{2\pi}\left(\sum_{k_1+k_2-k}A_{r_1;k_1;l}A_{r_2;k_2;l}\right)\mathrm{d}\boldsymbol{\theta}_1,\end{aligned} \tag{6.194}$$

$$\begin{aligned}\overline{a}_{r_1,v_1} = \overline{a}_{r_1,v_1}(\boldsymbol{I},\boldsymbol{\Psi})\big|_{\boldsymbol{\Psi}=\boldsymbol{\psi}} &= \frac{\varepsilon^2}{(2\pi)^{n-\alpha}}\int_0^{2\pi}\left(\sum_{i,j=1}^{n}\sum_{k=1}^{n_g}\sigma_{ik}\sigma_{jk}\frac{\partial I_{r_1}}{\partial p_i}\frac{\partial \psi_{v_1}}{\partial p_j}\right)\mathrm{d}\boldsymbol{\theta}_1\\ &\quad+\sum_{k=2}^{u}\varepsilon^{k}\frac{\lambda_l E[Y_l^k]}{(2\pi)^{n-\alpha}}\int_0^{2\pi}\left(\sum_{k_1+k_2=k}A_{r_1;k_1;l}C_{v_1;k_2;l}\right)\mathrm{d}\boldsymbol{\theta}_1,\end{aligned} \tag{6.195}$$

$$\begin{aligned}\overline{a}_{v_1,v_2} = \overline{a}_{v_1,v_2}(\boldsymbol{I},\boldsymbol{\Psi})\big|_{\boldsymbol{\Psi}=\boldsymbol{\psi}} &= \frac{\varepsilon^2}{(2\pi)^{n-\alpha}}\int_0^{2\pi}\left(\sum_{i,j=1}^{n}\sum_{k=1}^{n_g}\sigma_{ik}\sigma_{jk}\frac{\partial \psi_{v_1}}{\partial p_i}\frac{\partial \psi_{v_2}}{\partial p_j}\right)\mathrm{d}\boldsymbol{\theta}_1\\ &\quad+\sum_{k=2}^{u}\varepsilon^{k}\frac{\lambda_l E[Y_l^k]}{(2\pi)^{n-\alpha}}\int_0^{2\pi}\left(\sum_{k_1+k_2=k}C_{v_1;k_1;l}C_{v_2;k_2;l}\right)\mathrm{d}\boldsymbol{\theta}_1,\end{aligned} \tag{6.196}$$

$$\begin{aligned}\overline{a}_{r_1,r_2,r_3} &= \overline{a}_{r_1,r_2,r_3}(\boldsymbol{I},\boldsymbol{\Psi})\big|_{\boldsymbol{\Psi}=\boldsymbol{\psi}}\\ &= \sum_{k=3}^{u}\varepsilon^{k}\frac{\lambda_l E[Y_l^k]}{(2\pi)^{n-\alpha}}\int_0^{2\pi}\left(\sum_{k_1+k_2+k_3=k}A_{r_1;k_1;l}A_{r_2;k_2;l}A_{r_3;k_3;l}\right)\mathrm{d}\boldsymbol{\theta}_1,\end{aligned} \tag{6.197}$$

$$\overline{a}_{r_1,r_2,v_1} = \overline{a}_{r_1,r_2,v_1}(\boldsymbol{I},\boldsymbol{\Psi})\big|_{\boldsymbol{\Psi}=\boldsymbol{\psi}}$$

$$= \sum_{k=3}^{u} \varepsilon^k \frac{\lambda_l E[Y_l^k]}{(2\pi)^{n-\alpha}} \int_0^{2\pi} \left(\sum_{k_1+k_2+k_3=k} A_{r_1;k_1;l} A_{r_2;k_2;l} C_{v_1;k_3;l} \right) \mathrm{d}\boldsymbol{\theta}_1, \tag{6.198}$$

$$\overline{a}_{r_1,v_1,v_2} = \overline{a}_{r_1,v_1,v_2}(\boldsymbol{I}, \boldsymbol{\Psi})\big|_{\boldsymbol{\Psi}=\boldsymbol{\psi}}$$

$$= \sum_{k=3}^{u} \varepsilon^k \frac{\lambda_l E[Y_l^k]}{(2\pi)^{n-\alpha}} \int_0^{2\pi} \left(\sum_{k_1+k_2+k_3=k} A_{r_1;k_1;l} C_{v_1;k_2;l} C_{v_2;k_3;l} \right) \mathrm{d}\boldsymbol{\theta}_1 \tag{6.199}$$

$$\overline{a}_{v_1,v_2,v_3} = \overline{a}_{v_1,v_2,v_3}(\boldsymbol{I}, \boldsymbol{\Psi})\big|_{\boldsymbol{\Psi}=\boldsymbol{\psi}}$$

$$= \sum_{k=3}^{u} \varepsilon^k \frac{\lambda_l E[Y_l^k]}{(2\pi)^{n-\alpha}} \int_0^{2\pi} \left(\sum_{k_1+k_2+k_3=k} C_{v_1;k_1;l} C_{v_2;k_2;l} C_{v_3;k_3;l} \right) \mathrm{d}\boldsymbol{\theta}_1 \tag{6.200}$$

$$\overline{a}_{r_1,r_2,\cdots r_s,v_1,v_2,\cdots,v_{j-s}} = \overline{a}_{r_1,r_2,\cdots r_s,v_1,v_2,\cdots,v_{j-s}}(\boldsymbol{I}, \boldsymbol{\Psi})\big|_{\boldsymbol{\Psi}=\boldsymbol{\psi}}$$

$$= \sum_{k=j}^{u} \frac{\varepsilon^k \lambda_l E[Y_l^k]}{(2\pi)^{n-\alpha}} \int_0^{2\pi} \left(\sum_{k_1+k_2+\cdots+k_j=k} A_{r_1;k_1;l} \cdots A_{r_s;k_s;l} C_{v_1;k_{s+1};l} \cdots C_{v_{j-s};k_j;l} \right) \mathrm{d}\boldsymbol{\theta}_1, \tag{6.201}$$

$$j = 1, \cdots, u-1;\ s = 0, \cdots, j;\ r_i = 1, \cdots, n;\ v_i = 1, \cdots, \alpha;\ j = 4, \cdots, u,$$

in which $\boldsymbol{\psi} = [\psi_1, \psi_2, \cdots, \psi_\alpha]^T$, $A_{r;k;l} = A_{r;k;l}(\boldsymbol{I}, \boldsymbol{\psi}, \boldsymbol{\theta}_1)$, $C_{r;k;l} = C_{r;k;l}(\boldsymbol{I}, \boldsymbol{\psi}, \boldsymbol{\theta}_1)$. $\int_0^{2\pi} [\bullet] \mathrm{d}\boldsymbol{\theta}_1 = \int_0^{2\pi} \int_0^{2\pi} \cdots \int_0^{2\pi} [\bullet] \mathrm{d}\theta_1 \mathrm{d}\theta_2 \cdots \mathrm{d}\theta_\alpha$ represents $n-\alpha$-fold integral.

In Eq. (6.191), $p = p(\boldsymbol{I}, \boldsymbol{\psi}, t|\boldsymbol{I}_0, \boldsymbol{\psi}_0)$ is the transition PDF of the $n+\alpha$-dimensional process $[\boldsymbol{I}^\mathrm{T}, \boldsymbol{\Psi}^\mathrm{T}]^\mathrm{T}$ with initial condition

$$p(\boldsymbol{I}, \boldsymbol{\psi}, 0|\boldsymbol{I}_0, \boldsymbol{\psi}_0) = \delta(\boldsymbol{I} - \boldsymbol{I}_0)\delta(\boldsymbol{\psi} - \boldsymbol{\psi}_0) \tag{6.202}$$

or $p = p(\boldsymbol{I}, \boldsymbol{\psi}, 0)$ is the joint PDF of the $n+\alpha$-dimensional $[\boldsymbol{I}^\mathrm{T}, \boldsymbol{\Psi}^\mathrm{T}]^\mathrm{T}$ process with initial condition

$$p(\boldsymbol{I}, \boldsymbol{\psi}, 0) = p(\boldsymbol{I}, \boldsymbol{\psi}_0) \tag{6.203}$$

The boundary conditions for the averaged FPK Eq. (6.191) depend on the system's properties and constraints imposed on the system. In general, when $\boldsymbol{I}$ can take any values in the first quadrant of the n-dimensional space, the boundary conditions are

$$p = \text{finite};\quad \frac{\partial^k}{\partial I_r^k} p = \text{finite};\quad \text{when } I_r = 0. \tag{6.204}$$

$$p \to 0; \quad \frac{\partial^k}{\partial I_r^k} p \to 0; \quad \text{when } I_r \to \infty. \tag{6.205}$$

$$r = 1, \cdots, n; \quad k = 1, 2, \cdots$$

Equations (6.204) and (6.205) indicate that $I_r = 0$ is reflecting boundary and $I_r \to \infty$ is absorbing boundary. Since $p(\boldsymbol{I}, \boldsymbol{\psi})$ is a periodic function of $\boldsymbol{\psi}$, it satisfies the following periodic conditions

$$p\big|_{\psi_v+2n\pi} = p\big|_{\psi_v}; \quad \frac{\partial^k}{\partial \psi_v^k} p\bigg|_{\psi_v+2n\pi} = \frac{\partial^k}{\partial \psi_v^k} p\bigg|_{\psi_v}; \tag{6.206}$$

$$v = 1, \cdots, \alpha; \quad k = 1, 2, \cdots$$

Additionally, it must satisfy the normalization condition

$$\int_0^\infty \cdots \int_0^\infty \int_0^{2\pi} \cdots \int_0^{2\pi} p(\boldsymbol{I}, \boldsymbol{\psi}) \mathrm{d}\psi_1 \cdots \mathrm{d}\psi_\alpha \mathrm{d}I_1 \cdots \mathrm{d}I_n = 1 \tag{6.207}$$

The reduced averaged FPK equation of Eq. (6.191) can be solved using the finite difference methods and the successive over-relaxation iteration. The joint stationary PDF $p(\boldsymbol{q}, \boldsymbol{p})$ of generalized displacements and generalized momenta can be obtained as

$$\begin{aligned} p(\boldsymbol{q}, \boldsymbol{p}) &= p(\boldsymbol{I}, \boldsymbol{\psi}, \boldsymbol{\theta}_1) \left| \frac{\partial(\boldsymbol{I}, \boldsymbol{\psi}, \boldsymbol{\theta}_1)}{\partial(\boldsymbol{q}, \boldsymbol{p})} \right| \\ &= p(\boldsymbol{\theta}_1 | \boldsymbol{I}, \boldsymbol{\psi}) p(\boldsymbol{I}, \boldsymbol{\psi}) \left| \frac{\partial(\boldsymbol{I}, \boldsymbol{\psi}, \boldsymbol{\theta}_1)}{\partial(\boldsymbol{q}, \boldsymbol{p})} \right| \\ &= \frac{1}{(2\pi)^{n-\alpha}} p(\boldsymbol{I}, \boldsymbol{\psi}) \left| \frac{\partial(\boldsymbol{I}, \boldsymbol{\psi}, \boldsymbol{\theta}_1)}{\partial(\boldsymbol{q}, \boldsymbol{p})} \right|_{\boldsymbol{I}=\boldsymbol{I}(\boldsymbol{q},\boldsymbol{p}),\ \boldsymbol{\psi}=\boldsymbol{\psi}(\boldsymbol{q},\boldsymbol{p})}, \end{aligned} \tag{6.208}$$

where $\left|\partial(\boldsymbol{I}, \boldsymbol{\psi}, \boldsymbol{\theta}_1) / \partial(\boldsymbol{q}, \boldsymbol{p})\right|$ is the absolute value of the determinant of Jacobian matrix transforming from $\boldsymbol{q}, \boldsymbol{p}$ to $\boldsymbol{I}, \boldsymbol{\psi}, \boldsymbol{\theta}_1$.

Example 6.4 Consider two nonlinear damping oscillators coupled by linear damping and excited by both Gaussian and Poisson white noises (Jia and Zhu 2014a). The motion equations are

$$\begin{aligned} &\ddot{X}_1 + \omega_1^2 X_1 + (\alpha_{11} + \alpha_{12}\dot{X}_1^2)\dot{X}_1 + \beta_1 \dot{X}_2 \\ &= (c_{11} + c_{12}X_1)\big(W_{g1}(t) + W_{p1}(t)\big), \\ &\ddot{X}_2 + \omega_2^2 X_2 + (\alpha_{21} + \alpha_{22}\dot{X}_2^2)\dot{X}_2 + \beta_2 \dot{X}_1 \\ &= (c_{21} + c_{22}X_2)\big(W_{g2}(t) + W_{p2}(t)\big), \end{aligned} \tag{6.209}$$

where $\omega_2, \alpha_{11}, \alpha_{12}, \alpha_{21}, \alpha_{22}, \beta_1, \beta_2, c_{11}, c_{12}, c_{21}, c_{22}$ are constants; α_{ij}, β_i $(i, j = 1, 2)$ are small parameters of ε^2 order; $W_{gi}(t)$ $(i = 1, 2)$ are two independent Gaussian white noises with intensities $2\pi K_{ii}(i = 1, 2)$ of ε^2 order; $W_{pi}(t)(i = 1, 2)$ are two independent Poisson white noises with impulse average arrival rates λ_i (see Sect. 2.4.2), Gaussian distributed impulse amplitudes with zero mean value and small variances $\lambda_i E[Y_i^2]$ of ε^2 order; $W_{gi}(t)$ $(i = 1, 2)$ and $W_{pi}(t)(i = 1, 2)$ are independent.

Letting $Q_1 = X_1,\ P_1 = \dot{X}_1,\ Q_2 = X_2,\ P_2 = \dot{X}_2$, Eq. (6.209) can be written as the following SIDEs

$$\begin{aligned}
\mathrm{d}Q_1 &= P_1 \mathrm{d}t, \\
\mathrm{d}P_1 &= -\left[\omega_1^2 Q_1 + (\alpha_{11} + \alpha_{12} P_1^2) P_1 + \beta_1 P_2\right] \mathrm{d}t \\
&\quad + (c_{11} + c_{12} Q_1) \left(\sqrt{2\pi K_{11}} \mathrm{d}B_1(t) + \int_{\mathcal{Y}_1} Y_1 \mathcal{P}_1(\mathrm{d}t, \mathrm{d}Y_1) \right), \\
\mathrm{d}Q_2 &= P_2 \mathrm{d}t, \\
\mathrm{d}P_2 &= -\left[\omega_2^2 Q_2 + (\alpha_{21} + \alpha_{22} P_2^2) P_2 + \beta_2 P_1\right] \mathrm{d}t \\
&\quad + (c_{21} + c_{22} Q_2) \left(\sqrt{2\pi K_{22}} \mathrm{d}B_2(t) + \int_{\mathcal{Y}_2} Y_2 \mathcal{P}_2(\mathrm{d}t, \mathrm{d}Y_2) \right),
\end{aligned} \tag{6.210}$$

where $\mathcal{Y}_i$ $(i = 1, 2)$ are the Poisson symbol spaces.

The Hamiltonian associated with system (6.210) is

$$H = \sum_{i=1}^{2} \omega_i I_i, \tag{6.211}$$

where

$$I_i = \frac{1}{2\omega_i}(P_i^2 + \omega_i^2 Q_i^2), \quad i = 1, 2 \tag{6.212}$$

are action variables;

$$\Theta_i = -\arctan\left(\frac{P_i}{\omega_i Q_i}\right), \quad i = 1, 2 \tag{6.213}$$

are angle variables.

In non-resonant case, $r\omega_1 + s\omega_2 \neq 0$ where r, s are integers. Neglecting small terms of ε^4 and higher orders, the truncated averaged FPK equation is

$$
\begin{aligned}
\frac{\partial}{\partial t}p = &-\frac{\partial}{\partial I_1}(\overline{a}_1(I_1,I_2)p) - \frac{\partial}{\partial I_2}(\overline{a}_2(I_1,I_2)p) \\
&+ \frac{1}{2!}\frac{\partial^2}{\partial I_1^2}\big(\overline{a}_{1,1}(I_1,I_2)p\big) \\
&+ \frac{1}{2!}\frac{\partial^2}{\partial I_2^2}\big(\overline{a}_{2,2}(I_1,I_2)p\big) - \frac{1}{3!}\frac{\partial^3}{\partial I_1^3}\big(\overline{a}_{1,1,1}(I_1,I_2)p\big) \\
&- \frac{1}{3!}\frac{\partial^3}{\partial I_2^3}\big(\overline{a}_{2,2,2}(I_1,I_2)p\big) \\
&+ \frac{1}{4!}\frac{\partial^4}{\partial I_1^4}\big(\overline{a}_{1,1,1,1}(I_1,I_2)p\big) + \frac{1}{4!}\frac{\partial^4}{\partial I_2^4}\big(\overline{a}_{2,2,2,2}(I_1,I_2)p\big),
\end{aligned}
\tag{6.214}
$$

where the derivative moments are

$$
\begin{aligned}
\overline{a}_1(I_1,I_2) = &-\alpha_{11}I_1 - \frac{3}{2}\alpha_{12}\omega_1 I_1^2 \\
&+ \left(\frac{c_{11}^2}{2\omega_1} + \frac{c_{12}^2}{2\omega_1^2}I_1\right)\left(2\pi K_{11} + \lambda_1 E\left[Y_1^2\right]\right), \\
\overline{a}_2(I_1,I_2) = &-\alpha_{21}I_2 - \frac{3}{2}\alpha_{22}\omega_2 I_2^2 \\
&+ \left(\frac{c_{21}^2}{2\omega_2} + \frac{c_{22}^2}{2\omega_2^2}I_2\right)\left(2\pi K_{22} + \lambda_2 E\left[Y_2^2\right]\right),
\end{aligned}
\tag{6.215}
$$

$$
\begin{aligned}
\overline{a}_{1,1}(I_1,I_2) = &\left(\frac{c_{11}^2}{\omega_1}I_1 + \frac{c_{12}^2}{2\omega_1^2}I_1^2\right)\left(2\pi K_{11} + \lambda_1 E\left[Y_1^2\right]\right) \\
&+ \left(\frac{c_{11}^4}{4\omega_1^2} + \frac{3c_{12}^4}{8\omega_1^4}I_1^2 + \frac{3c_{11}^2c_{12}^2}{2\omega_1^3}I_1\right)\lambda_1 E\left[Y_1^4\right], \\
\overline{a}_{2,2}(I_1,I_2) = &\left(\frac{c_{21}^2}{\omega_2}I_2 + \frac{c_{22}^2}{2\omega_2^2}I_2^2\right)\left(2\pi K_{22} + \lambda_2 E\left[Y_2^2\right]\right) \\
&+ \left(\frac{c_{21}^4}{4\omega_2^2} + \frac{3c_{22}^4}{8\omega_2^4}I_2^2 + \frac{3c_{21}^2c_{22}^2}{2\omega_2^3}I_2\right)\lambda_2 E\left[Y_2^4\right],
\end{aligned}
\tag{6.216}
$$

$$
\begin{aligned}
\overline{a}_{1,1,1}(I_1,I_2) &= \left(\frac{3c_{11}^4}{2\omega_1^2}I_1 + \frac{3c_{12}^4}{4\omega_1^4}I_1^3 + \frac{9c_{11}^2c_{12}^2}{2\omega_1^3}I_1^2\right)\lambda_1 E\left[Y_1^4\right], \\
\overline{a}_{2,2,2}(I_1,I_2) &= \left(\frac{3c_{21}^4}{2\omega_2^2}I_2 + \frac{3c_{22}^4}{4\omega_2^4}I_2^3 + \frac{9c_{21}^2c_{22}^2}{2\omega_2^3}I_2^2\right)\lambda_2 E\left[Y_2^4\right],
\end{aligned}
\tag{6.217}
$$

$$
\overline{a}_{1,1,1,1}(I_1,I_2) = \left(\frac{3c_{11}^4}{2\omega_1^2}I_1^2 + \frac{3c_{12}^4}{8\omega_1^4}I_1^4 + \frac{3c_{11}^2c_{12}^2}{\omega_1^3}I_1^3\right)\lambda_1 E\left[Y_1^4\right],
$$

$$\overline{a}_{2,2,2,2}(I_1, I_2) = \left(\frac{3c_{21}^4}{2\omega_2^2}I_2^2 + \frac{3c_{22}^4}{8\omega_2^4}I_2^4 + \frac{3c_{21}^2c_{22}^2}{\omega_2^3}I_2^3\right)\lambda_2 E\left[Y_2^4\right]. \tag{6.218}$$

and all other derivative moments $\overline{a}_{r_1,r_2}(I_1, I_2)$, $\overline{a}_{r_1,r_2,r_3}(I_1, I_2)$ and $\overline{a}_{r_1,r_2,r_3,r_4}(I_1, I_2)$ are zero.

The stationary solution of the reduced averaged FPK of Eq. (6.214) can be obtained using perturbation methods. According to Eq. (6.145), the joint PDF $p(q_1, p_1, q_2, p_2)$ of the generalized displacements and generalized momenta of system (6.209) is given by

$$p(q_1, p_1, q_2, p_2) = \frac{1}{4\pi^2}p(I_1, I_2)|_{I_i=(p_i^2+\omega_i^2q_i^2)/(2\omega_i^2)} \tag{6.219}$$

Various statistics can be obtained from $p(q_1, p_1, q_2, p_2)$.

The approximate analytical solutions of $p(I_1)$, $p(I_2)$, $p(q_1, p_1)$, $p(q_2, p_2)$, $p(q_1)$, $p(p_1)$, $p(q_2)$, $p(p_2)$ and the results from Monte Carlo simulation are shown in Figs. 6.9–6.13. The good agreement between the two results demonstrates the good accuracy of the stochastic averaging method presented in this section. In order to investigate the effect of parameters α_{11} on system response, $p(I_1)$ and $p(q_1)$ for different α_{11} values are shown in Figs. 6.14 and 6.15. It is seen from these two figures that when α_{11} changes from positive to negative, the motion of the first oscillator is changed from random vibrations around the origin to a diffused limit cycle, i.e., a p-bifurcation occurs at $\alpha_{11} = 0$. Additionally, it can also be observed from these two figures that the approximate stationary solutions obtained using the stochastic averaging method agree well with the results from Monte Carlo simulation.

Example 6.5 Consider two linearly coupled nonlinear oscillators excited by both Gaussian and Poisson white noises (Jia et al. 2014). The motion equations are

$$\ddot{X}_1 + \omega_1^2 X_1 + (\alpha_{11} + \alpha_{12}\dot{X}_1^2)\dot{X}_1 + \beta_1\dot{X}_2 = W_{p1}(t) + W_{g1}(t),$$

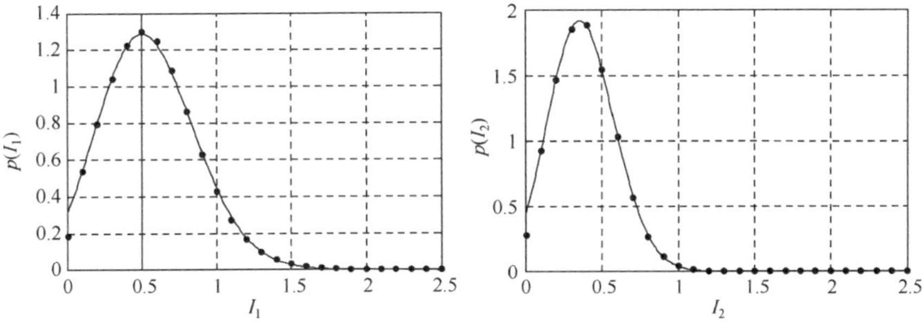

Fig. 6.9 $p(I_1)$ and $p(I_2)$ of system (6.209). System parameters are $\alpha_{11} = \alpha_{21} = -0.06$, $\alpha_{12} = \alpha_{22} = 0.08$, $\beta_1 = \beta_2 = 0.01$, $\omega_1 = 1.0$, $\omega_2 = \sqrt{2}$, $c_{11} = c_{12} = c_{21} = c_{22} = 1.0$, $2\pi K_{11} = 2\pi K_{22} = 0.01$, $\lambda_1 = \lambda_2 = 0.1$, $E[Y_1^2] = E[Y_2^2] = 0.1$. ——, result from the stochastic averaging method; ●, result from Monte Carlo simulation (Jia and Zhu 2014a)

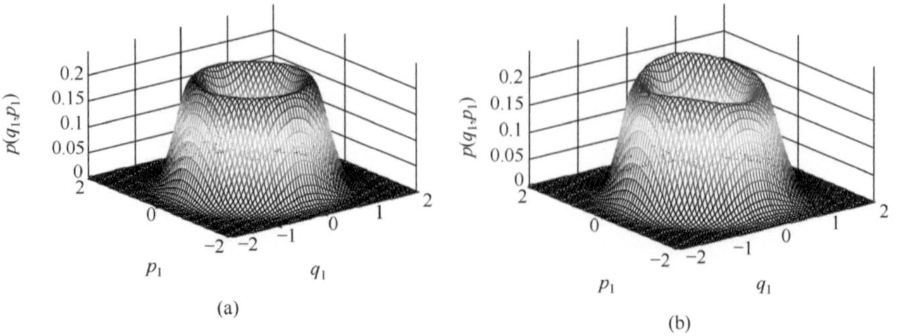

Fig. 6.10 $p(q_1, p_1)$ of system (6.209). **a** Result from the stochastic averaging method; **b** Result from Monte Carlo simulation. System parameters are the same as those in Fig. 6.9 (Jia and Zhu 2014a)

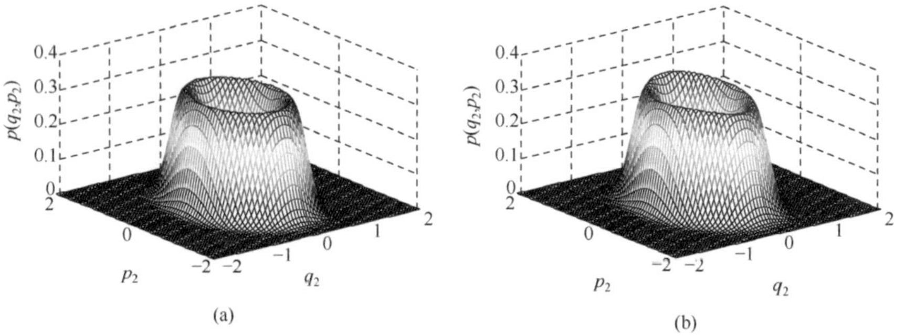

Fig. 6.11 $p(q_2, p_2)$ of system (6.209). **a** Result from the stochastic averaging method; **b** Result from Monte Carlo simulation. System parameters are the same as those in Fig. 6.9 (Jia and Zhu 2014a)

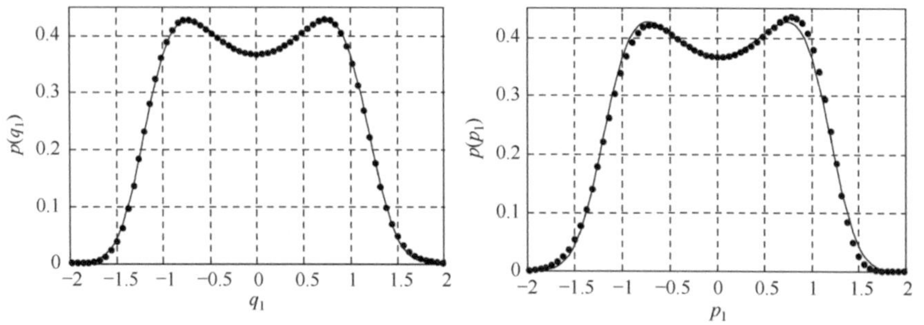

Fig. 6.12 $p(q_1)$ and $p(p_1)$ of system (6.209). System parameters are the same as those in Fig. 6.9. ——, result from the stochastic averaging method; ●, result from Monte Carlo simulation (Jia and Zhu 2014a)

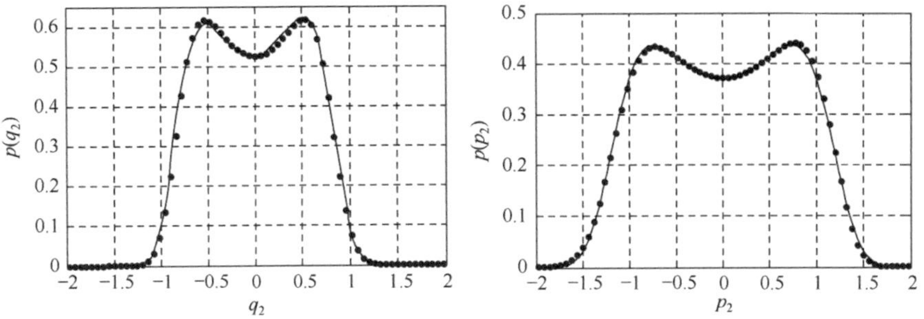

Fig. 6.13 $p(q_2)$ and $p(p_2)$ of system (6.209). System parameters are the same as those in Fig. 6.9. ——, result from the stochastic averaging method; ●, result from Monte Carlo simulation (Jia and Zhu 2014a)

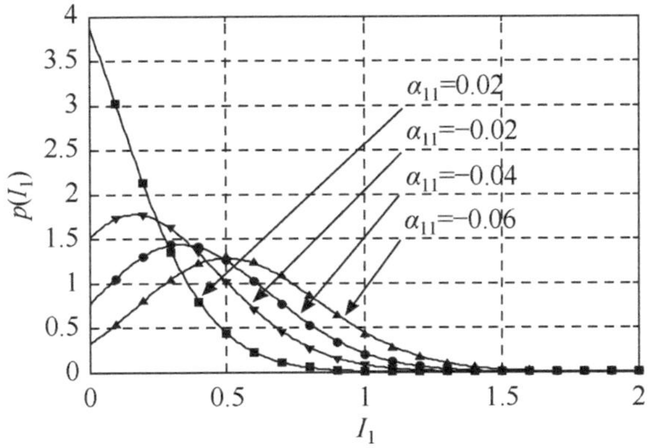

Fig. 6.14 $p(I_1)$ of system (6.209) with different α_{11}. The other system parameters are $\alpha_{12} = 0.08$, $\alpha_{21} = -0.06$, $\alpha_{22} = 0.08$, $\beta_1 = \beta_2 = 0.01$, $\omega_1 = 1.0$, $\omega_2 = \sqrt{2}$, $c_{11} = c_{12} = 1.0$, $c_{21} = c_{22} = 1.0$, $2\pi K_{11} = 2\pi K_{22} = 0.01$, $\lambda_1 = \lambda_2 = 1.0$, $E[Y_1^2] = E[Y_2^2] = 0.01$. ——, results from the stochastic averaging method; ▲■●▼, results from Monte Carlo simulation (Jia and Zhu 2014a)

Fig. 6.15 $p(q_1)$ of system (6.209) with different α_{11}. Other system parameters are the same as those in Fig. 6.14 ——, results from the stochastic averaging method; ▲■●▼, results from Monte Carlo simulation (Jia and Zhu 2014a)

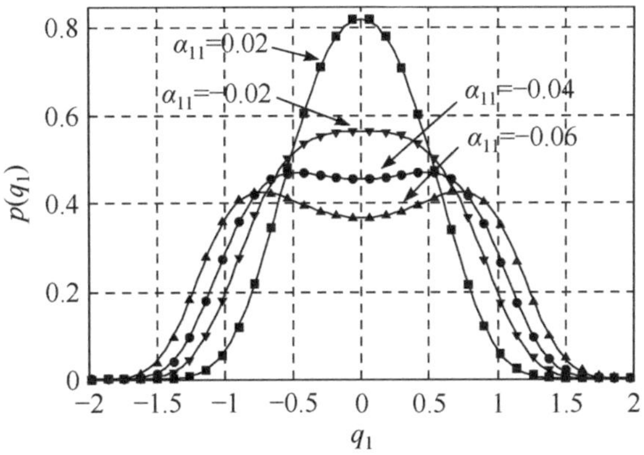

$$\ddot{X}_2 + \omega_2^2 X_2 + (\alpha_{21} + \alpha_{22}\dot{X}_2^2)\dot{X}_2 + \beta_2 \dot{X}_1 = W_{p2}(t) + W_{g2}(t). \tag{6.220}$$

where $\omega_i, \alpha_{ij}, \beta_i$ are constant; α_{ij}, β_i are small parameters of ε^2 order; $W_{gi}(t)$ are two independent Gaussian white noises with intensities $2\pi K_{ii}$ of ε^2 order; $W_{pi}(t)$ are two independent Poisson white noises with zero mean Gaussian distributed impulse amplitude Y_i and variances $\lambda_i E[Y_i^2]$ of ε^2 order; $W_{gi}(t)$ and $W_{pi}(t)$ are also independent.

Introducing the transformations $Q_1 = X_1, P_1 = \dot{X}_1, Q_1 = X_2, P_2 = \dot{X}_2$, the SIDEs associated with system (6.220) are

$$\begin{aligned}
\mathrm{d}Q_1 &= P_1 \mathrm{d}t, \\
\mathrm{d}P_1 &= -\left[\omega_1^2 Q_1 + (\alpha_{11} + \alpha_{12} P_1^2) P_1 + \beta_1 P_2\right]\mathrm{d}t \\
&\quad + \sqrt{2\pi K_{11}}\mathrm{d}B_1(t) + \int_{\mathcal{Y}_1} Y_1 \mathcal{P}_1(\mathrm{d}t, \mathrm{d}Y_1), \\
\mathrm{d}Q_2 &= P_2 \mathrm{d}t, \\
\mathrm{d}P_2 &= -\left[\omega_2^2 Q_2 + (\alpha_{21} + \alpha_{22} P_2^2) P_2 + \beta_2 P_1\right]\mathrm{d}t \\
&\quad + \sqrt{2\pi K_{22}}\mathrm{d}B_2(t) + \int_{\mathcal{Y}_2} Y_2 \mathcal{P}_2(\mathrm{d}t, \mathrm{d}Y_2).
\end{aligned} \tag{6.221}$$

The Hamiltonian associated with system (6.221) is

$$H = \sum_{i=1}^{2} \omega_i I_i; \quad I_i = \frac{1}{2\omega_i}(P_i^2 + \omega_i^2 Q_i^2) \tag{6.222}$$

Introduce the angle variables

$$\Theta_i = -\arctan\left(\frac{P_i}{\omega_i Q_i}\right); \quad i = 1, 2 \tag{6.223}$$

Using the stochastic jump-diffusion chain rule (Hanson 2007), the following SIDEs for I_r and Θ_r are obtained from Eq. (6.221).

$$\begin{aligned}
\mathrm{d}I_1 &= \left[-(\alpha_{11} + \alpha_{12}P_1^2)\frac{P_1^2}{\omega_1} - \frac{\beta_1}{\omega_1}P_1 P_2 + \frac{\pi K_{11}}{\omega_1}\right]\mathrm{d}t + \frac{\sqrt{2\pi K_{11}}}{\omega_1} P_1 \mathrm{d}B_1(t) \\
&\quad + \frac{1}{2\omega_1}\int_{\mathcal{Y}_1} (Y_1^2 + 2P_1 Y_1)\mathcal{P}_1(\mathrm{d}t, \mathrm{d}Y_1), \\
\mathrm{d}I_2 &= \left[-(\alpha_{21} + \alpha_{22}P_2^2)\frac{P_2^2}{\omega_2} - \frac{\beta_2}{\omega_2}P_1 P_2 + \frac{\pi K_{22}}{\omega_2}\right]\mathrm{d}t + \frac{\sqrt{2\pi K_{22}}}{\omega_2} P_2 \mathrm{d}B_2(t)
\end{aligned}$$

$$+\frac{1}{2\omega_2}\int\limits_{\mathcal{Y}_2}(Y_2^2+2P_2Y_2)\mathcal{P}_2(\mathrm{d}t,\mathrm{d}Y_2),$$

$$\mathrm{d}\Theta_1=\left\{\omega_1+(\alpha_{11}+\alpha_{12}P_1^2)\frac{\omega_1P_1Q_1}{\omega_1^2Q_1^2+P_1^2}+\beta_1\frac{\omega_1Q_1P_2}{\omega_1^2Q_1^2+P_1^2}+\frac{2\omega_1\pi K_{11}Q_1P_1}{\left(\omega_1^2Q_1^2+P_1^2\right)^2}\right\}\mathrm{d}t$$
$$-\frac{\sqrt{2\pi K_{11}}\omega_1Q_1}{\omega_1^2Q_1^2+P_1^2}\mathrm{d}B_1(t)-\int\limits_{\mathcal{Y}_1}\left[\arctan\left(\frac{P_1+Y_1}{\omega_1Q_1}\right)-\arctan\left(\frac{P_1}{\omega_1Q_1}\right)\right]\mathcal{P}_1(\mathrm{d}t,\mathrm{d}Y_1),$$

$$\mathrm{d}\Theta_2=\left\{\omega_2+(\alpha_{21}+\alpha_{22}P_2^2)\frac{\omega_2P_2Q_2}{\omega_2^2Q_2^2+P_2^2}+\beta_2\frac{\omega_2Q_2P_1}{\omega_2^2Q_2^2+P_2^2}+\frac{2\omega_2\pi K_{22}Q_2P_2}{\left(\omega_2^2Q_2^2+P_2^2\right)^2}\right\}\mathrm{d}t$$
$$-\frac{\sqrt{2\pi K_{22}}\omega_2Q_2}{\omega_2^2Q_2^2+P_2^2}\mathrm{d}B_2(t)-\int\limits_{\mathcal{Y}_2}\left[\arctan\left(\frac{P_2+Y_2}{\omega_2Q_2}\right)-\arctan\left(\frac{P_2}{\omega_2Q_2}\right)\right]\mathcal{P}_2(\mathrm{d}t,\mathrm{d}Y_2).\quad(6.224)$$

In the case of primary resonance, $\omega_1=\omega_2$, let $\Psi=\Theta_1-\Theta_2$. The following SIDE for Ψ can be obtained by linearly combining the SIDEs for Θ_i, expanding $\arctan\left(\frac{P_i}{\omega_iQ_i}+\frac{Y_i}{\omega_iQ_i}\right)-\arctan\left(\frac{P_i}{\omega_iQ_i}\right)$ to Taylor series and neglecting small terms of ε^4 and higher orders.

$$\mathrm{d}\Psi=\left\{(\alpha_{11}+\alpha_{12}P_1^2)\frac{\omega_1P_1Q_1}{\omega_1^2Q_1^2+P_1^2}+\beta_1\frac{\omega_1Q_1P_2}{\omega_1^2Q_1^2+P_1^2}+\frac{2\omega_1\pi K_{11}Q_1P_1}{(\omega_1^2Q_1^2+P_1^2)^2}\right.$$
$$\left.-(\alpha_{21}+\alpha_{22}P_2^2)\frac{\omega_2P_2Q_2}{\omega_2^2Q_2^2+P_2^2}-\beta_2\frac{\omega_2Q_2P_1}{\omega_2^2Q_2^2+P_2^2}-\frac{2\omega_2\pi K_{22}Q_2P_2}{(\omega_2^2Q_2^2+P_2^2)^2}\right\}\mathrm{d}t$$
$$-\frac{\sqrt{2\pi K_{11}}\omega_1Q_1}{\omega_1^2Q_1^2+P_1^2}\mathrm{d}B_1(t)+\frac{\sqrt{2\pi K_{22}}\omega_2Q_2}{\omega_2^2Q_2^2+P_2^2}\mathrm{d}B_2(t)$$
$$-\int\limits_{\mathcal{Y}_1}\left[\frac{\omega_1Q_1Y_1}{P_1^2+\omega_1^2Q_1^2}-\frac{\omega_1Q_1P_1}{(P_1^2+\omega_1^2Q_1^2)^2}Y_1^2+\frac{1}{3}\frac{(3P_1^2-\omega_1^2Q_1^2)\omega_1Q_1}{(P_1^2+\omega_1^2Q_1^2)^3}Y_1^3\right.$$
$$\left.-\frac{P_1(P_1^2-\omega_1^2Q_1^2)\omega_1Q_1}{(P_1^2+\omega_1^2Q_1^2)^4}Y_1^4\right]\mathcal{P}_1(\mathrm{d}t,\mathrm{d}Y_1)$$
$$+\int\limits_{\mathcal{Y}_2}\left[\frac{\omega_2Q_2}{P_2^2+\omega_2^2Q_2^2}Y_2-\frac{\omega_2Q_2P_2}{(P_2^2+\omega_2^2Q_2^2)^2}Y_2^2\right.$$
$$\left.+\frac{1}{3}\frac{(3P_2^2-\omega_2^2Q_2^2)\omega_2Q_2}{(P_2^2+\omega_2^2Q_2^2)^3}Y_2^3-\frac{P_2(P_2^2-\omega_2^2Q_2^2)\omega_2Q_2}{(P_2^2+\omega_2^2Q_2^2)^4}Y_2^4\right]\mathcal{P}_2(\mathrm{d}t,\mathrm{d}Y_2).\quad(6.225)$$

The new equations of system include the first three equations in Eq. (6.224) and Eq. (6.225). In this new system, I_1, I_2, Ψ are slowly-varying processes while Θ_2 is fast-varying process. The following truncated averaged FPK equation similar to Eq. (6.191) is obtained by applying the stochastic averaging.

$$
\begin{aligned}
0 = & -\frac{\partial}{\partial I_1}(\overline{a}_1 p) - \frac{\partial}{\partial I_2}(\overline{a}_2 p) - \frac{\partial}{\partial \psi}(\overline{a}_3 p) \\
& + \frac{1}{2}\frac{\partial^2}{\partial I_1^2}(\overline{a}_{1,1} p) + \frac{1}{2}\frac{\partial^2}{\partial I_2^2}(\overline{a}_{2,2} p) \\
& + \frac{1}{2}\frac{\partial^2}{\partial \psi^2}(\overline{a}_{3,3} p) - \frac{1}{3!}\frac{\partial^3}{\partial I_1^3}(\overline{a}_{1,1,1} p) \\
& - \frac{1}{3!}\frac{\partial^3}{\partial I_2^3}(\overline{a}_{2,2,2} p) - \frac{1}{3!}\frac{\partial^3}{\partial \psi^3}(\overline{a}_{3,3,3} p) \\
& + \frac{1}{4!}\frac{\partial^4}{\partial I_1^4}(\overline{a}_{1,1,1,1} p) + \frac{1}{4!}\frac{\partial^4}{\partial I_2^4}(\overline{a}_{2,2,2,2} p) \\
& + \frac{1}{4!}\frac{\partial^4}{\partial \psi^4}(\overline{a}_{3,3,3,3} p) + \frac{6}{4!}\frac{\partial^4}{\partial I_1^2 \partial \psi^2}(\overline{a}_{1,1,3,3} p) \\
& + \frac{6}{4!}\frac{\partial^2}{\partial I_2^2 \partial \psi^2}(\overline{a}_{2,2,3,3} p),
\end{aligned}
\tag{6.226}
$$

where the derivative moments are

$$
\begin{aligned}
\overline{a}_1 = & -\alpha_{11} I_1 - \frac{3}{2}\alpha_{12}\omega_1 I_1^2 \\
& - \beta_1 \sqrt{I_1 I_2}\cos\psi + \frac{2\pi K_{11} + \lambda_1 E[Y_1^2]}{2\omega_1}, \\
\overline{a}_2 = & -\alpha_{21} I_2 - \frac{3}{2}\alpha_{22}\omega_2 I_2^2 \\
& - \beta_2 \sqrt{I_1 I_2}\cos\psi + \frac{2\pi K_{22} + \lambda_2 E[Y_2^2]}{2\omega_2}, \\
\overline{a}_3 = & \left(\frac{\beta_1}{2}\sqrt{\frac{I_2}{I_1}} + \frac{\beta_2}{2}\sqrt{\frac{I_1}{I_2}}\right)\sin\psi,
\end{aligned}
\tag{6.227}
$$

$$
\begin{aligned}
\overline{a}_{1,1} = & \frac{2\pi K_{11} + \lambda_1 E[Y_1^2]}{\omega} I_1 + \frac{\lambda_1 E[Y_1^4]}{4\omega^2}, \\
\overline{a}_{2,2} = & \frac{2\pi K_{22} + \lambda_2 E[Y_2^2]}{\omega} I_2 + \frac{\lambda_2 E[Y_2^4]}{4\omega^2}, \\
\overline{a}_{3,3} = & \frac{1}{4\omega}\left(\frac{2\pi K_{11} + \lambda_1 E[Y_1^2]}{I_1} + \frac{2\pi K_{22} + \lambda_2 E[Y_2^2]}{I_2}\right) \\
& + \frac{1}{32\omega^2}\left(\frac{\lambda_1 E[Y_1^4]}{I_1^2} + \frac{\lambda_2 E[Y_2^4]}{I_2^2}\right),
\end{aligned}
\tag{6.228}
$$

$$
\begin{aligned}
\overline{a}_{1,1,1} = & \frac{3}{2}\frac{\lambda_1 E[Y_1^4]}{\omega^2} I_1, \\
\overline{a}_{2,2,2} = & \frac{3}{2}\frac{\lambda_2 E[Y_2^4]}{\omega^2} I_2,
\end{aligned}
$$

$$\overline{a}_{3,3,3} = 0, \tag{6.229}$$

$$\begin{aligned}
\overline{a}_{1,1,1,1} &= \frac{3}{2\omega^2} I_1^2 \lambda_1 E[Y_1^4], \\
\overline{a}_{2,2,2,2} &= \frac{3}{2\omega^2} I_2^2 \lambda_2 E[Y_2^4], \\
\overline{a}_{3,3,3,3} &= \frac{3}{32} \frac{\lambda_1 E[Y_1^4]}{\omega^2 I_1^2} + \frac{3}{32} \frac{\lambda_2 E[Y_2^4]}{\omega^2 I_2^2}, \\
\overline{a}_{1,1,3,3} &= \frac{1}{8\omega^2} \lambda_1 E[Y_1^4], \quad \overline{a}_{2,2,3,3} = \frac{1}{8\omega^2} \lambda_2 E[Y_2^4],
\end{aligned} \tag{6.230}$$

and all other derivative moments $\overline{a}_{r_1,r_2}, \overline{a}_{r_1,r_2,r_3}, \overline{a}_{r_1,r_2,r_3,r_4}$ are zero.

The truncated averaged FPK Eq. (6.226) can be solved numerically using the finite differences method and the over-relaxation iteration. After obtaining the approximately stationary solution $p(I_1, I_2, \psi)$, the joint stationary PDF $p(q_1, p_1, q_2, p_2)$ of the generalized displacements and generalized momenta of the original system (6.220) can be calculated using (6.208) as follows

$$p(q_1, p_1, q_2, p_2) = \frac{1}{2\pi} p(I_1, I_2, \psi) \bigg|_{I_1 = I_1(\boldsymbol{q},\boldsymbol{p}), I_2 = I_2(\boldsymbol{q},\boldsymbol{p}), \psi = \psi(\boldsymbol{q},\boldsymbol{p})} \tag{6.231}$$

Other statistics for system (6.220) can be obtained by integrating $p(I_1, I_2, \psi)$ or $p(q_1, p_1, q_2, p_2)$.

The stationary PDFs $p(I_1), p(I_2)$ of action variables and $p(q_1), p(p_1)$ of the generalized displacement and generalized momentum, as well as $p(\psi)$ of angle variable combination for system (6.220) are shown in Figs. 6.16–6.18. It is seen from these figures that the results from the stochastic averaging method and those from Monte Carlo simulation are in good agreement. In order to investigate the effect of parameter α_{11} on system response, the stationary PDF of the generalized displacement q_1 for different values of α_{11} is shown in Fig. 6.19. It is seen that the size of the diffused limit cycle increases as α_{11} decreases.

6.4 Quasi-Partially Integrable Hamiltonian Systems

Assume that the Hamiltonian system associated with system (6.5) is partially integrable with r $(1 < r < n)$ independent and involutive first integrals. Through a canonical transformation, a partially integrable Hamiltonian system can be transformed into a system consisting of $r - 1$ integrable Hamiltonian subsystems and a non-integrable Hamiltonian subsystem, i.e., the Hamiltonian is of the form

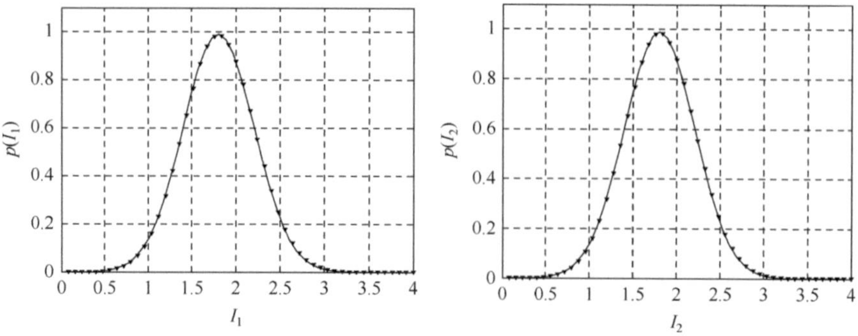

Fig. 6.16 $p(I_1)$ and $p(I_2)$ of system (6.220). System parameters are $\alpha_{11}=\alpha_{21}=-0.1$, $\alpha_{12}=\alpha_{22}=0.04$, $\omega_1=\omega_2=1.0$, $\beta_1=\beta_2=0.01$, $2\pi K_1 = 2\pi K_2 = 0.015$, $\lambda_1 = \lambda_2 = 2.5$, $E[Y_1^2] = E[Y_2^2] = 0.002$. ——, result from the stochastic averaging method; ▼, result from Monte Carlo simulation (Jia et al. 2014)

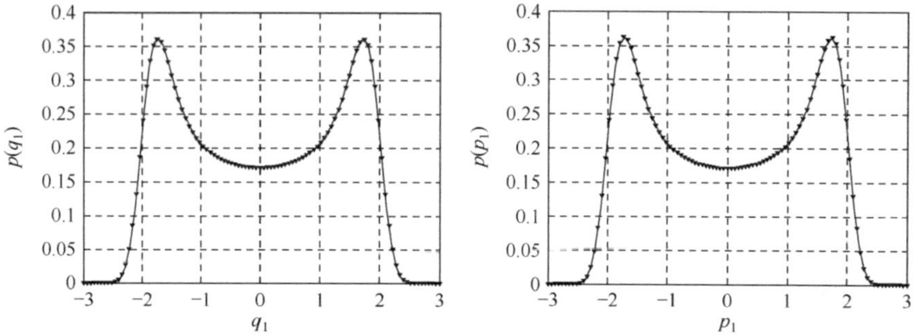

Fig. 6.17 $p(q_1)$ and $p(p_1)$ of system (6.220). System parameters are the same as those in Fig. 6.16. ——, result from the stochastic averaging method; ▼, result from Monte Carlo simulation (Jia et al. 2014)

Fig. 6.18 $p(\psi)$ of system (6.220) in primary resonant case. System parameters the same as those in Fig. 6.16. ——, result from the stochastic averaging method; ▼, result from Monte Carlo simulation (Jia et al. 2014)

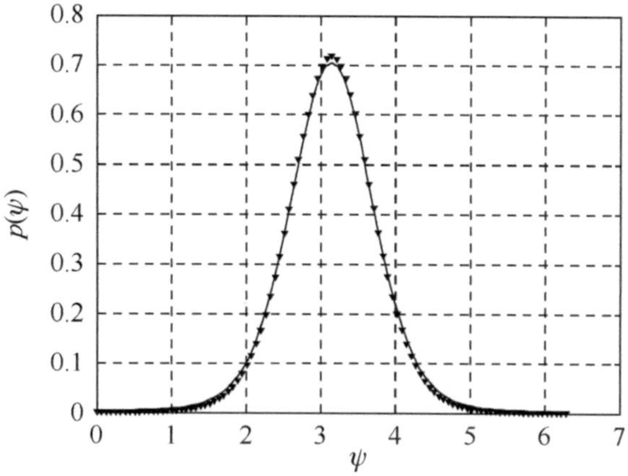

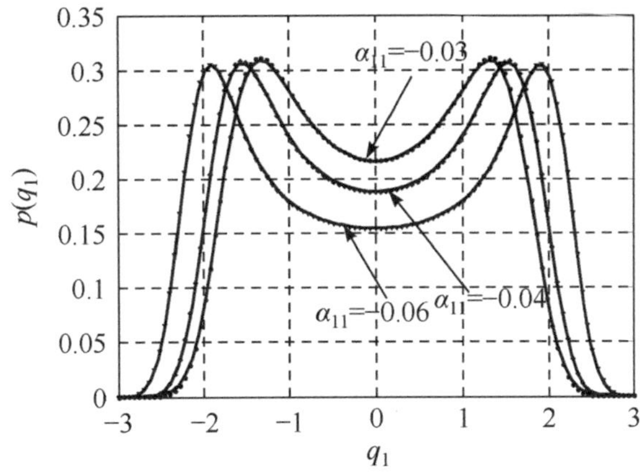

Fig. 6.19 $p(q_1)$ of system (6.220) with different α_{11}. Other system parameters are $\alpha_{12} = 0.02, \alpha_{12} = 0.02$, $\alpha_{21} = -0.1, \alpha_{22}=0.04$, $\omega_1=\omega_2=1.0, \beta_1=\beta_2=0.01$, $2\pi K_{11} = 2\pi K_{22} = 0.015$, $\lambda_1 = \lambda_2 = 2.5$, $E[Y_1^2] = E[Y_2^2] = 0.002$. ——, results from the stochastic averaging method; ▼•●, results from Monte Carlo simulation (Jia et al. 2014)

$$H(\boldsymbol{q},\boldsymbol{p}) = \sum_{\eta=1}^{r-1} H_\eta(\boldsymbol{q}_1,\boldsymbol{p}_1) + H_r(\boldsymbol{q}_2,\boldsymbol{p}_2), \tag{6.232}$$

where $\boldsymbol{q}_1 = [q_1, q_2, \cdots, q_{r-1}]^{\mathrm{T}}$, $\boldsymbol{p}_1 = [p_1, p_2, \cdots, p_{r-1}]^{\mathrm{T}}$, $\boldsymbol{q}_2 = [q_r, q_{r+1}, \cdots, q_n]^{\mathrm{T}}$, $\boldsymbol{p}_2 = [p_r, p_{r+1}, \cdots, p_n]^{\mathrm{T}}$. Introduce the following action-angle variables for the integrable Hamiltonian subsystems,

$$I_\eta = I_\eta(\boldsymbol{q}_1,\boldsymbol{p}_1), \quad \theta_\eta = \theta_\eta(\boldsymbol{q}_1,\boldsymbol{p}_1), \quad \eta = 1, 2, \cdots, r-1. \tag{6.233}$$

Equation (6.232) can be rewritten as

$$H(\boldsymbol{q},\boldsymbol{p}) = \sum_{\eta=1}^{r-1} I_\eta(\boldsymbol{q}_1,\boldsymbol{p}_1) + H_r(\boldsymbol{q}_2,\boldsymbol{p}_2), \tag{6.234}$$

Introduce the following transformations for Eq. (6.5)

$$\begin{aligned} &I_\eta = I_\eta(\boldsymbol{Q}_1, \boldsymbol{P}_1), \quad \Theta_\eta = \Theta_\eta(\boldsymbol{Q}_1, \boldsymbol{P}_1), \quad H_r = H_r(\boldsymbol{Q}_2, \boldsymbol{P}_2), \\ &\eta = 1, 2, \cdots, r-1. \end{aligned} \tag{6.235}$$

Using the stochastic jump-diffusion chain rule (Hanson 2007), the following SIDEs for I_η, Θ_η, and H_r can be derived from (6.5) (Jia and Zhu 2014b)

$$\begin{aligned} \mathrm{d}I_\eta = \varepsilon^2 &\left(-\sum_{j=1}^{n}\sum_{\eta'=1}^{r-1} m_{\eta' j}\frac{\partial H}{\partial P_j}\frac{\partial I_\eta}{\partial P_{\eta'}} + \frac{1}{2}\sum_{\eta',\eta''=1}^{r-1}\sum_{k=1}^{n_g}\sigma_{\eta' k}\sigma_{\eta'' k}\frac{\partial^2 I_\eta}{\partial P_{\eta'}\partial P_{\eta''}} \right)\mathrm{d}t \\ &+ \varepsilon \sum_{\eta'=1}^{r-1}\sum_{k=1}^{n_g}\frac{\partial I_\eta}{\partial P_{\eta'}}\sigma_{\eta' k}\mathrm{d}B_k(t) \end{aligned}$$

$$+\sum_{l=1}^{n_p}\int_{\mathcal{Y}_l}[I_\eta(\boldsymbol{Q},\boldsymbol{P}+\hat{\boldsymbol{\gamma}}_l)-I_\eta(\boldsymbol{I}',\boldsymbol{\Theta}',H_r,q_r,\cdots,q_n,p_{r+1},\cdots,p_n)]\mathcal{P}_l(\mathrm{d}t,\mathrm{d}Y_l), \quad (6.236)$$

$$\mathrm{d}\Theta_\eta=\left(\omega_\eta-\varepsilon^2\sum_{j=1}^{n}\sum_{\eta'=1}^{r-1}m_{\eta'j}\frac{\partial H}{\partial P_j}\frac{\partial\Theta_\eta}{\partial P_{\eta'}}+\frac{\varepsilon^2}{2}\sum_{\eta',\eta''=1}^{r-1}\sum_{k=1}^{n_g}\sigma_{\eta'k}\sigma_{\eta''k}\frac{\partial^2\Theta_\eta}{\partial P_{\eta'}\partial P_{\eta''}}\right)\mathrm{d}t$$
$$+\varepsilon\sum_{\eta'=1}^{r-1}\sum_{k=1}^{n_g}\frac{\partial\Theta_\eta}{\partial P_{\eta'}}\sigma_{\eta'k}\mathrm{d}B_k(t)$$
$$+\sum_{l=1}^{n_p}\int_{\mathcal{Y}_l}[\Theta_\eta(\boldsymbol{Q},\boldsymbol{P}+\hat{\boldsymbol{\gamma}}_l)-\Theta_\eta(\boldsymbol{I}',\boldsymbol{\Theta}',H_r,q_r,\cdots,q_n,p_{r+1},\cdots,p_n)]\mathcal{P}_l(\mathrm{d}t,\mathrm{d}Y_l), \quad (6.237)$$

$$\mathrm{d}H_r=\varepsilon^2\left(-\sum_{j=1}^{n}\sum_{\rho=r}^{n}m_{\rho j}\frac{\partial H}{\partial P_j}\frac{\partial H_r}{\partial P_\rho}+\frac{1}{2}\sum_{\rho,\rho'=r}^{n}\sum_{k=1}^{n_g}\sigma_{\rho k}\sigma_{\rho' k}\frac{\partial^2 H_r}{\partial P_\rho\partial P_{\rho'}}\right)\mathrm{d}t$$
$$+\varepsilon\sum_{\rho=r}^{n}\sum_{k=1}^{n_g}\frac{\partial H_r}{\partial P_\rho}\sigma_{\rho k}\mathrm{d}B_k(t)$$
$$+\sum_{l=1}^{n_p}\int_{\mathcal{Y}_l}[H_r(\boldsymbol{Q},\boldsymbol{P}+\hat{\boldsymbol{\gamma}}_l)-H_r(\boldsymbol{I}',\boldsymbol{\Theta}',H_r,q_r,\cdots,q_n,p_{r+1},\cdots,p_n)]\mathcal{P}_l(\mathrm{d}t,\mathrm{d}Y_l),$$
$$\eta=1,2,\cdots,r-1, \quad (6.238)$$

where $\boldsymbol{I}'=[I_1,I_2,\cdots,I_{r-1}]^{\mathrm{T}}$, $\boldsymbol{\Theta}'=[\Theta_1,\Theta_2,\cdots,\Theta_{r-1}]^{\mathrm{T}}$, $\hat{\boldsymbol{\gamma}}_l$ represent the l-th column of the matrix $[\gamma_{il}]_{n\times n_p}$ given in Eq. (6.6).

Expanding the last term in Eqs. (6.236) to (6.238) into Taylor series, see Appendix C in Reference (Jia and Zhu 2014b), results in

$$\mathrm{d}I_\eta=\varepsilon^2\left(-\sum_{j=1}^{n}\sum_{\eta'=1}^{r-1}m_{\eta'j}\frac{\partial H}{\partial P_j}\frac{\partial I_\eta}{\partial P_{\eta'}}+\frac{1}{2}\sum_{\eta',\eta''=1}^{r-1}\sum_{k=1}^{n_g}\sigma_{\eta'k}\sigma_{\eta''k}\frac{\partial^2 I_\eta}{\partial P_{\eta'}\partial P_{\eta''}}\right)\mathrm{d}t$$
$$+\varepsilon\sum_{\eta'=1}^{r-1}\sum_{k=1}^{n_g}\frac{\partial I_\eta}{\partial P_{\eta'}}\sigma_{\eta'k}\mathrm{d}B_k(t)$$
$$+\sum_{l=1}^{n_p}\int_{\mathcal{Y}_l}\left[\sum_{k=1}^{\infty}\varepsilon^kY_l^kA_{\eta;k;l}(\boldsymbol{I}',\boldsymbol{\Theta}',H_r,q_r,\cdots,q_n,p_{r+1},\cdots,p_n)\right]\mathcal{P}_l(\mathrm{d}t,\mathrm{d}Y_l), \quad (6.239)$$

$$\mathrm{d}\Theta_\eta=\left(\omega_\eta-\varepsilon^2\sum_{j=1}^{n}\sum_{\eta'=1}^{r-1}m_{\eta'j}\frac{\partial H}{\partial P_j}\frac{\partial\Theta_\eta}{\partial P_\eta}+\frac{\varepsilon^2}{2}\sum_{\eta',\eta''=1}^{r-1}\sum_{k=1}^{n_g}\sigma_{\eta'k}\sigma_{\eta''k}\frac{\partial^2\Theta_\eta}{\partial P_{\eta'}\partial P_{\eta''}}\right)\mathrm{d}t$$
$$+\varepsilon\sum_{\eta'=1}^{r-1}\sum_{k=1}^{n_g}\frac{\partial\Theta_\eta}{\partial P_{\eta'}}\sigma_{\eta'k}\mathrm{d}B_k(t)$$

$$+\sum_{l=1}^{n_p}\int_{\mathcal{Y}_l}\left[\sum_{k=1}^{\infty}\varepsilon^k Y_l^k D_{\eta;k;l}(\boldsymbol{I}',\boldsymbol{\Theta}',H_r,q_r,\cdots,q_n,p_{r+1},\cdots,p_n)\right]\mathcal{P}_l(\mathrm{d}t,\mathrm{d}Y_l),\tag{6.240}$$

$$\begin{aligned}\mathrm{d}H_r&=\varepsilon^2\left(-\sum_{j=1}^{n}\sum_{\rho=r}^{n}m_{\rho j}\frac{\partial H}{\partial P_j}\frac{\partial H_r}{\partial P_\rho}+\frac{1}{2}\sum_{\rho,\rho'=r}^{n}\sum_{k=1}^{n_g}\sigma_{\rho k}\sigma_{\rho' k}\frac{\partial^2 H_r}{\partial P_\rho\partial P_{\rho'}}\right)\mathrm{d}t\\&+\varepsilon\sum_{\rho=r}^{n}\sum_{k=1}^{n_g}\frac{\partial H_r}{\partial P_\rho}\sigma_{\rho k}\mathrm{d}B_k(t)\\&+\sum_{l=1}^{n_p}\int_{\mathcal{Y}_l}\left[\sum_{k=1}^{\infty}\varepsilon^k Y_l^k B_{r;k;l}\big(\boldsymbol{I}',\boldsymbol{\Theta}',H_r,q_r,\cdots,q_n,p_{r+1},\cdots,p_n\big)\right]\mathcal{P}_l(\mathrm{d}t,\mathrm{d}Y_l),\\&\eta=1,2,\cdots,r-1.\end{aligned}\tag{6.241}$$

The new SIDEs of the system consist of $r-1$ equations of I_η in Eq. (6.239), $r-1$ equations of Θ_η in Eq. (6.240), one equation of H_r in Eq. (6.241), $n-r+1$ equations of Q_i and $n-r$ equations of P_i in Eq. (6.5). The dimension and form of the averaged SIDEs and the averaged FPK equation depend on whether the $(r-1)$ DOF integrable Hamiltonian subsystem has resonance.

6.4.1 Non-Internal Resonant Case

In non-internal resonant cases, $I_1,\cdots,I_{r-1}$ and H_r are slowly-varying process while $\Theta_1,\cdots,\Theta_{r-1}$, $Q_r,\cdots,Q_n$, $P_{r+1},\cdots,P_n$, are fast-varying processes. According to the stochastic averaging principle (Khasminskii 1968; Xu et al. 2011), $\left[\boldsymbol{I}^{\mathrm{T}},H_r\right]^{\mathrm{T}}$ weakly converges to a r-dimensional vector Markov process as $\varepsilon\to 0$. The averaged SIDEs for by this r-dimensional vector Markov process can be obtained by time averaging of Eqs. (6.239) and (6.241). Note that in the case of non-internal resonance, $(r-1)$-DOF integrable Hamiltonian subsystem is ergodic on a $(r-1)$-dimensional torus while the non-integrable Hamiltonian subsystem is ergodic on $(2n-2r+1)$–dimensional constant H_r surface. Therefore, the time averaging can be replaced by spatial averaging in Eq. (5.149). Since the SIDEs (6.239) and (6.241) for slowly-varying variables contain infinite terms, truncation is necessary to obtain closed-form averaged SIDEs. By neglecting terms of order ε^u and higher, the following approximate averaged SIDEs for $I_1,\cdots,I_{r-1},H_r$ are derived

$$\mathrm{d}I_\eta=\left(\varepsilon^2\overline{m}_\eta(\boldsymbol{I}',H_r)+\sum_{i=0}^{u-2}\varepsilon^{i+2}U_{\eta,i}(\boldsymbol{I}',H_r)\right)\mathrm{d}t$$

$$+\sum_{k=1}^{n_g}\overline{\sigma}_{\eta k}\big(\boldsymbol{I}', H_r\big)\mathrm{d}B_k(t)+\sum_{l=1}^{u-1}G_{\eta,l} \tag{6.242}$$

$$\mathrm{d}H_r=\left(\varepsilon^2\overline{m}_r\big(\boldsymbol{I}', H_r\big)+\sum_{i=0}^{u-2}\varepsilon^{i+2}U_{r,i}\big(\boldsymbol{I}', H_r\big)\right)\mathrm{d}t$$

$$+\sum_{k=1}^{n_g}\overline{\sigma}_{rk}\big(\boldsymbol{I}', H_r\big)\mathrm{d}B_k(t)+\sum_{l=1}^{u-1}G_{r,l} \tag{6.243}$$

where

$$\overline{m}_\eta\big(\boldsymbol{I}', H_r\big)=\left\langle-\sum_{\eta'=1}^{r-1}\sum_{j=1}^{n}m_{\eta' j}\frac{\partial H}{\partial P_j}\frac{\partial I_\eta}{\partial P_{\eta'}}\right.$$

$$\left.+\frac{1}{2}\sum_{\eta',\eta''=1}^{r-1}\sum_{k=1}^{n_g}\sigma_{\eta' k}\sigma_{\eta'' k}\frac{\partial^2 I_\eta}{\partial P_{\eta'}\partial P_{\eta''}}\right\rangle_t \tag{6.244}$$

$$U_{\eta,0}\big(\boldsymbol{I}', H_r\big)=\sum_{l=1}^{n_p}\lambda_l E\big[Y_l^2\big]\langle A_{\eta;2;l}\rangle_t, \tag{6.245}$$

$$U_{\eta,k}\big(\boldsymbol{I}', H_r\big)=\sum_{l=1}^{n_p}\lambda_l E\big[Y_l^{k+2}\big]\langle A_{\eta;k+2;l}\rangle_t, \tag{6.246}$$

$$\overline{m}_r\big(\boldsymbol{I}', H_r\big)=\left\langle-\sum_{\rho'=r}^{n}\sum_{j=1}^{n}m_{\rho' j}\frac{\partial H}{\partial P_j}\frac{\partial H_r}{\partial P_{\rho'}}+\frac{1}{2}\sum_{\rho',\rho''=r}^{n}\sum_{k=1}^{n_g}\sigma_{\rho' k}\sigma_{\rho'' k}\frac{\partial^2 H_r}{\partial P_{\rho'}\partial P_{\rho''}}\right\rangle_t, \tag{6.247}$$

$$U_{r,0}\big(\boldsymbol{I}', H_r\big)=\sum_{l=1}^{n_p}\lambda_l E\big[Y_l^2\big]\langle B_{r;2;l}\rangle_t, \tag{6.248}$$

$$U_{r,k}\big(\boldsymbol{I}', H_r\big)=\sum_{l=1}^{n_p}\lambda_l E\big[Y_l^{k+2}\big]\langle B_{r;k+2;l}\rangle_t, \tag{6.249}$$

$$\sum_{k=1}^{n_g}\overline{\sigma}_{\eta_1 k}\big(\boldsymbol{I}', H_r\big)\overline{\sigma}_{\eta_2 k}\big(\boldsymbol{I}', H_r\big)=\left\langle\sum_{\eta',\eta''=1}^{r-1}\sum_{k=1}^{n_g}\sigma_{\eta' k}\sigma_{\eta'' k}\frac{\partial I_{\eta_1}}{\partial P_{\eta'}}\frac{\partial I_{\eta_2}}{\partial P_{\eta''}}\right\rangle_t \tag{6.250}$$

$$\sum_{k=1}^{n_g}\overline{\sigma}_{\eta k}\big(\boldsymbol{I}', H_r\big)\overline{\sigma}_{rk}\big(\boldsymbol{I}', H_r\big)=\left\langle\sum_{\eta'=1}^{r-1}\sum_{\rho'=r}^{n}\sum_{k=1}^{n_g}\sigma_{\eta' k}\sigma_{\rho' k}\frac{\partial I_\eta}{\partial P_{\eta'}}\frac{\partial H_r}{\partial P_{\rho'}}\right\rangle_t \tag{6.251}$$

$$\sum_{k=1}^{n_g} \overline{\sigma}_{rk}(\boldsymbol{I}', H_r)\overline{\sigma}_{rk}(\boldsymbol{I}', H_r) = \left\langle \sum_{\rho,\rho'=r}^{n} \sum_{k=1}^{n_g} \frac{\partial H_r}{\partial P_\rho} \frac{\partial H_r}{\partial P_{\rho'}} \sigma_{\rho k}\sigma_{\rho' k} \right\rangle_t \tag{6.252}$$

$$G_{\eta,l} = \sum_{s_1=1}^{r} \sum_{s_2=1}^{s_1} \cdots \sum_{s_{u-l+1}=1}^{s_{u-l}} \left[\int\limits_{\mathcal{Y}_{s_1,\cdots,s_{u-l+1};l}} \left(\sum_{k=1}^{l} \varepsilon^k V_{\eta;s_1,\cdots,s_{u-l+1};l,k}(\boldsymbol{I}', H_r) \right) \right.$$
$$\left. Y_{s_1,\cdots,s_{u-l+1};l} \mathcal{P}_{s_1,\cdots,s_{u-l+1};l}\left(\mathrm{d}t, \mathrm{d}Y_{s_1,\cdots,s_{u-l+1};l}\right) \right], \tag{6.253}$$

$$G_{r,l} = \sum_{s_1=1}^{r} \sum_{s_2=1}^{s_1} \cdots \sum_{s_{u-l+1}=1}^{s_{u-l}} \left[\int\limits_{\mathcal{Y}_{s_1,\cdots,s_{u-l+1};l}} \left(\sum_{k=1}^{l} \varepsilon^k V_{r;s_1,\cdots,s_{u-l+1};l,k}(\boldsymbol{I}', H_r) \right) \right.$$
$$\left. Y_{s_1,\cdots,s_{u-l+1};l} \mathcal{P}_{s_1,\cdots,s_{u-l+1};l}\left(\mathrm{d}t, \mathrm{d}Y_{s_1,\cdots,s_{u-l+1};l}\right) \right]. \tag{6.254}$$

The expression for the function $V_{\eta;s_1,s_2,\cdots,s_{u-k+1};l;k} = V_{\eta;s_1,s_2,\cdots,s_{u-k+1};l;k}(\boldsymbol{I}', H_r)$ can be determined by equating same ε order terms as follows

$$\sum_{k=1}^{u-1} \sum_{\eta=j}^{j\times k} \varepsilon^\eta \left\{ \sum_{s_1=1}^{r} \sum_{s_2=1}^{s_1} \cdots \sum_{s_{u-k+1}=1}^{s_{u-k}} \left(\sum_{k_1+\cdots+k_j=\eta} V_{\eta_1;s_1,s_2,\cdots,s_{u-k+1};k;k_1} V_{\eta_2;s_1,s_2,\cdots,s_{u-k+1};k;k_2} \right.\right.$$
$$\left. \cdots V_{\eta_s;s_1,s_2,\cdots,s_{u-k+1};k;k_s} \times V_{r;s_1,s_2,\cdots,s_{u-k+1};k;k_{s+1}} \cdots V_{r;s_1,s_2,\cdots,s_{u-k+1};k;k_j} \right)\Big]$$
$$\left. \lambda_{s_1,s_2,\cdots,s_{u-k+1};k} E\left[Y^j_{s_1,s_2,\cdots,s_{u-k+1};k} \right] \right\}, = \sum_{k=j}^{u} \varepsilon^k$$
$$\sum_{l=1}^{n_p} \left\{ \lambda_l E\left[Y_l^k\right] \left\langle \sum_{k_1+k_2+\cdots+k_j=k} A_{\eta_1;k_1;l} A_{\eta_2;k_2;l} \cdots A_{\eta_3;k_s;l} B_{r;k_{s+1};l} \cdots B_{r;k_j;l} \right\rangle_t \right\}$$
$$+ \mathrm{O}(\varepsilon^{u+1}), \tag{6.255}$$

$$s = 0, \cdots, u,$$

$$\langle[\bullet]\rangle_t = \frac{1}{(2\pi)^{r-1} T(H_r)} \int\limits_{\Omega'} \int\limits_0^{2\pi} \left[[\bullet] \Big/ \frac{\partial H_r}{\partial p_r} \right] \mathrm{d}\boldsymbol{\theta}' \mathrm{d}q_r \cdots \mathrm{d}q_n \mathrm{d}p_{r+1} \cdots \mathrm{dp}_n \tag{6.256}$$

$$T(H_r) = \int\limits_{\Omega'} \left(1 \Big/ \frac{\partial H_r}{\partial p_r} \right) \mathrm{d}q_r \cdots \mathrm{d}q_n \mathrm{d}p_{r+1} \cdots \mathrm{d}p_n \tag{6.257}$$

and $\boldsymbol{\theta}' = [\theta_1, \theta_2, \cdots, \theta_{r-1}]^{\mathrm{T}}$,

$$A_{\eta;k;l} = A_{\eta;k;l}(\boldsymbol{I}', \boldsymbol{\Theta}', H_r, q_r, \cdots, q_2, p_{r+1}, \cdots, p_n),$$

$$B_{r;k;l} = B_{r;k;l}(\boldsymbol{I}', \boldsymbol{\Theta}', H_r, q_r, \cdots, q_2, p_{r+1}, \cdots, p_n), \tag{6.258}$$

$$\Omega' = \{(q_r, \cdots, q_n, p_{r+1}, \cdots, p_n) | H_r(q_r, \cdots, q_n, 0, p_{r+1}, \cdots, p_n) \le H_r\} \tag{6.259}$$

where $\int_0^{2\pi} [\bullet]\, \mathrm{d}\boldsymbol{\theta}' = \int_0^{2\pi}\int_0^{2\pi} \cdots \int_0^{2\pi} [\bullet]\, \mathrm{d}\theta_1 \mathrm{d}\theta_2 \cdots \mathrm{d}\theta_{r-1}$ represents $(r-1)$-fold integral. In (6.253) and (6.254), $\mathcal{P}_{s_1,s_2,\cdots,s_{u-l+1};l}(\mathrm{d}t, \mathrm{d}Y_{s_1,s_2,\cdots,s_{u-l+1};l})$ are independent Poisson random measure with the following property

$$\lambda_{r_1,r_2,\cdots,r_{u-l+1};l} E[Y^r_{r_1,r_2,\cdots,r_{u-l+1};l}] = \begin{cases} 0 & r = 1, \\ \overline{M}_{r_1,r_2,\cdots,r_{u-l+1};s,r} & r = 2, 3, \cdots, u-l+1, \\ \overline{m}_{r_1,r_2,\cdots,r_{u-l+1};s,r} & r = u-l+2, u-l+3, \cdots, \end{cases} \tag{6.260}$$

$$\overline{M}_{r_1,r_2,\cdots,r_{u-l+1};l,r} \gg \varepsilon,\ \overline{m}_{r_1,r_2,\cdots,r_{u-l+1};l,r} < \varepsilon^u.$$

The truncated averaged FPK equation associated with truncated averaged SIDEs (6.242) and (6.243) is

$$\begin{aligned} \frac{\partial p}{\partial t} &= -\sum_{\eta_1=1}^{r-1} \frac{\partial}{\partial I_{\eta_1}}(\overline{a}_{\eta_1} p) - \frac{\partial}{\partial h_r}(\overline{a}_r p) \\ &+ \frac{1}{2!} \sum_{\eta_1,\eta_2=1}^{r-1} \frac{\partial^2}{\partial I_{\eta_1} \partial I_{\eta_2}} (\overline{a}_{\eta_1,\eta_2} p) \\ &+ \frac{C_2^1}{2!} \sum_{\eta_1=1}^{r-1} \frac{\partial^2}{\partial I_{\eta_1} \partial h_r} (\overline{a}_{\eta_1,r} p) + \frac{1}{2!} \frac{\partial^2}{\partial h_r^2} (\overline{a}_{r,r} p) + \cdots + \\ &\sum_{j=3}^{u} \sum_{s=0}^{j} \sum_{\eta_1,\eta_2,\cdots,\eta_s=1}^{r-1} (-1)^j \frac{C_j^s}{j!} \frac{\partial^j}{\partial I_{\eta_1} \partial I_{\eta_2} \cdots \partial I_{\eta_s} \partial h_r^{j-s}} \left(\overline{a}_{\eta_1,\eta_2,\cdots,\eta_s,\underbrace{r,\cdots,r}_{(j-s)-\text{ fold}}} p \right), \end{aligned} \tag{6.261}$$

where derivative moments are

$$\begin{aligned} \overline{a}_{\eta_1} = \overline{a}_{\eta_1}(\boldsymbol{I}', h_r) = \varepsilon^2 \Big\langle & -\sum_{j=1}^{n} \sum_{\rho=1}^{r-1} \frac{\partial H}{\partial p_j} \frac{\partial I_{\eta_1}}{\partial p_\rho} \\ & + \frac{1}{2} \sum_{\rho,\rho'=1}^{r-1} \sum_{k=1}^{n_g} \sigma_{\rho k} \sigma_{\rho' k} \frac{\partial^2 I_{\eta_1}}{\partial p_\rho \partial p_{\rho'}} \Big\rangle_t \end{aligned}$$

$$+\sum_{k=2}^{u}\varepsilon^k\sum_{l=1}^{n_p}\left\{\lambda_l E\left[Y_l^k\right]\left\langle A_{\eta_1;k;l}\right\rangle_t\right\}, \tag{6.262}$$

$$\begin{aligned}\overline{a}_r=\overline{a}_r\left(\boldsymbol{I}',h_r\right)=\varepsilon^2\Bigg\langle&\sum_{j=1}^{n}\sum_{\rho=r}^{n}\frac{\partial H}{\partial p_j}\frac{\partial H_r}{\partial p_\rho}\\&+\frac{1}{2}\sum_{\rho,\rho'=r}^{n}\sum_{k=1}^{n_g}\sigma_{\rho k}\sigma_{\rho' k}\frac{\partial^2 H_r}{\partial p_\rho\partial p_{\rho'}}\Bigg\rangle_t\\&+\sum_{k=2}^{u}\varepsilon^k\sum_{l=1}^{n_p}\left\{\lambda_l E\left[Y_l^k\right]\left\langle B_{r;k;l}\right\rangle_t\right\},\end{aligned} \tag{6.263}$$

$$\begin{aligned}\overline{a}_{\eta_1,\eta_2}=\overline{a}_{\eta_1,\eta_2}\left(\boldsymbol{I}',h_r\right)=&\,\varepsilon^2\left\langle\sum_{\rho,\rho'=1}^{r-1}\sum_{k=1}^{n_g}\sigma_{\rho k}\sigma_{\rho' k}\frac{\partial I_{\eta_1}}{\partial p_\rho}\frac{\partial I_{\eta_2}}{\partial p_{\rho'}}\right\rangle_t\\&+\sum_{k=2}^{u}\varepsilon^k\sum_{l=1}^{n_p}\left\{\lambda_l E\left[Y_l^k\right]\left\langle\sum_{k_1+k_2=k}A_{\eta_1;k_1;l}A_{\eta_2;k_2;l}\right\rangle_t\right\},\end{aligned} \tag{6.264}$$

$$\begin{aligned}\overline{a}_{\eta_1,r}=\overline{a}_{\eta_1,r}\left(\boldsymbol{I}',h_r\right)=&\,\varepsilon^2\left\langle\sum_{\eta'=1}^{r-1}\sum_{\rho'=r}^{n}\sum_{k=1}^{n_g}\sigma_{\eta' k}\sigma_{\rho' k}\frac{\partial I_{\eta_1}}{\partial p_{\eta'}}\frac{\partial H_r}{\partial p_{\rho'}}\right\rangle_t\\&+\sum_{k=2}^{u}\varepsilon^k\sum_{l=1}^{n_p}\left\{\lambda_l E\left[Y_l^k\right]\left\langle\sum_{k_1+k_2=k}A_{\eta_1;k_1;l}B_{r;k_2;l}\right\rangle_t\right\},\end{aligned} \tag{6.265}$$

$$\begin{aligned}\overline{a}_{r,r}=\overline{a}_{r,r}\left(\boldsymbol{I}',h_r\right)=&\,\varepsilon^2\left\langle\sum_{\rho,\rho'=r}^{n}\sum_{k=1}^{n_g}\sigma_{\rho k}\sigma_{\rho' k}\frac{\partial H_r}{\partial p_\rho}\frac{\partial H_r}{\partial p_{\rho'}}\right\rangle_t\\&+\sum_{k=2}^{u}\varepsilon^k\sum_{l=1}^{n_p}\left\{\lambda_l E\left[Y_l^k\right]\left\langle\sum_{k_1+k_2=k}B_{r;k_1;l}B_{r;k_2;l}\right\rangle_t\right\},\end{aligned} \tag{6.266}$$

$$\begin{aligned}\overline{a}_{\eta_1,\eta_2,\cdots,\eta_j}&=\overline{a}_{\eta_1,\eta_2,\cdots,\eta_j}\left(\boldsymbol{I}',h_r\right)\\&=\sum_{k=j}^{u}\varepsilon^k\sum_{l=1}^{n_p}\left\{\lambda_l E\left[Y_l^k\right]\left\langle\sum_{k_1+\cdots+k_j=k}A_{\eta_1;k_1;l}\cdots A_{\eta_j;k_j;l}\right\rangle_t\right\},\end{aligned} \tag{6.267}$$

$$\begin{aligned}&\overline{a}_{\eta_1,\cdots,\eta_s,\underbrace{r,\cdots,r}_{(j-s)-\text{fold}}}=\overline{a}_{\eta_1,\cdots,\eta_s,\underbrace{r,\cdots,r}_{(j-s)-\text{fold}}}\left(\boldsymbol{I}',h_r\right)\\&=\sum_{k=j}^{u}\varepsilon^k\sum_{l=1}^{n_p}\left\{\lambda_l E\left[Y_l^k\right]\left\langle\sum_{k_1+\cdots+k_j=k}A_{\eta_1;k_1;l}\cdots A_{\eta_s;k_s;l}B_{r;k_{s+1};l}\cdots B_{r;k_j;l}\right\rangle_t\right\},\end{aligned} \tag{6.268}$$

$$\overline{a}_{\underbrace{r,\cdots,r}_{j\text{-fold}}}=\overline{a}_{\underbrace{r,\cdots,r}_{j\text{-fold}}}(\boldsymbol{I}',h_r)=\sum_{k=j}^{u}\varepsilon^k\sum_{l=1}^{n_p}\left\{\lambda_l E\left[Y_l^k\right]\left\langle\sum_{k_1+\cdots+k_j=k}B_{r;k_1;l}\cdots B_{r;k_j;l}\right\rangle_t\right\},\tag{6.269}$$

$$j=3,\cdots,u.$$

In Eqs. (6.262) thought (6.269), $A_{\eta;k;l}=A_{\eta;k;l}(\boldsymbol{I}',\boldsymbol{\theta}',h_r,q_r,\cdots,q_2,p_{r+1},\cdots,p_n)$ and $B_{r;k;l}=B_{r;k;l}(\boldsymbol{I}',\boldsymbol{\theta}',h_r,q_r,\cdots,q_2,p_{r+1},\cdots,p_n)$.

In (6.261), $p=p(\boldsymbol{I}',h_r,t|\boldsymbol{I}'_0,h_{r0})$ is the transition PDF of $[\boldsymbol{I}'^{\mathrm{T}},H_r]^{\mathrm{T}}$ with initial condition

$$p(\boldsymbol{I}',h_r,0|\boldsymbol{I}'_0,h_{r0})=\delta(\boldsymbol{I}'-\boldsymbol{I}'_0)\delta(h_r-h_{r0}),\tag{6.270}$$

or $p=p(\boldsymbol{I}',h_r,t)$ is the joint PDF of $[\boldsymbol{I}'^{\mathrm{T}},H_r]^{\mathrm{T}}$ with initial condition

$$p(\boldsymbol{I}',h_r,0)=p(\boldsymbol{I}'_0,h_{r0}).\tag{6.271}$$

When I_η, h_r can vary within $[0,\infty)$, the boundary conditions of averaged FPK Eq. (6.261) are

$$p=\text{finite},\quad \frac{\partial^k}{\partial I_n^k}p=\text{finite},\quad \frac{\partial^k}{\partial h_r^k}p=\text{finite},\quad I_\eta=0\ \text{or}\ h_r=0.\tag{6.272}$$

$$p\to 0,\quad \frac{\partial^k}{\partial I_\eta^k}p\to 0,\quad \frac{\partial^k}{\partial H_r^k}p\to 0,\quad \text{when}\ |\boldsymbol{I}'|\to\infty\ \text{or}\ h_r\to\infty.\tag{6.273}$$

Equation (6.272) indicate that the boundaries $I_\eta=0$ and $h_r=0$ are reflective. Equation (6.273) indicate that the infinity boundary is absorbing. In addition, Eq. (6.261) must satisfy the following normalization condition

$$\int_0^\infty\cdots\int_0^\infty\int_0^\infty p(\boldsymbol{I}',h_r)\,\mathrm{d}I_1\cdots\mathrm{d}I_{r-1}\mathrm{d}h_r=1\tag{6.274}$$

According to Eq. (5.156), the stationary solution $p(\boldsymbol{I}',h_r)$ of Eq. (6.261) can yield approximate stationary PDF for the generalized displacements and generalized momenta of the original system, i.e.,

$$p(\boldsymbol{q},\boldsymbol{p})=\left.\frac{p(\boldsymbol{I}',h_r)}{(2\pi)^{r-1}T(h_r)}\right|_{\boldsymbol{I}'=\boldsymbol{I}'(\boldsymbol{q}_1,\boldsymbol{p}_1),h_r=H_r(\boldsymbol{q}_2,\boldsymbol{p}_2)}\tag{6.275}$$

6.4.2 Internal Resonant Case

Suppose that the quasi-partially integrable Hamiltonian system has the following $\beta\,(1 \le \beta \le r-2)$ weak internal resonant relations between $r-1$ frequencies ω_η

$$\sum_{\eta=1}^{r-1} k_\eta^v \omega_\eta = \mathrm{O}_v(\varepsilon^2), \quad v = 1, 2, \cdots, \beta \tag{6.276}$$

Introduce the following β combinations of angle variables

$$\Psi_v = \sum_{\eta=1}^{r-1} k_\eta^v \Theta_\eta, \quad v = 1, 2, \cdots, \beta \tag{6.277}$$

The SIDEs for Ψ_v can be obtained from equations of Θ_η in Eq. (6.240) using relations in Eq. (6.277) as follow

$$\begin{aligned} \mathrm{d}\Psi_v = &\left(O_v(\varepsilon^2) - \varepsilon^2 \sum_{j=1}^{n} \sum_{\eta'=1}^{r-1} m_{\eta' j} \frac{\partial \Psi_v}{\partial P_{\eta'}} \frac{\partial H}{\partial P_j} + \frac{\varepsilon^2}{2} \sum_{\eta',\eta''=1}^{r-1} \sum_{k=1}^{n_g} \sigma_{\eta' k} \sigma_{\eta'' k} \frac{\partial^2 \Psi_v}{\partial P_{\eta'} \partial P_{\eta''}} \right) \mathrm{d}t \\ &+ \varepsilon \sum_{\eta'=1}^{r-1} \sum_{k=1}^{n_g} \frac{\partial \Psi_v}{\partial P_{\eta'}} \sigma_{\eta' k} \mathrm{d}B_k(t) + \sum_{l=1}^{n_p} \int_{\mathcal{Y}_l} \big[\Psi_v(\boldsymbol{Q}, \boldsymbol{P} + \hat{\boldsymbol{\gamma}}_l) \\ &- \Psi_v(\boldsymbol{I}', H_r, \boldsymbol{\Psi}, \theta_1, \cdots, \theta_{r-1-\beta}, Q_r, \cdots, Q_n, P_{r+1}, \cdots, P_n) \big] \mathcal{P}_l(\mathrm{d}t, \mathrm{d}Y_l). \end{aligned} \tag{6.278}$$

Expanding $\Psi_v(\boldsymbol{Q}, \boldsymbol{P} + \hat{\gamma}_l) - \Psi_v(\boldsymbol{I}', H_r, \boldsymbol{\Psi}, \Theta_1, \cdots, \Theta_{r-1-\beta}, Q_r, \cdots, Q_n, P_{r+1}, \cdots, P_n)$ into Taylor series, Eq. (6.278) becomes

$$\begin{aligned} \mathrm{d}\Psi_v = &\left(O_v(\varepsilon^2) - \varepsilon^2 \sum_{j=1}^{n} \sum_{\eta'=1}^{r-1} m_{\eta' j} \frac{\partial \Psi_v}{\partial P_{\eta'}} \frac{\partial H}{\partial P_j} \right. \\ &\left. + \frac{\varepsilon^2}{2} \sum_{\eta',\eta''=1}^{r-1} \sum_{k=1}^{n_g} \sigma_{\eta' k} \sigma_{\eta'' k} \frac{\partial^2 \Psi_v}{\partial P_{\eta'} \partial P_{\eta''}} \right) \mathrm{d}t \\ &+ \varepsilon \sum_{\eta'=1}^{r-1} \sum_{k=1}^{n_g} \frac{\partial \Psi_v}{\partial p_{\eta'}} \sigma_{\eta' k} \mathrm{d}B_k(t) \\ &+ \sum_{l=1}^{n_p} \int_{\mathcal{Y}_l} \left[\sum_{k=1}^{\infty} \varepsilon^k Y_l^k C_{v;k;l}(\boldsymbol{I}', H_r, \Psi, \Theta_1, \cdots, \right. \\ &\Theta_{r-1-\beta}, Q_r, \cdots, Q_n, P_{r+1}, \cdots, P_n) \big] \mathcal{P}_l(\mathrm{d}t, \mathrm{d}Y_l), \end{aligned}$$

$$v = 1, 2, \cdots, \beta. \tag{6.279}$$

The new system consist of $r-1$ equations of I_η in Eq. (6.239), $r-1-\beta$ equations of Θ_η in Eq. (6.240), one equation of H_r in Eq. (6.241), β equations of Ψ_v in Eq. (6.279), $n-r+1$ equations of Q_i and $n-r$ equations of P_i in Eq. (6.5). I_η $(\eta = 1, 2, \cdots, r-1)$, H_r and Ψ_v $(v = 1, 2, \cdots, \beta)$ are slowly-varying processes while Θ_η $(\eta = \beta + 1, \beta + 2, \cdots, r-1)$,$Q_i$ $(i = r, \cdots, n)$ and P_i $(i = r+1, \cdots, n)$ are fast-varying processes. According to the stochastic averaging principle (Khasminskii 1968; Xu et al. 2011), $[\boldsymbol{I}^{\mathrm{T}}, \boldsymbol{\Psi}^{\mathrm{T}}, H_r]^{\mathrm{T}}$ converges weakly to a $(r+\beta)$-dimensional vector Markov process as $\varepsilon \to 0$. The SIDEs for this vector Markov process can be obtained by time averaging of Eqs. (6.239), (6.241), and (6.279). Due to that the integrable Hamiltonian subsystem is ergodic on the $(r-\beta-1)$-dimensional sub-torus and the non-integrable Hamiltonian subsystem is ergodic on the $[2(n-r)+1]$-dimensional constant H_r surface, time averaging can be replaced by spatial averaging in Eq. (5.173).

Due to infinite terms in the SIDEs for slowly-varying processes in Eqs. (6.239), (6.241) and (6.279), truncation is necessary. Ignoring the terms of order ε^u and higher lead to the following approximate truncated averaged SIDEs

$$\mathrm{d}I_\eta = \left[\varepsilon^2 \overline{m}_\eta(\boldsymbol{I}', \boldsymbol{\Psi}', H_r) + \sum_{i=0}^{u-2} \varepsilon^{i+2} U_{\eta,i}(\boldsymbol{I}', \boldsymbol{\Psi}', H_r)\right]\mathrm{d}t + \sum_{k=1}^{n_g} \overline{\sigma}_{\eta k}(\boldsymbol{I}', \boldsymbol{\Psi}', H_r)\mathrm{d}B_k(t) + \sum_{l=1}^{u-1} G_{\eta,l}(\boldsymbol{I}', \boldsymbol{\Psi}', H_r), \tag{6.280}$$

$$\mathrm{d}\Psi_v = \left[\varepsilon^2 \overline{m}_v(\boldsymbol{I}', \boldsymbol{\Psi}', H_r) + \sum_{i=0}^{u-2} \varepsilon^{i+2} U_{v,i}(\boldsymbol{I}', \boldsymbol{\Psi}', H_r)\right]\mathrm{d}t + \sum_{k=1}^{n_g} \overline{\sigma}_{vk}(\boldsymbol{I}', \boldsymbol{\Psi}', H_r)\mathrm{d}B_k(t) + \sum_{l=1}^{u-1} G_{v,l}(\boldsymbol{I}', \boldsymbol{\Psi}', H_r), \tag{6.281}$$

$$\mathrm{d}H_r = \left[\varepsilon^2 \overline{m}_r(\boldsymbol{I}', \boldsymbol{\Psi}', H_r) + \sum_{i=0}^{u-2} \varepsilon^{i+2} U_{r,i}(\boldsymbol{I}', \boldsymbol{\Psi}', H_r)\right]\mathrm{d}t + \sum_{k=1}^{n_g} \overline{\sigma}_{rk}(\boldsymbol{I}', \boldsymbol{\Psi}', H_r)\mathrm{d}B_k(t) + \sum_{l=1}^{u-1} G_{r,l}(\boldsymbol{I}', \boldsymbol{\Psi}', H_r), \tag{6.282}$$

$$\eta = 1, 2, \cdots, r-1; \quad v = 1, 2, \cdots, \beta$$

where $\boldsymbol{\Psi}' = \left[\Psi_1, \Psi_2, \cdots, \Psi_\beta\right]^{\mathrm{T}}$,

$$\overline{m}_\eta(\boldsymbol{I}', \boldsymbol{\Psi}', H_r) = \Bigg\langle -\sum_{\eta'=1}^{r-1}\sum_{j=1}^{n} m_{\eta' j}\frac{\partial I_\eta}{\partial P_{\eta'}}\frac{\partial H}{\partial P_j}$$

$$+\frac{1}{2}\sum_{\eta',\eta''=1}^{r-1}\sum_{k=1}^{n_g}\sigma_{\eta' k}\sigma_{\eta'' k}\frac{\partial^2 I_\eta}{\partial P_{\eta'}\partial P_{\eta''}}\Bigg\rangle_t, \tag{6.283}$$

$$U_{\eta,i}(\boldsymbol{I}',\boldsymbol{\Psi}',H_r)=\sum_{l=1}^{n_p}\lambda_l E\big[Y_l^{i+2}\big]\langle A_{\eta;i+2;l}\rangle_t, \tag{6.284}$$

$$\overline{m}_v(\boldsymbol{I}',\boldsymbol{\Psi}',H_r)=\Bigg\langle O_v(\varepsilon^2)-\sum_{\rho'=1}^{r-1}\sum_{j=1}^{n}m_{\rho' j}\frac{\partial \Psi_v}{\partial P_{\rho'}}\frac{\partial H}{\partial P_j}$$
$$+\frac{1}{2}\sum_{\rho',\rho''=1}^{r-1}\sum_{k=1}^{n_g}\sigma_{\rho' k}\sigma_{\rho'' k}\frac{\partial^2 \Psi_v}{\partial P_{\rho'}\partial P_{\rho''}}\Bigg\rangle_t, \tag{6.285}$$

$$U_{v,i}(\boldsymbol{I}',\boldsymbol{\Psi}',H_r)=\sum_{l=1}^{n_p}\lambda_l E\big[Y_l^{i+2}\big]\langle C_{v;i+2;l}\rangle_t, \tag{6.286}$$

$$\overline{m}_r(\boldsymbol{I}',\boldsymbol{\Psi}',H_r)=\Bigg\langle -\sum_{\rho'=r}^{n}\sum_{j=1}^{n}m_{\rho' j}\frac{\partial H_r}{\partial P_{\rho'}}\frac{\partial H}{\partial P_j}$$
$$+\frac{1}{2}\sum_{\rho',\rho''=r}^{n}\sum_{k=1}^{n_g}\sigma_{\rho' k}\sigma_{\rho'' k}\frac{\partial^2 H_r}{\partial P_{\rho'}\partial P_{\rho''}}\Bigg\rangle_t, \tag{6.287}$$

$$U_{r,i}(\boldsymbol{I}',\boldsymbol{\Psi}',H_r)=\sum_{l=1}^{n_p}\lambda_l E\big[Y_l^{i+2}\big]\langle B_{r;i+2;l}\rangle_t, \tag{6.288}$$

$\eta=1,\cdots,r-1;\quad v=1,\cdots,\beta;\quad i=0,1,\cdots,u-2,$

$$\sum_{k=1}^{n_g}\overline{\sigma}_{\eta_1 k}(\boldsymbol{I}',\boldsymbol{\Psi}',H_r)\overline{\sigma}_{\eta_2 k}(\boldsymbol{I}',\boldsymbol{\Psi}',H_r)=\Bigg\langle\sum_{\rho,\rho'=1}^{r-1}\sum_{k=1}^{n_g}\sigma_{\rho k}\sigma_{\rho' k}\frac{\partial I_{\eta_1}}{\partial P_\rho}\frac{\partial I_{\eta_2}}{\partial P_{\rho'}}\Bigg\rangle_t, \tag{6.289}$$

$$\sum_{k=1}^{n_g}\overline{\sigma}_{v_1 k}(\boldsymbol{I}',\boldsymbol{\Psi}',H_r)\overline{\sigma}_{v_2 k}(\boldsymbol{I}',\boldsymbol{\Psi}',H_r)$$
$$=\Bigg\langle\sum_{\rho,\rho'=1}^{r-1}\sum_{k=1}^{n_g}\sigma_{\rho k}\sigma_{\rho' k}\frac{\partial \Psi_{v_1}}{\partial P_\rho}\frac{\partial \Psi_{v_2}}{\partial P_{\rho'}}\Bigg\rangle_t, \tag{6.290}$$

$$\sum_{k=1}^{n_g}\overline{\sigma}_{rk}(\boldsymbol{I}',\boldsymbol{\Psi}',H_r)\overline{\sigma}_{rk}(\boldsymbol{I}',\boldsymbol{\Psi}',H_r)=\Bigg\langle\sum_{\rho,\rho'=r}^{n}\sum_{k=1}^{n_g}\sigma_{\rho k}\sigma_{\rho' k}\frac{\partial H_r}{\partial P_\rho}\frac{\partial H_r}{\partial P_{\rho'}}\Bigg\rangle_t, \tag{6.291}$$

$$\sum_{k=1}^{n_g}\overline{\sigma}_{\eta k}(\boldsymbol{I}',\boldsymbol{\Psi}',H_r)\overline{\sigma}_{vk}(\boldsymbol{I}',\boldsymbol{\Psi}',H_r)=\Bigg\langle\sum_{\rho,\rho'=1}^{r-1}\sum_{k=1}^{n_g}\sigma_{\rho k}\sigma_{\rho' k}\frac{\partial I_\eta}{\partial P_\rho}\frac{\partial \Psi_v}{\partial P_{\rho'}}\Bigg\rangle_t, \tag{6.292}$$

$$\sum_{k=1}^{n_g}\overline{\sigma}_{\eta k}(\boldsymbol{I}',\boldsymbol{\Psi}',H_r)\overline{\sigma}_{rk}(\boldsymbol{I}',\boldsymbol{\Psi}',H_r)=\left\langle\sum_{\eta'=1}^{r-1}\sum_{\rho'=r}^{n}\sum_{k=1}^{n_g}\sigma_{\eta' k}\sigma_{\rho' k}\frac{\partial I_\eta}{\partial P_{\eta'}}\frac{\partial H_r}{\partial P_{\rho'}}\right\rangle_t, \tag{6.293}$$

$$\sum_{k=1}^{n_g}\overline{\sigma}_{rk}(\boldsymbol{I}',\boldsymbol{\Psi}',H_r)\overline{\sigma}_{vl}(\boldsymbol{I}',\boldsymbol{\Psi}',H_r)=\left\langle\sum_{\rho=r}^{n}\sum_{\eta'=1}^{r-1}\sum_{k=1}^{n_g}\sigma_{\rho k}\sigma_{\eta' k}\frac{\partial H_r}{\partial P_{\rho}}\frac{\partial \Psi_v}{\partial P_{\eta'}}\right\rangle_t, \tag{6.294}$$

$$G_{\eta,l}(\boldsymbol{I}',\boldsymbol{\Psi}',H_r)=\sum_{s_1=1}^{r+\beta}\sum_{s_2=1}^{s_1}\cdots\sum_{s_{u-l+1}=1}^{s_{u-l}}\left[\int_{\mathcal{Y}_{s_1,\cdots,s_{u-l+1}}}\left(\sum_{k=1}^{l}\varepsilon^k V_{\eta;s_1,\cdots,s_{u-l+1};l;k}(\boldsymbol{I}',\boldsymbol{\Psi}',H_r)\right)\right.$$
$$\left.\times Y_{s_1,\cdots,s_{u-l+1};l;k}\mathcal{P}_{s_1,\cdots,s_{u-l+1};l}\left(\mathrm{d}t,\mathrm{d}Y_{s_1,\cdots,s_{u-l+1};l}\right)\right], \tag{6.295}$$

$$G_{v,l}(\boldsymbol{I}',\boldsymbol{\Psi}',H_r)=\sum_{s_1=1}^{r+\beta}\sum_{s_2=1}^{s_1}\cdots\sum_{s_{u-l+1}=1}^{s_{u-l}}\left[\int_{\mathcal{Y}_{s_1,\cdots,s_{u-l+1}}}\left(\sum_{k=1}^{l}\varepsilon^k V_{v;s_1,\cdots,s_{u-l+1};l;k}(\boldsymbol{I}',\boldsymbol{\Psi}',H_r)\right)\right.$$
$$\left.\times Y_{s_1,\cdots,s_{u-l+1};l;k}\mathcal{P}_{s_1,\cdots,s_{u-l+1};l}\left(\mathrm{d}t,\mathrm{d}Y_{s_1,\cdots,s_{u-l+1};l}\right)\right], \tag{6.296}$$

$$G_{r,l}(\boldsymbol{I}',\boldsymbol{\Psi}',H_r)=\sum_{s_1=1}^{r+\beta}\sum_{s_2=1}^{s_1}\cdots\sum_{s_{u-l+1}=1}^{s_{u-l}}\left[\int_{\mathcal{Y}_{s_1,\cdots,s_{u-l+1}}}\left(\sum_{k=1}^{l}\varepsilon^k V_{r;s_1,\cdots,s_{u-l+1};l;k}(\boldsymbol{I}',\boldsymbol{\Psi}',H_r)\right)\right.$$
$$\left.\times Y_{s_1,\cdots,s_{u-l+1};l;k}\mathcal{P}_{s_1,\cdots,s_{u-l+1};l}\left(\mathrm{d}t,\mathrm{d}Y_{s_1,\cdots,s_{u-l+1};l}\right)\right], \tag{6.297}$$

where functions $V_{\eta;s_1,\cdots,s_{u-l+1};l;k}(\boldsymbol{I}',\boldsymbol{\Psi}',H_r)$ can be obtained by setting the sum of same ε order terms equal to zero as follows

$$\sum_{k=1}^{u-1}\sum_{\eta=j}^{j\times k}\varepsilon^{\eta}\left\{\sum_{s_1=1}^{r+\beta}\sum_{s_2=1}^{s_1}\cdots\sum_{s_{u-k+1}=1}^{s_{u-k}}\left[\sum_{k_1+\cdots+k_j=\eta}V_{\eta_1;s_1,\cdots,s_{u-k+1};k;k_1}(\boldsymbol{I}',\boldsymbol{\Psi}',H_r)\times\cdots\right.\right.$$
$$V_{\eta_{\rho_1};s_1,\cdots,s_{u-k+1};k;k_{\rho_1}}(\boldsymbol{I}',\boldsymbol{\Psi}',H_r)V_{\eta_r;s_1,\cdots,s_{u-k+1};k;k_{\rho_1+1}}(\boldsymbol{I}',\boldsymbol{\Psi}',H_r)\times\cdots$$
$$V_{\eta_r;s_1,\cdots,s_{u-k+1};k;k_{\rho_1+\rho_2}}(\boldsymbol{I}',\boldsymbol{\Psi}',H_r)V_{\eta_{r+v_1};s_1,\cdots,s_{u-k+1};k;k_{\rho_1+\rho_2+1}}(\boldsymbol{I}',\boldsymbol{\Psi}',H_r)\times\cdots$$
$$\left.\left.V_{r+v_{\rho_3};s_1,\cdots,s_{u-k+1};k;k_{\rho_1+\rho_2+\rho_3}}(\boldsymbol{I}',\boldsymbol{\Psi}',H_r)\lambda_{s_1,s_2,\cdots,s_{u-k+1};k}E\left[Y^j_{s_1,s_2,\cdots,s_{u-k+1};k}\right]\right]\right\}$$
$$=\sum_{k=j}^{u}\varepsilon^k\sum_{l=1}^{n_p}\{\lambda_l E[Y_l^k]\left\langle\sum_{k_1+k_2+\cdots+k_j=k}A_{\eta_1;k_1;l}\cdots A_{\eta_{\rho_1};k_{\rho_1};l}B_{r;k_{\rho_1+1};l}\cdots\right.$$
$$\left.B_{r;k_{\rho_1+\rho_2};l}C_{v_1;k_{\rho_1+\rho_2+1};l}\cdots C_{v_{\rho_3};k_{\rho_1+\rho_2+\rho_3};l}\right\rangle\}, \tag{6.298}$$

$$\rho_1+\rho_2+\rho_3=j;\quad \rho_i=0,\cdots,j;\quad v_i=1,\cdots,\beta;\quad \eta_i=1,\cdots,r-1.$$

where

$$A_{r;k;l}=A_{r;k;l}(\boldsymbol{I}',\boldsymbol{\Psi}',H_r,\theta_1,\cdots,\theta_{r-1-\beta},q_r,\cdots,q_n,p_{r+1},\cdots,p_n)$$

$$B_{r;k;l} = B_{r;k;l}\big(\boldsymbol{I}', \boldsymbol{\Psi}', H_r, \theta_1, \cdots, \theta_{r-1-\beta}, q_r, \cdots, q_n, p_{r+1}, \cdots, p_n\big)$$
$$C_{r;k;l} = C_{r;k;l}\big(\boldsymbol{I}', \boldsymbol{\Psi}', H_r, \theta_1, \cdots, \theta_{r-1-\beta}, q_r, \cdots, q_n, p_{r+1}, \cdots, p_n\big) \tag{6.299}$$

$\mathcal{P}_{s_1,s_2,\cdots,s_{u-l+1};l}(\mathrm{d}t, \mathrm{d}Y_{s_1,s_2,\cdots,s_{u-l+1};l})$ are independent Poisson random measure with the following properties

$$\lambda_{r_1,r_2,\cdots,r_{u-s+1};s}E[Y^r_{r_1,r_2,\cdots,r_{u-s+1};s}]$$
$$= \begin{cases} 0 & r = 1, \\ \overline{M}_{r_1,r_2,r_{u-s+1};s,r} & r = 2, 3, \ldots, u-s+1, \\ \overline{m}_{r_1,r_2,r_{u-s+1};s,r} & r = u-s+2, u-s+3, \ldots, \end{cases} \tag{6.300}$$

and $\overline{M}_{r_1,r_2,\cdots,r_{u-s+1};s,r} \gg \varepsilon$, $\overline{m}_{r_1,r_2,\cdots,r_{u-s+1};s,r} < \varepsilon^u$.

The truncated averaged FPK equation associated with SIDEs (6.280)–(6.282) is

$$\begin{aligned}
\frac{\partial p}{\partial t} =& -\sum_{\eta_1=1}^{r-1}\frac{\partial}{\partial I_{\eta_1}}(\overline{a}_{\eta_1}p) - \sum_{v_1=1}^{\beta}\frac{\partial}{\partial \psi_{v_1}}(\overline{a}_{v_1}p) - \frac{\partial}{\partial h_r}(\overline{a}_r p) \\
&+ \frac{1}{2}\sum_{\eta_1,\eta_2=1}^{r-1}\frac{\partial^2}{\partial I_{\eta_1}\partial I_{\eta_2}}(\overline{a}_{\eta_1,\eta_2}p) \\
&+ \frac{1}{2}\sum_{v_1,v_2=1}^{\beta}\frac{\partial^2}{\partial \psi_{v_1}\partial \psi_{v_2}}(\overline{a}_{v_1,v_2}p) + \frac{1}{2}\frac{\partial^2}{\partial h_r^2}(\overline{a}_{r,r}p) \\
&+ \frac{C_2^1}{2}\sum_{\eta_1=1}^{r-1}\sum_{v_1=1}^{\beta}\frac{\partial^2}{\partial I_{\eta_1}\partial \psi_{v_1}}(\overline{a}_{\eta_1,v_1}p) \\
&+ \frac{C_2^1}{2}\sum_{\eta_1=1}^{r-1}\frac{\partial^2}{\partial I_{\eta_1}\partial h_r}(\overline{a}_{\eta_1,r}p) + \frac{C_2^1}{2}\sum_{v_1=1}^{\beta}\frac{\partial^2}{\partial h_r \partial \psi_{v_1}}(\overline{a}_{r,v_1}p) \\
&+ \cdots + \sum_{j=3}^{u}\sum_{\rho_1+\rho_2+\rho_3=j}\sum_{\eta_1,\cdots,\eta_{\rho_1}=1}^{r-1}\sum_{v_1,\cdots,v_{\rho_3}=1}^{\beta}\frac{(-1)^j C_j^{\rho_1}C_{j-\rho_1}^{\rho_2}}{j!} \\
&\times \frac{\partial^j}{\partial I_{\eta_1}\cdots\partial I_{\eta_{\rho_1}}\partial h_r^{\rho_2}\partial\psi_{v_1}\cdots\partial\psi_{v_{\rho_3}}}(\overline{a}_{\eta_1,\cdots,\eta_{\rho_1},r,\cdots,r,v_1,\cdots,v_{\rho_3}}p) + \mathrm{O}(\varepsilon^{u+1}),
\end{aligned} \tag{6.301}$$

where derivative moments are

$$\overline{a}_{\eta_1} = \varepsilon^2\left\langle -\sum_{\eta'=1}^{r-1}\sum_{j=1}^{n}\frac{\partial I_{\eta_1}}{\partial p_{\eta'}}\frac{\partial h}{\partial p_j} + \frac{1}{2}\sum_{\rho,\rho'=1}^{n}\sum_{k=1}^{n_g}\sigma_{\rho k}\sigma_{\rho' k}\frac{\partial^2 I_{\eta_1}}{\partial p_\rho \partial p_{\rho'}}\right\rangle_t$$

$$+\sum_{i=2}^{u}\varepsilon^{i}\sum_{l=1}^{n_p}\left\{\lambda_l E[Y_l^i]\langle A_{\eta_1;i;l}\rangle_t\right\}, \tag{6.302}$$

$$\overline{a}_{v_1}=\varepsilon^2\left\langle-\sum_{\eta'=1}^{r-1}\sum_{j=1}^{n}\frac{\partial\psi_{v_1}}{\partial p_{\eta'}}\frac{\partial h}{\partial p_j}+\frac{1}{2}\sum_{\rho,\rho'=1}^{n}\sum_{k=1}^{n_g}\sigma_{\rho k}\sigma_{\rho' k}\frac{\partial^2\psi_{v_1}}{\partial p_\rho\partial p_{\rho'}}\right\rangle_t$$
$$+\sum_{i=2}^{u}\varepsilon^{i}\sum_{l=1}^{n_p}\left\{\lambda_l E[Y_l^i]\langle C_{v_1;i;l}\rangle_t\right\}, \tag{6.303}$$

$$\overline{a}_{r}=\varepsilon^2\left\langle-\sum_{\rho=r}^{n}\sum_{j=1}^{n}\frac{\partial h_r}{\partial p_{\rho}}\frac{\partial h}{\partial p_j}+\frac{1}{2}\sum_{\rho,\rho'=1}^{n}\sum_{k=1}^{n_g}\sigma_{\rho k}\sigma_{\rho' k}\frac{\partial^2 h_r}{\partial p_\rho\partial p_{\rho'}}\right\rangle_t$$
$$+\sum_{i=2}^{u}\varepsilon^{i}\sum_{l=1}^{n_p}\left\{\lambda_l E[Y_l^i]\langle B_{r;i;l}\rangle_t\right\}, \tag{6.304}$$

$$\overline{a}_{\eta_1,\eta_2}=\varepsilon^2\left\langle\sum_{\eta',\eta''=1}^{r-1}\sum_{k=1}^{n_g}\sigma_{\eta' k}\sigma_{\eta'' k}\frac{\partial I_{\eta_1}}{\partial p_{\eta'}}\frac{\partial I_{\eta_2}}{\partial p_{\eta''}}\right\rangle_t$$
$$+\sum_{i=2}^{u}\varepsilon^{i}\sum_{l=1}^{n_p}\left\{\lambda_l E[Y_l^i]\left\langle\sum_{i_1+i_2=i}A_{\eta_1;i_1;l}A_{\eta_2;i_2;l}\right\rangle_t\right\}, \tag{6.305}$$

$$\overline{a}_{v_1,v_2}=\varepsilon^2\left\langle\sum_{\eta',\eta''=1}^{r-1}\sum_{k=1}^{n_g}\sigma_{\eta' k}\sigma_{\eta'' k}\frac{\partial \psi_{v_1}}{\partial p_{\eta'}}\frac{\partial \psi_{v_2}}{\partial p_{\eta''}}\right\rangle_t$$
$$+\sum_{i=2}^{u}\varepsilon^{i}\sum_{l=1}^{n_p}\left\{\lambda_l E[Y_l^i]\left\langle\sum_{i_1+i_2=i}C_{v_1;i_1;l}C_{v_2;i_2;l}\right\rangle_t\right\}, \tag{6.306}$$

$$\overline{a}_{r,r}=\varepsilon^2\left\langle\sum_{\rho,\rho'=r}^{n}\sum_{k=1}^{n_g}\sigma_{\rho k}\sigma_{\rho' k}\frac{\partial h_r}{\partial p_{\rho}}\frac{\partial h_r}{\partial p_{\rho'}}\right\rangle_t$$
$$+\sum_{i=2}^{u}\varepsilon^{i}\sum_{l=1}^{n_p}\left\{\lambda_l E[Y_l^i]\left\langle\sum_{i_1+i_2=i}B_{r;i_1;l}B_{r;i_2;l}\right\rangle_t\right\}, \tag{6.307}$$

$$\overline{a}_{\eta_1 r}=\varepsilon^2\left\langle\sum_{\eta'=1}^{r-1}\sum_{\rho'=r}^{n}\sum_{k=1}^{n_g}\sigma_{\eta' k}\sigma_{\rho' k}\frac{\partial I_{\eta_1}}{\partial p_{\eta'}}\frac{\partial h_r}{\partial p_{\rho'}}\right\rangle_t$$
$$+\sum_{i=2}^{u}\varepsilon^{i}\sum_{l=1}^{n_p}\left\{\lambda_l E\left[Y_l^i\right]\left\langle\sum_{i_1+i_2=i}A_{\eta_1;i_1;l}B_{r;i_2;l}\right\rangle_t\right\} \tag{6.308}$$

$$\begin{aligned}\overline{a}_{\eta_1,v_1} &= \varepsilon^2\left\langle \sum_{\eta',\eta''=1}^{r-1}\sum_{k=1}^{n_g}\sigma_{\eta'k}\sigma_{\eta''k}\frac{\partial\psi_{v_1}}{\partial p_{\eta'}}\frac{\partial I_{\eta_1}}{\partial p_{\eta''}}\right\rangle_t \\ &+\sum_{i=2}^{u}\varepsilon^i\sum_{l=1}^{n_p}\left\{\lambda_l E[Y_l^i]\left\langle\sum_{i_1+i_2=i}C_{v_1;i_1;l}A_{\eta_1;i_2;l}\right\rangle_t\right\} \end{aligned}\tag{6.309}$$

$$\begin{aligned}\overline{a}_{r,v_1} &= \varepsilon^2\left\langle \sum_{\eta'=1}^{r-1}\sum_{\rho'=r}^{n}\sum_{k=1}^{n_g}\sigma_{\eta'k}\sigma_{\rho'k}\frac{\partial\psi_{v_1}}{\partial p_{\eta'}}\frac{\partial h_r}{\partial p_{\rho'}}\right\rangle_t \\ &+\sum_{i=2}^{u}\varepsilon^i\sum_{l=1}^{n_p}\left\{\lambda_l E[Y_l^i]\left\langle\sum_{i_1+i_2=i}C_{v_1;i_1;l}B_{r;i_2;l}\right\rangle_t\right\} \end{aligned}\tag{6.310}$$

$$\begin{aligned}\overline{a}_{\eta_1,\cdots,\eta_{\rho_1},\underbrace{r,\cdots,r}_{\rho_2-\text{fold}},v_1,\cdots,v_{\rho_3}} &= \sum_{i=j}^{u}\varepsilon^i\sum_{l=1}^{n_p}\left\{\left\langle\sum_{i_1+\cdots+i_j=i}A_{\eta_1;i_1;l}\cdots A_{\eta_{\rho_1};i_{\rho_1};l}\right.\right. \\ &\times B_{r;i_{\rho_1+1};l}\cdots B_{r;i_{\rho_1+\rho_2};l}C_{v_1;i_{\rho_1+\rho_2+1};l}\cdots \left.\left. C_{v_{\rho_3};i_j;l}\right\rangle_t\right\}, \end{aligned}\tag{6.311}$$

$$v_i=1,\ldots,\beta;\quad \eta_i=1,\ldots,r-1;\quad \rho_1+\rho_2+\rho_3=j;\quad \rho_i=1,\ldots,j.$$

and

$$\langle[\cdot]\rangle_t=\frac{1}{(2\pi)^{r-\beta-1}T(h_r)}\int_{\Omega'}\int_0^{2\pi}\left([\cdot]/\frac{\partial h_r}{\partial p_r}\right)\mathrm{d}\boldsymbol{\theta}''\mathrm{d}q_r\cdots\,\mathrm{d}q_n\,\mathrm{d}p_{r+1}\cdots\,\mathrm{d}p_n,\tag{6.312}$$

$$T(h_r)=\int_{\Omega'}\left(1\bigg/\frac{\partial h_r}{\partial p_r}\right)\mathrm{d}q_r\cdots\mathrm{d}q_n\,\mathrm{d}p_{r+1}\cdots\mathrm{d}p_n,\tag{6.313}$$

$$\Omega'=\{(q_r,\ldots,q_n,p_{r+1},\ldots,p_n)|h_r(q_r,\ldots,q_n,0,p_{r+1},\ldots,p_n)\le h_r\},\tag{6.314}$$

in which

$$\begin{aligned}&\boldsymbol{\theta}'=\left[\theta_1,\theta_2,\cdots,\theta_{r-\beta-1}\right]^{\mathrm{T}}\\ &A_{\eta;i;l}=A_{\eta;i;l}\left(\boldsymbol{I}',\psi',h_r,\theta_1,\cdots,\theta_{r-1-\beta},q_r,\cdots,q_n,p_{r+1},\cdots,p_n\right)\\ &B_{r;i;l}=B_{r;i;l}\left(\boldsymbol{I}',\psi',h_r,\theta_1,\cdots,\theta_{r-1-\beta},q_r,\cdots,q_n,p_{r+1},\cdots,p_n\right)\\ &C_{v;i;l}=C_{v;i;l}\left(\boldsymbol{I}',\psi',h_r,\theta_1,\cdots,\theta_{r-1-\beta},q_r,\cdots,q_n,p_{r+1},\cdots,p_n\right)\end{aligned}\tag{6.315}$$

In Eq. (6.301), $p = p\left(\boldsymbol{I}',\boldsymbol{\Psi}',h_r,t\middle|\boldsymbol{I}_0',\boldsymbol{\Psi}_0',h_{r0}\right)$ is the transition PDF of $\left[\mathbf{I}'^{\mathrm{T}},\psi'^{\mathrm{T}},h_r\right]^{\mathrm{T}}$ with initial condition

$$p\left(\boldsymbol{I}',\boldsymbol{\Psi}',h_r,0\middle|\boldsymbol{I}_0',\boldsymbol{\Psi}_0'\right)=\delta\left(\boldsymbol{I}'-\boldsymbol{I}_0'\right)\delta\left(\boldsymbol{\Psi}'-\boldsymbol{\Psi}_0'\right)\delta\left(h_r-h_{r0}\right)\tag{6.316}$$

or $p = p(\boldsymbol{I}', \boldsymbol{\Psi}', h_r)$ is the PDF of $[\boldsymbol{I}'^{\mathrm{T}}, \boldsymbol{\Psi}'^{\mathrm{T}}, h_r]^{\mathrm{T}}$ with initial condition

$$p(\boldsymbol{I}', \boldsymbol{\Psi}', h_r, 0) = p(\boldsymbol{I}', \boldsymbol{\Psi}'_0, h_{r0}) \tag{6.317}$$

The boundary conditions of Eq. (6.301) depend on the system's characteristics and constraints. When I_η $(\eta = 1, 2, \ldots, r-1)$ and h_r vary on $[0, \infty)$, the boundary conditions are

$$\begin{aligned} p &= \text{finite}, \quad \frac{\partial^k p}{\partial I_\eta^k} = \text{finite}, \quad \text{when } I_\eta = 0; \\ p &= \text{finite}, \quad \frac{\partial^k p}{\partial h_r^k} = \text{finite}, \quad h_r = 0. \end{aligned} \tag{6.318}$$

$$\begin{aligned} &p \to 0, \quad \frac{\partial^k}{\partial I_n^k} p \to 0, \quad \text{when } I_n \to \infty; \quad p \to 0, \quad \frac{\partial^k}{\partial h_r^k} p \to 0, \quad h_r \to \infty, \\ &\eta = 1, \ldots, r-1; \; k = 1, 2, \cdots \end{aligned} \tag{6.319}$$

$$\begin{aligned} &p\big|_{\psi_v \pm 2n\pi} = p\big|_{\psi_v}; \quad \frac{\partial^k}{\partial \psi_v^k} p\bigg|_{\psi_v \pm 2n\pi} = \frac{\partial^k}{\partial \psi_v^k} p\bigg|_{\psi_v}; \\ &v = 1, 2, \ldots, \beta \end{aligned} \tag{6.320}$$

The averaged FPK equation must also satisfy the normalization condition

$$\int_0^\infty \cdots \int_0^\infty \int_0^{2\pi} \cdots \int_0^{2\pi} p(\boldsymbol{I}', \boldsymbol{y}', h_r) \mathrm{d}I_1 \cdots \mathrm{d}I_{r-1} \mathrm{d}\psi_1 \cdots \mathrm{d}\psi_\beta \mathrm{d}h_r = 1 \tag{6.321}$$

Similar to Eq. (5.178), the relationship between the stationary solution of the averaged FPK Eq. (6.300) and the approximate stationary PDF of the generalized displacements and generalized momenta is

$$p(\boldsymbol{q}, \boldsymbol{p}) = \frac{1}{(2\pi)^{r-\beta-1} T(h_r)} p(\boldsymbol{I}', \boldsymbol{\Psi}', h_r) \left| \frac{\partial(\boldsymbol{I}', \boldsymbol{\Psi}', \boldsymbol{\theta}'')}{\partial(\boldsymbol{q}_1, \boldsymbol{p}_1)} \right| \tag{6.322}$$

where $|\partial(\boldsymbol{I}', \boldsymbol{\Psi}', \boldsymbol{\theta}'')/\partial(\boldsymbol{q}_1, \boldsymbol{p}_1)|$ is the determinant of the Jacobian matrix for the transformation from $\boldsymbol{q}_1, \boldsymbol{p}_1$ to $\boldsymbol{I}', \boldsymbol{\Psi}', \boldsymbol{\theta}''$. It is a linear combination of $|\partial(\boldsymbol{I}', \boldsymbol{\theta}')/\partial(\boldsymbol{q}_1, \boldsymbol{p}_1)|$ and affects only the normalization constant.

Example 6.6 Consider a 4-DOF nonlinear system excited by Gaussian and Poisson white noises. The motion equations are (Jia and Zhu 2014b).

$$\begin{aligned} &\ddot{X}_1 + \dot{X}_1[\alpha_{10} + \alpha_{11}\dot{X}_1^2 + \alpha_{12}\dot{X}_2^2 + \alpha_{13}\dot{X}_3^2 \\ &+ \alpha_{14}\dot{X}_4^2 + (\alpha_{13} + \alpha_{14})U(X_3, X_4)] \end{aligned}$$

$$
\begin{aligned}
&+\omega_1^2 X_1 = W_{g1}(t) + W_{p1}(t),\\
&\ddot{X}_2 + \dot{X}_2[\alpha_{20} + \alpha_{21}\dot{X}_2^2 + \alpha_{22}\dot{X}_2^2 + \alpha_{23}\dot{X}_3^2\\
&\quad + \alpha_{24}\dot{X}_4^2 + (\alpha_{23} + \alpha_{24})U(X_3, X_4)]\\
&\quad + \omega_2^2 X_2 = W_{g2}(t) + W_{p2}(t),
\end{aligned}
$$

$$
\begin{aligned}
&\ddot{X}_3 + \dot{X}_3[\alpha_{30} + \alpha_{31}\dot{X}_1^2 + \alpha_{32}\dot{X}_2^2 + \alpha_{33}\dot{X}_3^2\\
&\quad + \alpha_{34}\dot{X}_4^2 + \frac{1}{2}(3\alpha_{33} + \alpha_{34})U(X_3, X_4)]\\
&\quad + \frac{\partial U}{\partial X_3} = W_{g3}(t) + W_{p3}(t),\\
&\ddot{X}_4 + \dot{X}_4[\alpha_{40} + \alpha_{41}\dot{X}_1^2 + \alpha_{42}\dot{X}_2^2 + \alpha_{43}\dot{X}_3^2\\
&\quad + \alpha_{44}\dot{X}_4^2 + \frac{1}{2}(\alpha_{43} + 3\alpha_{44})U(X_3, X_4)]\\
&\quad + \frac{\partial U}{\partial X_4} = W_{g4}(t) + W_{p4}(t),
\end{aligned}
\tag{6.323}
$$

where

$$
\begin{aligned}
U = U(X_3, X_4) &= \frac{1}{2}(\omega_3^2 X_3^2 + \omega_4^2 X_4^2)\\
&\quad + \frac{b}{4}(\omega_3^2 X_3^2 + \omega_4^2 X_4^2)^2,
\end{aligned}
\tag{6.324}
$$

α_{ij}, b are constant; $W_{gi}(t)$ $(i = 1, 2, 3, 4)$ are independent Gaussian white noises with excitation intensities $2\pi K_{ii}$; $W_{pi}(t)$ $(i = 1, 2, 3, 4)$ are zero-mean independent Poisson white noises with Gaussian distributed amplitudes Y_i and variances $\lambda_i E[Y_i^2]$. α_{ij}, K_{ii} and $\lambda_i E[Y_i^2]$ are small quantities of ε^2 order.

Introduce transformation $Q_1 = X_1$, $P_1 = \dot{X}_1$, $Q_2 = X_2$, $P_2 = \dot{X}_2$, $Q_3 = X_3$, $P_3 = \dot{X}_3$, $Q_4 = X_4$, $P_4 = \dot{X}_4$. The SIDEs for system (6.323) are

$$
\begin{aligned}
&\mathrm{d}Q_1 = P\mathrm{d}t,\\
&\mathrm{d}P_1 = \Big\{-\omega_1^2 Q_1 - P_1\Big[\alpha_{10} + \alpha_{11}P_1^2 + \alpha_{12}P_2^2 + \alpha_{13}P_3^2 + \alpha_{14}P_4^2 + (\alpha_{13} + \alpha_{14})U(Q_3, Q_4)\Big]\Big\}\mathrm{d}t\\
&\quad + \sqrt{2\pi K_{11}}\mathrm{d}B_1(t) + \int_{Y_1} Y_1 \mathcal{P}_1(\mathrm{dt}, \mathrm{dY}_1),\\
&\mathrm{d}Q_2 = P_2\mathrm{d}t,\\
&\mathrm{d}P_2 = \Big\{-\omega_2^2 Q_2 - P_2\Big[\alpha_{20} + \alpha_{21}P_1^2 + \alpha_{22}P_2^2 + \alpha_{23}P_3^2 + \alpha_{24}P_4^2 + (\alpha_{23} + \alpha_{24})U(Q_3, Q_4)\Big]\Big\}\mathrm{d}t\\
&\quad + \sqrt{2\pi K_{22}}\mathrm{d}B_2(t) + \int_{\nu_2} Y_2 \mathcal{P}_2(\mathrm{dt}, \mathrm{dY}_2),
\end{aligned}
$$

$$
\mathrm{d}Q_3 = P\mathrm{d}t,
$$

$$
\begin{aligned}
\mathrm{d}P_3 = & \left\{ -\frac{\partial U}{\partial Q_3} - P_3 \left[\alpha_{30} + \alpha_{31}P_1^2 + \alpha_{32}P_2^2 + \alpha_{33}P_3^2 + \alpha_{34}P_4^2 \right.\right. \\
& \left.\left. + \frac{1}{2}(3\alpha_{33} + \alpha_{34})U(Q_3, Q_4) \right] \right\} \mathrm{d}t + \sqrt{2\pi K_{33}} \mathrm{d}B_3(t) + \int_{y_3} Y_3 \mathcal{P}_3(\mathrm{dt}, \mathrm{dY}_3), \\
\mathrm{d}Q_4 = & P_4 \mathrm{d}t, \\
\mathrm{d}P_4 = & \left\{ -\frac{\partial U}{\partial Q_4} - P_4 \left[\alpha_{40} + \alpha_{41}P_1^2 + \alpha_{42}P_2^2 + \alpha_{43}P_3^2 + \alpha_4 P_+^2 \right.\right. \\
& \left.\left. + \frac{1}{2}(\alpha_{43} + 3\alpha_4)U(Q_3, Q_4) \right] \right\} \mathrm{d}t + \sqrt{2\pi K_{44}} \mathrm{d}B_4(t) + \int_{y_4} Y_4 \mathcal{P}_4(\mathrm{dt}, \mathrm{dY}_4),
\end{aligned} \tag{6.325}
$$

The Hamiltonian associated with Eq. (6.325) is

$$
H = H_1 + H_2 + H_3 = \omega_1 I_1 + \omega_2 I_2 + H_3 \tag{6.326}
$$

where

$$
I_1 = \frac{P_1^2 + \omega_1^2 Q_1^2}{2\omega_1}; \quad I_2 = \frac{P_2^2 + \omega_2^2 Q_2^2}{2\omega_2}; \quad H_3 = \frac{P_3^2 + P_4^2}{2} + U(Q_3, Q_4), \tag{6.327}
$$

$$
U(Q_3, Q_4) = \frac{1}{2}(\omega_3^2 Q_3^2 + \omega_4^2 Q_4^2) + \frac{b}{4}(\omega_3^2 Q_3^2 + \omega_4^2 Q_4^2)^2 \tag{6.328}
$$

Noting that $U(Q_3, Q_4)$ is inseparable. It is seen from Eqs. (6.325) to (6.328) that the Hamiltonian system associated with Eq. (6.325) is partially integrable. The SIDEs for I_1, I_2, H_3 can be derived from Eq. (6.325) by using the stochastic jump-diffusion chain rule

$$
\begin{aligned}
\mathrm{d}I_1 = & \left\{ -\left[\alpha_{10} + \alpha_{11}P_1^2 + \alpha_{12}P_2^2 + \alpha_{13}P_3^2 + \alpha_{14}P_4^2 \right.\right. \\
& \left.\left. + (\alpha_{13} + \alpha_{14})U(Q_3, Q_4) \right] \frac{P_1^2}{\omega_1} + \frac{\pi K_{11}}{\omega_1} \right\} \mathrm{d}t \\
& + \frac{\sqrt{2\pi K_{11}}}{\omega_1} P_1 \, \mathrm{d}B_1(t) + \frac{1}{2\omega_1} \int_{\mathcal{Y}_1} (Y_1^2 + 2P_1 Y_1) \mathcal{P}_1(\mathrm{d}t, \mathrm{d}Y_1), \\
\mathrm{d}I_2 = & \left\{ -\left[\alpha_{20} + \alpha_{21}P_1^2 + \alpha_{22}P_2^2 + \alpha_{23}P_3^2 + \alpha_{24}P_4^2 \right.\right. \\
& \left.\left. + (\alpha_{23} + \alpha_{24})U(Q_3, Q_4) \right] \frac{P_2^2}{\omega_2} + \frac{\pi K_{22}}{\omega_2} \right\} \mathrm{d}t \\
& + \frac{\sqrt{2\pi K_{22}}}{\omega_2} P_2 \, \mathrm{d}B_2(t) + \frac{1}{2\omega_2} \int_{\mathcal{Y}_2} (Y_2^2 + 2P_2 Y_2) \mathcal{P}_2(\mathrm{d}t, \mathrm{d}Y_2),
\end{aligned}
$$

$$
\begin{aligned}
\mathrm{d}H_3 = & \Big\{-\Big[\alpha_{30} + \alpha_{31}P_1^2 + \alpha_{32}P_2^2 + \alpha_{33}P_3^2 + \alpha_{34}P_4^2 \\
& + \frac{1}{2}(\alpha_{34} + 3\alpha_{33})U(Q_3, Q_4)\Big]P_3^2 + \pi K_{33} \\
& - \Big[\alpha_{40} + \alpha_{41}P_1^2 + \alpha_{42}P_2^2 + \alpha_{43}P_3^2 + \alpha_{44}P_4^2 \\
& + \frac{1}{2}(\alpha_{43} + 3\alpha_{44})U(Q_3, Q_4)\Big]P_4^2 + \pi K_{44}\Big\}\mathrm{d}t \\
& + \sqrt{2\pi K_{33}}P_3\,\mathrm{d}B_3(t) + \sqrt{2\pi K_{44}}P_4\mathrm{d}B_4(t) \\
& + \frac{1}{2}\int_{\mathcal{Y}_3}(Y_3^2 + 2P_3Y_3)\mathcal{P}_3(\mathrm{d}t, \mathrm{d}Y_3) \\
& + \frac{1}{2}\int_{\mathcal{Y}_4}(Y_4^2 + 2P_4Y_4)\mathcal{P}_4(\mathrm{d}t, \mathrm{d}Y_4).
\end{aligned}
\tag{6.329}
$$

The form and dimension of the averaged SIDEs depend on whether system (6.275) has internal resonance.

(i) **Non-internal resonant case**

In this case, the equations of the new system consist of Eq. (6.329) and the equations for Q_1, Q_2, Q_3, Q_4 and P_4 in Eq. (6.325). In these equations, P_1, P_2, P_3 are replaced by I_1, I_2, H_3 via Eq. (6.327). In the new system, I_1, I_2, H_3 are slowly-varying processes while Q_1, Q_2, Q_3, Q_4, P_4 are fast-varying processes. Applying the stochastic averaging method described in Sect. 6.4.1 to the new system yields following reduced averaged FPK equation

$$
\begin{aligned}
0 = & -\frac{\partial}{\partial I_1}(\overline{a}_1 p) - \frac{\partial}{\partial I_2}(\overline{a}_2 p) - \frac{\partial}{\partial h_3}(\overline{a}_3 p) + \frac{1}{2}\frac{\partial^2}{\partial I_1^2}(\overline{a}_{1,1}p) \\
& + \frac{1}{2}\frac{\partial^2}{\partial I_2^2}(\overline{a}_{2,2}p) + \frac{1}{2}\frac{\partial^2}{\partial h_3^2}(\overline{a}_{3,3}p) \\
& - \frac{1}{3!}\frac{\partial^3}{\partial I_1^3}(\overline{a}_{1,1,1}p) - \frac{1}{3!}\frac{\partial^3}{\partial I_2^3}(\overline{a}_{2,2,2}p) - \frac{1}{3!}\frac{\partial^3}{\partial h_3^3}(\overline{a}_{3,3,3}p) \\
& + \frac{1}{4!}\frac{\partial^4}{\partial I_1^4}(\overline{a}_{1,1,1,1}p) + \frac{1}{4!}\frac{\partial^4}{\partial I_1^4}(\overline{a}_{2,2,2,2}p) \\
& + \frac{1}{4!}\frac{\partial^4}{\partial h_3^4}(\overline{a}_{3,3,3,3}p),
\end{aligned}
\tag{6.330}
$$

where derivative moments are

$$
\begin{aligned}
\overline{a}_1 & = \overline{a}_1(I_1, I_2, h_3) \\
& = -\Big[\alpha_{10}I_1 + \frac{3}{2}\omega_1\alpha_{11}I_1^2 + \omega_2\alpha_{12}I_1I_2 + (\alpha_{13} + \alpha_{14})I_1h_3\Big]
\end{aligned}
$$

$$+\frac{2\pi K_{11}+\lambda_1 E\left[Y_1^2\right]}{2\omega_1}, \tag{6.331}$$

$$\begin{aligned}\overline{a}_2 &= \overline{a}_2(I_1, I_2, h_3)\\ &= -\left[\alpha_{20}I_2+\frac{3}{2}\omega_2\alpha_{22}I_2^2+\omega_1\alpha_{21}I_1I_2+(\alpha_{23}+\alpha_{24})I_2h_3\right]\\ &+\frac{2\pi K_{22}+\lambda_2 E\left[Y_2^2\right]}{2\omega_2},\end{aligned} \tag{6.332}$$

$$\begin{aligned}\overline{a}_3 &= \overline{a}_3(I_1, I_2, h_3)\\ &= -[(\alpha_{30}+\alpha_{40})+(\alpha_{31}+\alpha_{41})\omega_1 I_1+(\alpha_{32}+\alpha_{42})\omega_2 I_2\\ &+\frac{1}{2}(3\alpha_{33}+3\alpha_{44}+\alpha_{34}+\alpha_{43})I_1h_3]\times S_1(H_3)\\ &+\frac{2\pi K_{33}+\lambda_3E[Y_3^2]+2\pi K_{44}+\lambda_4E[Y_4^2]}{2},\end{aligned} \tag{6.333}$$

$$\overline{a}_{1,1}=\overline{a}_{1,1}(I_1, I_2, h_3)=\frac{2\pi K_{11}+\lambda_1E[Y_1^2]}{\omega_1}I_1+\frac{\lambda_1E[Y_1^4]}{4\omega_1^2}, \tag{6.334}$$

$$\overline{a}_{2,2}=\overline{a}_{2,2}(I_1, I_2, h_3)=\frac{2\pi K_{22}+\lambda_2E[Y_2^2]}{\omega_2}I_2+\frac{\lambda_2E[Y_2^4]}{4\omega_2^2}, \tag{6.335}$$

$$\begin{aligned}\overline{a}_{3,3} &= \overline{a}_{3,3}(I_1, I_2, h_3)\\ &= (2\pi K_{33}+\lambda_3E[Y_3^2]+2\pi K_{44}+\lambda_4E[Y_4^2])S_1(h_3)\\ &+\frac{\lambda_3E[Y_3^4]+\lambda_4E[Y_4^4]}{4},\end{aligned} \tag{6.336}$$

$$\overline{a}_{1,1,1}=\overline{a}_{1,1,1}(I_1, I_2, h_3)=\frac{3}{2}\frac{\lambda_1E[Y_1^4]}{\omega_1^2}I_1, \tag{6.337}$$

$$\overline{a}_{2,2,2}=\overline{a}_{2,2,2}(I_1, I_2, h_3)=\frac{3}{2}\frac{\lambda_2E[Y_2^4]}{\omega_2^2}I_2, \tag{6.338}$$

$$\overline{a}_{3,3,3}=\overline{a}_{3,3,3}(I_1, I_2, h_3)=\frac{3}{2}(\lambda_3E[Y_3^4]+\lambda_4E[Y_4^4])S_1(h_3), \tag{6.339}$$

$$\overline{a}_{1,1,1,1}=\overline{a}_{1,1,1,1}(I_1, I_2, h_3)=\frac{3}{2}\frac{\lambda_1E[Y_1^4]}{\omega_1^2}I_1^2, \tag{6.340}$$

$$\overline{a}_{2,2,2,2}=\overline{a}_{2,2,2,2}(I_1, I_2, h_3)=\frac{3}{2}\frac{\lambda_2E[Y_2^4]}{\omega_2^2}I_2^2, \tag{6.341}$$

$$\begin{aligned}\overline{a}_{3,3,3,3} &= \overline{a}_{3,3,3,3}(I_1, I_2, h_3)\\ &= (\lambda_3 E[Y_3^4] + \lambda_4 E[Y_4^4])\frac{32b^2h_3^2 + 9bh_3 + 1 - (7bh_3 + 1)\sqrt{1 + 4bh_3}}{40b^2},\end{aligned} \tag{6.342}$$

in which

$$S_1(h_3) = \frac{1 + 8bh_3 - \sqrt{1 + 4bh_3}}{12b} \tag{6.343}$$

The stationary solution $p(I_1, I_2, h_3)$ of the reduced averaged FPK Eq. (6.330) can be obtained by using the finite difference method and the successive over-relaxation iteration. The joint stationary PDF of generalized displacements and generalized momenta can be obtained from $p(I_1, I_2, h_3)$ according to Eq. (6.275) as follows

$$p(q_1, q_2, q_3, q_4, p_1, p_2, p_3, p_4) = \left.\frac{p(I_1, I_2, h_3)}{\sqrt{1 - 4h_3 b} - 1}\right|_{I_i = I_i(q_i, p_i), h_3 = H_3(q_3, p_3, q_4, p_4)} \tag{6.344}$$

The other statistics of the system can be obtained by integrating $p(q_1, q_2, q_3, q_4, p_1, p_2, p_3, p_4)$ or $p(I_1, I_2, h_3)$.

The stationary PDFs $p(I_1, I_2)$, $p(I_1, h_3)$ and $p(I_2, h_3)$ for system (6.323) in non-resonant case obtained from solving the reduced FPK Eq. (6.330) and Monte Carlo simulations are shown in Figs. 6.20–6.22. The stationary PDFs $p(q_1, p_1)$, $p(p_1)$ and $p(q_1)$ for system (6.323) in non-resonant case calculated by using Eq. (6.344) and from Monte Carlo simulation are shown in Figs. 6.23–6.25. In all cases, good agreement between the results from the stochastic averaging method and those from Monte Carlo simulation can be seen from Figs. 6.20–6.25.

(ii) **Primary internal resonant case**

Assume $\omega_1 \approx \omega_2$. Let

$$\Psi = \Theta_1 - \Theta_2 \tag{6.345}$$

where

$$\Theta_i = -\arctan\left(\frac{P_i}{\omega_i Q_i}\right);\ i = 1, 2 \tag{6.346}$$

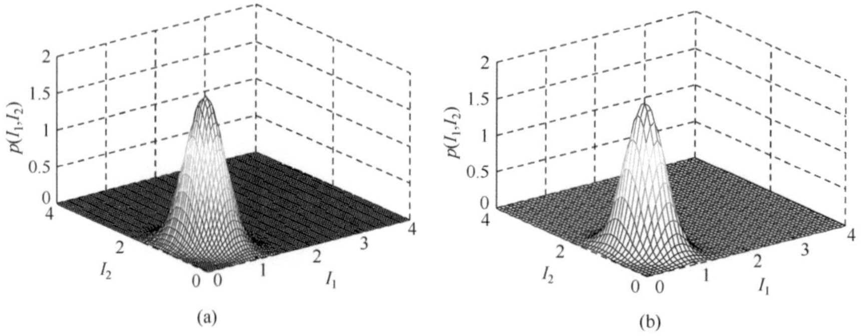

Fig. 6.20 $p(I_1, I_2)$ of system (6.323) in non-resonant case. **a** Result from Monte Carlo simulation; **b** results from the stochastic averaging method. System parameters are $\omega_1 = 1.0$, $\omega_2 = 1.414$, $\omega_3 = 1.1$, $\omega_4 = 1.732$; $\alpha_{10} = \alpha_{20} = \alpha_{30} = \alpha_{40} = -0.1$; $\alpha_{11} = \alpha_{22} = \alpha_{33} = 0.04$; $\alpha_{44} = \alpha_{34} = \alpha_{43} = 0.04$; $\alpha_{12} = \alpha_{21} = \alpha_{31} = 0.02$; $\alpha_{41} = \alpha_{32} = \alpha_{42} = 0.02$; $\alpha_{13} = \alpha_{14} = 0.01$; $\alpha_{23} = \alpha_{24} = 0.01$; $2\pi K_{11} = 2\pi K_{22} = 0.008$; $2\pi K_{33} = 2\pi K_{44} = 0.008$; $\lambda_1 = \lambda_2 = 2.0$; $\lambda_3 = \lambda_4 = 2.0$; $E[Y_1^2] = E[Y_2^2] = 0.002$; $E[Y_3^2] = E[Y_4^2] = 0.002$ (Jia and Zhu 2014b)

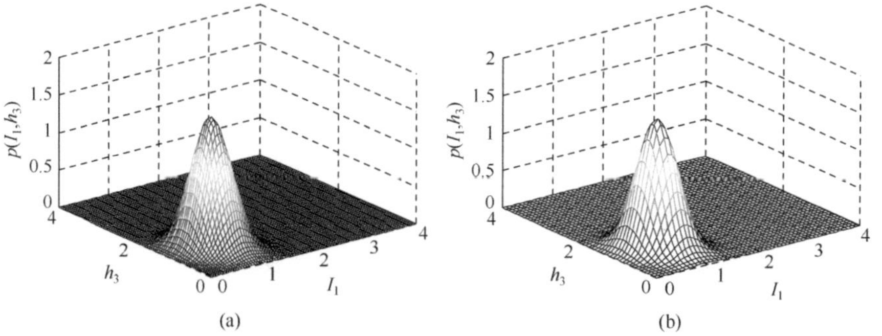

Fig. 6.21 $p(I_1, h_3)$ of system (6.323) in non-resonant case. **a** Result from Monte Carlo simulation; **b** results from the stochastic averaging method. System parameters are the same as those in Fig. 6.22 (Jia and Zhu 2014b) (Fig. 6.21)

represent the angle variables of the first two oscillators in system (6.322). Applying transformations (6.345) and (6.346) and then using the stochastic jump-diffusion chain rule result in the SIDE for Ψ as follows

$$
\begin{aligned}
\mathrm{d}\Psi = \Big\{ & O(\varepsilon^2) + \Big[\alpha_{10} + \alpha_{11}P_1^2 + \alpha_{12}P_2^2 + \alpha_{13}P_3^2 + \alpha_{14}P_4^2 + (\alpha_{13} + \alpha_{14})U(Q_3, Q_4)\Big] \\
& - \frac{\omega_1 Q_1 P_1}{P_1^2 + \omega_1^2 Q_1^2}\Big[\alpha_{20} + \alpha_{21}P_1^2 + \alpha_{22}P_2^2 + \alpha_{23}P_3^2 + \alpha_{24}P_4^2 + (\alpha_{23} + \alpha_{24})U(Q_3, Q_4)\Big] \\
& \frac{\omega_2 Q_2 P_2}{P_2^2 + \omega_2^2 Q_2^2} + 2\pi K_{11}\frac{\omega_1 Q_1 P_1}{(P_1^2 + \omega_1^2 Q_1^2)^2} - 2\pi K_{22}\frac{\omega_2 Q_2 P_2}{(P_2^2 + \omega_2^2 Q_2^2)^2}\Big\}\mathrm{d}t
\end{aligned}
$$

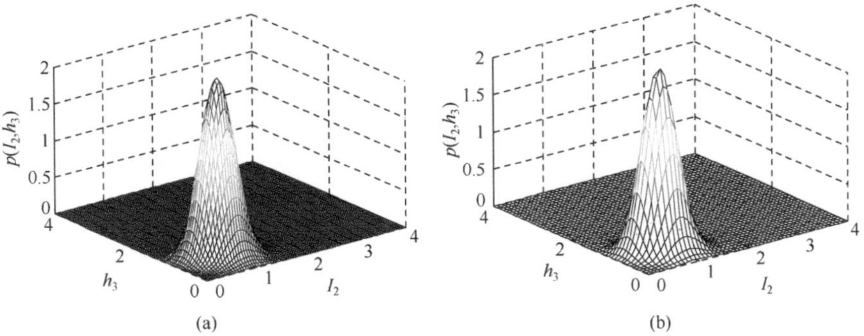

Fig. 6.22 $p(I_2, h_3)$ of system (6.323) in non-resonant case. **a** Result from Monte Carlo simulation; **b** results from the stochastic averaging method. System parameters are the same as those in Fig. 6.22 (Jia and Zhu 2014b)

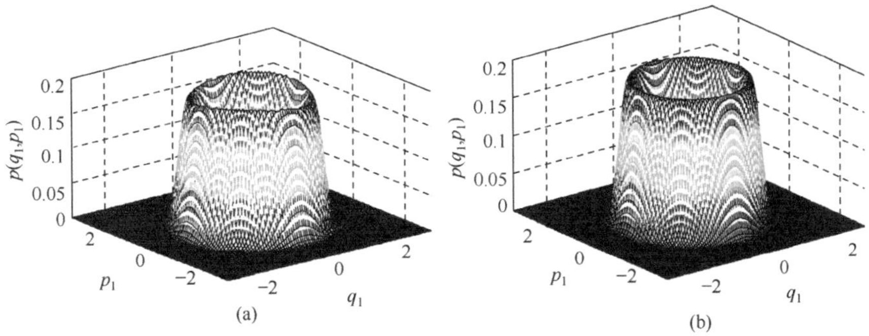

Fig. 6.23 $p(q_1, p_1)$ of system (6.323) in non-resonant case. **a** Result from Monte Carlo simulation; **b** results from the stochastic averaging method. System parameters are the same as those in Fig. 6.22 (Jia and Zhu 2014b)

Fig. 6.24 $p(p_1)$ of system (6.323) in non-resonant case. ——, result from the stochastic averaging method; ●, result from Monte Carlo simulation. System parameters are the same as those in Fig. 6.22 (Jia and Zhu 2014b) (Fig. 6.24)

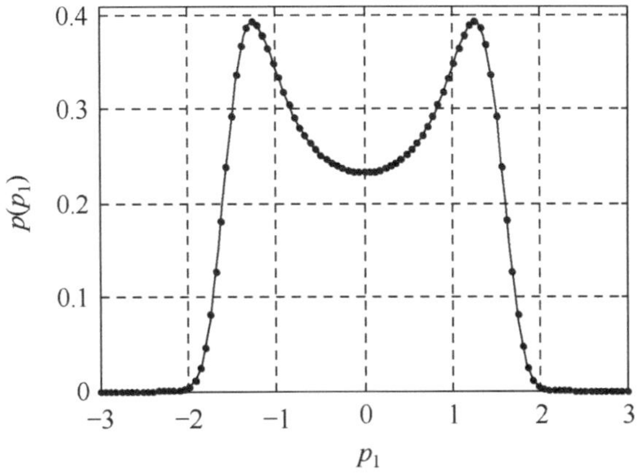

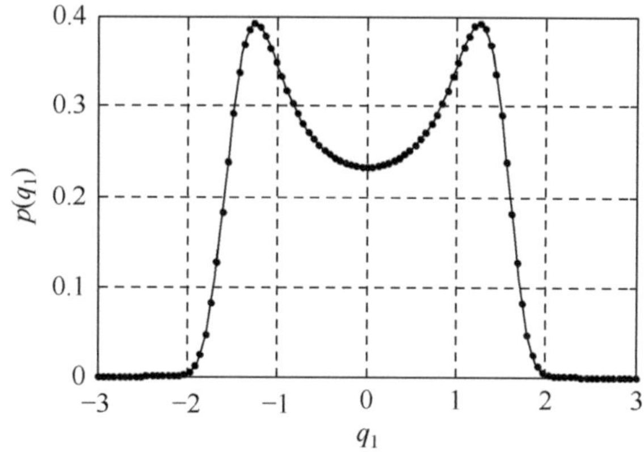

Fig. 6.25 $p(q_1)$ of system (6.323) in non-resonant case. ——, result from the stochastic averaging method; ●, result from Monte Carlo simulation. System parameters are the same as those in Fig. 6.22 (Jia and Zhu 2014b)

$$
\begin{aligned}
&-\sqrt{2\pi K_{11}}\frac{\omega_1 Q_1}{P_1^2+\omega_1^2 Q_1^2}\mathrm{d}B_1(t)+\sqrt{2\pi K_{22}}\frac{\omega_2 Q_2}{P_2^2+\omega_2^2 Q_2^2}\mathrm{d}B_2(t)\\
&-\int\limits_{\mathcal{Y}_1}\left[\frac{\omega_1 Q_1}{P_1^2+\omega_1^2 Q_1^2}Y_1-\frac{\omega_1 Q_1 P_1}{\left(P_1^2+\omega_1^2 Q_1^2\right)^2}Y_1^2+\frac{1}{3}\frac{\omega_1 Q_1\left(3P_1^2-\omega_1^2 Q_1^2\right)}{\left(P_1^2+\omega_1^2 Q_1^2\right)^3}Y_1^3\right.\\
&\quad\left.-\frac{\omega_1 Q_1 P_1\left(P_1^2-\omega_1^2 Q_1^2\right)}{\left(P_1^2+\omega_1^2 Q_1^2\right)^4}Y_1^4\right]\mathcal{P}_1(\mathrm{dt},\mathrm{dY}_1)\\
&\quad+\int\limits_{y_2}\left[\frac{\omega_2 Q_2}{P_2^2+\omega_2^2 Q_2^2}Y_2-\frac{\omega_2 Q_2 P_2}{\left(P_2^2+\omega_2^2 Q_2^2\right)^2}Y_2^2+\frac{1}{3}\frac{\omega_2 Q_2\left(3P_2^2-\omega_2^2 Q_2^2\right)}{\left(P_2^2+\omega_2^2 Q_2^2\right)^3}Y_2^3\right.\\
&\quad\left.-\frac{\omega_2 Q_2 P_2\left(P_2^2-\omega_2^2 Q_2^2\right)}{\left(P_2^2+\omega_2^2 Q_2^2\right)^4}Y_2^4\right]\mathcal{P}_2(\mathrm{dt},\mathrm{dY}_2).
\end{aligned}
\tag{6.347}
$$

Applying the stochastic averaging method described in Sect. 6.4.2 to Eqs. (6.329) and (6.347), with a truncation parameter $u = 4$, yields the following truncated averaged FPK equation

$$
\begin{aligned}
0=&-\frac{\partial}{\partial I_1}(\overline{a}_1 p)-\frac{\partial}{\partial I_2}(\overline{a}_2 p)-\frac{\partial}{\partial\psi}(\overline{a}_4 p)-\frac{\partial}{\partial h_3}(\overline{a}_3 p)+\frac{1}{2}\frac{\partial^2}{\partial I_1^2}(\overline{a}_{1,1}p)\\
&+\frac{1}{2}\frac{\partial^2}{\partial I_2^2}(\overline{a}_{2,2}p)+\frac{1}{2}\frac{\partial^2}{\partial\psi^2}\left(\overline{a}_{4,4}p\right)+\frac{1}{2}\frac{\partial^2}{\partial h_3^2}(\overline{a}_{3,3}p)-\frac{1}{3!}\frac{\partial^3}{\partial I_1^3}(\overline{a}_{1,1,1}p)\\
&-\frac{1}{3!}\frac{\partial^3}{\partial I_2^3}(\overline{a}_{2,2,2}p)-\frac{1}{3!}\frac{\partial^3}{\partial\psi^3}(\overline{a}_{4,4,4}p)-\frac{1}{3!}\frac{\partial^3}{\partial h_3^3}(\overline{a}_{3,3,3}p)+\frac{1}{4!}\frac{\partial^4}{\partial I_1^4}(\overline{a}_{1,1,1,1}p)\\
&+\frac{1}{4!}\frac{\partial^4}{\partial I_1^4}(\overline{a}_{2,2,2,2}p)+\frac{1}{4!}\frac{\partial^4}{\partial\psi^4}(\overline{a}_{4,4,4,4}p)+\frac{1}{4!}\frac{\partial^4}{\partial h_3^4}(\overline{a}_{3,3,3,3}p)
\end{aligned}
$$

$$+\frac{C_4^2}{4!}\frac{\partial^4}{\partial I_1^2\partial\psi^2}(\overline{a}_{1,1,4,4}p)+\frac{C_4^2}{4!}\frac{\partial^4}{\partial I_2^2\partial\psi^2}(\overline{a}_{2,2,4,4}p), \tag{6.348}$$

where derivative moments are

$$p=p(I_1,I_2,\psi,h_3),$$

$$\overline{a}_1(I_1,I_2,\psi,h_3)=-\left[\alpha_{10}I_1+\frac{3}{2}\alpha_{11}\omega_1I_1^2+\alpha_{12}\omega_2I_1I_2\left(\frac{1}{2}\cos 2\psi+1\right)\right.$$
$$\left.+(\alpha_{13}+\alpha_{14})I_1h_3\right]+\frac{2\pi K_{11}+\lambda_1E[Y_1^2]}{2\omega_1}, \tag{6.349}$$

$$\overline{a}_2(I_1,I_2,\psi,h_3)=-\left[\alpha_{20}I_2+\frac{3}{2}\alpha_{22}\omega_2I_2^2+\alpha_{21}\omega_1I_1I_2\left(\frac{1}{2}\cos 2\psi+1\right)\right.$$
$$\left.+(\alpha_{23}+\alpha_{24})I_2h_3\right]+\frac{2\pi K_{22}+\lambda_2E[Y_2^2]}{2\omega_2}, \tag{6.350}$$

$$\overline{a}_3(I_1,I_2,\psi,h_3)=-\left[(\alpha_{30}+\alpha_{40})+(\alpha_{31}+\alpha_{41})\omega_1I_1+(\alpha_{32}+\alpha_{42})\omega_2I_2\right.$$
$$\left.+\frac{1}{2}(3\alpha_{33}+3\alpha_{44}+\alpha_{34}+\alpha_{43})I_1h_3\right]S_1(h_3)$$
$$+\frac{2\pi K_{33}+\lambda_3E[Y_3^2]+2\pi K_{44}+\lambda_4E[Y_4^2]}{2}, \tag{6.351}$$

$$\overline{a}_4(I_1,I_2,\psi,h_3)=\frac{1}{4}(\alpha_{12}\omega_2I_2+\alpha_{21}\omega_1I_1)\sin 2\psi, \tag{6.352}$$

$$\overline{a}_{1,1}(I_1,I_2,\psi,h_3)=\frac{2\pi K_{11}+\lambda_1E[Y_1^2]}{\omega_1}I_1+\frac{1}{4}\frac{\lambda_1E[Y_1^4]}{\omega_1^2}, \tag{6.353}$$

$$\overline{a}_{2,2}(I_1,I_2,\psi,h_3)=\frac{2\pi K_{22}+\lambda_2E[Y_2^2]}{\omega_2}I_2+\frac{1}{4}\frac{\lambda_2E[Y_2^4]}{\omega_2^2}, \tag{6.354}$$

$$\overline{a}_{3,3}(I_1,I_2,\psi,h_3)=\left(2\pi K_{33}+\lambda_3E[Y_3^2]+2\pi K_{44}+\lambda_4E[Y_4^2]\right)S_1(h_3)$$
$$+\frac{1}{4}\left(\lambda_3E[Y_3^4]+\lambda_4E[Y_4^4]\right), \tag{6.355}$$

$$\overline{a}_{4,4}(I_1,I_2,\psi,h_3)=\frac{2\pi K_{11}+\lambda_1E[Y_1^2]}{4\omega_1I_1}+\frac{2\pi K_{22}+\lambda_2E[Y_2^2]}{4\omega_2I_2}$$
$$+\frac{\lambda_1E[Y_1^4]}{32\omega_1^2I_1^2}+\frac{\lambda_2E[Y_2^4]}{32\omega_2^2I_2^2}, \tag{6.356}$$

$$\overline{a}_{1,1,1}(I_1,I_2,\psi,h_3)=\frac{3}{2}\frac{\lambda_1E[Y_1^4]}{\omega_1^2}I_1 \tag{6.357}$$

$$\overline{a}_{2,2,2}(I_1, I_2, \psi, h_3) = \frac{3}{2}\frac{\lambda_2 E[Y_2^4]}{\omega_2^2} I_2 \tag{6.358}$$

$$\overline{a}_{3,3,3}(I_1, I_2, \psi, h_3) = \frac{3}{2}\left(\lambda_3 E[Y_3^4] + \lambda_4 E[Y_4^4]\right) S_1(h_3) \tag{6.359}$$

$$\overline{a}_{4,4,4}(I_1, I_2, \psi, h_3) = 0 \tag{6.360}$$

$$\overline{a}_{1,1,1,1}(I_1, I_2, \psi, h_3) = \frac{3}{2}\frac{\lambda_1 E[Y_1^4]}{\omega_1^2} I_1^2 \tag{6.361}$$

$$\overline{a}_{2,2,2,2}(I_1, I_2, \psi, h_3) = \frac{3}{2}\frac{\lambda_2 E[Y_2^4]}{\omega_2^2} I_2^2 \tag{6.362}$$

$$\begin{aligned}\overline{a}_{3,3,3,3}(I_1, I_2, \psi, h_3) = {} & \frac{32b^2h_3^2 + 9bh_3 + 1 - (7bh_3 + 1)\sqrt{1 + 4bh_3}}{40b^2} \\ & \times \left(\lambda_3 E[Y_3^4] + \lambda_4 E[Y_4^4]\right),\end{aligned} \tag{6.363}$$

$$\overline{a}_{4,4,4,4}(I_1, I_2, \psi, h_3) = \frac{3}{32}\frac{\lambda_1 E[Y_1^4]}{\omega_1^2 I_1^2} + \frac{3}{32}\frac{\lambda_2 E[Y_2^4]}{\omega_2^2 I_2^2} \tag{6.364}$$

$$\overline{a}_{1,1,4,4}(I_1, I_2, \psi, h_3) = \frac{1}{8}\frac{\lambda_1 E[Y_1^4]}{\omega_1^2} \tag{6.365}$$

$$\overline{a}_{2,2,4,4}(I_1, I_2, \psi, h_3) = \frac{1}{8}\frac{\lambda_2 E[Y_2^4]}{\omega_2^2} \tag{6.366}$$

in which

$$S_1(h_3) = \frac{1 + 8bh_3 - \sqrt{1 + 4bh_3}}{12b}. \tag{6.367}$$

Solving the reduced averaged FPK Eq. (6.348) by using the finite difference method yields the stationary PDF $p(I_1, I_2, \psi, h_3)$ of system (6.323). The joint stationary PDF of generalized displacements and generalized momenta can be obtained according to Eq. (6.322) as follows

$$\begin{aligned} & p(q_1, q_2, q_3, q_4, p_1, p_2, p_3, p_4) \\ & = \left.\frac{p(I_1, I_2, \psi, h_3)}{\sqrt{1 - 4h_3 b} - 1}\right|_{I_i = I_i(q_i, p_i), h_3 = H_3(q_3, p_3, q_4, p_4), \psi = \psi(q_1, q_2, p_1, p_2)} \end{aligned} \tag{6.368}$$

The stationary PDFs $p(I_1, I_2)$, $p(I_1, h_3)$, $p(I_2, h_3)$, $p(\psi)$, $p(I_1)$, $p(I_2)$, $p(h_3)$ for system (6.323) in primary resonant case obtained by using the stochastic averaging method and from Monte Carlo simulations are shown in Figs. 6.26–6.32. It is seen

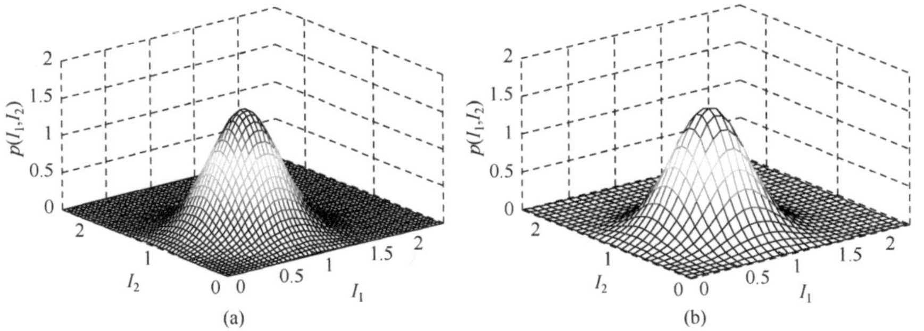

Fig. 6.26 $p(I_1, I_2)$ of system (6.323) in primary resonant case. **a** Result from Monte Carlo simulation; **b** results from the stochastic averaging method. System parameters are $\omega_1 = 1.0,\ \omega_2 = 1.0$, $\omega_3 = 1.11,\ \omega_4 = 1.732$, $\alpha_{10} = \alpha_{20} = \alpha_{30} = \alpha_{40} = -0.1$, $\alpha_{11} = \alpha_{22} = \alpha_{33} = \alpha_{44} = \alpha_{34} = \alpha_{43} = 0.04$, $\alpha_{12} = \alpha_{21} = \alpha_{31} = \alpha_{41} = \alpha_{32} = \alpha_{42} = 0.02$, $\alpha_{13} = \alpha_{14} = \alpha_{23} = \alpha_{24} = 0.01$, $2\pi K_{11} = 2\pi K_{22} = 2\pi K_{33} = 2\pi K_{44} = 0.008$, $\lambda_1 = \lambda_2 = \lambda_3 = \lambda_4$=2.0, $E[Y_1^2] = E[Y_2^2] = E[Y_3^2] = E[Y_4^2] = 0.002$ (Jia and Zhu 2014b)

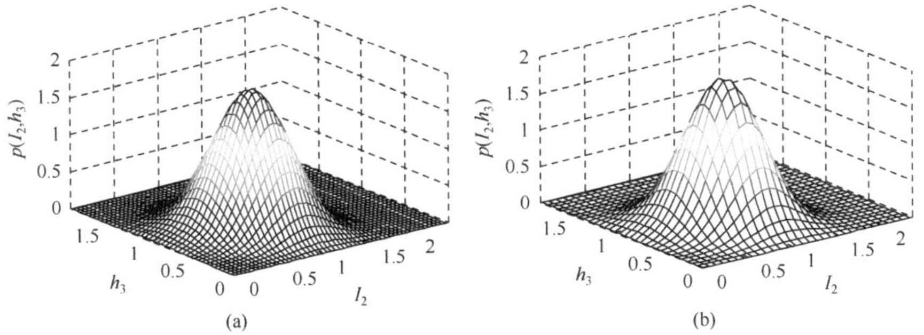

Fig. 6.27 $p(I_2, h_3)$ of system (6.323) in primary resonant case. **a** Result from Monte Carlo simulation; **b** results from the stochastic averaging method. System parameters are the same as those in Fig. 6.26 (Jia and Zhu 2014b)

that the results obtained using the stochastic averaging method and those from Monte Carlo simulation are in good agreement.

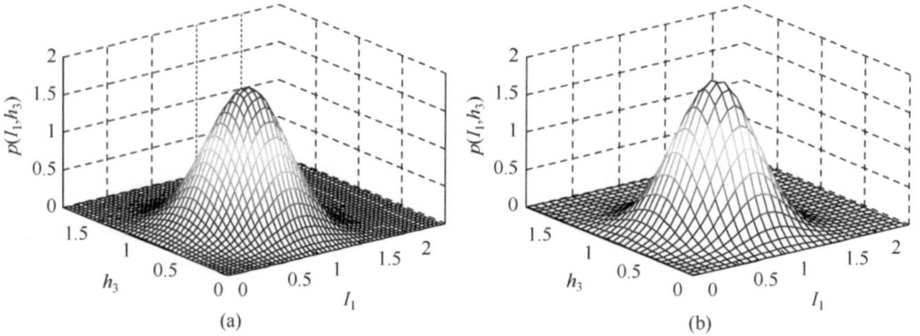

Fig. 6.28 $p(I_1, h_3)$ of system (6.323) in primary resonant case. **a** Result from Monte Carlo simulation; **b** results from the stochastic averaging method. System parameters are the same as those in Fig. 6.26 (Jia and Zhu 2014b)

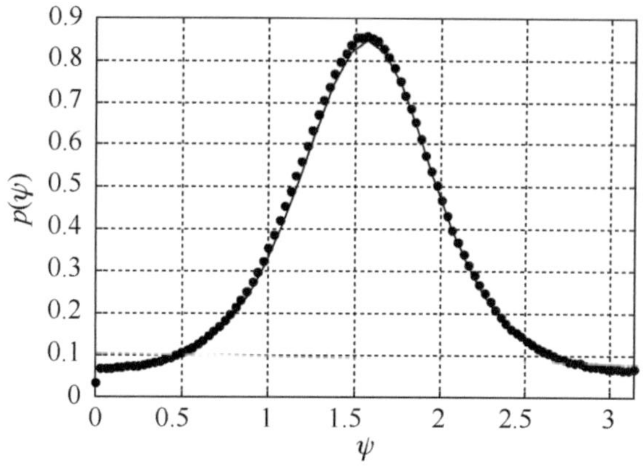

Fig. 6.29 $p(\psi)$ of system (6.323) in primary resonant case. ——, result from the stochastic averaging method; ●, result from Monte Carlo simulation. System parameters are the same as those in Fig. 6.26 (Jia and Zhu 2014b)

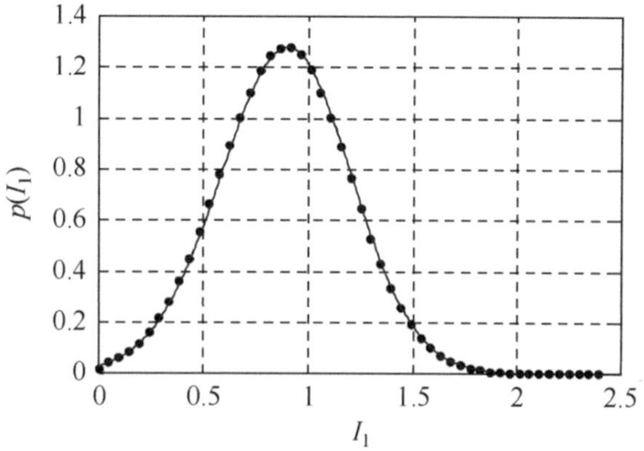

Fig. 6.30 $p(I_1)$ of system (6.323) in primary resonant case. ——, result from the stochastic averaging method; ●, result from Monte Carlo simulation. System parameters are the same as those in Fig. 6.26 (Jia and Zhu 2014b)

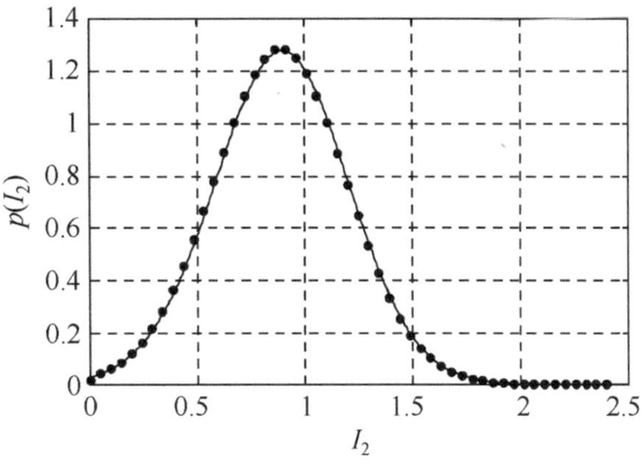

Fig. 6.31 $p(I_2)$ of system (6.323) in primary resonant case. ——, result from the stochastic averaging method; ●, result from Monte Carlo simulation. System parameters are the same as those in Fig. 6.26 (Jia and Zhu 2014b)

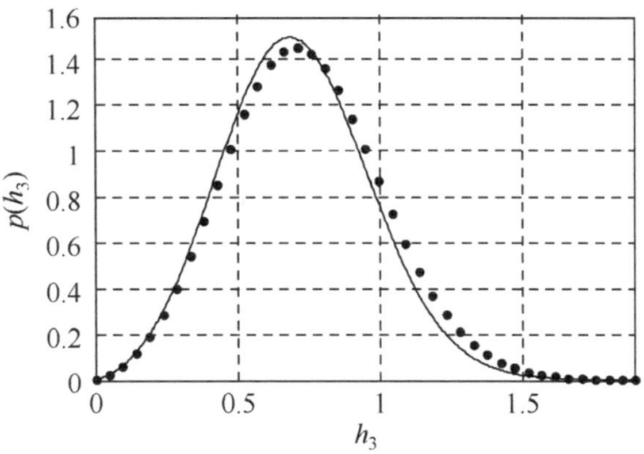

Fig. 6.32 $p(h_3)$ of system (6.323) in primary resonant case. ——, result from the stochastic averaging method; ●, result from Monte Carlo simulation. System parameters are the same as those in Fig. 6.26 (Jia and Zhu 2014b)

References

Binney J J, Dowrick N J, Fisher A J, Newman M. 1992. The Theory of Critical Phenomena: An Introduction to the Renormalization Group. New York: Oxford University Press, Inc.

Di Paola M, Falsone G. 1993a. Itô and Stratonovich integrals for delta-correlated processes. Probabilistic engineering mechanics, 8(3): 197–208.

Di Paola M, Falsone G. 1993b. Stochastic dynamics of nonlinear systems driven by non-normal delta-correlated processes. Journal of Applied Mechanics, 60(1): 141–148.

Hanson F B. 2007. Applied stochastic processes and control for Jump-diffusions: modeling, analysis, and computation. Philadelphia: SIAM.

Jia W T. 2014. Stochastic Averaging Methods for Quasi-Hamiltonian Systems under Combined Gaussian and Poisson White Noise Excitations. Ph.D. Dissertation. Xi'an: Northwestern Polytechnical University. (in Chinese).

Jia W T, Zhu W Q. 2014a. Stochastic averaging of quasi-integrable and non-resonant Hamiltonian systems under combined Gaussian and Poisson white noise excitations. Nonlinear Dynamics, 76(2): 1271–1289.

Jia W T, Zhu W Q. 2014b. Stochastic averaging of quasi-partially integrable Hamiltonian systems under combined Gaussian and Poisson white noise excitations. Physica A: Statistical Mechanics and its Applications, 398: 125–144.

Jia W T, Zhu W Q, Xu Y. 2013. Stochastic averaging of quasi-non-integrable Hamiltonian systems under combined Gaussian and Poisson white noise excitations. International Journal of Non-Linear Mechanics, 51: 45–53.

Jia W T, Zhu W Q, Xu Y, Liu W Y. 2014. Stochastic averaging of quasi-integrable and resonant Hamiltonian systems under combined Gaussian and Poisson white noise excitations. ASME Journal of Applied Mechanics, 81(4): 041009.

Khasminskii R Z. 1968. On the averaging principle for Itô stochastic differential equations. Kibernetika, 3(4): 260–279. (in Russian)

Xu Y, Duan J Q, Xu W. 2011. An averaging principle for stochastic dynamical systems with Lévy noise. Physica D-Nonlinear Phenomena, 240(17): 1395–1401.

Zeng Y. 2010. Stochastic Averaging Methods for Nonlinear Systems Subject to Non-Gaussian Random Excitation. Ph.D. Dissertation. Hangzhou: Zhejiang University. (in Chinese).

Zeng Y, Zhu W Q. 2011. Stochastic averaging of quasi-nonintegrable-Hamiltonian systems under Poisson white noise excitation. ASME Journal of Applied Mechanics, 78: 021002.

Zhu W Q, Huang Z L, Suzuki Y. 2002. Stochastic averaging and Lyapunov exponent of quasi partially integrable Hamiltonian systems. International Journal of Non-Linear Mechanics, 37(3): 419–437.

Chapter 7
Stochastic Averaging Methods of Quasi-Hamiltonian Systems Excited by Fractional Gaussian Noises

It is learned from Sect. 2.5.3 that the fractional Gaussian noise (fGn) with Hurst index $\mathcal{H}$ in range of $(1/2,\ 1)$ is a color noise with certain correlation function and power spectral density. Its main feature is long-range dependence or long memory, making it a power mathematical model for various random phenomena in economics, finance, science, and engineering. Although the response of a nonlinear system excited by fGn is not a Markov process, stochastic averaging method still can reduce the system dimension, thereby reducing the time required for Monte Carlo simulation in predicting the system response. Thus, in this chapter, the stochastic averaging method of quasi-Hamiltonian systems excited by fGn (Deng and Zhu 2016a) is presented. Since the power spectral density of fGn varies slowly in the medium to high-frequency range, the fGn in this frequency range can be treated as a wideband process and the stochastic averaging method of quasi-integrable Hamiltonian systems excited by wideband noise is applicable. It will be discussed in Sect. 1.2 of vol. 2.

7.1 Quasi-Hamiltonian Systems Excited by FGns

Consider a n-DOF quasi-Hamiltonian system excited by fGns governed by the following equations

$$\dot{Q}_i = \frac{\partial H}{\partial P_i},$$
$$\dot{P}_i = -\frac{\partial H}{\partial Q_i} - \varepsilon \sum_{j=1}^{n} m_{ij}(\boldsymbol{Q}, \boldsymbol{P}) \frac{\partial H}{\partial P_j} + \varepsilon^{1/2} \sum_{l=1}^{m} g_{il}(\boldsymbol{Q}, \boldsymbol{P}) W_{Hl}(t),$$

W.-Q. Zhu et al., *Stochastic Averaging*, https://doi.org/10.1007/978-981-96-3829-1_7

$$i = 1, 2, \ldots, n. \tag{7.1}$$

where $\boldsymbol{Q} = [Q_1, Q_2, \ldots, Q_n]^T$ and $\boldsymbol{P} = [P_1, P_2, \ldots, P_n]^T$ are the vectors of generalized displacements and generalized momenta, respectively; ε is a small parameter; $H = H(\boldsymbol{Q}, \boldsymbol{P})$ represent twice-differentiable Hamiltonian function; $\varepsilon m_{ij}(\boldsymbol{Q}, \boldsymbol{P})$ represent weak damping coefficients; $\varepsilon^{1/2} g_{il}(\boldsymbol{Q}, \boldsymbol{P})$ represent small excitation amplitudes; $W_{H1}(t), W_{H2}(t), \ldots, W_{Hm}(t)$ are m independent unit fGns with Hurst indexes $1/2 < \mathcal{H}_1, \mathcal{H}_2, \ldots, \mathcal{H}_m < 1$, respectively. The auto-correlation functions $R_1(\tau), R_2(\tau), \ldots, R_m(\tau)$ and power spectral densities $S_1(\omega), S_2(\omega), \ldots, S_m(\omega)$ of fGns are determined by Eqs. (2.217) and (2.219), respectively.

According to the stochastic integration of fractional Brownian motion (fBm) introduced in Sect. 2.5, the motion Eq. (7.1) can be transformed into the following fractional stochastic differential equations (SDEs) defined in the type of pathwise symmetric integral, i.e.,

$$\begin{aligned}
\mathrm{d}Q_i &= \frac{\partial H}{\partial P_i}\mathrm{d}t, \\
\mathrm{d}P_i &= \left(-\frac{\partial H}{\partial Q_i} - \varepsilon \sum_{j=1}^{n} m_{ij}\frac{\partial H}{\partial P_j}\right)\mathrm{d}t + \varepsilon^{1/2}\sum_{l=1}^{m} g_{il}\mathrm{d}^{\circ}B_{Hl}(t), \\
i &= 1, 2, \ldots, n.
\end{aligned} \tag{7.2}$$

where $B_{H1}(t), B_{H2}(t), \ldots, B_{Hm}(t)$ are m independent unit fBms with Hurst indexes $\mathcal{H}_1, \mathcal{H}_2, \ldots, \mathcal{H}_m$, respectively. Equation (7.2) can be further transformed into the following fractional SDEs defined in the type of pathwise forward integral, i.e.,

$$\begin{aligned}
\mathrm{d}Q_i &= \frac{\partial H}{\partial P_i}\mathrm{d}t, \\
\mathrm{d}P_i &= \left(-\frac{\partial H}{\partial Q_i} - \varepsilon \sum_{j=1}^{n} m_{ij}\frac{\partial H}{\partial P_j}\right)\mathrm{d}t + \varepsilon^{1/2}\sum_{l=1}^{m} g_{il}\mathrm{d}B_{Hl}(t), \\
i &= 1, 2, \ldots, n.
\end{aligned} \tag{7.3}$$

Note that, according to Eq. (2.223), there is no Wong-Zakai correction term from Eqs. (7.2) to (7.3). Hence, Eq. (7.3) is different from Eq. (5.5) for the case of Gaussian white noise excitation and not applicable to the case of $\mathcal{H}_1, \mathcal{H}_2, \ldots, \mathcal{H}_m = 1/2$. In the following, the averaged fractional SDEs for five cases of quasi-Hamiltonian systems excited by fGns will be derived from Eq. (7.3).

7.2 Quasi-Non-Integrable Hamiltonian Systems

Assume that the Hamiltonian system associated with Eq. (7.3) is non-integrable. According to the fBm stochastic differentials rule described in Sect. 2.5.4, the fractional SDE governing Hamiltonian $H(\boldsymbol{Q}, \boldsymbol{P})$ can be derived as follows

$$dH = -\varepsilon \sum_{j,k=1}^{n} m_{jk} \frac{\partial H}{\partial P_j} \frac{\partial H}{\partial P_k} dt + \varepsilon^{1/2} \sum_{j=1}^{n} \sum_{l=1}^{m} \frac{\partial H}{\partial P_j} g_{jl} dB_{Hl}(t). \tag{7.4}$$

In the case of $\mathcal{H}_1, \mathcal{H}_2, \ldots, \mathcal{H}_m = \mathcal{H}$, Eq. (7.4) is equivalent to the following fractional SDE in the sense of the same Gaussian distribution with zero mean value and the same mean-square (Deng and Zhu 2016b)

$$\begin{aligned} dH = &- \varepsilon \sum_{j,k=1}^{n} m_{jk} \frac{\partial H}{\partial P_j} \frac{\partial H}{\partial P_k} dt \\ &+ \varepsilon^{1/2} \left(\sum_{j,k=1}^{n} \sum_{l=1}^{m} g_{jl} g_{kl} \frac{\partial H}{\partial P_j} \frac{\partial H}{\partial P_k} \right)^{1/2} dB_H(t). \end{aligned} \tag{7.5}$$

where $B_H(t)$ is unit fBm with Hurst index $\mathcal{H}$. Now, the new system is composed of Eq. (7.5) and $2n - 1$ equations in Eq. (7.3) except that for $dP_1(t)$. The Hamiltonian $H(t)$ is a slowly-varying process while the displacements $Q_j(t)$ and momenta $P_j(t)$ are fast-varying processes. According to the averaging principles of fBm (Xu et al. 2014a, b), as $\varepsilon \to 0$, $H(t)$ converges in the mean-square sense to the following averaged fractional stochastic process

$$dH = m(H)dt + \sigma(H)dB_H(t), \tag{7.6}$$

where the coefficients $m(H)$, $\sigma^2(H)$ are determined from the coefficients in Eq. (7.5) by time averaging,

$$\begin{aligned} &\lim_{T\to\infty} \frac{1}{T} \int_0^T \left| -\varepsilon \sum_{j,k=1}^{n} m_{jk} \frac{\partial H}{\partial P_j} \frac{\partial H}{\partial P_k} - m(H) \right| dt = 0, \\ &\lim_{T\to\infty} \frac{1}{T} \int_0^T \left| \left(\varepsilon \sum_{j,k=1}^{n} \sum_{l=1}^{m} g_{jl} g_{kl} \frac{\partial H}{\partial P_j} \frac{\partial H}{\partial P_k} \right)^{1/2} - \sigma(H) \right|^2 dt = 0. \end{aligned} \tag{7.7}$$

Since the states of n-DOF non-integrable Hamiltonian system are uniformly distributed on $(2n - 1)$-dimensional isoenergetic surface, similar to Eqs. (5.10) to (5.12), the time averaging can be replaced by spatial averaging, i.e.,

$$m(H) = \frac{\varepsilon}{T(H)} \int\limits_{\Omega} \left(-\sum_{j,k=1}^{n} m_{jk} \frac{\partial H}{\partial p_j} \frac{\partial H}{\partial p_k} \Big/ \frac{\partial H}{\partial p_1} \right) \mathrm{d}q_1 \ldots \mathrm{d}q_n \mathrm{d}p_2 \ldots \mathrm{d}p_n,$$

$$\sigma^2(H) = \frac{\varepsilon}{T(H)} \int\limits_{\Omega} \left(\sum_{j,k=1}^{n} \sum_{l=1}^{m} g_{jl} g_{kl} \frac{\partial H}{\partial p_j} \frac{\partial H}{\partial p_k} \Big/ \frac{\partial H}{\partial p_1} \right) \mathrm{d}q_1 \ldots \mathrm{d}q_n \mathrm{d}p_2 \ldots \mathrm{d}p_n,$$

$$T(H) = \int\limits_{\Omega} \left(1 \Big/ \frac{\partial H}{\partial p_1} \right) \mathrm{d}q_1 \ldots \mathrm{d}q_n \mathrm{d}p_2 \ldots \mathrm{d}p_n, \tag{7.8}$$

where the integral domain is

$$\Omega = \{(q_1, \ldots, q_n, p_2, \ldots, p_n) | H(q_1, \ldots, q_n, 0, p_2, \ldots, p_n) \le H\}. \tag{7.9}$$

Due to the non-Markovian nature of the process governed by Eq. (7.6), the PDF and statistics of the response can be obtained only by using Monte Carlo simulation. Once the stationary PDF $p(H)$ is obtained, the stationary $p(\boldsymbol{q}, \boldsymbol{p})$ of the generalized displacements and generalized momenta can be obtained approximately by using Eq. (5.18), i.e.,

$$p(\boldsymbol{q}, \boldsymbol{p}) = \left. \frac{p(h)}{T(h)} \right|_{h=H(\boldsymbol{q}, \boldsymbol{p})}. \tag{7.10}$$

The stochastic averaging method of quasi-non-integrable Hamiltonian system excited by fGns presented here may save a significant amount of computation time, see Example 7.1.

Example 7.1 Considers a 2-DOF quasi-non-integrable Hamiltonian system excited by fGns governed by the motion equations (Deng and Zhu 2016b)

$$\begin{aligned} \ddot{X}_1 + \gamma \dot{X}_1 + \frac{\partial U(X_1, X_2)}{\partial X_1} &= \sqrt{2D_1} W_{H1}(t), \\ \ddot{X}_2 + \gamma \dot{X}_2 + \frac{\partial U(X_1, X_2)}{\partial X_2} &= \sqrt{2D_2} W_{H2}(t), \end{aligned} \tag{7.11}$$

where $W_{H1}(t)$ and $W_{H2}(t)$ are two independent unit fGns with Hurst index $\mathcal{H} \in (1/2, 1)$, $2D_1$ and $2D_2$ are excitation intensities, γ is linear damping constant, γ, D_1, D_2 are small quantities of ε order. Let $Q_1 = X_1, Q_2 = X_2, P_1 = \dot{X}_1, P_2 = \dot{X}_2$. The Hamiltonian associated with Eq. (7.11) is

$$\begin{aligned} H &= p_1^2/2 + p_2^2/2 + U(q_1, q_2), \\ U(q_1, q_2) &= \frac{1}{2}(\omega_1^2 q_1^2 + \omega_2^2 q_2^2) + \frac{1}{4}\lambda(\omega_1^2 q_1^2 + \omega_2^2 q_2^2)^2, \end{aligned} \tag{7.12}$$

where ω_1 and ω_2 are the natural frequencies of the linearized system of system (7.11), and λ represents the nonlinearity intensity. The pathwise forward type fractional SDE governing the Hamiltonian process is

$$dH = -\gamma(P_1^2 + P_2^2)dt + \sqrt{2D_1}P_1 dB_{H1}(t) + \sqrt{2D_2}P_2 dB_{H2}(t). \tag{7.13}$$

This equation is equivalent to

$$dH = -\gamma(P_1^2 + P_2^2)dt + \sqrt{2D_1P_1^2 + 2D_2P_2^2}dB_H(t). \tag{7.14}$$

Applying the stochastic averaging method of quasi-non-integrable Hamiltonian systems excited by fGns presented in Sect. 7.1, the following one-dimensional averaged fractional SDE can be obtained

$$dH = m(H)dt + \sigma(H)dB_H(t). \tag{7.15}$$

According to Eq. (7.8), the coefficients are derived as

$$\begin{aligned} m(H) &= -2\gamma\left(H - \frac{1}{4}R - \frac{\lambda}{12}R^2\right), \\ \sigma^2(H) &= 2(D_1 + D_2)\left(H - \frac{1}{4}R - \frac{\lambda}{12}R^2\right), \\ R &= \frac{1}{\lambda}\left(\sqrt{1 + 4\lambda H} - 1\right). \end{aligned} \tag{7.16}$$

The stationary PDF $p(h)$ and statistics $E[H], E[H^2]$ obtained from Monte Carlo simulation of Eq. (7.15) and the original system (7.11), respectively, are shown in Figs. 7.1 and 7.2. According to Eq. (5.18), the approximate stationary PDF $p(q_1, q_2, p_1, p_2)$, marginal stationary PDF $p(q), p(p)$, and mean-square value $E[Q^2], E[P^2]$ of the original system (7.11) can then be calculated as follows

$$\begin{aligned} p(q_1, q_2, p_1, p_2) &= \left.\frac{p(h)}{T(h)}\right|_{h=(p_1^2+p_2^2)/2+U(q_1,q_2)}, \\ p(q_1, q_2) &= \int_{-\infty}^{\infty}\int_{-\infty}^{\infty} p(q_1, q_2, p_1, p_2)dp_1dp_2, \\ E[Q_1^2] &= \int_{-\infty}^{\infty}\int_{-\infty}^{\infty} q_1^2 p(q_1, q_2)dq_1dq_2, \\ E[Q_2^2] &= \int_{-\infty}^{\infty}\int_{-\infty}^{\infty} q_2^2 p(q_1, q_2)dq_1dq_2. \end{aligned} \tag{7.17}$$

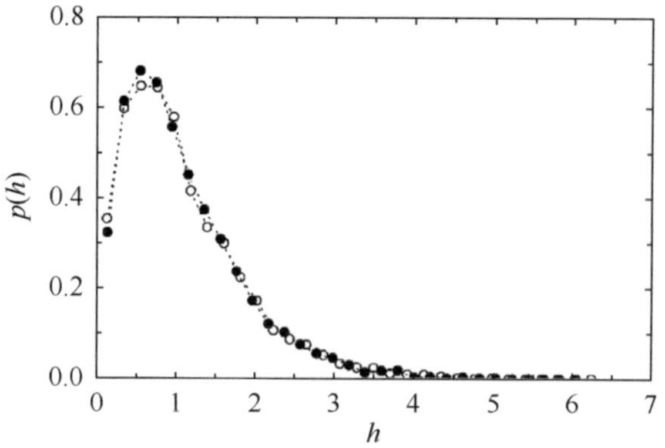

Fig. 7.1 Stationary PDFs $p(h)$ obtained from simulation of averaged fractional SDE (7.15) and original system (7.11). System parameters are $\omega_1 = 1.414$, $\omega_2 = 2$, $\lambda = 1$, $\gamma = 0.01$, $D_1 = D_2 = 0.01$, $\mathcal{H} = 0.75$. $\cdots\circ\cdots$, results from simulation of system (7.11); $\cdots\bullet\cdots$, results from simulation of Eq. (7.15) (Deng and Zhu 2016b)

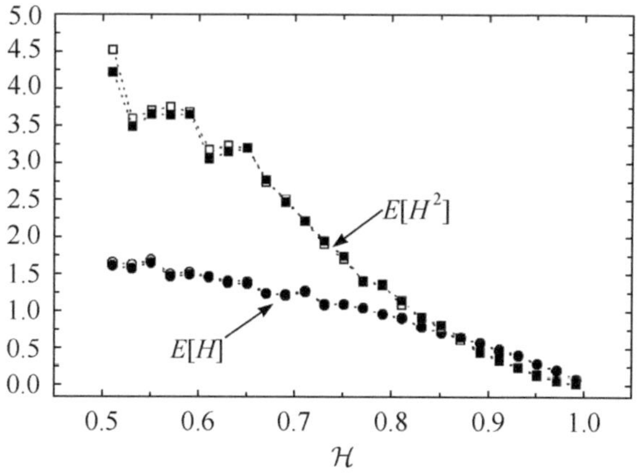

Fig. 7.2 $E[H]$, $E[H^2]$ obtained from simulation of averaged fractional SDE (7.15) and original system (7.11). System parameters are the same as those in Fig. 7.1. $\cdots\circ\cdots$, $\cdots\square\cdots$, results from simulation of system (7.11); $\cdots\bullet\cdots$, $\cdots\blacksquare\cdots$, results from simulation of Eq. (7.15) (Deng and Zhu 2016b)

The contour plots of the joint stationary PDF $p(q_1, q_2)$ and mean-square value $E[Q_1^2]$, $E[Q_2^2]$ obtained from Monte Carlo simulation of the averaged fractional SDE (7.15) and the original system (7.11) are shown in Figs. 7.3 and 7.4. It is seen from Figs. 7.1–7.4 that the results from Monte Carlo simulation of averaged Eq. (7.15) and those from the original system (7.11) are nearly identical. However, the computation time for the former is much short than that for the later.

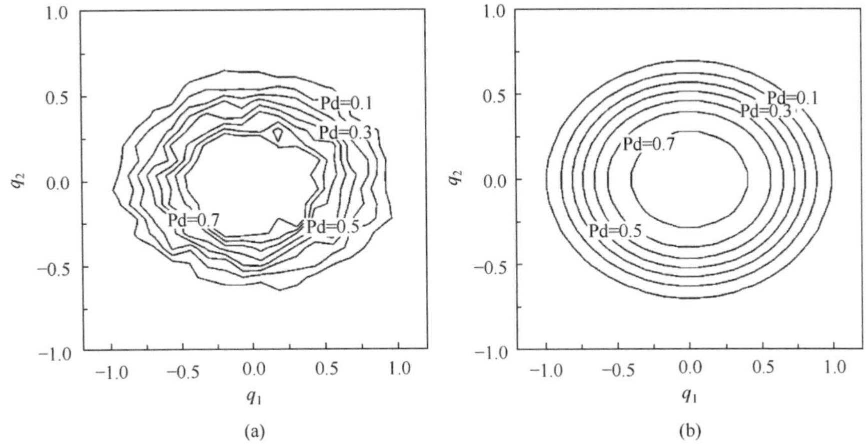

Fig. 7.3 Contours of $p(q_1, q_2)$ obtained from: **a** simulation of the averaged fractional SDE (7.15) and **b** simulation of the original system (7.11). System parameters are the same as those in Fig. 7.1 (Deng and Zhu 2016b)

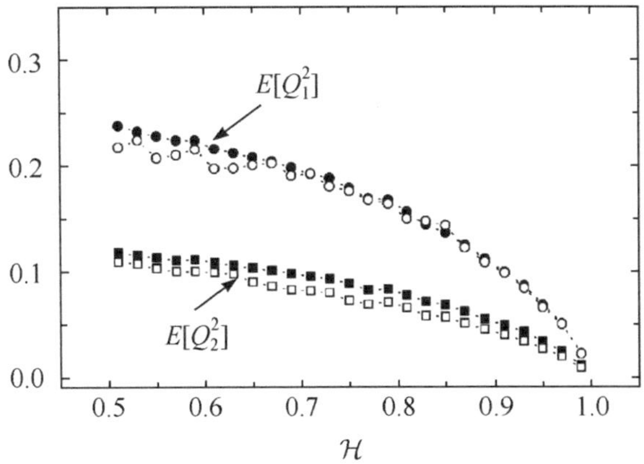

Fig. 7.4 $E[Q_1^2]$, $E[Q_2^2]$ obtained from simulation of averaged fractional SDE (7.15) and original system (7.11). System parameters are the same as those in Fig. 7.1. $\cdots\circ\cdots$, $\cdots\square\cdots$, results from simulation of system (7.11); $\cdots\bullet\cdots$, $\cdots\blacksquare\cdots$, results from simulation of Eq. (7.15) (Deng and Zhu 2016b)

7.3 Quasi-Integrable Hamiltonian Systems

7.3.1 Non-Internal Resonant Case

Now assume that the Hamiltonian system (7.3) is integrable and non-resonant with n independent first integrals in involution. According to the fBm stochastic differential rules in Sect. 2.5.4, the fractional SDEs governing the n first integrals are obtained as follows (Deng and Zhu 2018)

$$\mathrm{d}H_i = -\varepsilon \sum_{j,k=1}^{n} m_{jk} \frac{\partial H}{\partial P_j} \frac{\partial H_i}{\partial P_k} \mathrm{d}t + \varepsilon^{1/2} \sum_{j=1}^{n} \sum_{l=1}^{m} g_{jl} \frac{\partial H_i}{\partial P_j} \mathrm{d}B_{Hl}(t),$$
$$i = 1, 2, \ldots, n. \tag{7.18}$$

In Eq. (7.18), the generalized momenta P_i must be replaced with H_i and Q_i according to $H_i = H_i(\boldsymbol{Q}, \boldsymbol{P})$. Then, the system consists of n equations for Q_i in Eq. (7.3) and n equations for H_i in Eq. (7.18). In this new system, $\boldsymbol{H}(t) = [H_1(t), H_2(t), \ldots, H_n(t)]^{\mathrm{T}}$ is a slowly-varying vector process while $\boldsymbol{Q}(t) = [Q_1(t), Q_2(t), \ldots, Q_n(t)]^{\mathrm{T}}$ is a fast-varying vector process. According to the stochastic averaging principle (Xu et al. 2014a, b), as $\varepsilon \to 0$, $\boldsymbol{H}(t)$ converges in the mean-square sense to a vector stochastic process governed by the following averaged fractional SDEs

$$\mathrm{d}H_i = m_i(\boldsymbol{H})\mathrm{d}t + \sum_{l=1}^{m} \sigma_{il}(\boldsymbol{H})\mathrm{d}B_{Hl}(t),$$
$$i = 1, 2, \ldots, n. \tag{7.19}$$

The coefficients $m_i(\boldsymbol{H}), \sigma_{il}(\boldsymbol{H})$ can be obtained from time averaging of the coefficients of Eq. (7.18) as follows

$$m_i(\boldsymbol{H}) = \varepsilon \left\langle -\sum_{j,k=1}^{n} m_{jk} \frac{\partial H}{\partial P_j} \frac{\partial H_i}{\partial P_k} \right\rangle_t,$$
$$\sum_{l=1}^{m} \sigma_{il}(\boldsymbol{H})\sigma_{rl}(\boldsymbol{H}) = \varepsilon \left\langle \sum_{j,k=1}^{n} \sum_{l=1}^{m} g_{jl} g_{kl} \frac{\partial H_i}{\partial P_j} \frac{\partial H_r}{\partial P_k} \right\rangle_t,$$
$$i, r = 1, 2, \ldots, n. \tag{7.20}$$

If the n first integrals of the Hamiltonian system associated with Eq. (7.3) are separable, namely, $H = \sum_{\mu=1}^{n} H_\mu$, $H_\mu = H_\mu(q_\mu, p_\mu)$, $\mu = 1, 2, \ldots, n$ and the motion of each Hamiltonian subsystem is periodic, then similar to Eqs. (5.94) to (5.96), the time averaging in Eq. (7.20) can be replaced by spatial averaging with respect to $q_1, q_2, \ldots, q_n$, i.e.,

$$m_i(\boldsymbol{H}) = \frac{\varepsilon}{T(\boldsymbol{H})} \oint \left(-\sum_{j,k=1}^{n} m_{jk} \frac{\partial H}{\partial p_j} \frac{\partial H_i}{\partial p_k} \right) \prod_{\mu=1}^{n} \left(1 \Big/ \frac{\partial H_\mu}{\partial p_\mu} \right) \mathrm{d}q_\mu,$$
$$\sum_{l=1}^{m} \sigma_{il}(\boldsymbol{H})\sigma_{rl}(\boldsymbol{H}) = \frac{\varepsilon}{T(\boldsymbol{H})} \oint \left(\sum_{j,k=1}^{n} \sum_{l=1}^{m} g_{jl} g_{kl} \frac{\partial H_i}{\partial p_j} \frac{\partial H_r}{\partial p_k} \right) \prod_{\mu=1}^{n} \left(1 \Big/ \frac{\partial H_\mu}{\partial p_\mu} \right) \mathrm{d}q_\mu,$$

$$T(\boldsymbol{H}) = \oint \prod_{\mu=1}^{n} \left(1 \Big/ \frac{\partial H_\mu}{\partial p_\mu}\right) \mathrm{d}q_\mu, \quad i, r = 1, 2, \ldots, n. \tag{7.21}$$

Since the $\boldsymbol{H}(t)$ governed by Eq. (7.19) is not Markov process, its probability and statistics can be obtained only by Monte Carlo simulation. Once its stationary PDF $p(\boldsymbol{h})$ is obtained, the approximate stationary PDF for the original system (7.1) can be derived using Eq. (5.106) as follows

$$p(\boldsymbol{q}, \boldsymbol{p}) = \left. \frac{p(\boldsymbol{h})}{T(\boldsymbol{h})} \right|_{\boldsymbol{h}=\boldsymbol{H}(\boldsymbol{q},\boldsymbol{p})}. \tag{7.22}$$

If the action-angle variables I_i and Θ_i $i = 1, 2, \ldots, n$ of the integrable Hamiltonian system associated with Eq. (7.3) are available, then the fBm stochastic differential rule described in Sect. 2.5.4 can be used to derive the governing fractional SDEs for I_i, Θ_i $(i = 1, 2, \ldots, n)$ as follows

$$\begin{aligned} \mathrm{d}I_i &= -\varepsilon \sum_{j,k=1}^{n} m_{jk} \frac{\partial I_i}{\partial P_j} \frac{\partial H}{\partial P_k} \mathrm{d}t + \varepsilon^{1/2} \sum_{j=1}^{n} \sum_{l=1}^{m} g_{jl} \frac{\partial I_i}{\partial P_j} \mathrm{d}B_{Hl}(t), \\ \mathrm{d}\Theta_i &= \left(\omega_i - \varepsilon \sum_{j,k=1}^{n} m_{jk} \frac{\partial \Theta_i}{\partial P_j} \frac{\partial H}{\partial P_k} \right) \mathrm{d}t + \varepsilon^{1/2} \sum_{j=1}^{n} \sum_{l=1}^{m} g_{jl} \frac{\partial \Theta_i}{\partial P_j} \mathrm{d}B_{Hl}(t), \\ & i = 1, 2, \ldots, n. \end{aligned} \tag{7.23}$$

where Q_j, P_j $(j = 1, 2, \ldots, n)$ must be replaced by I_i, Θ_i $(i = 1, 2, \ldots, n)$ according to $I_i = I_i(\boldsymbol{Q}, \boldsymbol{P})$, $\Theta_i = \Theta_i(\boldsymbol{Q}, \boldsymbol{P})$. In the non-resonant case, $\boldsymbol{I}(t) = [I_1(t), I_2(t), \ldots, I_n(t)]^{\mathrm{T}}$ is a slowly-varying vector process while $\boldsymbol{\Theta}(t) = [\Theta_1(t), \Theta_2(t), \ldots, \Theta_n(t)]^{\mathrm{T}}$ is a fast-varying vector process. According to the fractional stochastic averaging principle (Xu et al. 2014a, b), as $\varepsilon \to 0$, $I_i(t)$ converge in mean-square sense to processes governed by the following averaged fractional SDEs

$$\mathrm{d}I_i = m_i(\boldsymbol{I})\mathrm{d}t + \sum_{l=1}^{m} \sigma_{il}(\boldsymbol{I})\mathrm{d}B_{Hl}(t), \quad i = 1, 2, \ldots, n. \tag{7.24}$$

where the coefficients are obtained from time averaging of the coefficients of Eq. (7.23). Due to the ergodicity of an n-DOF integrable and non-resonant Hamiltonian system in the n–dimensional torus with constant I_i, the time averaging can be replaced by spatial averaging with respect to the angle variables. Similar to the derivation from Eq. (5.79) to Eq. (5.80), $m_i(\boldsymbol{I})$ and $\sigma_{il}(\boldsymbol{I})$ can be obtained as follows

$$m_i(\boldsymbol{I}) = \frac{\varepsilon}{(2\pi)^n} \int_0^{2\pi} \left(-\sum_{j,k=1}^{n} m_{jk} \frac{\partial I_i}{\partial p_j} \frac{\partial H}{\partial p_k} \right) \mathrm{d}\boldsymbol{\theta},$$

$$\sum_{l=1}^{m}\sigma_{il}(\boldsymbol{I})\sigma_{rl}(\boldsymbol{I}) = \frac{\varepsilon}{(2\pi)^n}\int_0^{2\pi}\left(\sum_{j,k=1}^{n}\sum_{l=1}^{m} g_{jl}g_{kl}\frac{\partial I_i}{\partial p_j}\frac{\partial I_r}{\partial p_k}\right)\mathrm{d}\boldsymbol{\theta},$$
$$i, r = 1, 2, \ldots, n. \tag{7.25}$$

where $\int_0^{2\pi}(\cdot)\mathrm{d}\boldsymbol{\theta}$ represents n-fold integral with respect to $\theta_1, \theta_2, \ldots, \theta_n$.

The stationary PDF $p(\boldsymbol{I})$ can be obtained by performing Monte Carlo simulation on the averaged fractional SDEs (7.24). Similar to Eq. (5.104), the approximate stationary PDF $p(\boldsymbol{q}, \boldsymbol{p})$ for the original system (7.1) can then be obtained as follows

$$p(\boldsymbol{q}, \boldsymbol{p}) = \left.\frac{p(\boldsymbol{I})}{(2\pi)^n}\right|_{\boldsymbol{I}=\boldsymbol{I}(\boldsymbol{q},\boldsymbol{p})}. \tag{7.26}$$

Example 7.2 Consider two linear oscillators coupled by linear and nonlinear dampings and excited by fGns (Deng and Zhu 2018). The governing equations are

$$\begin{aligned}
&\ddot{X}_1 + \alpha_{11}\dot{X}_1 + \alpha_{12}\dot{X}_2 + \beta_1(X_1^2 + X_2^2)\dot{X}_1 \\
&\quad + \omega_1^2 X_1 = \sqrt{2D_1}W_{H1}(t), \\
&\ddot{X}_2 + \alpha_{21}\dot{X}_1 + \alpha_{22}\dot{X}_2 + \beta_2(X_1^2 + X_2^2)\dot{X}_2 \\
&\quad + \omega_2^2 X_2 = \sqrt{2D_2}W_{H2}(t),
\end{aligned} \tag{7.27}$$

where α_{ij}, β_i, ω_i and $2D_i$ $(i, j = 1, 2)$ are constants; α_{ij}, β_i, D_i are small quantities of order ε; $W_{H1}(t)$, $W_{H2}(t)$ are independent unit fGns with Hurst index $\mathcal{H} \in (1/2, 1)$.

Let $X_1 = Q_1, X_2 = Q_2, \dot{X}_1 = P_1, \dot{X}_2 = P_2$. Equation (7.27) can be rewritten in the following quasi-Hamiltonian system excited by fGns

$$\begin{aligned}
&\dot{Q}_1 = P_1, \\
&\dot{P}_1 = -\omega_1^2 Q_1 - \left[\alpha_{11} + \beta_1(Q_1^2 + Q_2^2)\right]P_1 - \alpha_{12}P_2 + \sqrt{2D_1}W_{H1}(t), \\
&\dot{Q}_2 = P_2, \\
&\dot{P}_2 = -\omega_2^2 Q_2 - \left[\alpha_{22} + \beta_2(Q_1^2 + Q_2^2)\right]P_2 - \alpha_{21}P_1 + \sqrt{2D_2}W_{H2}(t).
\end{aligned} \tag{7.28}$$

The Hamiltonian associated with system (7.28) is

$$H = H_1 + H_2, \quad H_i = \frac{1}{2}(p_i^2 + \omega_i^2 q_i^2), \quad i = 1, 2. \tag{7.29}$$

Note that H can be separated, and the motion of two oscillators associated with H_1, H_2 are periodic in whole planes (q_1, p_1) and (q_2, p_2), respectively. If the frequencies of the two oscillators are not in weak resonant relation, i.e., $k_1\omega_1 + k_2\omega_2 \neq O(\varepsilon)$, k_1, k_2 are integers. The system (7.28) is a quasi-integrable and non-internal resonant Hamiltonian system. Using the stochastic averaging method described in Sect. 7.3.1, the following averaged fractional SDEs can be obtained

$$dH_i = m_i(H_1, H_2)dt + \sum_{l=1}^{2} \sigma_{il}(H_1, H_2)dB_{Hl}(t), \quad i = 1, 2. \tag{7.30}$$

where the coefficients can be obtained according to (7.21) as follows

$$\begin{aligned}
m_1 &= -\alpha_{11}H_1 - \frac{\beta_1}{2\omega_1^2}H_1^2 - \frac{\beta_1}{\omega_2^2}H_1H_2, \\
m_2 &= -\alpha_{22}H_2 - \frac{\beta_2}{2\omega_2^2}H_2^2 - \frac{\beta_2}{\omega_1^2}H_1H_2, \\
\sigma_{11}^2 &= 2D_1H_1, \quad \sigma_{22}^2 = 2D_2H_2, \quad \sum_{l=1}^{2} \sigma_{1l}\sigma_{2l} = 0.
\end{aligned} \tag{7.31}$$

Monte Carlo simulation of Eq. (7.30) with coefficients (7.31) can be carried out to obtain the stationary PDF $p(h_1, h_2)$. Then the joint stationary PDF $p(q_1, q_2, p_1, p_1)$, marginal PDFs $p(q_1, q_2)$ and statistical moments $E[Q_1^2]$, $E[Q_2^2]$ can be calculated using Eq. (7.22) as follows

$$\begin{aligned}
p(q_1, q_2, p_1, p_2) &= \left. \frac{\omega_1\omega_2 p(h_1, h_2)}{4\pi^2} \right|_{\substack{h_1=(1/2)(p_1^2+\omega_1^2q_1^2), \\ h_2=(1/2)(p_2^2+\omega_2^2q_2^2)}} \\
p(q_1, q_2) &= \int_{-\infty}^{\infty}\int_{-\infty}^{\infty} p(q_1, q_2, p_1, p_2)dp_1dp_2, \\
E[Q_1^2] &= \int_{-\infty}^{\infty}\int_{-\infty}^{\infty} q_1^2 p(q_1, q_2)dq_1dq_2, \\
E[Q_2^2] &= \int_{-\infty}^{\infty}\int_{-\infty}^{\infty} q_2^2 p(q_1, q_2)dq_1dq_2.
\end{aligned} \tag{7.32}$$

The stationary PDFs $p(h_1), p(h_2), p(h_1, h_2)$ and mean value $E[H_1]$, $E[H_2]$ obtained from Monte Carlo simulation of Eq. (7.30) and original system (7.28) are shown in Figs. 7.5–7.7. It is seen that the two results from Eq. (7.30) and from Eq. (7.28) are in good agreement. The joint stationary PDFs $p(q_1), p(q_2), p(q_1, q_2)$ of displacements and mean-square $E[Q_1^2]$, $E[Q_2^2]$ obtained from simulation of Eq. (7.30) and Eq. (7.32) and from simulation of Eq. (7.28) are shown in Figs. 7.8-7.10. It is seen that for a Hurst index closing to 1/2, the results from the averaged equation and from the original system agree well. However, the error between the two results increases as the Hurst index increases. This is because the correlation time of fGn increases as the Hurst index increases, while relaxation time of system remains unchanged. That makes the applicability of the stochastic averaging method deteriorates as the Hurst index increases.

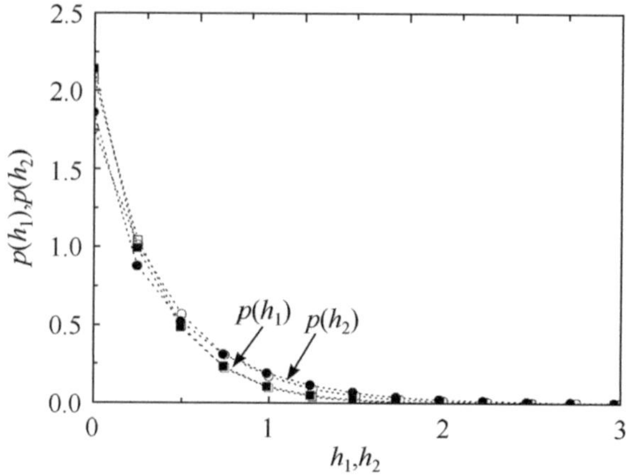

Fig. 7.5 $p(h_1), p(h_2)$ obtained from simulation of averaged fractional SDE (7.30) and original system (7.28). System parameters are $\alpha_{11} = 0.1$, $\alpha_{12} = 0.0$, $\alpha_{21} = 0.0$, $\alpha_{22} = 0.2$, $\beta_1 = 0.1$, $\beta_2 = 0.2$, $\omega_1 = 1.0$, $\omega_2 = 1.414$, $D_1 = 0.05$, $D_2 = 0.15$, H = 0.75. ···○···, ···□···, results from simulation of system (7.28); ···●···, ···■···, results from simulation of Eq. (7.30) (Deng and Zhu 2018)

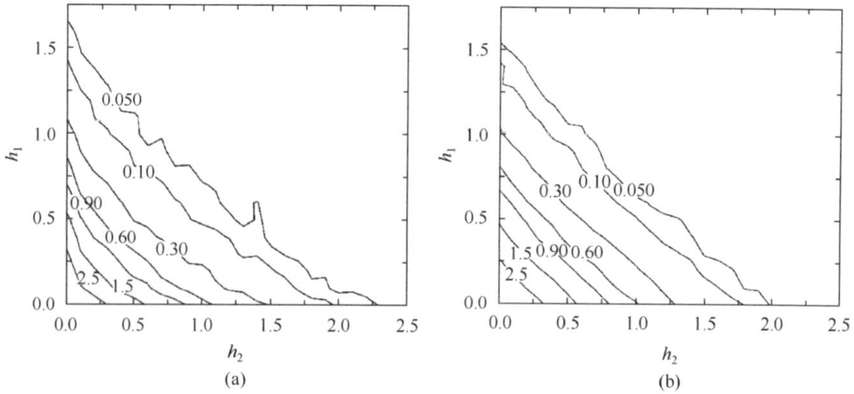

Fig. 7.6 Contours of $p(h_1, h_2)$ obtained from: **a** simulation of the averaged fractional SDE (7.30) and **b** simulation of the original system (7.28). System parameters are the same as those in Fig. 7.5 (Deng and Zhu 2018)

Example 7.3 Consider a coupled van der Pol oscillator and Duffing oscillator excited by fGns (Deng and Zhu 2018). The motion equations are

$$\begin{aligned}\ddot{X}_1 + (-\beta_1 + \alpha_1 X_1^2 + \alpha_2 X_2^4 + \alpha_3 \dot{X}_2^2)\dot{X}_1 + \omega^2 X_1 &= \sqrt{2D_1} W_{H1}(t),\\ \ddot{X}_2 + (\beta_2 + \alpha_4 X_1^2)\dot{X}_2 + kX_2^3 &= \sqrt{2D_2} W_{H2}(t), \end{aligned} \tag{7.33}$$

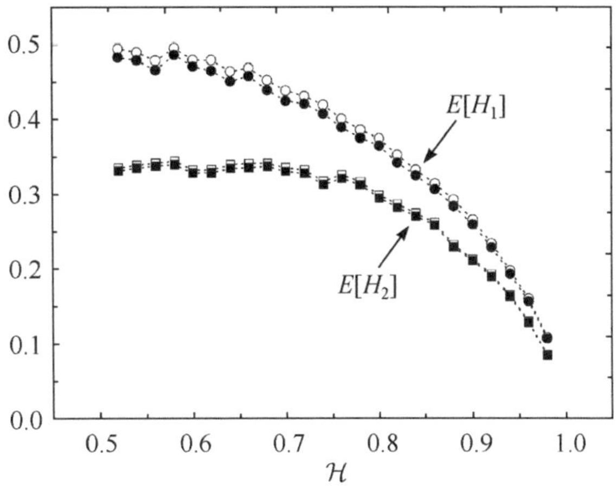

Fig. 7.7 $E[H]$, $E[H^2]$ versus $\mathcal{H}$. Results are obtained from simulation of averaged fractional SDE (7.30) and original system (7.28). System parameters are the same as those in Fig. 7.5. ···○···, ···□···, results from simulation of system (7.28); ···●···, ···■···, results from simulation of Eq. (7.30) (Deng and Zhu 2016b)

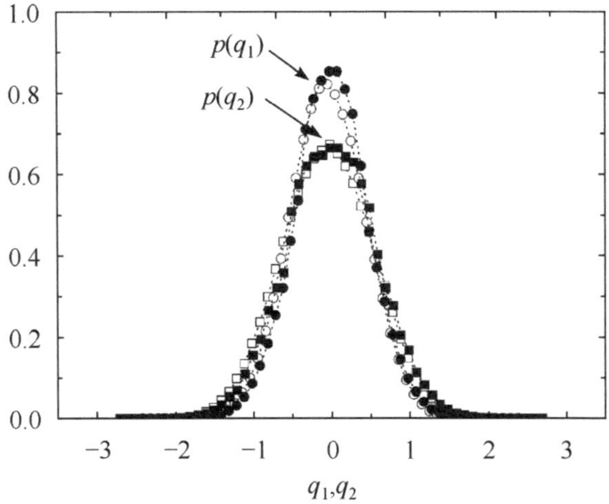

Fig. 7.8 $p(q_1)$, $p(q_2)$ obtained from simulation of averaged fractional SDE (7.30) and original system (7.28). System parameters are the same as those in Fig. 7.5. ···○···, ···□···, results from simulation of system (7.28); ···●···, ···■···, results from simulation of Eq. (7.30) (Deng and Zhu 2018)

where $\alpha_1, \alpha_2, \alpha_3, \alpha_4, \beta_1, \beta_2, \omega, k$ are positive constants; $W_{H1}(t), W_{H2}(t)$ are independent unit fGns with Hurst index $\mathcal{H} \in (1/2, 1)$ and excitation intensities $2D_1$ and $2D_2$, respectively; α_j $(j = 1, 2, 3, 4)$, β_i, D_i $(i = 1, 2)$ are small quantities of order ε.

Letting $X_1 = Q_1, X_2 = Q_2, \dot{X}_1 = P_1, \dot{X}_2 = P$, system (7.33) can be written in the following quasi-Hamiltonian system excited by fGns

$$\begin{aligned}
\dot{Q}_1 &= P_1,\\
\dot{P}_1 &= -\omega^2 Q_1 - (-\beta_1 + \alpha_1 Q_1^2 + \alpha_2 Q_2^4 + \alpha_3 P_2^2)P_1 + \sqrt{2D_1}W_{H1}(t),\\
\dot{Q}_2 &= P_2,\\
\dot{P}_2 &= -kQ_2^3 - (\beta_2 + \alpha_4 Q_1^2)P_2 + \sqrt{2D_2}W_{H2}(t).
\end{aligned} \tag{7.34}$$

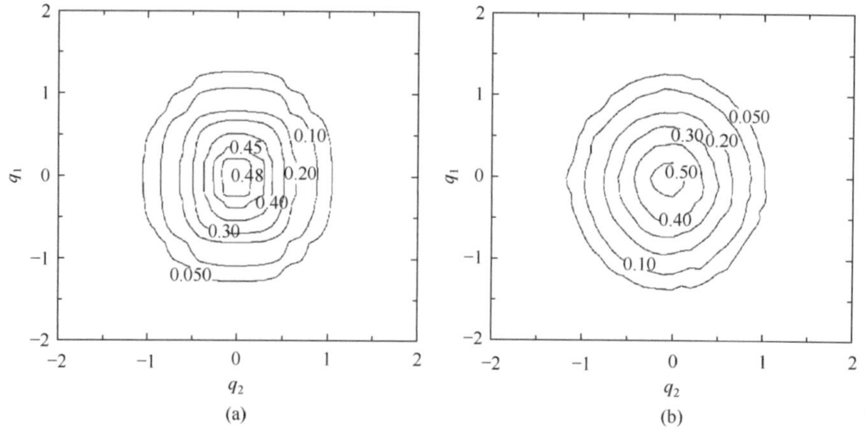

Fig. 7.9 Contours of $p(q_1, q_2)$ obtained from: **a** simulation of the averaged fractional SDE (7.30) and **b** simulation of the original system (7.28). System parameters are the same as those in Fig. 7.5 (Deng and Zhu 2018)

Fig. 7.10 $E[Q_1^2], E[Q_2^2]$ versus $\mathcal{H}$. Results are obtained from simulation of averaged fractional SDE (7.30) (computation time 503 s) and of original system (7.28) (computation time 941 s). System parameters are the same as those in Fig. 7.5. $\cdots\circ\cdots$, $\cdots\square\cdots$, results from simulation of system (7.28); $\cdots\bullet\cdots$, $\cdots\blacksquare\cdots$, results from simulation of Eq. (7.30) (Deng and Zhu 2018)

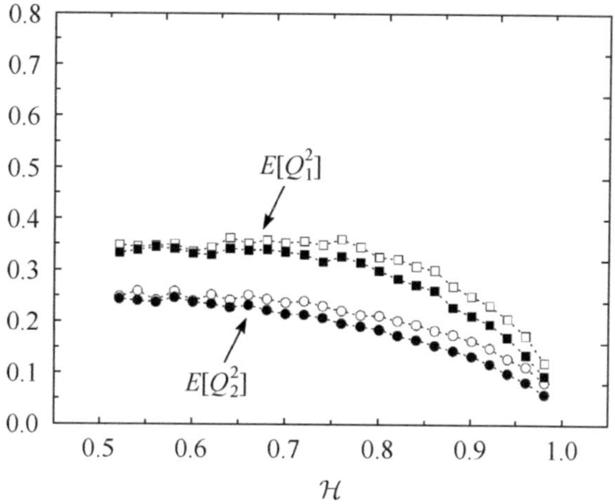

System (7.34) can further be converted into the following fractional SDEs

$$
\begin{aligned}
\mathrm{d}Q_1 &= P_1\mathrm{d}t,\\
\mathrm{d}P_1 &= \left[-\omega^2 Q_1 - (-\beta_1 + \alpha_1 Q_1^2 + \alpha_2 Q_2^4 + \alpha_3 P_2^2)P_1\right]\mathrm{d}t + \sqrt{2D_1}\mathrm{d}B_{H1}(t),\\
\mathrm{d}Q_2 &= P_2\mathrm{d}t,\\
\mathrm{d}P_2 &= \left[-kQ_2^3 - (\beta_2 + \alpha_4 Q_1^2)P_2\right]\mathrm{d}t + \sqrt{2D_2}\mathrm{d}B_{H2}(t).
\end{aligned}
\tag{7.35}
$$

The Hamiltonian associated with system (7.35) is

$$H = H_1 + H_2,$$
$$H_1 = \frac{1}{2}(p_1^2 + \omega^2 q_1^2), \quad H_2 = \frac{1}{2}p_2^2 + \frac{1}{4}kq_2^4. \tag{7.36}$$

It is seen that H can be separated, and the motion of two Hamiltonian subsystems associated with H_1 and H_2 are periodic. The Hamiltonian system associated with Eq. (7.35) is integrable and non-resonant. Similar to Example 7.2, the averaged fractional SDE (7.30) governing $H_1(t)$, $H_2(t)$ can be obtained from Eq. (7.35). The coefficients are derived from Eq. (7.21) as follows

$$m_1 = \beta_1 H_1 - \frac{\alpha_1}{2w^2}H_1^2 - \frac{4}{3}\left(\frac{\alpha_2}{k} + \alpha_3\right)H_1 H_2,$$
$$m_2 = -\frac{4\beta_2}{3}H_2 - \frac{4\alpha_4}{3\omega^2}H_1 H_2,$$
$$\sigma_{11}^2 = 2D_1 H_1, \quad \sigma_{22}^2 = \frac{8}{3}D_2 H_2, \quad \sum_{l=1}^{2}\sigma_{1l}\sigma_{2l} = 0,$$
$$T(H_1, H_2) = \frac{4\pi \mathrm{B}(1/4, 1/2)}{\sqrt{2}\omega}(4kH_2)^{-1/4}. \tag{7.37}$$

where $\mathrm{B}(\cdot, \cdot)$ is the Beta function

Performing Monte Carlo simulation on the SDE (7.30) with coefficients (7.37) yield $p(h_1, h_2)$, and then using Eq. (7.22) and $p(h_1, h_2)$ yield the approximate joint stationary PDF $p(q_1, q_2, p_1, p_1)$ for the original system (7.34) as follows

$$p(q_1, q_2, p_1, p_2) = \left.\frac{p(h_1, h_2)}{T(h_1, h_2)}\right|_{\substack{h_1=(1/2)(p_1^2+\omega_1^2 q_1^2),\\ h_2=p_2^2/2+kq_2^4/4}}. \tag{7.38}$$

The stationary PDF $p(q_1), p(q_2)$ and the mean-square value $E[Q_1^2], E[Q_2^2]$ can then be obtained from $p(q_1, q_2, p_1, p_2)$.

The stationary PDFs $p(h_1)$, $p(h_2)$, $p(h_1, h_2)$ and mean-value $E[H_1], E[H_2]$ obtained from Monte Carlo simulation of Eqs. (7.30) and (7.37 and of original system (7.34) are plotted in Figs. 7.11, 7.12 and 7.13. It is seen that the two results are in good agreement. The stationary joint PDFs $p(q_1), p(q_2), p(q_1, q_2)$ of displacements and the mean-square value $E[Q_1^2], E[Q_2^2]$ are shown in Figs. 7.14, 7.15 and 7.16. The error between the results from simulation of Eq. (7.30) and results from simulation of original system (7.34) is small.

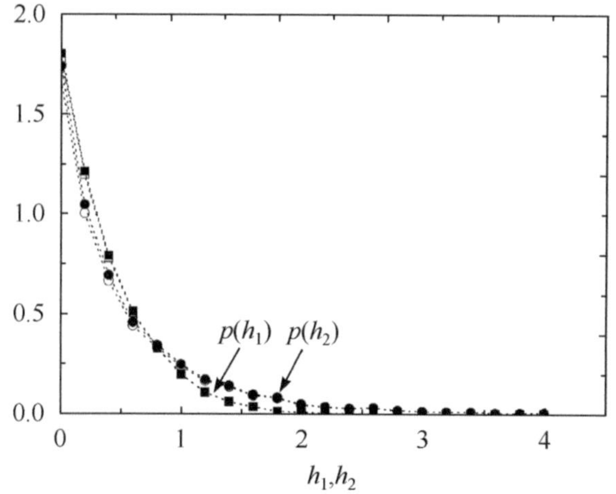

Fig. 7.11 $p(h_1), p(h_2)$ obtained from simulation of averaged fractional SDE (7.30) and original system (7.33). System parameters are $\alpha_1 = 0.2$, $\alpha_2 = 0.1$, $\alpha_3 = 0.1$, $\alpha_4 = 0.4$, $\beta_1 = 0.05$, $\beta_2 = 0.1$, $\omega = 1.0$, $k = 1.0$, $D_1 = 0.05$, $D_2 = 0.15$, $\mathcal{H} = 0.75$.$\cdots\circ\cdots$, $\cdots\square\cdots$,, results from simulation of system (7.33); $\cdots\bullet\cdots$, $\cdots\blacksquare\cdots$, results from simulation of Eq. (7.30) (Deng and Zhu 2018)

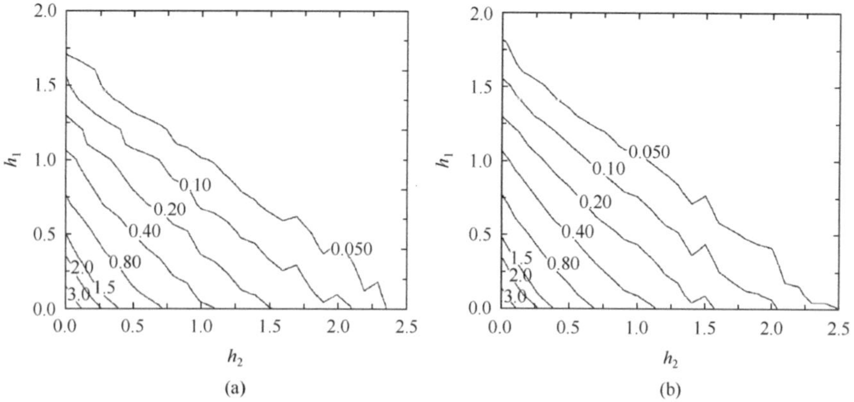

Fig. 7.12 Contours of $p(h_1, h_2)$ obtained from: **a** simulation of the averaged fractional SDE (7.30) and **b** simulation of the original system (7.33). System parameters are the same as those in Fig. 7.11 (Deng and Zhu 2018)

7.3.2 Internal Resonant Case

Assume that the Hamiltonian system associated with Eq. (7.3) is integrable and internal resonant, i.e., $\sum_{i=1}^{n} k_i^u \omega_i = O(\varepsilon)$ $(u = 1, 2, \ldots, \alpha, \quad 1 \le \alpha < n)$, k_i^u are integers. Introduce the following α combinations of angle variables

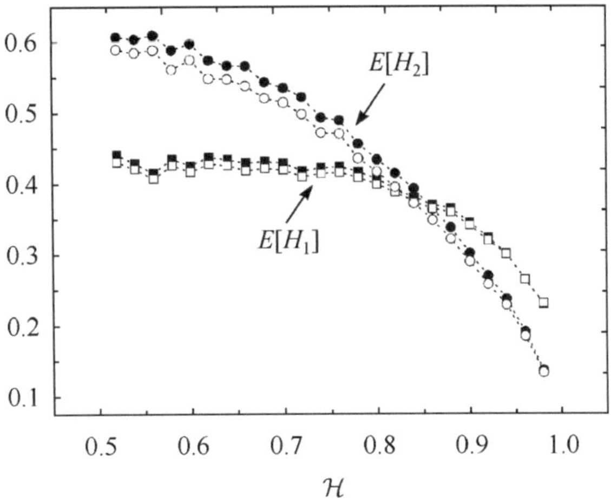

Fig. 7.13 $E[H_1], E[H_2]$ versus $\mathcal{H}$. Results arc obtained from simulation of averaged fractional SDE (7.30) and original system (7.33). System parameters are the same as those in Fig. 7.11. ···○···, ···□···, results from simulation of system (7.33); ···●···, ···■···, results from simulation of Eq. (7.30) (Deng and Zhu 2018)

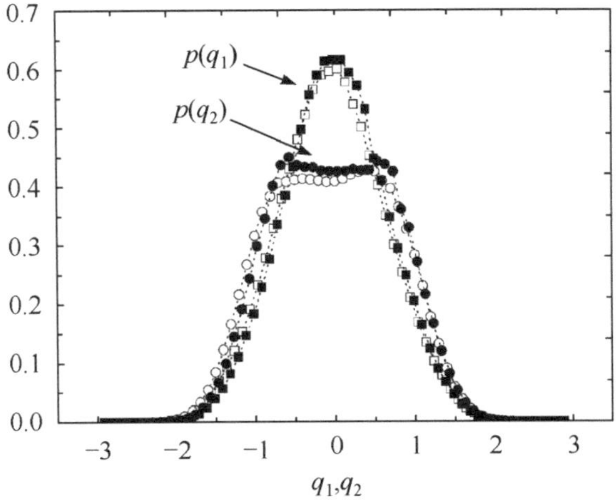

Fig. 7.14 $p(q_1), p(q_2)$ obtained from simulation of averaged fractional SDE (7.30) and original system (7.33). System parameters are the same as those in Fig. 7.11. ···○···, ···□···, results from simulation of system (7.33); ···●···, ···■···, results from simulation of Eq. (7.30) (Deng and Zhu 2018)

$$\Psi_u = \sum_{i=1}^{n} k_i^u \Theta_i, \quad u = 1, 2, \ldots, \alpha. \tag{7.39}$$

The fractional SDEs governing Ψ_u can be obtained from the equations for Θ_r in Eq. (7.23) according to Eq. (7.39) as follows

$$\mathrm{d}\Psi_u = \left[O_u(\varepsilon) - \varepsilon \sum_{j,k=1}^{n} m_{jk} \frac{\partial \Psi_u}{\partial P_j} \frac{\partial H}{\partial P_k} \right] \mathrm{d}t$$
$$+ \varepsilon^{1/2} \sum_{j=1}^{n} \sum_{l=1}^{m} g_{jl} \frac{\partial \Psi_u}{\partial P_j} \mathrm{d}B_{Hl}(t),$$

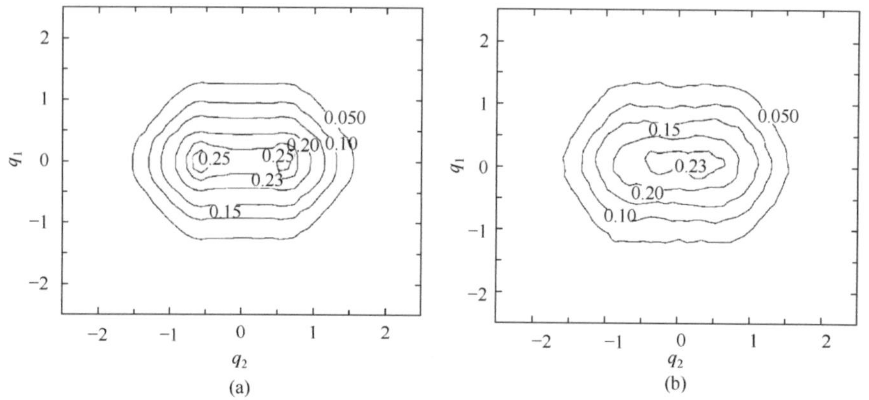

Fig. 7.15 Contours of $p(q_1, q_2)$ obtained from: **a** simulation of the averaged fractional SDE (7.30) and **b** simulation of the original system (7.33). System parameters are the same as those in Fig. 7.11 (Deng and Zhu 2018)

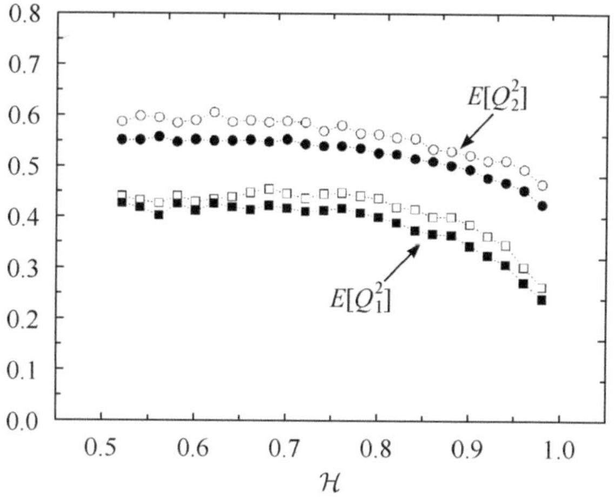

Fig. 7.16 $E[Q_1^2]$, $E[Q_2^2]$ versus $\mathcal{H}$. Results are obtained from simulation of averaged fractional SDE (7.30) (computation time 247 s) and original system (7.33) (computation time 852 s). System parameters are the same as those in Fig. 7.11. ···○···, ···□··· results from simulation of system (7.33); ···●···, ···■···, results from simulation of Eq. (7.30) (Deng and Zhu 2018)

$$u = 1, 2, \ldots, \alpha. \tag{7.40}$$

The new system consists of α equations for Ψ_u in Eq. (7.40), n equations for I_r and $(n-\alpha)$ equations for Θ_r from Eq. (7.23). $\boldsymbol{I}(t)$, $\boldsymbol{\Psi}(t) = [\Psi_1(t), \Psi_2(t), \ldots, \Psi_\alpha(t)]^{\mathrm{T}}$ are two slowly-varying vector processes while $\boldsymbol{\Theta}_1(t) = [\Theta_1(t), \Theta_2(t), \ldots, \Theta_{n-\alpha}(t)]^{\mathrm{T}}$ is a fast-varying vector process. According to the fractional stochastic averaging principle (Xu et al. 2014a, b), as $\varepsilon \to 0$, $[\boldsymbol{I}(t)^{\mathrm{T}}, \boldsymbol{\Psi}(t)^{\mathrm{T}}]^{\mathrm{T}}$ converges in the mean-square sense to a $(n+\alpha)$-dimensional vector process governed by the following averaged fractional SDEs

$$
\begin{aligned}
\mathrm{d}I_i &= m_i(\boldsymbol{I}, \boldsymbol{\Psi})\mathrm{d}t + \sum_{l=1}^{m} \sigma_{il}(\boldsymbol{I}, \boldsymbol{\Psi})\mathrm{d}B_{Hl}(t),\\
\mathrm{d}\Psi_u &= m_u(\boldsymbol{I}, \boldsymbol{\Psi})\mathrm{d}t + \sum_{l=1}^{m} \sigma_{ul}(\boldsymbol{I}, \boldsymbol{\Psi})\mathrm{d}B_{Hl}(t),\\
&i = 1, 2, \ldots, n; \quad u = 1, 2, \ldots, \alpha.
\end{aligned} \tag{7.41}
$$

where the coefficients are determined by the time averaging of the coefficients for I_i in Eq. (7.23) and of the coefficients for Ψ_u in Eq. (7.40). Due to the ergodicity of an integrable and resonant Hamiltonian system in a $(n-\alpha)$-dimensional subtorus, time averaging can be replaced by spatial averaging with respect to $(n-\alpha)$-dimensional vector $\boldsymbol{\theta}_1$ of phase angles, i.e.,

$$
\begin{aligned}
m_r(\boldsymbol{I}, \boldsymbol{\Psi}) &= \frac{\varepsilon}{(2\pi)^{n-\alpha}} \int_0^{2\pi} \left(-\sum_{j,k=1}^{n} m_{jk} \frac{\partial I_r}{\partial p_k} \frac{\partial H}{\partial p_j} \right) \mathrm{d}\boldsymbol{\theta}_1,\\
m_u(\boldsymbol{I}, \boldsymbol{\Psi}) &= O_u(\varepsilon) + \frac{\varepsilon}{(2\pi)^{n-\alpha}} \int_0^{2\pi} \left(-\sum_{j,k=1}^{n} m_{jk} \frac{\partial \Psi_u}{\partial p_k} \frac{\partial H}{\partial p_j} \right) \mathrm{d}\boldsymbol{\theta}_1,\\
\sum_{l=1}^{m} \sigma_{rl}\sigma_{il}(\boldsymbol{I}, \boldsymbol{\Psi}) &= \frac{\varepsilon}{(2\pi)^{n-\alpha}} \int_0^{2\pi} \left(\sum_{j,k=1}^{n} \sum_{l=1}^{m} g_{jl} g_{kl} \frac{\partial I_r}{\partial p_j} \frac{\partial I_i}{\partial p_k} \right) \mathrm{d}\boldsymbol{\theta}_1,\\
\sum_{l=1}^{m} \sigma_{rl}\sigma_{ul}(\boldsymbol{I}, \boldsymbol{\Psi}) &= \frac{\varepsilon}{(2\pi)^{n-\alpha}} \int_0^{2\pi} \sum_{j,k=1}^{n} \sum_{l=1}^{m} g_{jl} g_{kl} \frac{\partial I_r}{\partial p_j} \frac{\partial \Psi_u}{\partial p_k} \mathrm{d}\boldsymbol{\theta}_1,\\
\sum_{l=1}^{m} \sigma_{ul}\sigma_{vl}(\boldsymbol{I}, \boldsymbol{\Psi}) &= \frac{\varepsilon}{(2\pi)^{n-\alpha}} \int_0^{2\pi} \sum_{j,k=1}^{n} \sum_{l=1}^{m} g_{jl} g_{kl} \frac{\partial \Psi_u}{\partial p_j} \frac{\partial \Psi_v}{\partial p_k} \mathrm{d}\boldsymbol{\theta}_1,\\
&r, i = 1, 2, \ldots, n; \quad u, v = 1, 2, \ldots, \alpha.
\end{aligned} \tag{7.42}
$$

where $\int_0^{2\pi}(\cdot)d\boldsymbol{\theta}_1$ represents $(n-\alpha)$-fold integral with respect to $\boldsymbol{\theta}_1 = [\theta_1, \theta_2, \ldots, \theta_{n-\alpha}]^{\mathrm{T}}$. Note that the dimension of the averaged fractional SDEs (7.41) is the sum of the number n of action variables and the number α of resonance relations. The total dimension is less than dimension $2n$ of the original system. Monte Carlo simulation of the averaged fractional SDEs (7.41) can yield the stationary PDF $p(\boldsymbol{I}, \boldsymbol{\psi})$. According to (5.125), the approximate stationary PDF $p(\boldsymbol{q}, \boldsymbol{p})$ for the original system (7.1) can be obtained as follows

$$
p(\boldsymbol{q}, \boldsymbol{p}) = \frac{1}{(2\pi)^{n-\alpha}} p(\boldsymbol{I}, \boldsymbol{\psi}) \left| \frac{\partial(\boldsymbol{I}, \boldsymbol{\psi}, \boldsymbol{\theta}_1)}{\partial(\boldsymbol{q}, \boldsymbol{p})} \right|. \tag{7.43}
$$

Example 7.4 Consider the following quasi-integrable and internal resonant Hamiltonian systems excited by fGns (Lü et al. 2017a)

$$\begin{aligned}
&\dot{Q}_1 = P_1, \quad \dot{Q}_2 = P_2, \\
&\dot{P}_1 = -\omega_1^2 Q_1 - \alpha_{11} P_1 - \alpha_{12} P_2 + \sqrt{2D_1} W_{H1}(t), \\
&\dot{P}_2 = -\omega_2^2 Q_2 - \alpha_{22} P_2 - \alpha_{21} P_1 + \sqrt{2D_2} W_{H2}(t).
\end{aligned} \tag{7.44}$$

System (7.44) can be transformed into the following fractional SDE

$$\begin{aligned}
&\mathrm{d}Q_1 = P_1 \mathrm{d}t, \quad \mathrm{d}Q_2 = P_2 \mathrm{d}t, \\
&\mathrm{d}P_1 = \left(-\omega_1^2 Q_1 - \alpha_{11} P_1 - \alpha_{12} P_2\right)\mathrm{d}t + \sqrt{2D_1}\mathrm{d}B_{H1}(t), \\
&\mathrm{d}P_2 = \left(-\omega_2^2 Q_2 - \alpha_{22} P_2 - \alpha_{21} P_1\right)\mathrm{d}t + \sqrt{2D_2}\mathrm{d}B_{H2}(t).
\end{aligned} \tag{7.45}$$

The Hamiltonian system associated with Eq. (7.44) has the following Hamiltonian H, action variables I_i and angle variables θ_i

$$H = \sum_{i=1}^{2} \omega_i I_i, \quad I_i = \frac{1}{2\omega_i}(p_i^2 + \omega_i^2 q_i^2), \quad \theta_i = -tg^{-1}(p_i/\omega_i q_i), \ i = 1, 2. \tag{7.46}$$

Assume that $\omega_1 = \omega_2$. Introduce phase difference

$$\psi = \theta_1 - \theta_2. \tag{7.47}$$

Using the stochastic averaging method for quasi-integrable and resonant Hamiltonian systems described in Sect. 7.3.2, the following averaged fractional SDE can be obtained

$$\begin{aligned}
&\mathrm{d}I_1 = \overline{m}_1(I_1, I_2, \Psi)\mathrm{d}t + \overline{\sigma}_{11}(I_1, I_2, \Psi)\mathrm{d}B_{H1}(t), \\
&\mathrm{d}I_2 = \overline{m}_2(I_1, I_2, \Psi)\mathrm{d}t + \overline{\sigma}_{22}(I_1, I_2, \Psi)\mathrm{d}B_{H2}(t), \\
&\mathrm{d}\Psi = \overline{m}_3(I_1, I_2, \Psi)\mathrm{d}t + \overline{\sigma}_{33}(I_1, I_2, \Psi)\mathrm{d}B_{H3}(t),
\end{aligned} \tag{7.48}$$

where the coefficients can be obtained according to Eq. (7.42) as follows

$$\begin{aligned}
&\overline{m}_1(I_1, I_2, \psi) = -\alpha_{11} I_1 - \alpha_{12}(\omega_2/\omega_1)^{1/2}(I_1 I_2)^{1/2}\cos\psi, \\
&\overline{m}_2(I_1, I_2, \psi) = -\alpha_{22} I_2 - \alpha_{21}(\omega_1/\omega_2)^{1/2}(I_1 I_2)^{1/2}\cos\psi, \\
&\overline{m}_3(I_1, I_2, \psi) = \left\{\left[\alpha_{12}\left(I_2/I_1\right)^{1/2}\omega_2^2 + \alpha_{21}\left(I_1/I_2\right)^{1/2}\omega_1^2\right]\Big/ 2(\omega_1\omega_2)^{1/2}\right\}\sin\psi, \\
&\qquad \overline{\sigma}_{11}^2 = 2D_1 I_1/\omega_1, \ \overline{\sigma}_{22}^2 = 2D_2 I_2/\omega_2, \\
&\qquad \overline{\sigma}_{33}^2 = \left(\omega_2^2 D_1/I_1\omega_1 + \omega_1^2 D_2/I_2\omega_2\right)/2, \\
&\qquad \overline{\sigma}_{12}^2 = \overline{\sigma}_{13}^2 = \overline{\sigma}_{23}^2 = 0.
\end{aligned} \tag{7.49}$$

Monte Carlo simulation of Eq. (7.48) with coefficients in Eq. (7.49) yields the stationary PDF $p(I_1, I_2, \psi)$ and then $p(I_1, I_2)$ using integration of $p(I_1, I_2, \psi)$ with respect to ψ. Using (7.43),the joint PDF $p(q_1, q_2, p_1, p_1)$, marginal density $p(q_1, q_2)$, and other statistics $E[Q_1^2]$, $E[Q_2^2]$ can be obtained as follows

$$\begin{aligned} p(q_1, q_2, p_1, p_2) &= \left.\frac{\omega_1\omega_2 p(I_1, I_2)}{4\pi^2}\right|_{\substack{I_1=(p_1^2+\omega_1^2 q_1^2)/(2\omega_1), \\ I_2=(p_2^2+\omega_2^2 q_2^2)/(2\omega_2)}} \\ p(q_1, q_2) &= \int_{-\infty}^{\infty}\int_{-\infty}^{\infty} p(q_1, q_2, p_1, p_2)\mathrm{d}p_1\mathrm{d}p_2, \\ E[Q_1^2] &= \int_{-\infty}^{\infty}\int_{-\infty}^{\infty} q_1^2 p(q_1, q_2)\mathrm{d}q_1\mathrm{d}q_2, \\ E[Q_2^2] &= \int_{-\infty}^{\infty}\int_{-\infty}^{\infty} q_2^2 p(q_1, q_2)\mathrm{d}q_1\mathrm{d}q_2. \end{aligned} \tag{7.50}$$

It is shown in Fig. 7.17 that, I_1, I_2, Ψ in system (7.44) are slowly-varying processes, while Θ_1 is a fast-varying process. Figs. 7.18–7.21 provide contour plots of stationary PDF $p(I_1, I_2)$, mean values $p(I_1), p(I_2)$, and mean-square values $E[I_1^2]$, $E[I_2^2]$, obtained from Monte Carlo simulation of Eq. (7.44) and system (7.48), respectively. It is evident that the two results from Eq. (7.44) and from Eq. (7.48) are in good agreement. Figures 7.22–7.25show joint stationary PDFs $p(q_1, p_1)$, stationary PDFs $p(q_1), p(q_2), p(\psi)$, and mean square values $E[Q_1^2]$, $E[Q_2^2]$. It can be observed that the two results from the averaged fractional SDE (7.48) and the original system (7.44) agree well.

7.4 Quasi-Partially Integrable Hamiltonian Systems

7.4.1 Non-Internal Resonant Case

Assume that the Hamiltonian system associated with Eq. (7.3) is partially integrable and non-resonant. There are a total of r $(1 < r < n)$ independent, involutive first integrals $H_1, H_2, \ldots, H_r$, which can be separated. The Hamiltonian can be expressed as follows

$$H(\boldsymbol{q}, \boldsymbol{p}) = \sum_{\eta=1}^{r-1} H_\eta(\boldsymbol{q}_1, \boldsymbol{p}_1) + H_r(\boldsymbol{q}_2, \boldsymbol{p}_2), \tag{7.51}$$

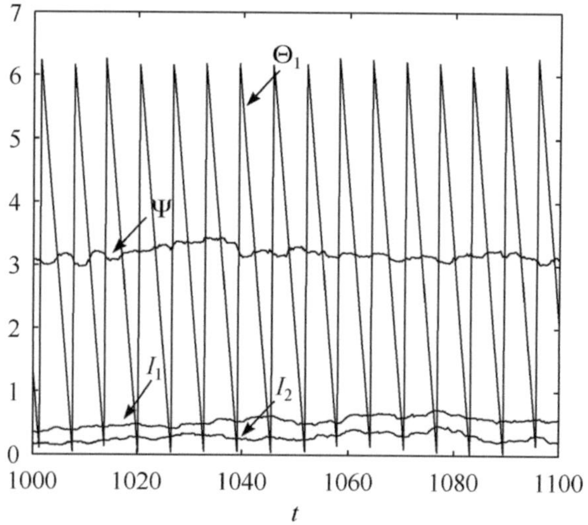

Fig. 7.17 Samples of the random process I_1, I_2, Ψ, Θ_1 in system (7.44). System parameter are $\alpha_{11} = 0.05, \alpha_{12} = 0.07, \alpha_{21} = 0.07, \alpha_{22} = 0.1, \omega_1 = 1.0, \omega_2 = 1.0, D_1 = 0.0005, D_2 = 0.001, \mathcal{H} = 0.7$ (Lü et al. 2017a)

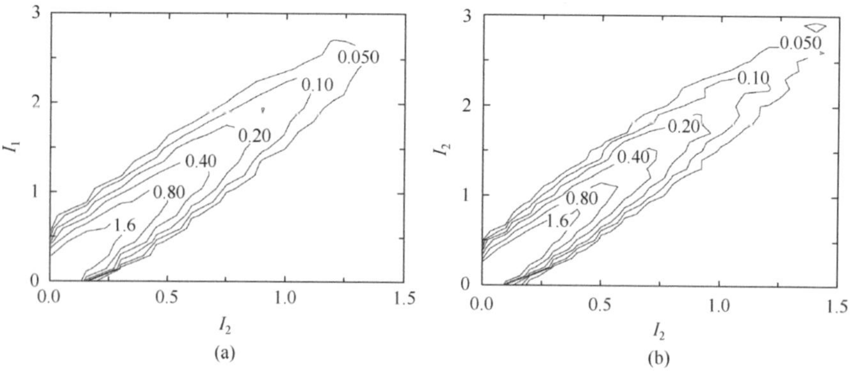

Fig. 7.18 Contours of $p(I_1, I_2)$ obtained from: **a** simulation of the averaged fractional SDE (7.48) and **b** simulation of the original system (7.44). System parameters are the same as those in Fig. 7.17 (Lü et al. 2017a)

where $\boldsymbol{q}_1 = [q_1, q_2, \ldots, q_{r-1}]^{\mathrm{T}}, \boldsymbol{p}_1 = [p_1, p_2, \ldots, p_{r-1}]^{\mathrm{T}}, \boldsymbol{q}_2 = [q_r, q_{r+1}, \ldots, q_n]^{\mathrm{T}}$, and $\boldsymbol{p}_2 = [p_r, p_{r+1}, \ldots, p_n]^{\mathrm{T}}$. Equation (7.51) indicates that the Hamiltonian of integrable subsystem is the sum of $H_1, H_2, \ldots, H_{r-1}$ while the Hamiltonian of non-integrable subsystem is H_r.

If it is possible to obtain action-angle variables for the integrable Hamiltonian subsystem, then Eq. (7.51) can be expressed as follows

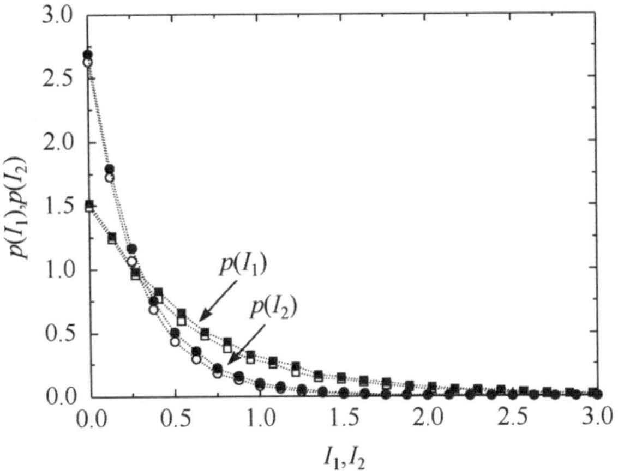

Fig. 7.19 $p(I_1), p(I_2)$ obtained from simulation of averaged fractional SDE (7.48) and original system (7.44). System parameters are the same as those in Fig. 7.17. For 40,000 samples, the computation time for the original system is 421 s, while the computation time for the averaged fractional SDE is 289 s. ···○···, ···□···, results from simulation of system (7.44); ···●···, ···■···, results from simulation of Eq. (7.48) (Lü et al. 2017a)

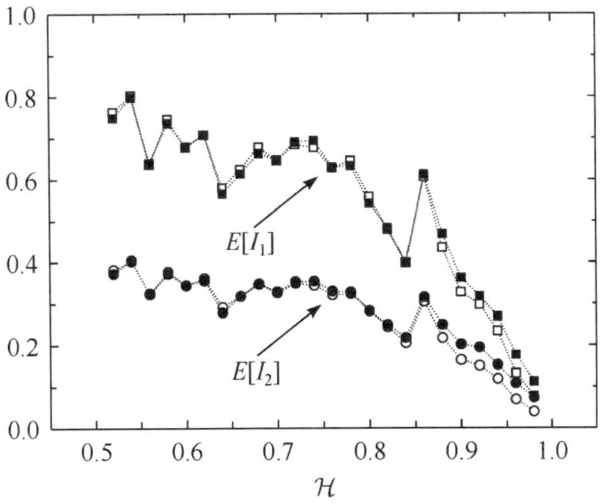

Fig. 7.20 $E[I_1], E[I_2]$ obtained from simulation of averaged fractional SDE (7.48) and original system (7.44). System parameters are the same as those in Fig. 7.17. ···○···, ···□···, results from simulation of system (7.44); ···●···, ···■···, results from simulation of Eq. (7.48) (Lü et al. 2017a)

$$H(\boldsymbol{I}', \boldsymbol{q}_2, \boldsymbol{p}_2) = \sum_{\eta=1}^{r-1} H_\eta(I_\eta) + H_r(\boldsymbol{q}_2, \boldsymbol{p}_2) \tag{7.52}$$

where $\boldsymbol{I}' = [I_1, I_2, \ldots, I_{r-1}]^{\mathrm{T}}$. Using fractional stochastic differential rules, the following fractional SDE can be derived from Eq. (7.3)

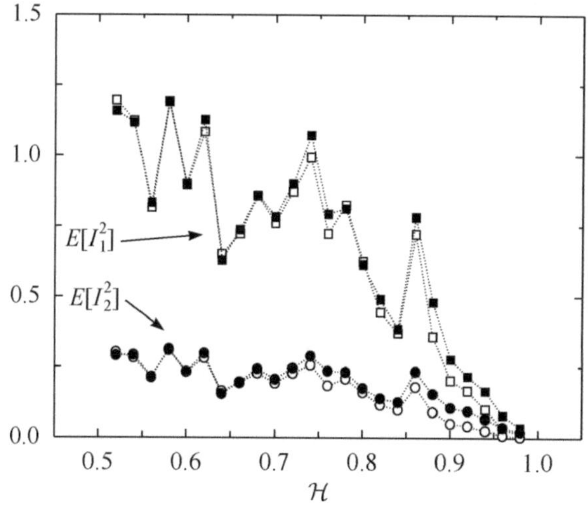

Fig. 7.21 $E[I_1^2]$, $E[I_2^2]$ obtained from simulation of averaged fractional SDE (7.48) and original system (7.44). System parameters are the same as those in Fig. 7.17. ···○···, ···□···, results from simulation of system (7.44); ···●···, ···■···, results from simulation of Eq. (7.48) (Lü et al. 2017a)

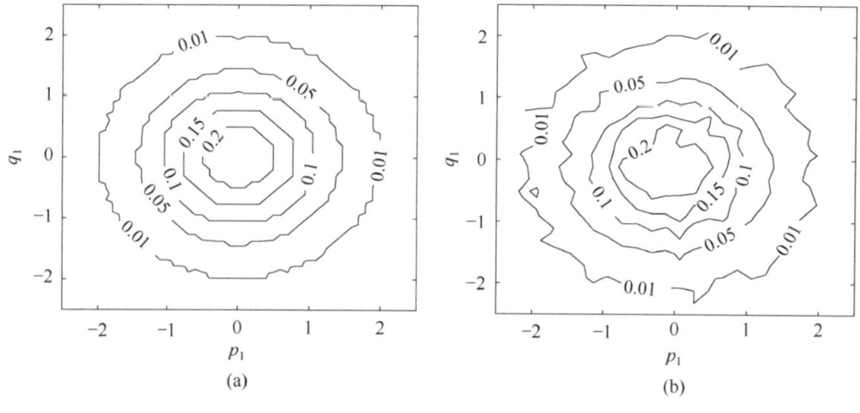

Fig. 7.22 Contours of $p(q_1, p_1)$ obtained from: **a** simulation of the averaged fractional SDE (7.48) and **b** simulation of the original system (7.44). System parameters are the same as those in Fig. 7.17 (Lü et al. 2017a)

$$
\begin{aligned}
\mathrm{d}I_\eta = \varepsilon \sum_{j=1}^{n} \sum_{\eta'=1}^{r-1} \left(-m_{\eta' j} \frac{\partial I_\eta}{\partial P_{\eta'}} \frac{\partial H}{\partial P_j} \right) \mathrm{d}t \\
+ \varepsilon^{1/2} \sum_{\eta'=1}^{r-1} \sum_{l=1}^{m} g_{\eta' l} \frac{\partial I_\eta}{\partial P_{\eta'}} \mathrm{d}B_{Hl}(t),
\end{aligned}
$$

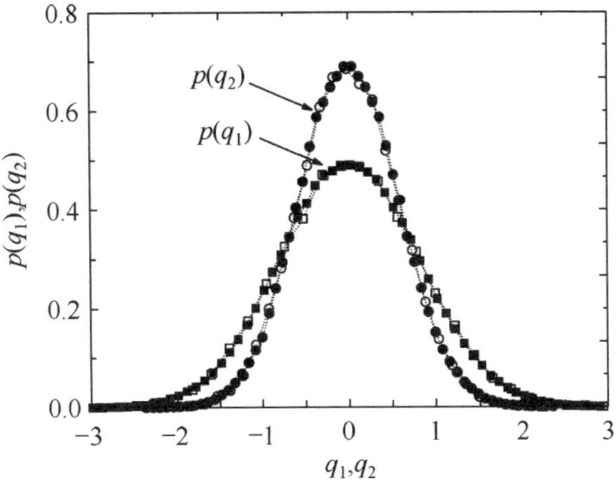

Fig. 7.23 $p(q_1), p(q_2)$ obtained from simulation of averaged fractional SDE (7.48) and original system (7.44). System parameters are the same as those in Fig. 7.17. ···○···, ···□···, results from simulation of system (7.44); ···●···, ···■···, results from simulation of Eq. (7.48) (Lü et al. 2017a)

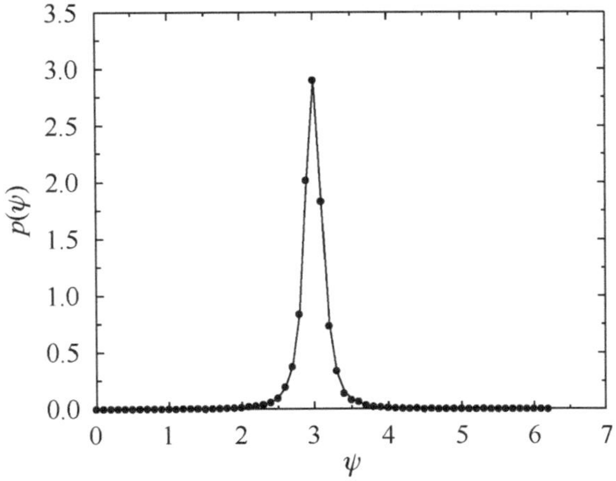

Fig. 7.24 $p(\psi)$ obtained from simulation of averaged fractional SDE (7.48) and original system (7.44). System parameters are the same as those in Fig. 7.17. —, results from simulation of system (7.44); ···●···, results from simulation of Eq. (7.48) (Lü et al. 2017a)

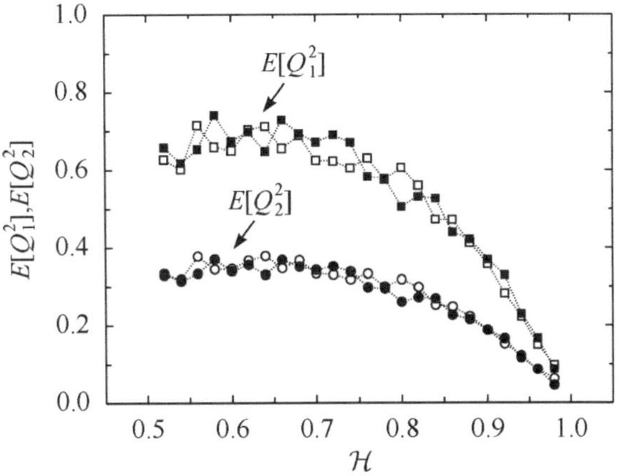

Fig. 7.25 $E[Q_1^2], E[Q_2^2]$ obtained from simulation of averaged fractional SDE (7.48) and original system (7.44). System parameters are the same as those in Fig. 7.17. ···○···, ···□···, results from simulation of system (7.44); ···●···, ···■···, results from simulation of Eq. (7.48) (Lü et al. 2017a)

$$
\begin{aligned}
\mathrm{d}\Theta_\eta &= \left[\omega_\eta - \varepsilon \sum_{j=1}^{n}\sum_{\eta'=1}^{r-1}\left(m_{\eta' j}\frac{\partial \Theta_\eta}{\partial P_{\eta'}}\frac{\partial H}{\partial P_j}\right)\right]\mathrm{d}t \\
&\quad + \varepsilon^{1/2}\sum_{\eta'=1}^{r-1}\sum_{l=1}^{m} g_{\eta' l}\frac{\partial \Theta_\eta}{\partial P_{\eta'}}\mathrm{d}B_{Hl}(t), \\
\mathrm{d}H_r &= \varepsilon \sum_{j}^{n}\sum_{\rho=r}^{n}\left(-m_{\rho j}\frac{\partial H_r}{\partial P_\rho}\frac{\partial H}{\partial P_j}\right)\mathrm{d}t \\
&\quad + \varepsilon^{1/2}\sum_{\rho=r}^{n}\sum_{l=1}^{m} g_{\rho l}\frac{\partial H_r}{\partial P_\rho}\mathrm{d}B_{Hl}(t), \\
\eta &= 1, 2, \ldots, r-1.
\end{aligned}
\tag{7.53}
$$

where $B_{Hl}(t)$ are independent unit fBms. Now the new system consists of all the equations in Eq. (7.53), $(n - r + 1)$ equations of Q_j and $(n - r)$ equations of P_j in Eq. (7.3). In this new system, I_η and H_r are slowly-varying processes, while $\Theta_1, \Theta_2, \ldots, \Theta_{r-1}, Q_r, \ldots, Q_n, P_{r+1}, \ldots, P_n$ are fast-varying processes. According to the principle of fractional stochastic averaging (Xu et al. 2014a, b), as $\varepsilon \to 0$, $\boldsymbol{I}'(t), H_r(t)$ converge in the sense of mean-square to the processes governed by the following averaged fractional SDEs

$$
\begin{aligned}
\mathrm{d}I_\eta &= m_\eta(\boldsymbol{I}', H_r)\mathrm{d}t + \sum_{l=1}^{m}\sigma_{\eta l}(\boldsymbol{I}', H_r)\mathrm{d}B_{Hl}(t), \\
\mathrm{d}H_r &= m_r(\boldsymbol{I}', H_r)\mathrm{d}t + \sum_{l=1}^{m}\sigma_{rl}(\boldsymbol{I}', H_r)\mathrm{d}B_{Hl}(t), \\
\eta &= 1, 2, \ldots, r-1.
\end{aligned}
\tag{7.54}
$$

The coefficients in Eq. (7.54) can be obtained from the time averaging of the coefficients of Eq. (7.53) as follows

$$
\begin{aligned}
m_\eta(I', H_r) &= \varepsilon\left\langle \sum_{j=1}^{n}\sum_{\eta'=1}^{r-1}\left(-m_{\eta' j}\frac{\partial I_\eta}{\partial P_{\eta'}}\frac{\partial H}{\partial P_j}\right)\right\rangle_t, \\
m_r(I', H_r) &= \varepsilon\left\langle \sum_{j=1}^{n}\sum_{\rho=r}^{n}\left(-m_{\rho j}\frac{\partial H_r}{\partial P_\rho}\frac{\partial H}{\partial P_j}\right)\right\rangle_t, \\
\sum_{l=1}^{m}\sigma_{\eta l}\sigma_{\bar{\eta} l}(I', H_r) &= \varepsilon\left\langle \sum_{\eta',\eta''=1}^{r-1}\sum_{l=1}^{m} g_{\eta' l}g_{\eta'' l}\frac{\partial I_\eta}{\partial P_{\eta'}}\frac{\partial I_{\bar{\eta}}}{\partial P_{\eta''}}\right\rangle_t, \\
\sum_{l=1}^{m}\sigma_{\eta l}\sigma_{rl}(I', H_r) &= \varepsilon\left\langle \sum_{\eta'=1}^{r-1}\sum_{\rho=r}^{n}\sum_{l=1}^{m} g_{\eta' l}g_{\rho l}\frac{\partial I_\eta}{\partial P_{\eta'}}\frac{\partial H_r}{\partial P_\rho}\right\rangle_t,
\end{aligned}
$$

$$\sum_{l=1}^{m} \sigma_{rl}\sigma_{rl}(I', H_r) = \varepsilon \left\langle \sum_{\rho,\rho'=r}^{n} \sum_{l=1}^{m} g_{\rho l} g_{\rho' s} \frac{\partial H_r}{\partial P_\rho} \frac{\partial H_r}{\partial P_{\rho'}} \right\rangle_t . \tag{7.55}$$

Since the integrable Hamiltonian subsystems is ergodicity on the $(r-1)$-dimensional torus and the non-integrable Hamiltonian subsystems is ergodicity on the $(2n-2r+1)$-dimensional isoenergetic surface, the time averaging in Eq. (7.55) can be replaced by the spatial averaging as follows

$$\begin{aligned}
\langle [\cdot] \rangle_t &= \frac{1}{(2\pi)^{r-1} T(H_r)} \int_{\Omega_1} \int_0^{2\pi} \left([\cdot] \Big/ \frac{\partial H_r}{\partial p_r} \right) \mathrm{d}\boldsymbol{\theta}' \mathrm{d}q_r \ldots \mathrm{d}q_n \mathrm{d}p_{r+1} \ldots \mathrm{d}p_n, \\
T(H_r) &= \int_{\Omega_1} \left(1 \Big/ \frac{\partial H_r}{\partial p_r} \right) \mathrm{d}q_r \ldots \mathrm{d}q_n \mathrm{d}p_{r+1} \ldots \mathrm{d}p_n, \\
\Omega_1 &= \{(q_r, \ldots, q_n, p_{r+1}, \ldots, p_n) | H_r(q_r, \ldots q_n, 0, p_{r+1}, \ldots p_n) \le H_r\}, \\
\boldsymbol{\theta}' &= [\theta_1, \theta_2, \ldots, \theta_{r-1}]^{\mathrm{T}}.
\end{aligned} \tag{7.56}$$

Monte Carlo simulation of the averaged fractional SDEs (7.54) yields the stationary PDF $p(\boldsymbol{I}', h_r)$ of the averaged system. According to Eq. (5.192), the approximate stationary PDF $p(\boldsymbol{q}, \boldsymbol{p})$ of the original system can be obtained as follows

$$p(\boldsymbol{q}, \boldsymbol{p}) = \left. \frac{p(\boldsymbol{I}', h_r)}{(2\pi)^{r-1} T(h_r)} \right|_{\boldsymbol{I}'=\boldsymbol{I}'(\boldsymbol{q}_1, \boldsymbol{p}_1), h_r = H_r(\boldsymbol{q}_2, \boldsymbol{p}_2)} . \tag{7.57}$$

The marginal PDFs and statistics of the generalized displacements and generalized momenta can be obtained from $p(\boldsymbol{q}, \boldsymbol{p})$.

If the action-angle variables I_η, θ_η cannot be found for the integrable Hamiltonian subsystem, H_η can be used to replace I_η in Eq. (7.53), which results in

$$\begin{aligned}
\mathrm{d}H_\eta &= \varepsilon \sum_{j=1}^{n} \sum_{\eta'=1}^{r-1} \left(-m_{\eta' j} \frac{\partial H_\eta}{\partial P_{\eta'}} \frac{\partial H}{\partial P_j} \right) \mathrm{d}t \\
&\quad + \varepsilon^{1/2} \sum_{\eta'=1}^{r-1} \sum_{l=1}^{m} g_{\eta' l} \frac{\partial H_\eta}{\partial P_{\eta'}} \mathrm{d}B_{Hl}(t), \\
\eta &= 1, 2, \ldots, r-1.
\end{aligned} \tag{7.58}$$

Then, the averaged fractional SDE (7.54) becomes

$$\mathrm{d}H_\eta = m_\eta(\boldsymbol{H}_1)\mathrm{d}t + \sum_{l=1}^{m} \sigma_{\eta l}(\boldsymbol{H}_1) \mathrm{d}B_{Hl}(t),$$

$$
\begin{aligned}
\mathrm{d}H_r &= m_r(\boldsymbol{H}_1)\mathrm{d}t + \sum_{l=1}^{m} \sigma_{rl}(\boldsymbol{H}_1)\mathrm{d}B_{Hl}(t),\\
\eta &= 1, 2, \ldots, r-1.
\end{aligned} \tag{7.59}
$$

where $\boldsymbol{H}_1 = [H_1, \ldots, H_{r-1}, H_r]^{\mathrm{T}}$. The coefficients in Eq. (7.59) can be obtained from the time averaging of the coefficient of H_r in Eq. (7.53) and the coefficients of H_η $(\eta = 1, 2, \ldots, r-1)$ in Eq. (7.58) as follows

$$
\begin{aligned}
m_\eta(\boldsymbol{H}_1) &= \varepsilon\left\langle \sum_{j=1}^{n}\sum_{\eta'=1}^{r-1}\left(-m_{\eta' j}\frac{\partial H_\eta}{\partial P_{\eta'}}\frac{\partial H}{\partial P_j}\right)\right\rangle_t, \quad \eta = 1, 2, \ldots, r-1,\\
m_r(\boldsymbol{H}_1) &= \varepsilon\left\langle \sum_{j=1}^{n}\sum_{\rho=r}^{n}\left(-m_{\rho j}\frac{\partial H_r}{\partial P_\rho}\frac{\partial H}{\partial P_j}\right)\right\rangle_t,\\
\sum_{l=1}^{m}\sigma_{\eta l}\sigma_{\overline{\eta} l}(\boldsymbol{H}_1) &= \varepsilon\left\langle \sum_{\eta',\eta''=1}^{r-1}\sum_{l=1}^{m} g_{\eta' l}g_{\eta'' l}\frac{\partial H_\eta}{\partial P_{\eta'}}\frac{\partial H_{\overline{\eta}}}{\partial P_{\eta''}}\right\rangle_t,\\
\sum_{l=1}^{m}\sigma_{\eta l}\sigma_{rl}(\boldsymbol{H}_1) &= \varepsilon\left\langle \sum_{\eta'=1}^{r-1}\sum_{\rho=r}^{n}\sum_{l=1}^{m} g_{\eta' l}g_{\rho l}\frac{\partial H_\eta}{\partial P_{\eta'}}\frac{\partial H_r}{\partial P_\rho}\right\rangle_t,\\
\sum_{l=1}^{m}\sigma_{rl}\sigma_{rl}(\boldsymbol{H}_1) &= \varepsilon\left\langle \sum_{\rho,\rho'=r}^{n}\sum_{l=1}^{m} g_{\rho l}g_{\rho' s}\frac{\partial H_r}{\partial P_\rho}\frac{\partial H_r}{\partial P_{\rho'}}\right\rangle_t.
\end{aligned} \tag{7.60}
$$

If each integrable Hamiltonian subsystem with Hamiltonian H_η $(\eta = 1, 2, \ldots, r-1)$ has a family of periodic solutions on the entire phase plane and the non-integrable Hamiltonian subsystem is ergodic on $(2n-2r+1)$-dimensional isoenergetic surface, time averaging in Eq. (7.60) can be replaced by the following spatial averaging

$$
\begin{aligned}
\langle[\cdot]\rangle_t &= \frac{1}{T(H_1)\ldots T(H_{r-1})T(H_r)}\\
&\quad \oint\left([\cdot]\Big/\prod_{\mu=1}^{r}\frac{\partial H_\mu}{\partial p_\mu}\right)\mathrm{d}q_1\mathrm{d}q_2\ldots\mathrm{d}q_n\mathrm{d}p_{r+1}\ldots\mathrm{d}p_n,\\
T(H_\eta) &= \int_{\Omega_\eta}\left(1\Big/\frac{\partial H_\eta}{\partial p_\eta}\right)\mathrm{d}q_\eta, \quad \Omega_\eta = \{q_\eta | H_\eta(q_r, 0) \le H_\eta\}, \quad \eta = 1, 2, \ldots, r-1,\\
T(H_r) &= \int_{\Omega_r}\left(1\Big/\frac{\partial H_r}{\partial p_r}\right)\mathrm{d}q_r\ldots\mathrm{d}q_n\mathrm{d}p_{r+1}\ldots\mathrm{d}p_n,\\
\Omega_r &= \{(q_r, \ldots, q_n, p_{r+1}, \ldots, p_n)|H_r(q_r, \ldots q_n, 0, p_{r+1}, \ldots p_n) \le H_r\}.
\end{aligned} \tag{7.61}
$$

Monte Carlo simulation of the averaged fractional SDEs (7.59) yields the stationary PDF $p(\boldsymbol{h}_1)$ of the averaged system. Following Eq. (5.3.25), the approximate stationary PDF of the original system can be obtained as follows

$$p(\boldsymbol{q}, \boldsymbol{p}) = \left. \frac{p(\boldsymbol{h}_1)}{T(h_1)\ldots T(h_r)} \right|_{h_\eta = H_\eta(q_\eta, p_\eta), h_r = H_r(\boldsymbol{q}_2, \boldsymbol{p}_2)} . \tag{7.62}$$

The marginal stationary PDFs and statistics of the generalized displacements and generalized momenta can be further derived from $p(\boldsymbol{q}, \boldsymbol{p})$ by integration.

The dimension of averaged Eqs. (7.54) and (7.59) is lower than that of the original system (7.1). So, the computation time for simulation of averaged system is much shorter than that for original system while both two simulations yield similar results. Therefore, the stochastic averaging method is useful for studying the dynamics of quasi-Hamiltonian systems excited by fGns. Two examples are presented below to illustrate the applications of the stochastic averaging method presented in this section.

Example 7.5 Consider a 4-DOF fGns excited quasi-Hamiltonian system governed by the following equations (Lü et al. 2017c)

$$\begin{aligned}
\dot{Q}_1 &= P_1, \\
\dot{P}_1 &= -\omega_1^2 Q_1 - \left[\alpha_{10} + \alpha_{11} P_1^2 + \alpha_{12} P_2^2 + \alpha_{13} P_3^2 + \alpha_{14} P_4^2 \right. \\
&\quad \left. + (\alpha_{13} + \alpha_{14}) U(Q_3, Q_4)\right] P_1 + \sqrt{2D_1} W_{H1}(t), \\
\dot{Q}_2 &= P_2, \\
\dot{P}_2 &= -\omega_2^2 Q_2 - \left[\alpha_{20} + \alpha_{21} P_1^2 + \alpha_{22} P_2^2 + \alpha_{23} P_3^2 + \alpha_{24} P_4^2 \right. \\
&\quad \left. + (\alpha_{23} + \alpha_{24}) U(Q_3, Q_4)\right] P_2 + \sqrt{2D_2} W_{H2}(t), \\
\dot{Q}_3 &= P_3, \\
\dot{P}_3 &= -\partial U / \partial Q_3 - \left[\alpha_{30} + \alpha_{31} P_1^2 + \alpha_{32} P_2^2 + \alpha_{33} P_3^2 + \alpha_{34} P_4^2 \right. \\
&\quad \left. + (1/2)(\alpha_{34} + 3\alpha_{33}) U(Q_3, Q_4)\right] P_3 + \sqrt{2D_3} W_{H3}(t), \\
\dot{Q}_4 &= P_4, \\
\dot{P}_4 &= -\partial U / \partial Q_4 - \left[\alpha_{40} + \alpha_{41} P_1^2 + \alpha_{42} P_2^2 + \alpha_{43} P_3^2 + \alpha_{44} P_4^2 \right. \\
&\quad \left. + (1/2)(\alpha_{43} + 3\alpha_{44}) U(Q_3, Q_4)\right] P_4 + \sqrt{2D_4} W_{H4}(t),
\end{aligned} \tag{7.63}$$

where

$$U(Q_3, Q_4) = k\left(\omega_3^2 Q_3^2 + \omega_4^2 Q_4^2\right)^3 / 6. \tag{7.64}$$

is the potential function; ω_j, α_{ij}, k are constants; $W_{Hl}(t)$ $(l = 1, 2, 3, 4)$ are independent unit fGns with Hurst index $\mathcal{H} \in (1/2, 1)$ and intensities $2D_l$; α_{ij} and D_l are small quantities of order ε. The Hamiltonian associated with system (7.63) is

$$H = H_1 + H_2 + H_3 = \omega_1 I_1 + \omega_2 I_2 + H_3. \tag{7.65}$$

where

$$\begin{aligned} &I_1 = \left(p_1^2 + \omega_1^2 q_1^2\right)/2\omega_1, \; I_2 = \left(p_2^2 + \omega_2^2 q_2^2\right)/2\omega_2, \\ &H_3 = \left(p_3^2 + p_4^2\right)/2 + U(q_3, q_4). \end{aligned} \tag{7.66}$$

The potential function $U(q_3, q_4)$ is inseparable, so system (7.63) is a quasi-partially integrable Hamiltonian system. When ω_1 and ω_2 do not satisfy weak resonant condition $\sum_{i=1}^{2} k_i^u \omega_i = O(\varepsilon)$, the system (7.63) is non-internal resonant. The fractional SDEs for $I_1(t), I_2(t), H_3(t)$ are of the form in Eq. (7.53). For system (7.63), only $I_1(t), I_2(t), H_3(t)$ are slowly-varying processes. They converge to a 3-dimensional vector process in mean-square sense governed by the following fractional SDEs

$$\begin{aligned} dI_1 &= m_1(I_1, I_2, H_3)dt + \sigma_{11}(I_1, I_2, H_3)dB_{H1}(t), \\ dI_2 &= m_2(I_1, I_2, H_3)dt + \sigma_{22}(I_1, I_2, H_3)dB_{H2}(t), \\ dH_3 &= m_3(I_1, I_2, H_3)dt + \sigma_{33}(I_1, I_2, H_3)dB_{H3}(t). \end{aligned} \tag{7.67}$$

According to Eqs. (7.55) and (7.56), the coefficients in Eq. (7.67) can be obtained as follows

$$\begin{aligned} m_1(I_1, I_2, H_3) &= -\left[\alpha_{10} I_1 + 3\alpha_{11}\omega_1 I_1^2/2 + \alpha_{12}\omega_2 I_1 I_2 + (\alpha_{13} + \alpha_{14}) I_1 H_3\right], \\ m_2(I_1, I_2, H_3) &= -\left[\alpha_{20} I_2 + 3\alpha_{22}\omega_2 I_2^2/2 + \alpha_{21}\omega_1 I_1 I_2 + (\alpha_{23} + \alpha_{24}) I_2 H_3\right], \\ m_3(I_1, I_2, H_3) &= -[(\alpha_{30} + \alpha_{40}) + (\alpha_{31} + \alpha_{41})\omega_1 I_1 \\ &\quad + (\alpha_{32} + \alpha_{42})\omega_2 I_2 + (3\alpha_{33} + 3\alpha_{44} + \alpha_{34} + \alpha_{43}) H_3/2] \frac{3H_3}{4}, \\ \sigma_{11}^2(I_1, I_2, H_3) &= 2D_1 I_1/\omega_1, \quad \sigma_{22}^2(I_1, I_2, H_3) = 2D_2 I_2/\omega_2, \\ \sigma_{33}^2(I_1, I_2, H_3) &= \frac{3}{2}(D_3 + D_4) H_3, \quad \sum_{l=1}^{3} \sigma_{1l}\sigma_{2l} = \sum_{l=1}^{3} \sigma_{1l}\sigma_{3l} = \sum_{l=1}^{3} \sigma_{2l}\sigma_{3l} = 0, \\ T(H_3) &= \frac{2\pi^2}{\omega_3\omega_4}\sqrt[3]{\frac{6H_3}{k}}. \end{aligned} \tag{7.68}$$

Monte Carlo simulation of the averaged Eq. (7.67) yields the stationary PDF $p(I_1, I_2, h_3)$. According to Eq. (7.57), the approximate stationary PDF $p(\boldsymbol{q}, \boldsymbol{p})$, the marginal PDFs and statistics for the original system (7.63) can be derived as follows

$$p(\boldsymbol{q},\boldsymbol{p}) = \left.\frac{p(I_1,I_2,h_3)}{4\pi^2 T(h_3)}\right|_{\substack{I_1=(p_1^2+\omega_1^2 q_1^2)/2\omega_1,\\ I_2=(p_2^2+\omega_2^2 q_2^2)/2\omega_2,\\ H_3=(p_3^2+p_4^2)/2+U(q_3,q_4)}},$$

$$p(q_1,q_2) = \int_{-\infty}^{\infty}\cdots\int_{-\infty}^{\infty} p(\boldsymbol{q},\boldsymbol{p})\mathrm{d}q_3\mathrm{d}q_4\mathrm{d}p_1\ldots\mathrm{d}p_4,$$

$$p(q_3,q_4) = \int_{-\infty}^{\infty}\cdots\int_{-\infty}^{\infty} p(\boldsymbol{q},\boldsymbol{p})\mathrm{d}q_1\mathrm{d}q_2\mathrm{d}p_1\ldots\mathrm{d}p_4,$$

$$E[Q_1^2] = \int_{-\infty}^{\infty}\cdots\int_{-\infty}^{\infty} q_1^2 p(\boldsymbol{q},\boldsymbol{p})\mathrm{d}q_1\ldots\mathrm{d}q_4\mathrm{d}p_1\ldots\mathrm{d}p_4,$$

$$E[Q_2^2] = \int_{-\infty}^{\infty}\cdots\int_{-\infty}^{\infty} q_2^2 p(\boldsymbol{q},\boldsymbol{p})\mathrm{d}q_1\ldots\mathrm{d}q_4\mathrm{d}p_1\ldots\mathrm{d}p_4,$$

$$E[Q_3^2] = \int_{-\infty}^{\infty}\cdots\int_{-\infty}^{\infty} q_3^2 p(\boldsymbol{q},\boldsymbol{p})\mathrm{d}q_1\ldots\mathrm{d}q_4\mathrm{d}p_1\ldots\mathrm{d}p_4,$$

$$E[Q_4^2] = \int_{-\infty}^{\infty}\cdots\int_{-\infty}^{\infty} q_4^2 p(\boldsymbol{q},\boldsymbol{p})\mathrm{d}q_1\ldots\mathrm{d}q_4\mathrm{d}p_1\ldots\mathrm{d}p_4. \tag{7.69}$$

The stationary PDF $p(I_1, I_2), p(I_1, h_3)$ and the mean-values $E[I_1], E[I_2], E[H_3]$ obtained from Monte Carlo simulation of the averaged Eq. (7.67) and the original system (7.63), respectively, are shown in Figs. 7.26–7.28. The stationary PDFs and mean-square values of the displacements are shown in Figs. 7.29–7.31. It is evident from these figures that, in the senses of probability and mean-square, the response of the original system can be predicted closely by using the averaged system. Furthermore, simulation of the averaged system is more efficient computationally than the simulation of original system. For example, the former simulation takes approximately 34 s for 10,000 samples, while the later simulation takes 81 s for the same number of samples.

Example 7.6 Consider a 3-DOF fGns excited quasi-Hamiltonian system governed by the following equations (Lü et al. 2017c)

$$\begin{aligned}
\dot{Q}_1 &= P_1,\\
\dot{P}_1 &= -\omega_1^2 Q_1 - P_1(\alpha_{10}+\alpha_{11}P_1^2+\alpha_{12}P_2^2+\alpha_{13}P_3^2) + \sqrt{2D_1}W_{H1}(t),\\
\dot{Q}_2 &= P_2,\\
\dot{P}_2 &= -\frac{\partial U}{\partial Q_2} - P_2(\alpha_{20}+\alpha_{21}P_1^2+\alpha_{22}P_2^2+\alpha_{23}P_3^2) + \sqrt{2D_2}W_{H2}(t),\\
\dot{Q}_3 &= P_3,
\end{aligned}$$

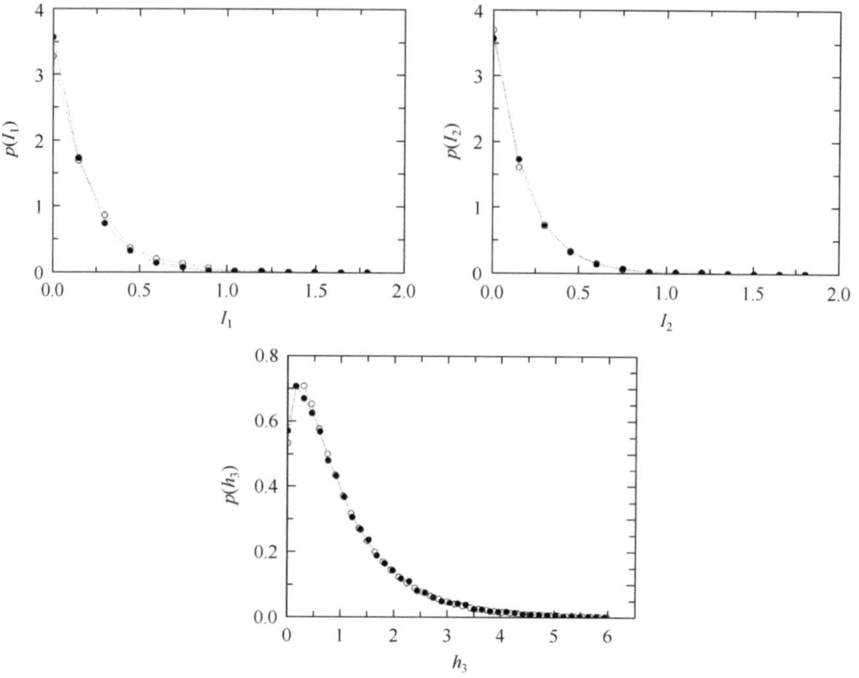

Fig. 7.26 $p(I_1), p(I_2), p(h_3)$ obtained from simulation of averaged Eq. (7.67) and original system (7.63). System parameters are $\alpha_{10} = \alpha_{20} = 0.05$; $\alpha_{30} = \alpha_{40} = 0.01$; $\alpha_{13} = \alpha_{14} = \alpha_{23} = \alpha_{24} = 0.01$; $\alpha_{31} = \alpha_{32} = \alpha_{41} = \alpha_{42} = 0.02$; $\alpha_{11} = \alpha_{22} = \alpha_{33} = \alpha_{44} = \alpha_{34} = \alpha_{43} = 0.0$; $\alpha_{12} = \alpha_{21} = 0.01$; $\omega_1^2 = 3$; $\omega_2^2 = 4$; $\omega_3^2 = 5$; $\omega_4^2 = 6$; $D_1 = D_2 = D_3 = D_4 = 0.04$; $k = 3$, $\mathcal{H} = 0.75$. $\cdots\circ\cdots$, results from simulation of system (7.63); $\cdots\bullet\cdots$, results from simulation of Eq. (7.67) (Lü et al. 2017c)

$$\dot{P}_3 = -\frac{\partial U}{\partial Q_3} - P_3(\alpha_{30} + \alpha_{31}P_1^2 + \alpha_{32}P_2^2 + \alpha_{33}P_3^2) + \sqrt{2D_3}W_{H3}(t). \tag{7.70}$$

where

$$U(Q_3, Q_4) = \left(\omega_3^2 Q_3^2 + \omega_4^2 Q_4^2\right)^3 \Big/ 2 + b\left(\omega_3^2 Q_3^2 + \omega_4^2 Q_4^2\right)^3 \Big/ 4. \tag{7.71}$$

is potential function; α_{ij} and $b > 0$ are constants; $W_{Hl}(t)$ $(l = 1, 2, 3)$ are independent unit fGns with Hurst index $\mathcal{H} \in (1/2, 1)$ and intensities $2D_k$; α_{ij} and D_k are small quantities of order ε. The Hamiltonian associated with system (7.71) is

$$H = H_1 + H_2 = \omega_1 I_1 + H_2, \tag{7.72}$$

where

$$I_1 = \left(p_1^2 + \omega_1^2 q_1^2\right) / 2\omega_1, \ H_2 = \left(p_2^2 + p_3^2\right) / 2 + U(Q_2, Q_3). \tag{7.73}$$

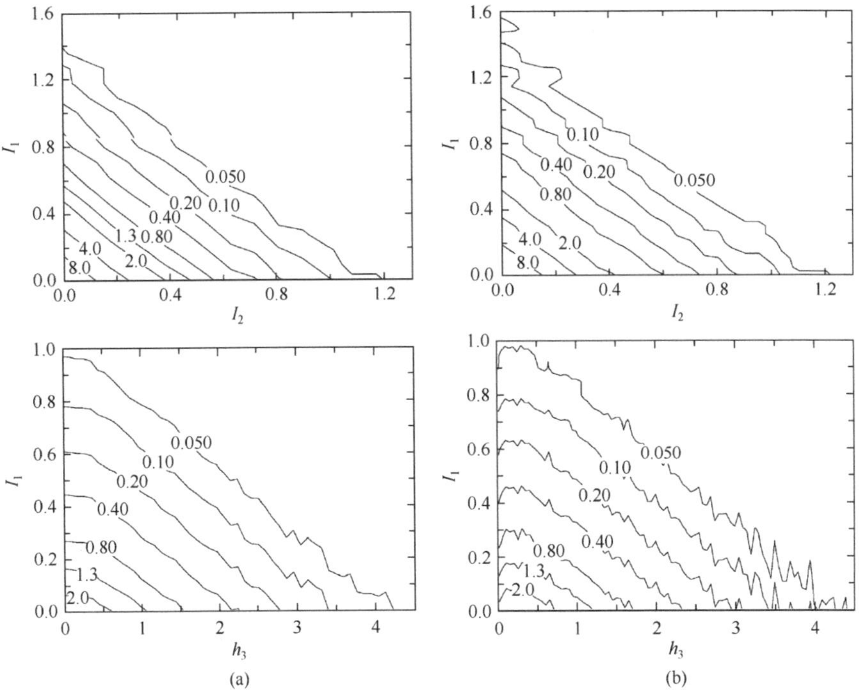

Fig. 7.27 Contours of $p(I_1, I_2)$, $p(I_1, h_3)$ obtained from: **a** simulation of the averaged Eq. (7.67) and **b** simulation of the original system (7.63). System parameters are the same as those in Fig. 7.26 (Lü et al. 2017c)

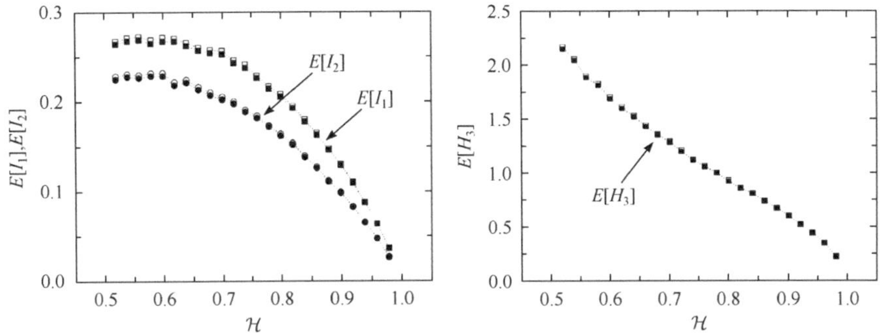

Fig. 7.28 $E[I_1]$, $E[I_2]$, $E[H_3]$ versus $\mathcal{H}$ are obtained from simulation of averaged Eq. (7.67) and original system (7.63). System parameters are the same as those in Fig. 7.26. ···○···, ···□···, results from simulation of system (7.63); ···●···, ···■···, results from simulation of Eq. (7.67) (Lü et al. 2017c)

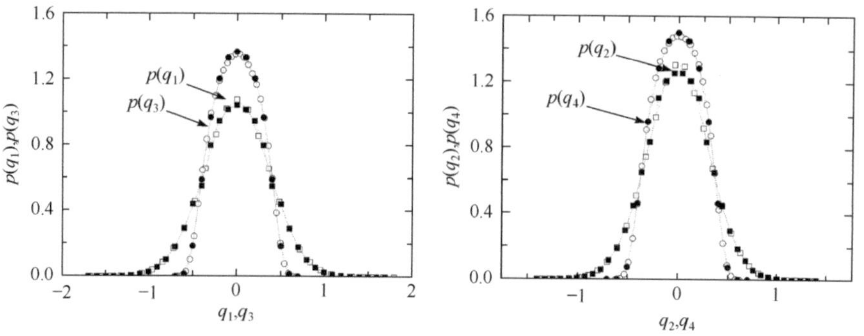

Fig. 7.29 $p(q_1), p(q_2), p(q_3), p(q_4)$ obtained from simulation of averaged Eq. (7.67) and original system (7.63). System parameters are the same as those in Fig. 7.26. ···○···, ···□···, results from simulation of system (7.63); ···●···, ···■···, results from simulation of Eq. (7.67) (Lü et al. 2017c)

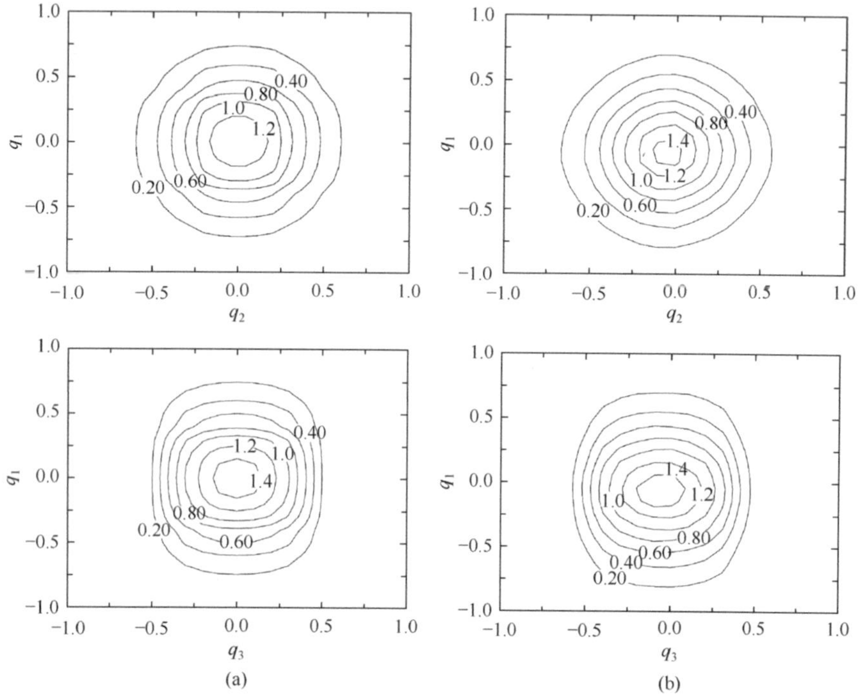

Fig. 7.30 Contours of $p(q_1, q_2), p(q_1, q_3)$ obtained from: **a** simulation of the averaged Eq. (7.67) and **b** simulation of the original system (7.63). System parameters are the same as those in Fig. 7.26 (Lü et al. 2017c)

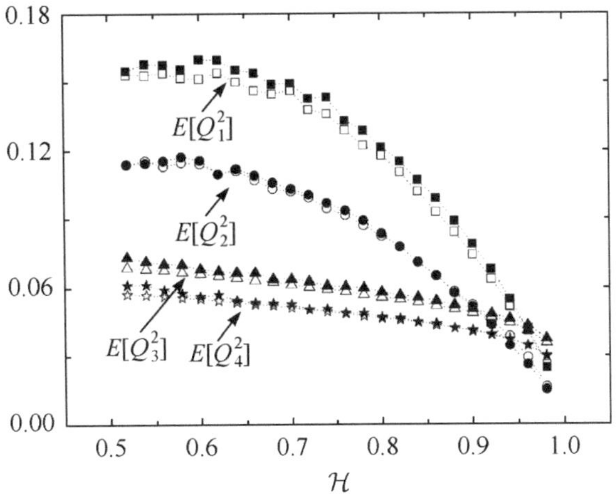

Fig. 7.31 $E[Q_1^2]$, $E[Q_2^2]$, $E[Q_3^2]$ versus $\mathcal{H}$ are obtained from simulation of averaged Eq. (7.67) and original system (7.63). System parameters are the same as those in Fig. 7.26. ···○···, ···□···, ···△···, ···☆···, results from simulation of system (7.63); ···●···, ···■···, ···▲···, ···★···, results from simulation of Eq. (7.67) (Lü et al. 2017c)

The system (7.70) is a quasi-partially integrable and non-resonant Hamiltonian system. Following the derivation from Eq. (7.53) to Eq. (7.56), the averaged fractional SDEs for system (7.70) are

$$\begin{aligned} \mathrm{d}I_1 &= m_1(I_1,H_2)\mathrm{d}t + \sigma_{11}(I_1,H_2)\mathrm{d}B_{H1}(t), \\ \mathrm{d}H_2 &= m_2(I_1,H_2)\mathrm{d}t + \sigma_{22}(I_1,H_2)\mathrm{d}B_{H2}(t). \end{aligned} \tag{7.74}$$

where the coefficients are

$$\begin{aligned} m_1(I_1, H_2) &= -\Bigg[\alpha_{10}I_1 + 3\alpha_{11}\omega_1 I_1^2/2 \\ &\quad + (\alpha_{12}+\alpha_{13})\left(H_2 - \frac{1}{4}R^2 - \frac{b}{12}R^4\right)\Bigg], \\ m_2(I_1, H_2) &= -(\alpha_{20} + \alpha_{21}\omega_1 I_1 + \alpha_{30} + \alpha_{31}\omega_1 I_1)\left(H_2 - \frac{1}{4}R^2 - \frac{b}{12}R^4\right) \\ &\quad - \frac{1}{8}(3\alpha_{22} + \alpha_{23} + \alpha_{32} + 3\alpha_{33}) \\ &\quad \left[4H_2^2 + 8H_2R^2\left(\frac{1}{4} + \frac{b}{12}R^2\right) + R^4\left(\frac{1}{3} + \frac{b^2}{20}R^4 + \frac{b}{4}R^2\right)\right], \\ \sigma_{11}^2(I_1, H_2) &= 2D_1 I_1/\omega_1, \quad \sigma_{22}^2(I_1, H_2) \\ &= 2(D_2 + D_3)\left(H_2 - \frac{1}{4}R^2 - \frac{b}{12}R^4\right), \end{aligned}$$

$$\sum_{l=1}^{3} \sigma_{1l}\sigma_{2l}(I_1, H_2) = 0, \tag{7.75}$$

in which

$$R = \left(\frac{\sqrt{1 + 4bH_2} - 1}{b} \right)^{\frac{1}{2}}. \tag{7.76}$$

Monte Carlo simulation of the averaged Eq. (7.74) yields the stationary PDF $p(I_1, h_2)$. Following Eq. (7.57), the approximate stationary PDF for the original system (7.70) can be obtained as follows

$$p(\boldsymbol{q}, \boldsymbol{p}) = \left. \frac{p(I_1, h_2)}{2\pi T(h_2)} \right|_{\substack{I_1=(p_1^2+\omega_1^2 q_1^2)/2\omega_1, \\ h_2=(p_2^2+p_3^2)/2+U(q_2,q_3)}}. \tag{7.77}$$

Integration of $p(\boldsymbol{q}, \boldsymbol{p})$ yields the stationary PDF $p(q_1, q_2), p(q_1, q_3)$ and statistics $E[Q_1^2], E[Q_2^2], E[Q_3^2]$. The numerical results for PDF $p(I_1, h_2), p(I_1), p(I_1, h_2)$ and mean-value $E[I_1], E[H_2]$ are depicted in Figs. 7.32–7.34. The numerical results for PDF $p(q_1, q_2), p(q_1, q_3)$ and mean-square value $E[Q_1^2], E[Q_2^2], E[Q_3^2]$ are shown in Figs. 7.35 and 7.36. It is evident from the figures that the simulation results of the averaged Eq. (7.74) closely match those of the original system (7.70). Simulation of the averaged system (7.74) takes 23 s for 10,000 samples, while simulation of the original system (7.70) requires approximately 62 s for the same number of samples. The stochastic averaging method presented in this section saves much computation time in response prediction.

7.4.2 Internal Resonant Case

Assume that the Hamiltonian system associated with Eq. (7.3) is partially integrable and resonant, and the Hamiltonian is of the form of Eq. (7.52). There are $\beta(1 \le \beta \le r-1)$ weak resonance relations

$$\sum_{\eta=1}^{r-1} k_\eta^u \omega_\eta = O_u(\varepsilon), \quad u = 1, 2, \ldots, \beta. \tag{7.78}$$

Introduce β combinations of angle variables

$$\Psi_u = \sum_{\eta=1}^{r-1} k_\eta^u \Theta_\eta, \quad u = 1, 2, \ldots, \beta. \tag{7.79}$$

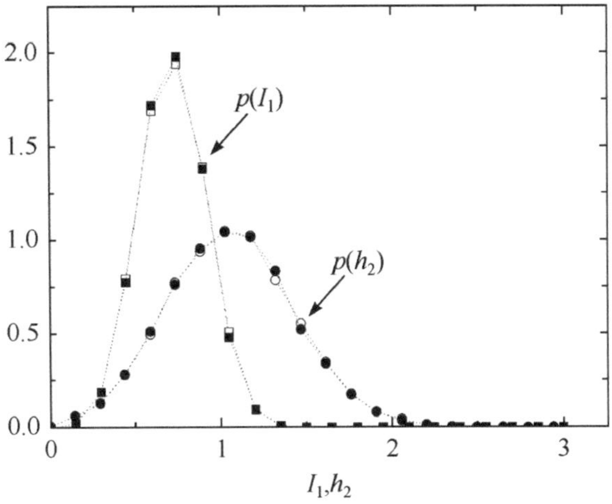

Fig. 7.32 $p(I_1), p(h_2)$ obtained from simulation of averaged Eq. (7.74) and original system (7.70). System parameters are $\alpha_{10} = \alpha_{20} = \alpha_{30} = -0.08$; $\alpha_{12} = \alpha_{13} = 0.01$; $\alpha_{21} = \alpha_{31} = 0.02$; $\alpha_{11} = \alpha_{22} = \alpha_{33} = 0.01$; $\alpha_{23} = \alpha_{32} = 0.04$; $\omega_1 = 1.414$; $\omega_2 = 1$; $\omega_3 = 1.732$; $b = 1$; $D_1 = 0.005$, $D_2 = 0.01$, $D_3 = 0.015$, $\mathcal{H} = 0.75$. $\cdots\circ\cdots$ $\cdots\square\cdots$, results from simulation of system (7.70); $\cdots\bullet\cdots$, $\cdots\blacksquare\cdots$, results from simulation of Eq. (7.74) (Lü et al. 2017c)

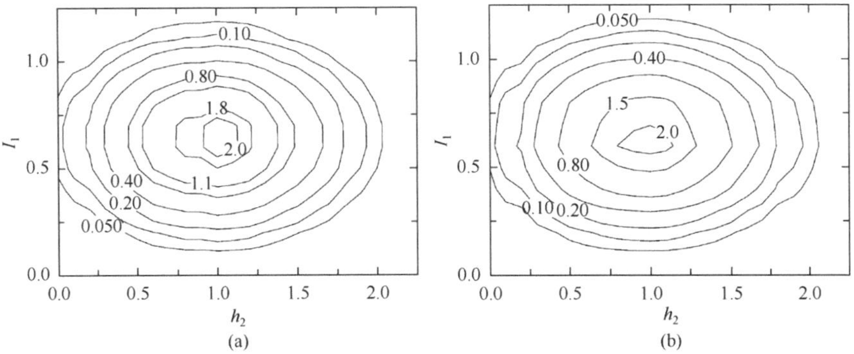

Fig. 7.33 Contours of $p(I_1, h_2)$ obtained from: **a** simulation of the averaged Eq. (7.74) and **b** simulation of the original system (7.70). System parameters are the same as those in Fig. 7.32 (Lü et al. 2017c)

The fractional SDEs for Ψ_u can be derived from Θ_η equations in Eq. (7.53) by combinations according to Eq. (7.79) as follows

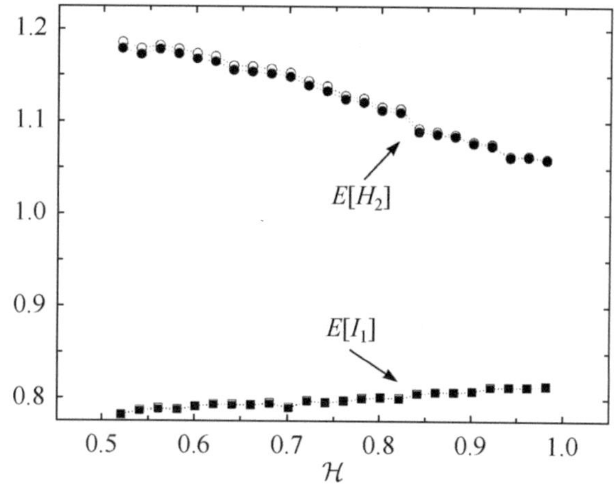

Fig. 7.34 $E[I_1]$, $E[H_2]$ with varying Hurst index $\mathcal{H}$. Results are obtained from simulation of averaged Eq. (7.74) and original system (7.70). System parameters are the same as those in Fig. 7.32. ···○···, ···□···, results from simulation of system (7.70); ···●···, ···■···, results from simulation of Eq. (7.74) (Lü et al. 2017c)

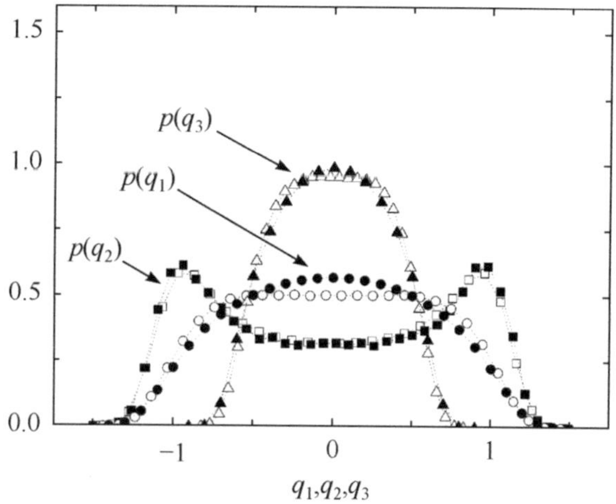

Fig. 7.35 $p(q_1), p(q_2), p(q_3)$ obtained from simulation of averaged Eq. (7.74) and original system (7.70). System parameters are the same as those in Fig. 7.32. ···○···, ···□···, ···△···, results from simulation of system (7.70); ···●···, ···■···, ···▲···, results from simulation of Eq. (7.74) (Lü et al. 2017c)

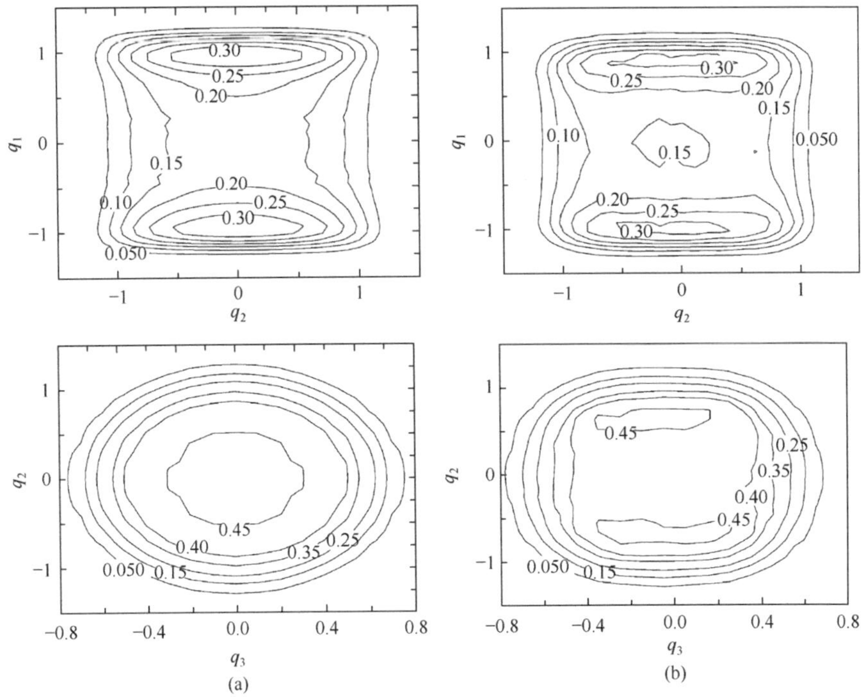

Fig. 7.36 Contours of $p(q_1, q_2)$, $p(q_2, q_3)$ obtained from: **a** simulation of the averaged Eq. (7.74) and **b** simulation of the original system (7.70). System parameters are the same as those in Fig. 7.32 (Lü et al. 2017c)

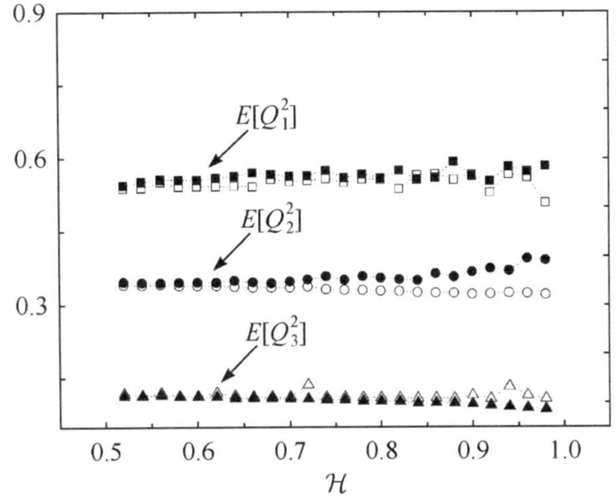

Fig. 7.37 $E[Q_1^2]$, $E[Q_2^2]$, $E[Q_3^2]$ versus $\mathcal{H}$ are obtained from simulation of averaged Eq. (7.74) and original system (7.70). System parameters are the same as those in Fig. 7.32. $\cdots\circ\cdots$, $\cdots\square\cdots$, $\cdots\triangle\cdots$ results from simulation of system (7.70); $\cdots\bullet\cdots$, $\cdots\blacksquare\cdots$, $\cdots\blacktriangle\cdots$, results from simulation of Eq. (7.74) (Lü et al. 2017c)

$$
\begin{aligned}
\mathrm{d}\Psi_u = & \left[O_u(\varepsilon) - \varepsilon \sum_{j=1}^{n} \sum_{\eta'=1}^{r-1} m_{\eta' j} \frac{\partial \Psi_u}{\partial P_{\eta'}} \frac{\partial H}{\partial P_j} \right] \mathrm{d}t \\
& + \varepsilon^{1/2} \sum_{\eta'=1}^{r-1} \sum_{l=1}^{m} g_{\eta' l} \frac{\partial \Psi_u}{\partial P_{\eta'}} \mathrm{d}B_{Hl}(t), \\
& u = 1, 2, \ldots, \beta.
\end{aligned} \tag{7.80}
$$

Now the system is composed by Eq. (7.80), $r-1$ equations of I_η, $r-1-\beta$ equations of Θ_η and one equation of H_r in Eq. (7.53), $n-r+1$ equations of Q_j and $n-r$ equations of P_j in Eq. (7.3). In this new system, $\boldsymbol{I}'(t)$, $\boldsymbol{\Psi}'(t)$ and $H_r(t)$ are slowly-varying processes while the others are fast-varying processes. According to the fractional stochastic averaging principle (Xu et al. 2014a, b), as $\varepsilon \to 0$, $\boldsymbol{I}'(t)$, $\boldsymbol{\Psi}'(t)$, $H_r(t)$ converge in mean-square sense to the processes governed by the following averaged fractional SDEs

$$
\begin{aligned}
\mathrm{d}I_\eta &= m_\eta(\boldsymbol{I}', \boldsymbol{\Psi}', H_r)\mathrm{d}t + \sum_{l=1}^{m} \sigma_{\eta l}(\boldsymbol{I}', \boldsymbol{\Psi}', H_r)\mathrm{d}B_{Hl}(t), \\
\mathrm{d}\Psi_u &= m_u(\boldsymbol{I}', \boldsymbol{\Psi}', H_r)\mathrm{d}t + \sum_{l=1}^{m} \sigma_{ul}(\boldsymbol{I}', \boldsymbol{\Psi}', H_r)\mathrm{d}B_{Hl}(t), \\
\mathrm{d}H_r &= m_r(\boldsymbol{I}', \boldsymbol{\Psi}', H_r)\mathrm{d}t + \sum_{l=1}^{m} \sigma_{rl}(\boldsymbol{I}', \boldsymbol{\Psi}', H_r)\mathrm{d}B_{Hl}(t), \\
\eta &= 1, 2, \ldots, r-1; \quad u = 1, 2, \ldots, \beta.
\end{aligned} \tag{7.81}
$$

where $\boldsymbol{\Psi}' = [\Psi_1, \Psi_2, \ldots, \Psi_\beta]^{\mathrm{T}}$. The coefficient in Eq. (7.81) are obtained from time averaging of the corresponding coefficients of Eqs. (7.53) and (7.80) as follows

$$
\begin{aligned}
m_\eta(\boldsymbol{I}', \boldsymbol{\Psi}', H_r) &= \varepsilon \left\langle \sum_{j=1}^{n} \sum_{\eta'=1}^{r-1} \left(-m_{\eta' j} \frac{\partial I_\eta}{\partial P_{\eta'}} \frac{\partial H}{\partial P_j} \right) \right\rangle_t, \\
m_u(\boldsymbol{I}', \boldsymbol{\Psi}', H_r) &= O(\varepsilon) + \varepsilon \left\langle \sum_{j=1}^{n} \sum_{\eta'=1}^{r-1} \left(-m_{\eta' j} \frac{\partial \Psi_u}{\partial P_{\eta'}} \frac{\partial H}{\partial P_j} \right) \right\rangle_t, \\
m_r(\boldsymbol{I}', \boldsymbol{\Psi}', H_r) &= \varepsilon \left\langle \sum_{j=1}^{n} \sum_{\rho=r}^{n} \left(-m_{\rho j} \frac{\partial H_r}{\partial P_\rho} \frac{\partial H}{\partial P_j} \right) \right\rangle_t, \\
\sum_{l=1}^{m} \sigma_{\eta l}\sigma_{\bar{\eta} l}(\boldsymbol{I}', \Psi', H_r) &= \varepsilon \left\langle \sum_{\eta', \eta''=1}^{r-1} \sum_{l=1}^{m} g_{\eta' l} g_{\eta'' l} \frac{\partial I_\eta}{\partial P_{\eta'}} \frac{\partial I_{\bar{\eta}}}{\partial P_{\eta''}} \right\rangle_t
\end{aligned}
$$

$$\sum_{l=1}^{m}\sigma_{\eta l}\sigma_{rl}(\boldsymbol{I}',\boldsymbol{\Psi}',H_r)=\varepsilon\left\langle\sum_{\eta'=1}^{r-1}\sum_{\rho=r}^{n}\sum_{l=1}^{m}g_{\eta' l}g_{\rho l}\frac{\partial I_\eta}{\partial P_{\eta'}}\frac{\partial H_r}{\partial P_\rho}\right\rangle_t,$$

$$\sum_{l=1}^{m}\sigma_{rl}\sigma_{rl}(\boldsymbol{I}',\boldsymbol{\Psi}',H_r)=\varepsilon\left\langle\sum_{\rho,\rho'=r}^{n}\sum_{l=1}^{m}g_{\rho l}g_{\rho' l}\frac{\partial H_r}{\partial P_\rho}\frac{\partial H_r}{\partial P_{\rho'}}\right\rangle_t,$$

$$\sum_{l=1}^{m}\sigma_{ul}\sigma_{vl}(\boldsymbol{I}',\boldsymbol{\Psi}',H_r)=\varepsilon\left\langle\sum_{\eta',\eta''=1}^{r-1}\sum_{l=1}^{m}g_{\eta' l}g_{\eta'' l}\frac{\partial \psi_u}{\partial P_{\eta'}}\frac{\partial \psi_v}{\partial P_{\eta''}}\right\rangle_t,$$

$$\sum_{l=1}^{m}\sigma_{\eta l}\sigma_{ul}(\boldsymbol{I}',\boldsymbol{\Psi}',H_r)=\varepsilon^{2\mathcal{H}}\left\langle\sum_{\eta',\eta''=1}^{r-1}\sum_{l=1}^{m}g_{\eta' l}g_{\eta'' l}\frac{\partial I_\eta}{\partial P_{\eta'}}\frac{\partial \psi_u}{\partial P_{\eta''}}\right\rangle_t, \tag{7.82}$$

$$\sum_{l=1}^{m}\sigma_{rl}\sigma_{ul}(\boldsymbol{I}',\boldsymbol{\Psi}',H_r)=\varepsilon\left\langle\sum_{\eta'=1}^{r-1}\sum_{\rho=r}^{n}\sum_{l=1}^{m}g_{\eta' l}g_{\rho l}\frac{\partial \psi_u}{\partial P_{\eta'}}\frac{\partial H_r}{\partial P_\rho}\right\rangle_t,$$

$$\eta,\overline{\eta}=1,2,\ldots,r-1;\quad u,v=1,2,\ldots,\beta.$$

Since the integrable Hamiltonian subsystem is ergodic on $(r-1-\beta)$-dimensional torus and the non-integrable Hamiltonian subsystems is ergodic on $[2(n-\beta)+1]$-dimensional isoenergetic surfaces, the time averaging can be replaced by the following spatial averaging

$$\langle[\cdot]\rangle_t=\frac{1}{(2\pi)^{r-\beta-1}T(H_r)}\int\limits_{\Omega_1}\int\limits_{0}^{2\pi}\left([\cdot]\Big/\frac{\partial H_r}{\partial p_r}\right)\mathrm{d}\theta''\mathrm{d}q_r\ldots\mathrm{d}q_n\mathrm{d}p_{r+1}\ldots\mathrm{d}p_n. \tag{7.83}$$

where $\int_0^{2\pi}(\cdot)\mathrm{d}\boldsymbol{\theta}''$ represents a $(r-\beta-1)$-fold integral with respect to $\boldsymbol{\theta}''=[\theta_1,\theta_2,\ldots,\theta_{r-\beta-1}]^{\mathrm{T}}$.

Monte Carlo simulation of the averaged fractional SDEs (7.81) yields the stationary PDF $p(\boldsymbol{I}',\boldsymbol{\psi}',h_r)$. According to Eq. (5.3.36), the approximate stationary PDF of the original system (7.1) is given by

$$p(\boldsymbol{q},\boldsymbol{p})=\frac{p(\boldsymbol{I},\boldsymbol{\psi}',h_r)}{(2\pi)^{r-\beta-1}T(h_r)}\left|\frac{\partial(\boldsymbol{I},\boldsymbol{\psi}',\boldsymbol{\theta}'')}{\partial(\boldsymbol{q}_1,\boldsymbol{p}_1)}\right|. \tag{7.84}$$

The dimension $r+\beta$ of the averaged Eq. (7.81) is lower than the dimension $2n$ of the original system (7.1). Therefore, simulation of the averaged system require less computation time than simulation of the original system.

Example 7.7 Consider fGns excited two linear oscillators and two nonlinear oscillators coupled with nonlinear damping forces. The governing equations are (Lü et al. 2017b)

$$\ddot{X}_1+\dot{X}_1\big[\alpha_{10}+\alpha_{11}\dot{X}_1^2+\alpha_{12}\dot{X}_2^2+\alpha_{13}\dot{X}_3^2+\alpha_{14}\dot{X}_4^2$$

$$
\begin{aligned}
&+(\alpha_{13}+\alpha_{14})U(X_3,X_4)]+\omega_1^2X_1=\sqrt{2D_1}W_{H1}(t),\\
&\ddot{X}_2+\dot{X}_2\big[\alpha_{20}+\alpha_{21}\dot{X}_1^2+\alpha_{22}\dot{X}_2^2+\alpha_{23}\dot{X}_3^2+\alpha_{24}\dot{X}_4^2\\
&\quad+(\alpha_{23}+\alpha_{34})U(X_3,X_4)]+\omega_2^2X_2=\sqrt{2D_2}W_{H2}(t),
\end{aligned}
$$

$$
\begin{aligned}
&\ddot{X}_3+\dot{X}_3\big[\alpha_{30}+\alpha_{31}\dot{X}_1^2+\alpha_{32}\dot{X}_2^2+\alpha_{33}\dot{X}_3^2+\alpha_{34}\dot{X}_4^2\\
&\quad+\frac{1}{2}(\alpha_{34}+3\alpha_{33})U(X_3,X_4)\Big]+\partial U/\partial X_3=\sqrt{2D_3}W_{H3}(t),\\
&\ddot{X}_4+\dot{X}_4\big[\alpha_{40}+\alpha_{41}\dot{X}_1^2+\alpha_{42}\dot{X}_2^2+\alpha_{43}\dot{X}_3^2+\alpha_{44}\dot{X}_4^2\\
&\quad+\frac{1}{2}(\alpha_{43}+3\alpha_{44})U(X_3,X_4)\Big]+\partial U/\partial X_4=\sqrt{2D_4}W_{H4}(t),
\end{aligned}
\tag{7.85}
$$

where

$$
U(X_3,X_4)=\frac{1}{2}\left(\omega_3^2X_3^2+\omega_4^2X_4^2\right)+\frac{b}{4}\left(\omega_3^2X_3^2+\omega_4^2X_4^2\right)^2,
\tag{7.86}
$$

ω_j, α_{ij}, b are positive constants; $W_{Hk}(t)$ are unit fGns with Hurst index $\mathcal{H} \in (1/2, 1)$ and intensities $2D_k$; α_{ij} and D_k are small quantities of order ε. Letting $Q_i = X_i,\quad P_i = \dot{X}_i,\quad i = 1, 2, 3, 4$, Eq. (7.85) can be rewritten as the following quasi-Hamiltonian system

$$
\begin{aligned}
\dot{Q}_1 =&P_1,\\
\dot{P}_1 =&-\omega_1^2Q_1-\big[\alpha_{10}+\alpha_{11}P_1^2+\alpha_{12}P_2^2+\alpha_{13}P_3^2+\alpha_{14}P_4^2\\
&+(\alpha_{13}+\alpha_{14})U(Q_3,Q_4)\big]P_1+\sqrt{2D_1}W_{H1}(t),\\
\dot{Q}_2 =&P_2,\\
\dot{P}_2 =&-\omega_2^2Q_2-\big[\alpha_{20}+\alpha_{21}P_1^2+\alpha_{22}P_2^2+\alpha_{23}P_3^2+\alpha_{24}P_4^2\\
&+(\alpha_{23}+\alpha_{24})U(Q_3,Q_4)\big]P_2+\sqrt{2D_2}W_{H2}(t),
\end{aligned}
$$

$$
\begin{aligned}
\dot{Q}_3 =&P_3,\\
\dot{P}_3 =&-\partial U/\partial Q_3-\big[\alpha_{30}+\alpha_{31}P_1^2+\alpha_{32}P_2^2+\alpha_{33}P_3^2+\alpha_{34}P_4^2\\
&+(1/2)(\alpha_{34}+3\alpha_{33})U(Q_3,Q_4)\big]P_3+\sqrt{2D_3}W_{H3}(t),\\
\dot{Q}_4 =&P_4,\\
\dot{P}_4 =&-\partial U/\partial Q_4-\big[\alpha_{40}+\alpha_{41}P_1^2+\alpha_{42}P_2^2+\alpha_{43}P_3^2+\alpha_{44}P_4^2\\
&+(1/2)(\alpha_{43}+3\alpha_{44})U(Q_3,Q_4)\big]P_4+\sqrt{2D_4}W_{H4}(t).
\end{aligned}
\tag{7.87}
$$

The Hamiltonian associated with system (7.87) is

$$
H=H_1+H_2+H_3,
$$

$$H_i = \left(p_i^2 + \omega_i^2 q_i^2\right)/2,\ i = 1, 2,\ H_3 = \left(p_3^2 + p_4^2\right)/2 + U(q_3, q_4). \tag{7.88}$$

H_1, H_2, H_3 are three independent and involutive first integrals. So, the Hamiltonian system is partially integrable, in which the subsystem with H_1, H_2 is integrable, and the subsystem with H_3 is non-integrable. Introduce action-angle variables

$$I_i = H_i/\omega_i,\ \theta_i = \tan^{-1}(p_i/\omega_i q_i),\ i = 1, 2. \tag{7.89}$$

Assume that $\omega_1 = \omega_2$, i.e., the integrable subsystem in Eq. (7.87) is in primary resonance. Introduce phase difference

$$\psi = \theta_1 - \theta_2. \tag{7.90}$$

Using the stochastic averaging method presented in Sect. 7.42, the following averaged fractional SDEs can be obtained

$$\begin{aligned}
\mathrm{d}I_1 &= \overline{m}_1(I_1,I_2,H_3,\ \Psi)\mathrm{d}t + \overline{\sigma}_{11}(I_1,I_2,H_3,\ \Psi)\mathrm{d}B_{H1}(t),\\
\mathrm{d}I_2 &= \overline{m}_2(I_1,I_2,H_3,\ \Psi)\mathrm{d}t + \overline{\sigma}_{22}(I_1,I_2,H_3,\ \Psi)\mathrm{d}B_{H2}(t),\\
\mathrm{d}\Psi &= \overline{m}_3(I_1,I_2,H_3,\ \Psi)\mathrm{d}t + \overline{\sigma}_{33}(I_1,I_2,H_3,\ \Psi)\mathrm{d}B_{H3}(t),\\
\mathrm{d}H_3 &= \overline{m}_4(I_1,I_2,H_3,\ \Psi)\mathrm{d}t + \overline{\sigma}_{44}(I_1,I_2,H_3,\ \Psi)\mathrm{d}B_{H4}(t).
\end{aligned} \tag{7.91}$$

where the coefficients can be obtained by using Eqs. (7.82) and (7.83) as follows

$$\begin{aligned}
\overline{m}_1(I_1,I_2,H_3,\ \psi) &= -\Bigg[\alpha_{10}I_1 + \frac{3\alpha_{11}\omega_1 I_1^2}{2} + \alpha_{12}\omega_2 I_1 I_2\\
&\quad \left(1 + \frac{1}{2}\cos 2\psi\right) + (\alpha_{13} + \alpha_{14})I_1 H_3\Bigg],\\
\overline{m}_2(I_1,I_2,H_3,\ \psi) &= -\Bigg[\alpha_{20}I_2 + \frac{3\alpha_{22}\omega_2 I_2^2}{2} + \alpha_{21}\omega_1 I_1 I_2\\
&\quad \left(1 + \frac{1}{2}\cos 2\psi\right) + (\alpha_{23} + \alpha_{24})I_2 H_3\Bigg],\\
\overline{m}_3(I_1,I_2,H_3,\ \psi) &= \frac{1}{4}(\alpha_{12}\omega_2 I_2 + \alpha_{21}\omega_1 I_1)\sin 2\psi,
\end{aligned}$$

$$\begin{aligned}
\overline{m}_4(I_1,I_2,H_3,\ \psi) =& -\big[(\alpha_{30} + \alpha_{40}) + (\alpha_{31} + \alpha_{41})\omega_1 I_1\\
&+ (\alpha_{32} + \alpha_{42})\omega_2 I_2 + (3\alpha_{33} + 3\alpha_{44} + \alpha_{34} + \alpha_{43})H_3/2\big]R(H_3),\\
\overline{\sigma}_{11}^2(I_1,I_2,H_3,\ \psi) =& 2D_1 I_1/\omega_1,\ \overline{\sigma}_{22}^2(I_1,I_2,H_3,\ \psi) = 2D_2 I_2/\omega_2,\\
\overline{\sigma}_{33}^2(I_1,I_2,H_3,\ \psi) =& \frac{D_1}{2\omega_1 I_1} + \frac{D_2}{2\omega_2 I_2},\ \overline{\sigma}_{44}^2(I_1,I_2,H_3,\ \psi) = 2(D_3 + D_4)R(H_3),\\
R(H_3) =& \left(1 + 8H_3 b - \sqrt{1 + 4H_3 b}\right)/12b.
\end{aligned} \tag{7.92}$$

The stationary PDF $p(I_1, I_2, \psi, h_3)$ can be obtained from Monte Carlo simulation of (7.91). The stationary joint PDF $p(\boldsymbol{q}, \boldsymbol{p})$ and statistics of original system (7.87) can be calculated according to Eq. (7.84) as follows

$$p(\boldsymbol{q},\boldsymbol{p}) = \left. \frac{p(I_1, I_2, \psi, h_3)}{2\pi T(h_3)} \right|_{\substack{I_1 = (p_1^2 + \omega_1^2 q_1^2)/2\omega_1, \\ I_2 = (p_2^2 + \omega_2^2 q_2^2)/2\omega_2, \\ h_3 = (p_3^2 + p_4^2)/2 + U(q_3, q_4) \\ \psi = \tan^{-1}(p_1/\omega_1 q_1) - \tan^{-1}(p_2/\omega_2 q_2)}},$$

$$E[Q_i^2] = \int_{-\infty}^{\infty} \cdots \int_{-\infty}^{\infty} q_i^2 p(\boldsymbol{q},\boldsymbol{p}) \mathrm{d}q_1 \ldots \mathrm{d}q_4 \mathrm{d}p_1 \ldots \mathrm{d}p_4, \quad i = 1, 2, 3, 4. \tag{7.93}$$

The parameter values used in the Monte Carlo simulation are $\alpha_{10} = \alpha_{20} = \alpha_{30} = \alpha_{40} = -0.08$, $\alpha_{13} = \alpha_{14} = \alpha_{23} = \alpha_{24} = 0.01$, $\alpha_{12} = \alpha_{21} = 0.02$, $\alpha_{31} = \alpha_{32} = \alpha_{41} = \alpha_{42} = 0.02$, $\alpha_{11} = \alpha_{22} = \alpha_{33} = \alpha_{44} = 0.04$, $\alpha_{34} = \alpha_{43} = 0.04$, $\omega_1 = 1$, $\omega_2 = 1$, $\omega_3 = 1.1$, $\omega_4 = 1.732$, $b = 1$, $D_1 = 0.04$, $D_2 = 0.02$, $D_3 = D_4 = 0.05$, and $\mathcal{H} = 0.7$. Samples of processes $I_1, I_2, H_3, \Psi, Q_3, Q_4, P_4, \Theta_2$ of system (7.87) are shown in Fig. 7.38. It can be seen that the former four processes are slowly-varying processes while the later four processes are fast-varying processes. The stationary PDFs $p(I_1)$, $p(h_3)$ and those of same system under Gaussian white noise excitations are shown in Fig. 7.39. Solid lines denote the results for Gaussian white noise excitation, while the other lines, long dashed, dash-dotted, and dotted, denote the results for fGn excitation with different Hurst index $\mathcal{H}$. It can be observed that when $\mathcal{H}$ approaches 1/2, the results for fGn excitation converge to those for Gaussian white noise excitation, confirming that when $\mathcal{H} \to 1/2$, fGn degenerates into Gaussian white noise. Figures 7.40–7.43 show the stationary PDF $p(I_1, I_2)$, $p(I_1, h_3), p(I_1), p(I_2), p(h_3)$, mean-value $E[I_1], E[I_2], E[H_3]$, and mean-square values $E[I_1^2], E[I_2^2], E[H_3^2]$. Figures 7.44–7.47 show the stationary PDF $p(q_1, q_2), p(q_1, q_3)$, $p(q_1), p(q_2), p(q_3), p(q_4), p(\psi)$ and mean-square values $E[Q_1^2], E[Q_2^2], E[Q_3^2], E[Q_4^2]$ of displacements. These figures demonstrate that the results from simulation of the averaged fractional SDE (7.91) closely match those from the simulation of original system (7.87).

For 10,000 samples, on a computer of i5-2400 CPU@3.10 GHz, the computation time for the original system (7.87) is 85.3 s, while that for the averaged fractional SDE (7.91) is 48.7 s. So, it is worthy to apply the stochastic averaging method of quasi-Hamiltonian systems excited by fGns presented in this section.

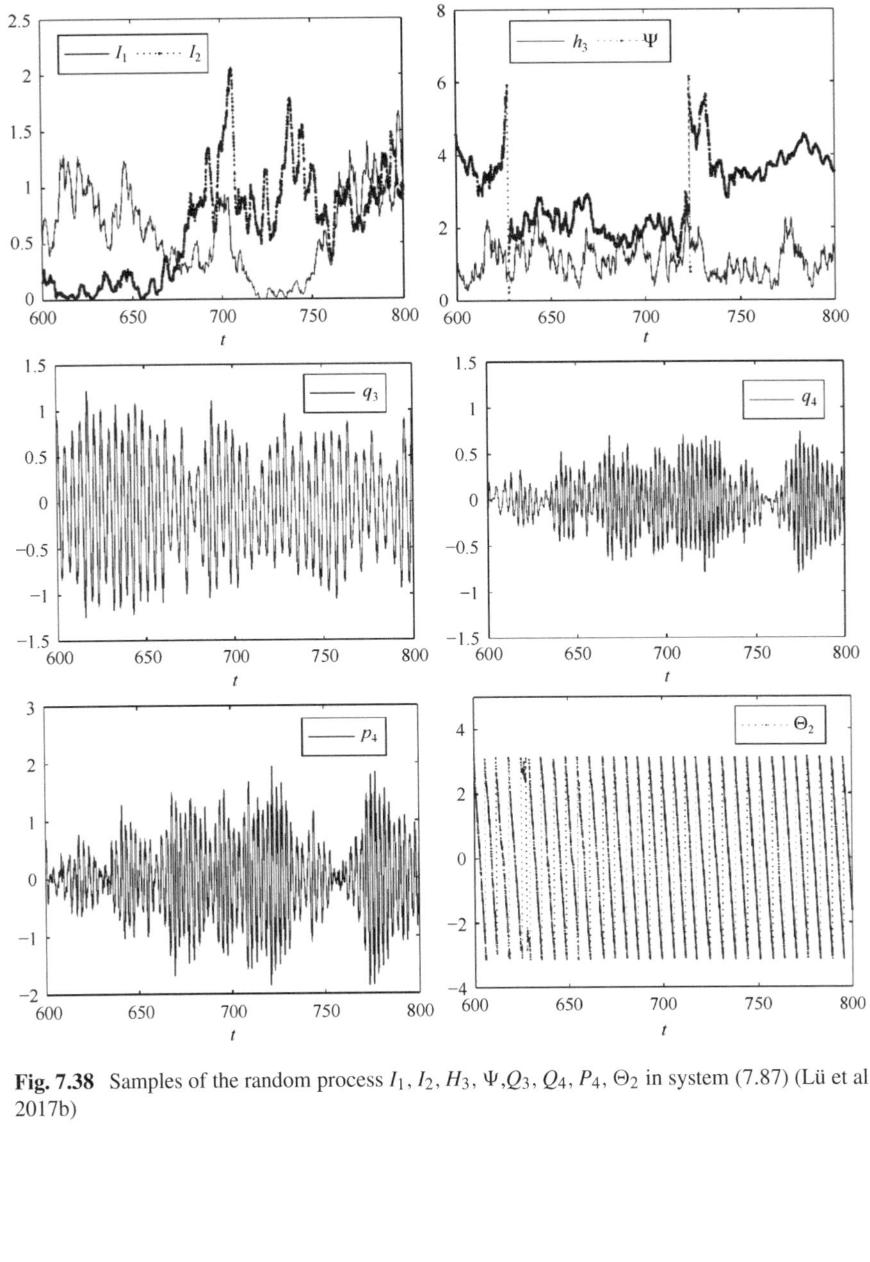

Fig. 7.38 Samples of the random process I_1, I_2, H_3, Ψ,Q_3, Q_4, P_4, Θ_2 in system (7.87) (Lü et al. 2017b)

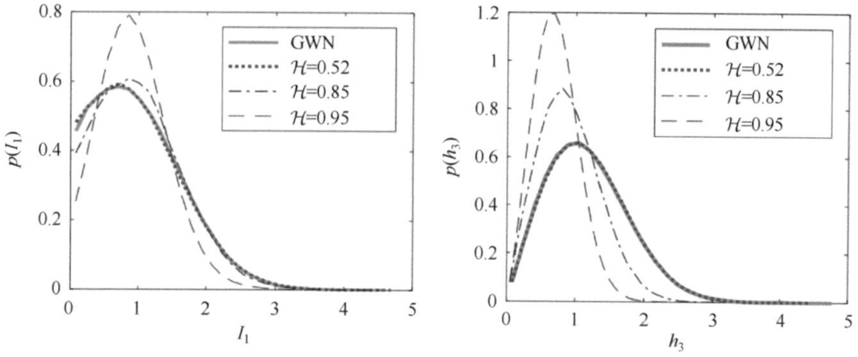

Fig. 7.39 $p(I_1), p(h_3)$ of system (7.87) excited by Gaussian white noise ($\mathcal{H} = 0.5$) and fGn, respectively (Lü et al. 2017b)

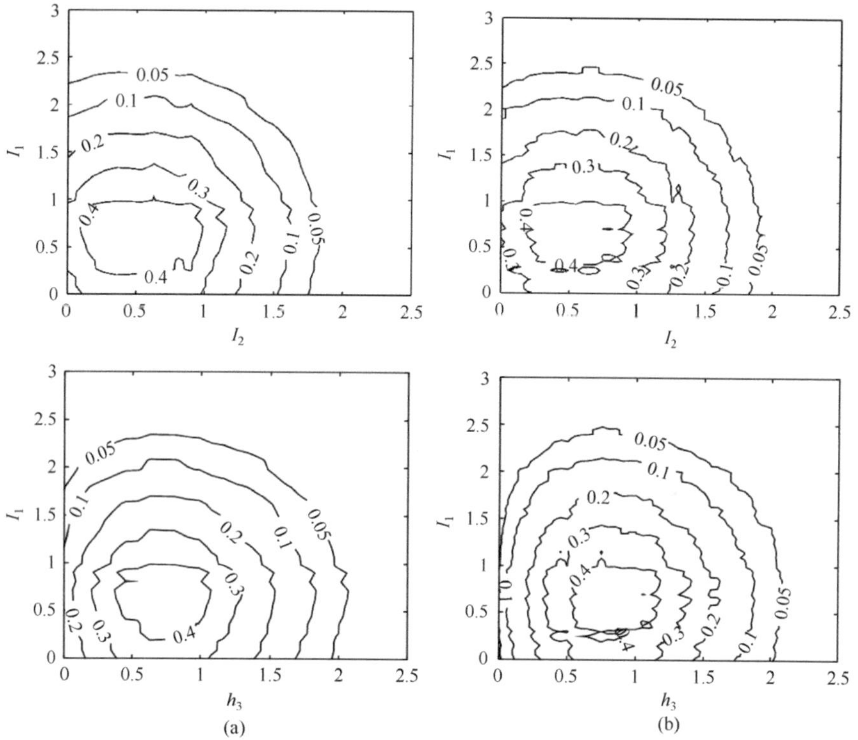

Fig. 7.40 Contours of $p(I_1, I_2), p(I_1, h_3)$ obtained from: **a** simulation of the averaged fractional SDE (7.91) and **b** simulation of the original system (7.87) (Lü et al. 2017b)

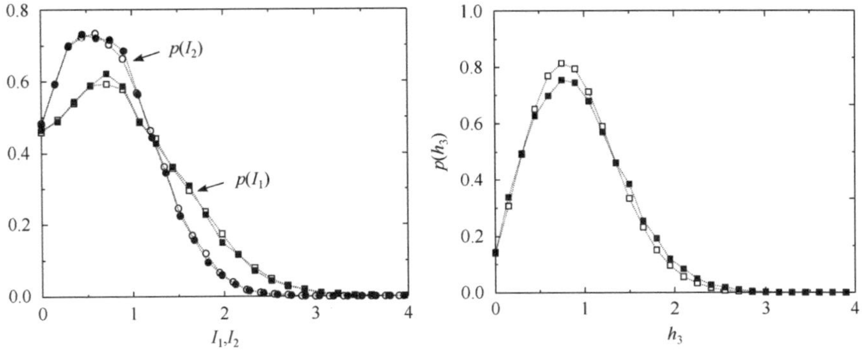

Fig. 7.41 $p(I_1), p(h_3)$ obtained from simulation of averaged fractional SDE (7.91) and original system (7.87). . ···○···, ···□···, results from simulation of system (7.87); ···●···, ···■···, results from simulation of Eq. (7.91) (Lü et al. 2017b)

Fig. 7.42
$E[I_1]$, $E[I_2]$, $E[H_3]$ versus $\mathcal{H}$ are obtained from simulation of averaged fractional SDE (7.91) and original system (7.87). . ···○···, ···□···, ···△···, results from simulation of system (7.87); ···●···, ···■···, ···▲···, results from simulation of Eq. (7.91) (Lü et al. 2017b)

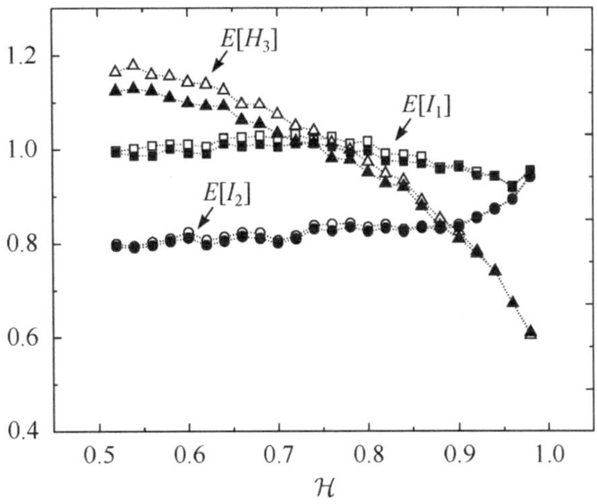

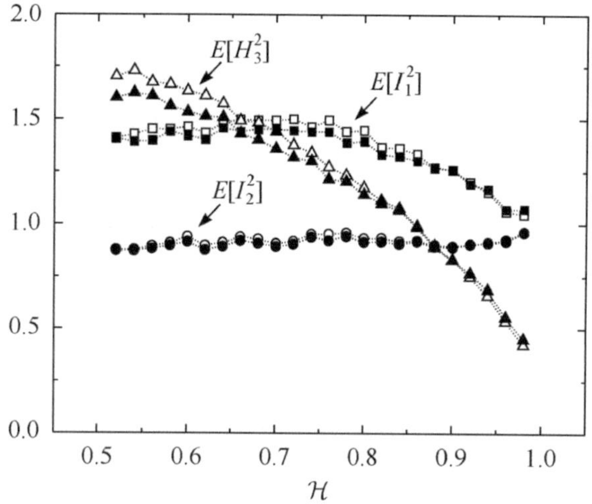

Fig. 7.43 $E[I_1^2]$, $E[I_2^2]$, $E[H_3^2]$ versus $\mathcal{H}$ are obtained from simulation of averaged fractional SDE (7.91) and original system (7.87). ···○···, ···□···, ···△···, results from simulation of system (7.87); ···●···, ···■···, ···▲···, results from simulation of Eq. (7.91) (Lü et al. 2017b)

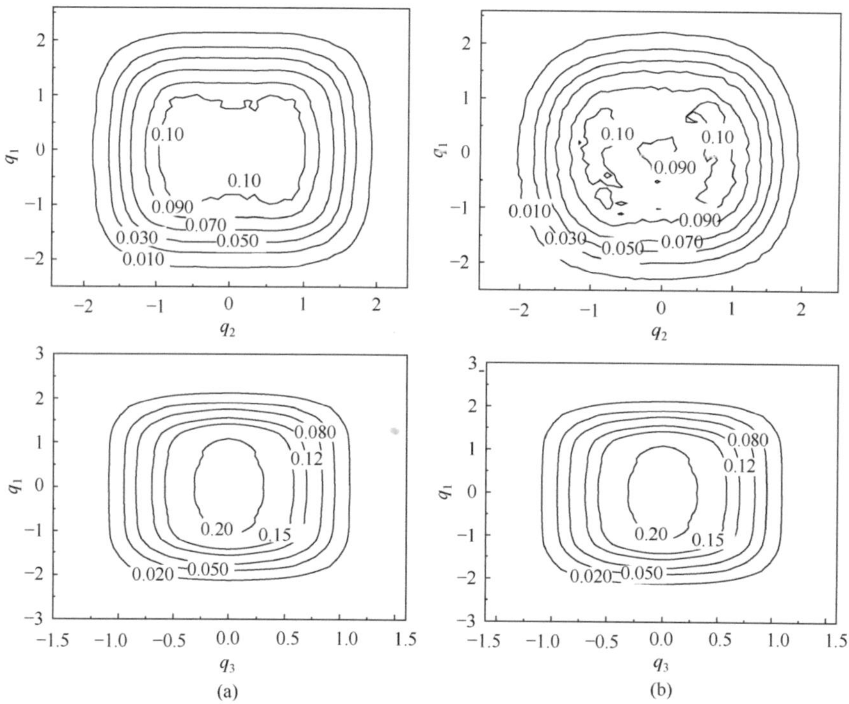

Fig. 7.44 Contours of $p(q_1, q_2)$, $p(q_1, q_3)$ obtained from: **a** simulation of the averaged fractional SDE (7.91) and **b** simulation of the original system (7.87) (Lü et al. 2017b)

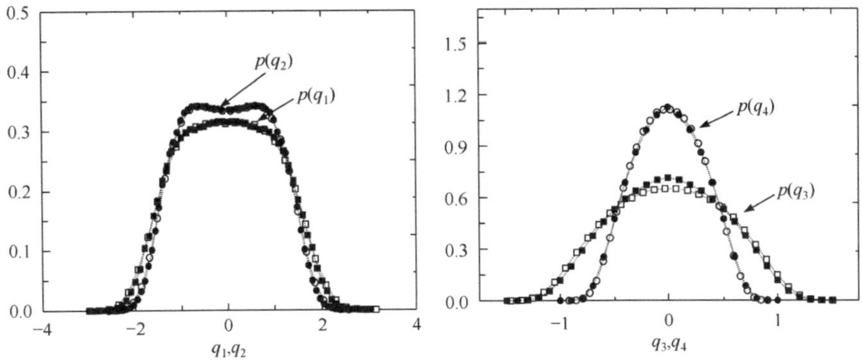

Fig. 7.45 $p(q_1), p(q_2), p(q_3), p(q_4)$ obtained from simulation of averaged fractional SDE (7.91) and original system (7.87). ···○···, ···□···, results from simulation of system (7.87); ···●···, ···■···, results from simulation of Eq. (7.91) (Lü et al. 2017b)

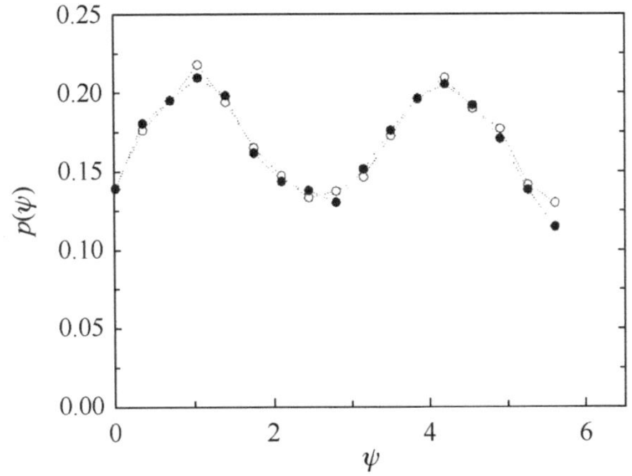

Fig. 7.46 $p(\psi)$ obtained from simulation of averaged fractional SDE (7.91) and original system (7.87). ···○···,results from simulation of system (7.87); ···●···, results from simulation of Eq. (7.91) (Lü et al. 2017b)

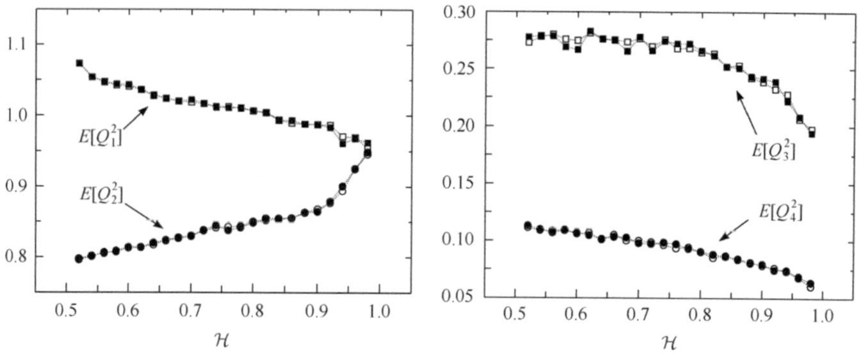

Fig. 7.47 $E[Q_1^2]$, $E[Q_2^2]$, $E[Q_3^2]$, $E[Q_4^2]$ versus $\mathcal{H}$ are obtained from simulation of averaged fractional SDE (7.91) and original system (7.87). $\cdots\circ\cdots$, $\cdots\square\cdots$, results from simulation of system (7.87); $\cdots\bullet\cdots$, $\cdots\blacksquare\cdots$, results from simulation of Eq. (7.91) (Lü et al. 2017b)

References

Deng M L, Zhu W Q. 2016a. Response of MDOF strongly nonlinear systems to fractional Gaussian noises. Chaos, 26: 084313.

Deng M L, Zhu W Q. 2016b. Stochastic averaging of quasi-non-integrable Hamiltonian systems under fractional Gaussian noise excitation. Nonlinear Dynamics, 83(1–2): 1015–1027.

Deng M L, Zhu W Q. 2018. Stochastic averaging of quasi integrable and non-resonant Hamiltonian systems excited by fractional Gaussian noise with Hurst index H∈(1/2,1). International Journal of Non-linear Mechanics, 98: 43–50.

Lü Q F, Deng M L, Zhu W Q. 2017a. Stochastic averaging of quasi integrable and resonant Hamiltonian systems excited by fractional Gaussian noise with Hurst index 1/2< H< 1. Acta Mechanica Solida Sinica, 30(1): 11–19.

Lü Q F, Deng M L, Zhu W Q. 2017b. Stationary response of multi-degree-of-freedom strongly nonlinear systems to fractional Gaussian noise. ASME Journal of Applied Mechanics, 84(10): 101001.

Lü Q F, Deng M L, Zhu W Q. 2017c. Stochastic averaging of quasi partially integrable Hamiltonian systems under fractional Gaussian noise. Journal of Zhejiang University-Science A, 18(9): 704–717.

Xu Y, Guo R, Xu W. 2014a. A limit theorem for the solutions of slow-fast systems with fractional Brownian motion. Theoretical and Applied Mechanics Letters, 4: 013003.

Xu Y, Guo R, Liu D, Zhang H Q, Duan J Q. 2014b. Stochastic averaging principle for dynamical systems with fractional Brownian motion. AIMS Discrete and Continuous Dynamical Systems-Series B, 19: 1197–1212.

Index

W.-Q. Zhu et al., *Stochastic Averaging*, https://doi.org/10.1007/978-981-96-3829-1

MIX
Papier aus verantwortungsvollen Quellen
Paper from responsible sources
FSC® C105338

FSC
www.fsc.org

If you have any concerns about our products,
you can contact us on
ProductSafety@springernature.com

In case Publisher is established outside the EU,
the EU authorized representative is:
Springer Nature Customer Service Center GmbH
Europaplatz 3, 69115 Heidelberg, Germany

Printed by Libri Plureos GmbH
in Hamburg, Germany